Manual of Geospatial Science and Technology

Manual of Geospatial Science and Technology

Edited by

John D. Bossler

Associate Editors:

John R. Jensen
Robert B. McMaster
Chris Rizos

London and New York

First published 2002 by Taylor & Francis
11 New Fetter Lane, London EC4P 4EE

Simultaneously published in the USA and Canada
by Taylor & Francis Inc.
29 West 35th Street,
New York, NY 10001

Taylor & Francis is an imprint of the Taylor & Francis Group

Typeset in Sabon by Deerpark Publishing Services Ltd
Printed and bound in Great Britain by TJ International Ltd, Padstow, Cornwall

Every effort has been made to ensure that the advice and information in this book is
true and accurate at the time of going to press. However, neither the publisher nor the
authors can accept any legal responsibility or liability for any errors or omissions that
may be made. In the case of drug administration, any medical procedure or the use of
technical equipment mentioned within this book, you are strongly advised to consult
the manufacturer's guidelines.

British Library Cataloguing in Publication Data
A catalogue record for this book is available from the British Library

Library of Congress Cataloging in Publication Data
Bossler, John D.
    Manual of geospatial science and technology / John D. Bossler.
        p. cm.
      1. Geographic information systems. 2. Remote sensing. 3. Global Positioning
System.
    I. Title.

    G70.212 .B67 2001
    910'.285–dc21

                                                        2001023762
ISBN 0-7484-0924-6

Contents

Contributors

John D. Bossler is a consultant specializing in GIS, GPS and remote sensing. He retired from the positions of Professor and Director of the Center for Mapping at The Ohio State University. Previously, he was Director of both the US National Geodetic Survey and the Coast and Geodetic Survey. He has authored over 100 papers in the fields mentioned. Employment: Private Consultant; e-mail: jbossler@aol.com.

Ayman Habib's research involves digital and analog photogrammetry. His research goal is to bridge the gap between the fields of computer vision and traditional photogrammetry by using theoretical work in computer vision. His research results have been published in the ASPRS journal and related publications. Employment: Assistant Professor, Department of Civil and Environmental Engineering and Geodetic Science, The Ohio State University; e-mail: habib.1@osu.edu.

Alan Saalfeld is foremost an applied mathematician/computer scientist who specializes in GIS applications such as conflation, generalization and related topics. He is well published in journals of mathematics, computer science and those covering GIS. Employment: Associate Professor, Department of Civil and Environmental Engineering and Geodetic Science, The Ohio State University; e-mail: saalfeld.1@osu.edu.

Carolyn J. Merry's research focuses on the application of remote sensing to land cover mapping, ice flow, water quality modeling, watershed management, and other engineering-related problems. Her research articles are published in several remote sensing journals. Employment: Professor, Department of Civil and Environmental Engineering and Geodetic Science, The Ohio State University; e-mail: merry.1@osu.edu.

Chris Rizos is leader of the Satellite Navigation and Positioning (SNAP) group, staff and graduate students of which are undertaking research into a variety of precise GPS/navigation technologies and applications. He is the author of over 100 journal and conference papers, and several books relating to GPS, surveying and geodesy. Employment: Professor, School of Geomatic Engineering, The University of New South Wales, Sydney, Australia; e-mail: c.rizos@unsw.edu.au.

Hasanuddin Abidin is head of the Geodetic Laboratory at ITB, where he conducts research into a range of geodetic problems using GPS technology. These include volcano monitoring, gravity and geoid studies. He has authored many papers, and written several books on GPS in the Bahasa Indonesian language. Employment: Lecturer, Department of Geodetic Engineering, Institute of Technology Bandung (ITB), Indonesia; e-mail: hzabidin@gd.itb.ac.id.

George Dedes is a graduate of surveying engineering from the Technical University of Athens, and a postgraduate in geodesy from the Ohio State University. Currently George is involved in the development and management of GPS software for surveying and GIS applications at Topcon-GeoComp. Employment: Vice President, Topcon; e-mail: dedes@iwaynet.net.

Dorota Grejner-Brzezinska's current research interest is integrated multisensor geospatial data acquisition systems, which involves kinematic positioning with GPS, GPS/INS integration, mobile mapping technology, and robust estimation techniques. Dorota has over 10 years experience working with the GPS technology and its engineering and scientific applications. Employment: Assistant Professor, Department of Civil and Environmental Engineering and Geodetic Science, The Ohio State University, Columbus, OH; e-mail: dorota@cfm.ohio-state.edu.

Mike Stewart After completing a Ph.D. in Physical Geodesy at the University of Edinburgh, Mike spent 4 years writing GPS software as a Senior Research Assistant at the Institute of Engineering Surveying and Space Geodesy in Nottingham. He migrated to Australia where his main research interests now lie in the fields of GPS algorithm development. Employment: Associate Professor, Department of Spatial Sciences, Curtin University of Technology, Perth, Australia; e-mail: stewart@vesta.curtin.edu.au.

Mike Stanoikovich is a partner and director of surveying and GPS with Woolpert, a top 100 engineering consulting firm. Over the past decade Mike has created one of the largest private surveying groups in the US. In addition, Mike has been closely involved with Trimble Navigation in the development and testing of Trimble's hardware and software products. Employment: Woolpert LLP, Dayton, OH; e-mail: mike.stanoikovich@woolpert.com.

Gérard Lachapelle has been involved with satellite and ground-based navigation systems since 1980, including 8 years of GPS development experience in industry. He has developed a number of precise GPS positioning methods and co-authored several related GPS software packages. He is widely published in GPS related publications. Employment: Professor and Head of Department, Department of Geomatics Engineering, The University of Calgary, Calgary, Canada; e-mail: lachapel@geomatics.ucalgary.ca.

Sam Ryan is the project manager for Navigation Systems in the Integrated Technical Support Directorate of the Canadian Coast Guard. Most recently he has been involved with the design and implementation of the Canadian Coast Guard's Marine System. Employment: Project Manager, Canadian Coast Guard, Ottawa, Canada; e-mail: ryans@dfo-mpo.gc.ca.

Don L. Light is a photogrammetrist who was employed by the Defense Mapping Agency and the USGS National Mapping Division from 1952 to 1995. After a distinguished career with the federal government, he joined Eastman Kodak as Manager of Business Development for Commercial Remote Sensing Systems. In 1998, he joined Landcare Aviation. Employment: Manager Business Development, Landcare Aviation, Inc., and PAR Government Systems, 6 Tawney Pointe, Rochester, NY 14626, USA; e-mail: dlight1@frontiernet.net.

John R. Jensen specializes in digital image processing of remote sensor data. He has authored over 100 journal articles and is the author of two well-known books on remote sensing and digital image processing. He is the director of the Center for Remote Sensing and GIS at USC and past-President and Fellow of ASPRS – the Imaging and Geospatial Information Society. Employment: Carolina Research Professor, University of South Carolina; e-mail: jrjensen@sc.edu.

Minhe Ji has major research interests in remote sensing land use characterization and biophysical modeling using artificial intelligence, accuracy assessment of thematic maps, and derivation of high-resolution DEMs using softcopy photogrammetry. Employment: Assistant Professor, Department of Geography, Box 305279, University of North Texas, Denton, TX, 76203, USA; e-mail: jminhe@unt.edu.

John D. Althausen Professional research interests include monitoring coastal habitats using remote sensing and geographic information systems, analyzing the impact of human populations on the natural landscape, and spatial analysis of business socio-economic patterns. Employment: Assistant Professor, Central Michigan University, Department of Geography, 296A Dow Science Building, Mount Pleasant, MI 48859, USA; e-mail: john.althausen@cmich.edu.

Sunil Narumalani Research interests focus on the use of remote sensing for the extraction of biophysical information from satellite data and the integration of geospatial data sets for ecological/natural resource mapping and monitoring. Employment: Associate Professor, School of Natural Resources Sciences, 13 Nebraska Hall, University of Nebraska, Lincoln, NB 68588-0517, USA; e-mail: snarumal@unlnotes.unl.edu.

Joseph T. Hlady received his B.Sc. in Geography from the University of Alberta, Canada in 1998 and is currently working on an M.A. in Geography in remote sensing at the University of Nebraska. Employment: Student, University of Nebraska, Lincoln, NB 68588, USA; e-mail: jhlady@calmit.unl.edu.

Ryan R. Jensen received his Ph.D. at the University of Florida in 2000. His research areas include the use of remote sensing for assessing biogeographical landscape characteristics and carbon cycling. Employment: Assistant Professor, Department of Geography and Geology and Anthropology, Indiana State University, Terre Haute, IN 47809, USA; e-mail: r-jensen@indstate.edu.

Russell G. Congalton is an expert in the practical application of remote sensing for forest resource management. He has developed the most widely adopted methods for assessing the accuracy of information derived from remotely sensed data. Employment: Professor, Department of Natural Resources, 215 James Hall, University of New Hampshire, Durham, NH 03824, USA; e-mail: russ.congalton@unh.edu.

Lucie C. Plourde received her M.Sc. in Natural Resources and Environmental Conservation from the University of New Hampshire in 2000. Her research focuses on sampling methods and logic to assess the accuracy of spatial products derived from remote sensor data. Employment: Remote Sensing Scientist, Complex Systems Research Center, University of New Hampshire, Durham, NH 03824, USA; e-mail: lucie@shooter.sr.unh.edu.

Donald C. Rundquist has a dual appointment in the School of Natural Resource Sciences and the Conservation and Survey Division, Institute of Agriculture and Natural Resources. He is the Director of the Center for Advanced Land Management Information Technologies, a research and development facility focused primarily on remote sensing and GIS. Employment: Professor, Center for Advanced Land Management Information Technologies (CALMIT), 113 Nebraska Hall, University of Nebraska, Lincoln, NE 68588-0517, USA; e-mail: dr1000@tan.unl.edu.

Maurice Nyquist has worked in the field of natural resource applications of geospatial technologies for over 25 years. He is the Chair of the Federal Geographic Data Committee's (FGDC) Biological Data Working Group. He is past President and Fellow of ASPRS – the Imaging and Geospatial Information Society. Employment: Government Scientist/Manager, US Geological Survey – Center for Biological Informatics, Box 25046, MS 302, DFC, Building 810, Rm 8000, Denver, CO 80225-0046, USA; e-mail: maury_nyquist@usgs.gov.

Thomas W. Owens is the coordinator for the USGS-NPS Vegetation Mapping Program. He has over 20 year experience in natural resource applications of remote sensing and geospatial technology. He has a Master of Science degree. Employment: Remote Sensing Specialist, US Geological Survey Center for Biological Informatics, Box 20046 MS 302, Denver, CO, 80225-0046, USA; e-mail: tom_owens@usgs.gov.

Robert B. McMaster's research interests include automated cartographic generalization including the development of practical algorithms. He is also known for work in environmental risk assessment and the history of US cartography. He has authored two books on generalization. Employment: Professor, University of Minnesota; e-mail: mcmaster@umn.edu.

William J. Craig is a past president of URISA and the University Consortium for Geographic Information Science (UCGIS). His research interests include the use and value of geographic information systems. He is Vice Chairman of the Minnesota's Governors Council on GIS for Minnesota. Employment: Associate Director of the Center for Urban and Regional Affairs, University of Minnesota; e-mail: wcraig@umn.edu.

David A. Bennett's research interests include geographic information science, spatial decision support systems, environmental modeling, and evolutionary computation. He serves as an associate editor for Cartography and Geographic Information Science published by the American Congress on Surveying and Mapping. Employment: Assistant Professor, University of Iowa; e-mail: david-bennett@uiowa.edu.

Marc P. Armstrong's research interests, within the broad area of geographic information science, include parallel and evolutionary computational analysis methods, spatial decision support systems and geographic visualization. He is a member of the Center for Global and Regional Environmental Science. Employment: Professor, University of Iowa; e-mail: marc-armstrong@uiowa.edu.

May Yuan received a Ph.D. degree in Geography from the State University of New York at Buffalo. Her research interest centers on geographic representation, query, analysis, and modeling and expands into geospatial data mining and knowledge discovery. She has published in various GIS journals. Employment: Associate Professor, University of Oklahoma; e-mail: myuan@ou.edu.

Francis J. Harvey's current research focuses on potentials and difficulties for developing the National Spatial Data Infrastructure in local governments. Previous projects have examined GIS implementation and design issues through a social constructivist approach that examined the relationships between private, public, and education sectors. Employment: Assistant Professor, University of Kentucky; e-mail: fharvey@pop.uky.edu.

Mark Lindberg is the Director of the University of Minnesota Cartography Laboratory and he is a co-director of the university's Masters in Geographic Information Science program. His primary interests focus on the use of GIS software in cartographic design and problems in contemporary map production with special emphasis on cross-platform issues. Employment: Director, University of Minnesota; e-mail: lindberg@atlas.socsci.umn.edu.

Michael F. Goodchild is chair of the Executive Committee, National Center for Geographic Information and Analysis (NCGIA) and Director of NCGIA's Center for Spatially Integrated Social Science. His current research interests center on geographic information science, spatial analysis, the future of the library, and uncertainty in geographic data. Employment: Professor, University of California at Santa Barbara; e-mail: good@geog.ucsb.edu.

Joel Morrison is the Director of the Center for Mapping at the Ohio State University and Professor in the Departments of Geography and, Civil and Environmental Engineering and Geodetic Science. Previously he was Division Chief, Geography Division, US Bureau of the Census. Employment: Professor/Director, The Ohio State University; e-mail: morrison@cfm.ohio-state.edu.

Rebecca Somers has more than 20 years experience helping government agencies, companies, and non-profit organizations develop GIS projects, programs, publications, and instructional resources. She has published dozens of articles on GIS planning, development, and management. She is also a prominent GIS workshop instructor. Employment: President, Somers-St. Claire; e-mail: SStCl@aol.com.

Susanna McMaster's research interests are in biophysical modeling, public participation GIS, and the history of US academic cartography. She obtained a Ph.D. in Environmental Science and Forestry from the SUNY College of Environmental Science and Forestry. Employment: Associate Program Director, Master of Geographic Information Science Program, University of Minnesota; e-mail: none at this time.

Zorica Nedović-Budić earned her Ph.D. at the University of North Carolina. Her research interests focus on the implementation of GIS in local governments. She is active in and has published extensively in the URISA journal and related publications. Employment: Associate Professor, University of Illinois at Urbana Champaign; e-mail: budic@uiuc.edu.

Lisa Warnecke is President of GeoManagement Associates, Inc. in Syracuse, New York and is a Senior Consultant at the National Academy of Public Administration in Washington, DC. She formerly served as a Town Manager in Colorado and GIS Coordinator for the States of Colorado and Wyoming. Employment: Consultant, GeoManagement Associates, Inc.; e-mail: lisaw@twcny.rr.com.

Drew Decker is Texas State Cartographer with the Texas Natural Resources Information System in Austin, Texas. He serves as co-chair of the Texas Geographic Information Council's Technical Advisory Committee and is the Project Manager of the Texas Strategic Mapping Program. Employment: State Cartographer, Texas Natural Resources Information System in Austin, Texas; e-mail: ddecker@twdb.state.tx.us,

John J. Moeller is the Staff Director for the Federal Geographic Data Committee (FGDC). He is responsible for coordination of the operations for the FGDC and for providing federal leadership in facilitating the implementation of the National Spatial Data Infrastructure (NSDI). Employment: US Geological Survey; e-mail: jmoeller@usgs.gov.

Mark E. Reichardt in 1999 was selected to establish and lead an international Spatial Data Infrastructure (SDI) program for the Federal Geographic Data Committee. He helped to establish globally compatible national and regional SDI practices in Africa, South America, Europe, and the Caribbean. He is now with the OpenGIS Consortium. Employment: Director of Marketing, Open GIS Consortium; e-mail: markoone@earthlink.net.

Robin Antenucci has 18 years experience in GIS project planning and management, photogrammeteric mapping and remote sensing and the design and development of GIS applications and databases. She has held the position of Senior Technical Consultant for the Convergent Group. Employment: Consultant, PlanGraphics, Inc.; e-mail: Robinantenucci@aol.com.

John Antenucci is an engineer, planner, management consultant, executive and author. He is founder and President of PlanGraphics Inc. a firm that specializes in the design and implementation of geographic information systems and related spatial technologies. He has served as president and director of the Geospatial Information Technology Association (GITA). Employment: President, Plan-Graphics, Inc.; e-mail: jantenucci@PlanGraphics.com.

Preface

Taylor & Francis is responsible for the concept of producing a manual useful for practitioners, covering geographic information systems. This initial concept evolved to a 'practical manual' covering geospatial science and technology, which is defined, somewhat arbitrarily, as a manual covering the Global Positioning System (GPS), Remote Sensing and Geographic Information Systems (GIS). It is a difficult task to produce such a manual, given that the authors invariably come from academe where the tendency is to produce an erudite journal paper as contrasted with a practical manual. The extent to which we have succeeded in producing a document valuable to the person involved in setting up a 'GIS project' will be found from a market analysis of the sales of this manual. We tried our best.

The manual is divided into five parts, Prerequisites, GPS, Remote Sensing , GIS and Applications. If the reader is somewhat familiar with these topics, it is suggested that he/she skip Part 1, Prerequisites, and proceed to the part of the manual they are most interested in. If the reader is unfamiliar with the three main topics, appropriate chapters in Part 1, along with the references provided, are highly recommended. The integration of these technologies is best found in Part 5, Applications. Actual case studies describing the integration of these technologies are best found in Chapter 37, which also adds the interesting aspect of cost savings as a result of the use of these technologies and their integration.

There are 43 authors and 37 chapters, with hundreds of references. The references in Parts 1 and 5 were kept to a minimum and textbooks were recommended wherever possible. In Parts 2–4 especially, a number of papers were provided as references because the topics are more specialized. The figures speak for themselves.

John D. Bossler
Editor

Acknowledgments

More than 50 people were involved in the creation of this manual and I would like to name and thank each of them. Obviously, that cannot be done but the geospatial community will, in years to come I believe, realize what a significant contribution they have made. Each of the 43 authors and especially the three associate editors, deserve special credit and thanks for their efforts. The monetary remuneration for writing a chapter in this manual is not sufficient motivation to contribute their valuable time and energy. Their motivation came from believing that they were providing a service to the community of users of geospatial data. For this service they deserve our sincere thanks. The three associate editors, Chris Rizos for Part 2, the Global Positioning System, John Jensen for Part 3, Remote Sensing and Bob McMaster for Part 4, Geographic Information Systems, deserve very special thanks because they had to coordinate, edit and organize all the chapters in their part of the manual in addition to writing several chapters.

The staff of Taylor & Francis, especially Tony Moore, provided patient and forthright advice throughout the nearly 2-year period that spanned the development of the manual.

It is my opinion that one of the most important factors in the acceptance and use of any technical document of this kind depends on the quality of the figures or illustrations. When I started this task, I ran into an old friend and colleague from my NOAA (National Oceanic and Atmospheric Administration) days, Robert (Bud) Hanson. Bud is retired but is still a highly capable computer graphics aficionado who, when begged to take on the task of editing, creating and organizing the figures in this manual, finally agreed. As a result, the more than 200 figures and illustrations (100 of which were drawn from scratch or extensively modified) are outstanding and we all owe Bud a huge debt. He spent hundreds of hours at this task, simply because 'a job well done becomes a pleasure'. He also generously helped in editing about 12 of the chapters. His advice was extremely valuable.

Part 1

Prerequisites

Chapter 1

An introduction to geospatial science and technology

John D. Bossler

1.1. About this manual

This manual is written for professionals working in the private sector, various levels of government, and government agencies who are faced with, or require knowledge of, the tasks involved in performing an area-wide mapping, inventory, and data conversion and analysis project. Such projects are commonly referred to as geographic information system projects.

The manual covers the three sciences and technologies needed to accomplish such a project – the Global Positioning System (GPS), remote sensing, and Geographic Information Systems (GIS). These three disciplines are sometimes called the geospatial information sciences. The manual covers the basic mathematics, computer science, and physics necessary to understand how these disciplines are used to accomplish mapping, inventory, and data conversion. Examples of the basics covered in Part 1 of this manual include co-ordinate systems, co-ordinate transformations, datums; computer architecture, operating systems, database structures; electromagnetic radiation, and atmospheric disturbances.

The manual addresses background, how-to, and frequently-asked questions. Questions such as 'What GPS equipment should I use?', 'What accuracy is required?', 'What scale should I use?', 'How do I buffer this river?', 'What satellite data is available?' and 'What spectral bands can I obtain, and which should I use?' are typical of the questions addressed in this manual. It contains 37 chapters written by 43 authors, most of whom are widely known experts in the field. The authors and editors hope that the manual will make decision-making easier and better for the professionals using this manual.

1.2. Geospatial data

Features shown on maps or those organized in a digital database that are tied to the surface of the earth by co-ordinates, addresses, or other means are collectively called geospatial data. These data are also called *spatial* or *geographic* data. Almost 80 per cent of all data are geospatial data. A house whose address is provided or a geodetic control monument with its latitude and longitude are examples of geospatial data. An example of data that are not geospatial is a budget for an organization.

Geospatial data can be acquired by digitizing maps, by traditional surveying, or by modern positioning methods using GPS. Geospatial data can also be acquired using remote sensing techniques, i.e. by airborne or spaceborne platforms. After such data are acquired, they must be organized and utilized. A GIS serves that purpose admirably by providing organizing capability through a database and utilizing query capability through sophisticated graphics software. The reader should now understand the intent of this manual – to help understand the process involved in modern map making and information age decision making.

If we define the process needed to make decisions using geospatial data as

1 acquiring data
2 analyzing and processing data
3 distributing data

then this manual provides significant material about the first two steps above.

The manual does not address the issue of distributing geospatial data.

1.3. Spatial data infrastructure

A large number of futurists believe that we are probably in the middle of the information age. The ramifications of this assertion require us to collect, process, manage, and distribute geospatial and other data. In the US, we have defined a *process* (*not an organization!*) called the National Spatial Data Infrastructure (NSDI), which is comprised of the people, policies, information, technology, and institutional support needed to utilize geospatial data for the enhancement of society (NAS/NRC, 1993). This manual describes the three nascent information technologies that are central to the NSDI process because they allow us to *acquire* digital information and to *process* and *analyze* these data.

The NSDI and in particular the technologies described in this manual, along with the Internet, are now essential for an extremely wide variety of applications. Today, one of the most important applications is probably the management of our environment. Other important applications include sustainable economic development, taxation, transportation, public safety, and many more. While we have attempted to describe these three technologies in a broad manner, we have deliberately focused our description of these technologies on natural resource questions, land use planning, and other issues concerned with the land and its use because in our opinion, there is significant 'GIS growth' associated with land issues, and we believe there is a need for educational material in this arena.

1.4. Relationship of geospatial science to other disciplines

The foundation of the geospatial sciences, especially those topics covered in this manual, is mathematics, computer science, physics, and engineering. Depending on the depth and breadth of study, the practitioner may find biology, cartography, geodetic science, geography, geology, and surveying very helpful.

Table 1.1 shows the basic educational needs when working with each of the topics covered in this manual. The table is highly subjective, and all topics require some knowledge of all the educational subjects shown. However, if the practitioner wants to acquire fundamental *background* information about the topics discussed in this manual, Table 1.1 provides reasonable general guidance. For continuing education

Table 1.1 Educational background needed for geospatial science and technology

Topic	Educational needs in approximate order of relevance
GIS	Computer science, geography, cartography, surveying, operations research
GPS	Geodetic science, mathematics, physics, surveying, celestial mechanics
Remote sensing	Physics, mathematics, computer science, engineering, geology/biology

purposes, most universities offer courses in these fundamentals and many provide specific courses in GIS, GPS and remote sensing. Consider, for example, the universities represented by the editor and the three associate editors of this manual. They all offer courses in these subjects.

1.5. Three important geospatial technologies

This manual covers the topics of GPS, GIS, and remote sensing. It is important therefore, that we discuss each of these briefly here and refer the reader to the individual part of this manual covering this material and to other writings.

1.5.1. GPS

It is probably true that no other technology has affected the surveying and mapping profession as profoundly as GPS. The purpose of this manual is to introduce practitioners to the principles of this technology so that they can understand its capability, accuracy, and limitations. Hopefully, this material will also allow the practitioner to use the equipment, if provided with adequate additional support from the receiver manufacturer.

The single most powerful feature related to GPS, which is not true of traditional surveying techniques, is that its use does not require a line of sight between adjacent surveyed points. This factor is very important in understanding the impact that GPS has had on the surveying, mapping, and GIS communities.

GPS has been used by the surveying and mapping community since the late 1970s, when only a few hours of satellite coverage were available. It was clear even then that centimeter-level accuracy was obtainable over very long baselines (hundreds of kilometers). In the early 1980s, users of GPS faced several problems: the cost of GPS receivers; poor satellite coverage, which resulted in long lengths of time at each survey location; and poor user-equipment interfaces. Today, instantaneous measurements with centimeter accuracy over tens of kilometers and with one part in 10^8 accuracy over nearly any distance greater than 10 km can be made. The cost of 'surveying and mapping-level' receivers in 1999 ranged from $10,000 to $25,000, and these costs are falling. Practitioners are developing numerous new applications in surveying, such as the use of GPS in a kinematic (real-time) mode to determine the elevation of terrain prior to grading (NAS/NRC 1995).

Traditional land surveying is increasingly being accomplished using GPS because of the continuous reduction in receiver costs, combined with an increase in user friendliness. This trend towards the use of GPS for surveying and mapping has enhanced the volume of survey receiver sales.

The standard surveying and mapping activities that are enhanced by GPS include surveying for subdivisions, highways, planned communities, etc. To date, GPS has not been used widely for single lot or mortgage surveys. This is because the accuracy is marginal for distances of 30 m or less and the cost of very local GPS operations is still slightly higher than traditional techniques. It is likely that in the future GPS will be used for such activities.

1.5.2. GIS

Prior to the decade of the 1960s there was no efficient way to manage natural or man-made resources associated with large areas of the land. During the 1960s and 1970s, computer capability increased significantly and graphics capability became feasible and even commonplace. Today GIS are used extensively to perform spatial analysis related to many of our important resources. Examples of the applications supporting natural resource management, equitable taxation, environmental monitoring, and civil infrastructure enhancement are numerous and are covered in Part 5. To quickly understand how a GIS can be used to benefit society, a short example is provided here: consider a demographic analysis of recent crime increases. The areas of increased crime can be tabulated, displayed graphically, and then increased police patrols can be dispatched to these areas.

Computer readable data and information can be easily shared within an organization, resulting in enhanced efficiency and effectiveness. This is especially important in public organizations such as cities, counties, and municipalities where a rapidly changing political environment can cause significant differences in the organization's modus operandi. Today's GIS, which are fundamentally a marriage of database management systems with graphics capability, are designed to allow for changes in the processes of individuals and organizations and changes in the data. Therefore, they are able to serve the complete spectrum of individuals, from political appointee to supervisor to computer programmer. It is now an indispensable tool for policy makers as well as technicians.

1.5.3. Remote sensing

With an increase in spatial and spectral image resolution and a decrease in cost of these images, the new millennium may finally fulfil the decades-old dream of remote sensing enthusiasts. This dream is that remotely sensed data and information will become a retail product, i.e. a commodity. The Internet will play a major role in making this happen. These accurate, high resolution images, in the form of dense pixels, are becoming cheap, easily accessible and most importantly, current. Modern remote sensing satellites are capable of acquiring images in dozens and even hundreds of spectral bands and this (at least panchromatic images) can now be done with a resolution of 1 m.

Our purpose in this manual is to alert the reader to the possibilities of usage of these data, and to provide enough background material to acquire, manage and understand such data and information.

It is intuitive that remote sensing provides us with an outstanding tool for global science studies such as greenhouse gas concentration, how the oceans interact with the atmosphere and the land masses, and the effects on ecosystems of acid rain, soil erosion and deforestation. This manual discusses the use of modern remote sensing systems for many of these applications.

Images can and are used as a base 'map.' Other information can be overlaid or used in conjunction with this 'map.' The images can also be used to determine the areal extent of the various land covers, e.g. forests, various crops and even some of the geological attributes. Moreover, and quite importantly, remote sensing can be used to inventory and determine the health of these land covers. In short, remote sensing is a powerful tool to tell us more about the land.

Remotely sensed data was a major driver in the development of GIS. The enormous quantity of data produced by remote sensors needed to be efficiently analyzed and managed and dealing with these data expedited the development of GIS. Therefore the relationship between GIS and remote sensing should be obvious. Noting that GPS is the most accurate and efficient way to relate images to the ground, either from GPS positions of the satellites themselves or positions of points on the ground that are in the image, it is easy to see that these three technologies provide a basis for the development of this manual.

References

National Research Council, National Academy of Sciences (NAS/NRC). *Toward a Co-ordinated Spatial Data Infrastructure for the Nation*. National Academy Press. Washington, DC 1993.

National Research Council, National Academy of Sciences (NAS/NRC). *The Global Positioning System, A Shared National Asset*. National Academy Press. Washington, DC 1995.

Chapter 2

Coordinates and coordinate systems

John D. Bossler

In the following sections, the fundamentals of coordinates and coordinate systems are described. A description of the most important and commonly used coordinate system associated with Global Positioning System (GPS), Geographic Information Systems (GIS), and Remote Sensing (RS) usage is provided.

2.1. Rectangular coordinates

Consider two straight lines OX and OY intersecting at a right angle (Figure 2.1). On each line we construct a scale using the point of intersection O as a common origin. Positive numbers are generally shown to the right, on horizontal line OX and upwards on OY. Consider the point P in the plane of the two axes. The distance of the point P from the y-axis (line OY) is called the abscissa of the point, denoted by x, and its distance from the x-axis (line OX) is called the ordinate and is denoted by y. In this case, the abscissa and ordinate together are called the *rectangular coordinates* of the point P. In Figure 2.1, P has coordinates $(-2,4)$, the abscissa, $x = -2$, and the ordinate, $y = +4$. The distance d is equal to

$$\sqrt{x^2 + y^2} = \sqrt{4 + 16} = \sqrt{20}$$

by trigonometry. The distance between any two points in general is equal to

$$\sqrt{(x_j - x_i)^2 + (y_j - y_i)^2}$$

where the subscripts j and i refer to point j and point i, respectively.

2.2. Polar coordinates

Rather than locating a point by its distances from two intersecting lines, we may locate it from its distance and direction from a fixed point. Consider a fixed point O, which is called the pole and a fixed line OX, which we call the polar axis (Figure 2.2).

Suppose now that an angle θ has been generated by the rotation of a line from initial coincidence with the polar axis and let P be a point on the terminal side of the angle. If we denote OP by ρ, the polar coordinates of P are given by ρ and θ, i.e. $P(\rho, \theta)$. Theta, θ, is called the vectorial angle and ρ the radius vector. If θ has been generated by a counter-clockwise rotation, it is usually regarded as positive; if it has been generated

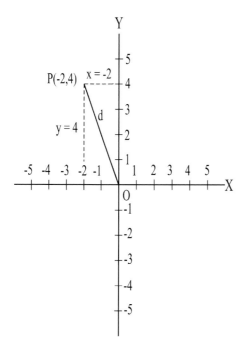

Figure 2.1   Rectangular coordinate system.

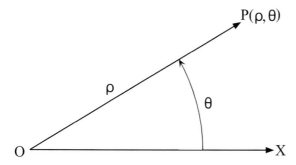

Figure 2.2   Polar coordinate system.

by a clockwise rotation, it will be regarded as negative. The radius vector will be regarded as positive or negative as the point P lies in the direction determined by the vectorial angle or backwards through the origin. If theta is allowed to take on any positive or negative value, greater or less than 360°, a point can have an infinity of polar coordinates, e.g. $(3, 60°)$, $(-3, 240°)$, and $(3, -300°)$ all represent the same point.

2.3. Transformation between polar and rectangular coordinates

If the polar axis coincides with the positive x-axis of a rectangular system of coordinates, the pole being at the origin, the relationship between the two coordinate systems is readily obtained using trigonometry (see Figure 2.3) as follows:

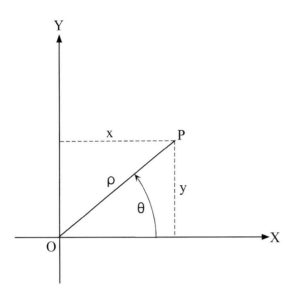

Figure 2.3   Relationship between rectangular and polar coordinates.

$$x = \rho\cos\theta, \quad y = \rho\sin\theta \tag{2.1}$$

$$\rho = \pm\sqrt{x^2 + y^2}, \quad \theta = \tan^{-1}y/x \tag{2.2}$$

2.4. Rectangular coordinates in three dimensions

To locate a point in space, three coordinates are necessary. Consider three mutually perpendicular planes that intersect in lines OX, OY, and OZ, shown in Figure 2.4. These lines are mutually perpendicular. The three planes are called the coordinate planes (*xy* plane, *xz* plane, *yz* plane); the three lines are called the coordinate axes (*x*-axis, *y*-axis, *z*-axis), and the point O is the origin. A point P (x, y, z) may be located by providing the orthogonal distances from the coordinate planes to the point P.

Again, the quantities x, y, and z are the rectangular coordinates of the point P. The positive and negative directions of the axes are somewhat arbitrary. Most mathematical textbooks use a left-handed coordinate system, but practitioners in the geospatial community use right-handed systems. Definition: a rectangular coordinate system in three dimensions that has the positive directions on the three axes (x, y, and z) defined in the following way: If the thumb of the right hand is imagined to point in the positive direction of the z-axis and the forefinger in the positive direction of the x-axis, then the middle finger, extended at right-angles to the thumb and forefinger, will point in the positive direction of the y-axis (Geodetic Glossary 1986).

If the coordinate system is left-handed, the middle finger will point in the negative direction of the y-axis.

The use of GPS and other geometric procedures has fostered the use of rectangular coordinate systems for geospatial endeavors.

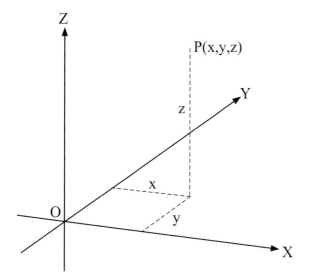

Figure 2.4 Rectangular coordinates in 3D.

2.5. Spherical coordinates

A point in space can also be located by the length ρ of its radius vector from the origin, the angle ϕ which this vector makes with the xy-plane and the angle λ from the x-axis to the projection of the radius vector in the xy-plane. The coordinates (ρ, ϕ, λ) are called spherical coordinates (see Figure 2.5).

The relationship between spherical and rectangular coordinates is easily derived from the figure as follows:

$$x = \rho \cos\phi \cos\lambda \qquad (2.3)$$

$$y = \rho \cos\phi \sin\lambda \qquad (2.4)$$

$$z = \rho \sin\phi \qquad (2.5)$$

The reader may recognize the angles ϕ, λ as the familiar latitude and longitude used on globes, maps, etc.

2.6. Ellipsoidal coordinates

A reasonable approximation of the surface of the earth is found by rotating an ellipse that is slightly flattened at the poles, around its shortest (minor) axis. Such a figure is called an ellipsoid of revolution or simply an ellipsoid. It is defined by its semi-major axis a and semi-minor axis b (see Figures 2.6 and 2.7).

The equation governing the ellipsoid obtained by rotating Figure 2.6 about axis b is:

$$\frac{x^2 + y^2}{a^2} + \frac{z^2}{b^2} = 1 \qquad (2.6)$$

The above equation follows directly from the definitions of an ellipse (Zwillinger

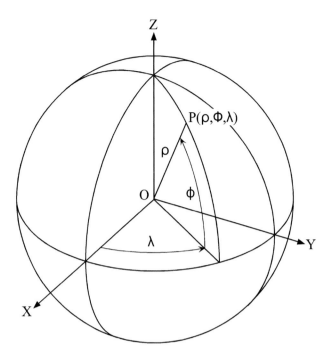

Figure 2.5 Spherical coordinate system.

1996; Swokowski 1998) and an ellipsoid with two of its three orthogonal axes equal in length. To facilitate the development of additional equations and relationships, it is convenient to define the following quantities:

$$f = \frac{a - b}{a}, \quad e = \sqrt{\frac{a^2 - b^2}{a^2}} = \sqrt{2f - f^2} \tag{2.7}$$

where f is defined as the flattening and e is defined as the first eccentricity.

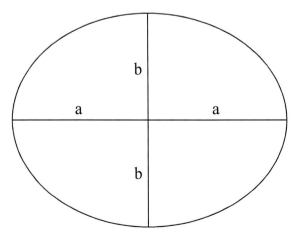

Figure 2.6 Ellipse.

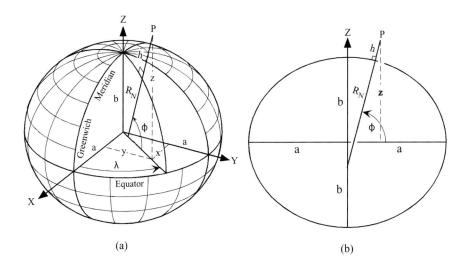

Figure 2.7 Ellipsoidal coordinate system.

Textbooks in geodesy develop the equations relating rectangular coordinates to ellipsoidal coordinates (Bomford 1980; Torge 1980), and this development will not be repeated here. The following equations refer to the right-handed coordinate system shown in Figure 2.7a. Figure 2.7b is a cross section of the ellipsoid shown in Figure 2.7a.

$$x = (R_N + h)\cos\phi\cos\lambda \tag{2.8}$$

$$y = (R_N + h)\cos\phi\cos\lambda \tag{2.9}$$

$$z = \left(\frac{b^2}{a^2}R_N + h\right)\sin\phi \tag{2.10}$$

where

$$R_N = \frac{a^2}{(a^2\cos^2\phi + b^2\sin^2\phi)^{1/2}} = \frac{a}{(1 - e^2\sin^2\phi)^{1/2}} \tag{2.11}$$

R_N is the radius of curvature in the prime vertical (Torge 1980: 50). The GRS 80 reference ellipsoid used in the North American Datum, NAD 83, has values of a and b as:

$$a = 6,378,137.0m, \quad b = 6,356,752.314m$$

This reference ellipsoid is now used throughout the world.

2.7. State plane coordinate systems (US)

Using the curved reference ellipsoid is awkward and time-consuming for carrying out mapping, surveying, and GPS projects, so most countries use plane coordinate systems for those activities. The equations and concepts in Section 2.1 can then be used.

In the US, the individual states are of such size that one or two plane coordinate systems can be used to cover an entire state, which allows plane surveying to be used over relatively great distances while maintaining the simplicity of plane rectangular x and y coordinates. With such a system, azimuths are referred to grid lines that are parallel to one another. Two types of projections are used in developing the state plane coordinate systems: the transverse Mercator projection and the Lambert conformal projection. The transverse Mercator projection employs a cylindrical surface (see Figure 2.8). The distortions in this system occur in an east-west direction; therefore, the projection is used for states, such as New Hampshire, that have relatively short east-west dimensions. The Lambert conformal projection uses a conical surface, the axis of which coincides with the ellipsoid's axis of rotation and which intersects the surface of the ellipsoid along two parallels of latitude that are approximately equidistant from a parallel lying in the center of the area to be projected (see Figure 2.9). The distortions occur in a north-south direction, and therefore, the projection is used for states, such as Tennessee, with relatively short north-south dimensions. One exception to this rule exists. The state of Alaska uses an oblique transverse Mercator.

2.8. Universal Transverse Mercator (UTM)

Because of its popularity world-wide and its use by geospatial practitioners, a brief discussion of this projection is appropriate. The UTM is simply a transverse Mercator projection to which specific parameters, such as central meridians, have been applied. The UTM covers the earth between latitudes 84° north and 80° south. The earth is

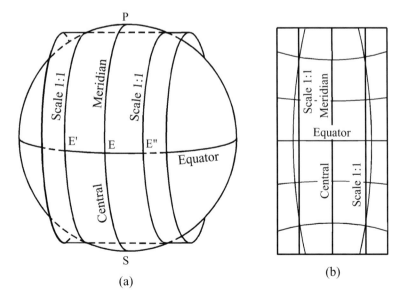

(a)

(b)

Figure 2.8 Transverse Mercator projection.

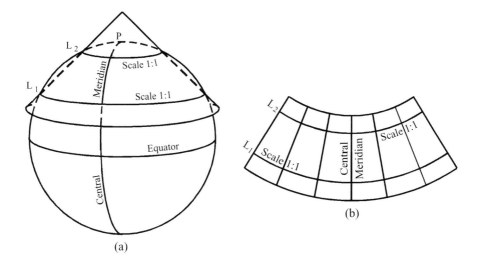

Figure 2.9 Lambert conformal projection.

divided into 60 zones each generally 6° wide in longitude. The bounding meridians are therefore, evenly divisible by 6° and zones are numbered from one to 60 proceeding east from the 180th meridian from Greenwich, with minor exceptions. Tables and other easy to use data and information for the UTM are widely available (Snyder 1987).

The equations governing state plane coordinates, both the Lambert conformal and the transverse Mercator, are provided in Chapter 4. Several books are valuable as references. To cover the background of the systems, tables of projection systems and their use in surveying, see Moffitt and Bossler (1998). To understand the development of the mapping equations, the reader is referred to Snyder (1987).

References

Bomford, G., *Geodesy* (4th ed.), Clarendon Press, Oxford, 1980.

Moffitt, F. and Bossler, J., *Surveying* (10th ed.), Addison Wesley Longman, Menlo Park, CA, 1998.

Snyder, J. P., *Map Projections – A Working Manual*, United States Government Printing Office. Washington, DC, 1987.

Swokowski, E. W., *Calculus and Analytic Geometry* (2nd alternate ed.), PWS Kent Publishing Co., Boston, MA, 1998.

Torge, W., *Geodesy*, Walter de Gruyter, New York, 1980.

Geodetic Glossary, National Geodetic Survey, United States Governmental Printing Office, Washington, DC, 1986.

Zwillinger, D. (ed.), *Standard Mathematical Tables and Formulae* (30th ed.), CRC Press, New York, 1996.

Chapter 3

Datums and geodetic systems

John D. Bossler

This chapter introduces and defines the concepts of *datums* and *reference systems*. It also illustrates various applications involving datums and reference systems and, in tabular form, provides a list of the major datums used in the world as of this writing.

A datum, in general, is any quantity or set of quantities that may serve as a referent or basis for the calculation of other quantities.

3.1. Geodetic datums

According to the (Geodetic Glossary, 1986), a geodetic datum is 'a set of constants specifying the coordinate system used for geodetic control, i.e. for calculating coordinates of points on the earth.'

3.1.1. Horizontal datums

If surveying, mapping and geodesy activities are carried out over large areas, the use of plane coordinates becomes impractical. Therefore, for geodetic purposes especially, the earth is considered as a whole body. Prior to about 1650, the earth was assumed to be spherical in shape. However, the results of using more accurate measuring instruments, combined with increased knowledge of earth physics, yielded the fact that an ellipsoid of revolution best approximated the figure of the earth (see Chapter 2, Section 2.6).

During the period from about 1700 to 1850 numerous measurements were made along arcs of meridians to verify the ellipsoidal assumption and then to determine the flattening of the ellipsoid. An accounting of these activities can be found in (Torge 1980). Once the ellipsoidal figure of the earth was accepted, it was logical to adopt an ellipsoidal coordinate system. Then, the problem was to establish these coordinates, i.e. to assign coordinates to a monument on the surface of the earth.

In the days before the 'deflection of the vertical' (Geodetic Glossary 1986) was known or understood, astronomic positions were used to establish these coordinates. Later, in the pre-satellite era, using classical geodetic techniques, the horizontal coordinates, geodetic latitude and longitude, defined a horizontal datum. This was usually accomplished on a country by country basis. A few examples of such datums are the North American Datum (NAD) 1927, European Datum (ED) 1950, Tokyo Datum (TD), Australian Geodetic Datum (AGD) 1966 and South American Datum (SAD) 1969.

At least eight quantities are needed to form a complete datum – three to specify the location of the origin of the coordinate system, three to specify the orientation of the coordinate system and two to specify the reference ellipsoid. In Moritz (1980) there is a discussion of recent determinations of the size of the reference ellipsoid.

The question of how many local horizontal datums defined under the classical definition are in use in the world may never be answered correctly. A very good attempt to list the datums in use today can be found in DIA (1971).

However, the problem with this list is that a number of datum names are either generic or extremely localized; in such cases it becomes practically impossible to associate them with a country or any region, e.g. Red Bridge Datum. In some cases, there are multiple names listed for the same datum, e.g. Leigon Pillar (GCS No. 1212), Accra, and Ghana and the listed names may not strictly relate to a geodetic datum or may refer to the 'approximate' versions of the same datum.

Due to 'localized' time saving readjustments, many local datums have been developed inadvertently within or in parallel to the original datums. These 'pseudo' datums then exist concurrently with other recognized coordinates, which are different for the same control points. Unless the two sets of coordinates indicate systematic biases and/or show very large differences, it would not be possible for users to suspect their existence. An interesting case is when two or more countries define datums with the same ellipsoid and name, but use different defining parameters. Fortunately this case is easy to identify and the chances of any mix up are almost nil.

3.1.1.1. Horizontal datums with multiple ellipsoids

In some cases, a local datum already defined by an ellipsoid was later redefined using a different ellipsoid. Two such datums are Old Hawaiian Datum and Timbalai Datum 1948.

In the case of Old Hawaiian Datum, the two ellipsoids, namely, Clarke 1866 and International 1924, were assumed to be tangent at an arbitrarily selected point. Then, polynomials were developed to establish transformations between the two coordinate sets based on the two reference ellipsoids.

3.1.2. Vertical datums

Heights[1] of surveyed stations or Bench Marks (BMs), which may or may not be part of the horizontal geodetic network, are used to adjust and define a vertical datum.

In keeping with the general definition of a datum, there must be a quantity that serves as a referent. In the case of a vertical datum this quantity is the geoid or, more practically, a surface which approximates the geoid. The geoid is defined as 'The equipotential surface of the earth's gravity field which best fits, in the least squares sense, mean sea level,' (Geodetic Glossary 1986). In the past, geodesists could not determine the geoid accurately enough over land areas for it to serve as a reliable reference surface. Hence, historically, Mean Sea Level (MSL) was used as the vertical datum. Sea level was (and still is) monitored and measured at Tidal Bench Marks (TBM) for a minimum period of 18.67 years to compute MSL. Today, determinations

---

[1] Strictly speaking, it would be better to use the term Orthometric Height (Geodetic Glossary 1986).

of the geoid by many scientists in different countries around the world have yielded adequate accuracy's so that the geoid can be used as a vertical datum over land areas (Lemoine *et al.* 1997).

In the past, such vertical datums were defined separately from the horizontal datum – even in the same country. A few examples are: National Geodetic Vertical Datum (NGVD) 1929, North American Vertical Datum (NAVD) 1988, Genoa 1942 and Bluff 1960.

The reference system for heights used in North America today is the NAVD 88. The NAVD 88 heights are the result of a mathematical (least squares) adjustment of the vertical control in North America. Over 500,000 permanent benchmarks are included in the datum. The datum surface is an equipotential surface that passes through a point on the International Great Lakes Datum. The datum closely corresponds with MSL along the coasts of the US. For further information about the NAVD 88 (see Zilkoski *et al.* 1992).

3.2. Geodetic reference systems

Soon after the launch of the first man-made satellite in 1957, geodesists defined geodetic coordinates with respect to a rectangular coordinate system whose origin is located at the center of mass (geocenter) of the earth. Using a mean earth (best fitting) ellipsoid, these coordinates can be converted to geodetic latitude, longitude and height above the ellipsoid using the inverse of equations (2.8)–(2.10). Figure 2.7a,b shows the relationship between rectangular coordinates and geodetic latitude and longitude.

The concept of a reference system is closely related to a datum. A geodetic reference system reflects the evolution of a datum given changes in technology. The introduction of Doppler and Global Positioning System (GPS) technologies led the way to a more global, rather than local, system. More importantly, these technologies demanded that the definitions of the coordinate system(s) be well-defined since the accuracy of GPS, for example, is 1 part in 10^8 of the distance between points. This implies that the orientation, size and scale of the (reference) coordinate system should be even more accurate.

Therefore, the modern geodetic reference system can be viewed as an evolved (updated) datum.

3.2.1. Modern geodetic reference systems

The practitioner in the geospatial sciences in most developed countries in the world will be faced with the possibility of using several reference systems. Therefore, it is important that the differences between these possibilities be discussed. North America will be used as an example. The fundamental concept behind the three reference systems that are discussed is nearly identical. However, the implementation (realization) of these coordinate systems is different (Snay and Soler 1999). The reason for this is straightforward. In order to implement these systems, measurements had to be obtained. Even if the scientists involved used the same instruments, times of observations, etc. they would not achieve the same results. The measurements would be contaminated by small errors and hence different results would be obtained.

The three reference systems that are in use in North America are the North American Datum of 1983 (NAD 83), the World Geodetic System of 1984 (WGS 84) and the International Terrestrial Reference System (ITRS). The practical result of the above discussion is that the coordinates of a point on the ground would be different in each of the above mentioned systems. The magnitude of this difference is as much as several meters, however, the difference in the distances between two points in each system would be quite small. In general, the differences between quantities that tend to be independent of the coordinate system will be very small.

For many applications in mapping and Geographic Information Systems (GIS) development, the coordinate differences between these systems may not be important. For geodetic control work, nearly all differences are important.

3.2.1.1. Reference system definition

Two broad aspects of the three reference systems will be discussed: the reference ellipsoid and the orientation of the system. The *reference ellipsoid* for each of the above reference systems is shown in Table 3.1.

The orientation of a reference system is complicated. This is because the rotation and wobble of the earth, at the level of centimeters, is somewhat irregular. For purposes of this manual, the z-axis of the rectangular coordinate system is aligned with the North Pole. However, because of the small irregularities of the earth's motions and the increased accuracy's of the instrumentation used to measure the motion, there is a need for something more precise than 'North Pole.'. This refinement is the International Reference Pole (IRP) as defined by an organization known as the International Earth Rotation Service (IERS). The IERS is headquartered in Paris, France. The x-axis of the right-handed system passes through the point of zero longitude located on the plane of the equator. The zero longitude is also defined by the IERS. This meridian passes through Greenwich, England, although this fact has more meaning historically today than it does scientifically.

The origin of the coordinate system is at the geocenter, and this has probably been accomplished to centimeter-level accuracy.

It should be pointed out that the land masses of the earth are in a state of constant motion and that even the center of mass moves slightly with respect to the solid earth. These motions are at the centimeter or millimeter per year level and are of interest, generally speaking, only to earth scientists. To them, not only are the coordinates important, but so are the velocities of the points. In Chapter 7, coordinates obtained

Table 3.1 Reference ellipsoid parameters for three reference systems

Reference system[a]	Reference ellipsoid semi-major axis (m)	Reference ellipsoid semi-minor axis (m)	Flattening, f ($f = (a - b)/a$)
NAD 83	6,378,137.0	6,356,752.314	1/298.257222101
WGS 84	6,378,137.0	6,356,752.314	1/298.257223563
ITRS	6,378,136.49	6,356,751.750	1/298.25645

[a] Note: the semi-major axis and the flattening are given quantities. The semi-minor axis is computed from them and rounded to the nearest millimeter.

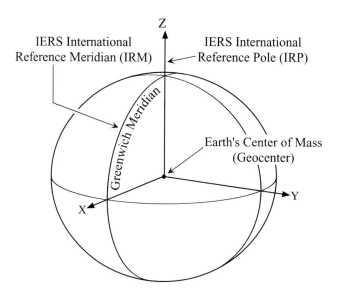

Figure 3.1 Geodetic reference system.

using GPS technology are discussed using the term Earth-Centered-Earth-Fixed (ECEF) coordinates. The three systems discussed in this chapter are ECEF systems. Figure 3.1 describes the coordinate system utilized by all three of the geodetic reference systems described above. The interested reader is referred to (McCarthy 1996; Kouba and Popelar 1994) for further details.

3.2.1.2. Usage of North American reference systems

It is important to understand the context of the general usage of each of these systems. The NAD 83 system is used by surveyors, geodesists, mappers, remote sensing users and those involved in GIS. Generally, these professionals occupy, with various instrumentation, NAD 83 geodetic monuments whose coordinates are known in the NAD 83 system. They make additional measurements that are computed using the NAD 83 system.

The WGS 84 reference system is very close to the NAD 83 system and provides the reference system for the GPS satellite system (see Section 7.2.1). The differences in the reference ellipsoids (Table 3.1) are negligible, but the coordinates may be different (<2 m in all cases) because of the different realization of the two systems. The fact that the two systems are nearly the same is evident in the (Federal Register 1995), wherein it is stated that 'all maps and charts produced for North America, at scales of 1:5,000 or smaller, that are based on *either* the NAD 83 or the WGS 84, should have the horizontal datum labeled as NAD 83/WGS 84.'

Almost all modern geodetic measurements are accomplished using GPS. The reason that geodesists and surveyors can occupy NAD 83 monuments, use NAD 83 coordinates and GPS observations (i.e., the WGS 84 system) is that they are generally using distances and angles derived from the GPS (WGS 84) system, which tend to be independent of the WGS 84 coordinate system.

The ITRS is the most frequently updated system and is used primarily by earth scientists (see Section 7.2.2). Both the NAD 83 and WGS 84 systems attempted to realize a system as close to the ITRS as possible at the last epoch at which they were realized. This implies that the technology, observations and supporting information was the best available at that time.

There are several important summary points to make. All three reference systems are very similar. They differ by parts in 10^8. They will probably all be revised on an aperiodic basis for decades to come (e.g. the ITRS is updated annually). Most of these future changes will not affect practitioners involved in the geospatial sciences, but will be important for earth scientists. It should be noted that many countries during the last decade or so, have redefined their datums to be 'close' to WGS 84. The principles of relating the three reference systems discussed herein are essentially the same.

3.2.2. Transformations between systems

This general topic is covered in detail in Chapter 4, where both traditional datum transformations and 3D transformations are discussed. In the case of North America, the National Geodetic Survey (NGS) in the US has developed transformation software that is freely available to all via the Internet. The URL is: http://www.ngs.noaa.gov/ tools/Htdp/Htdp.html. This software enables users to transform individual positions entered interactively or a collection of positions entered as a formatted file. Also, if users expect to transform only a few positions, they may exercise the Horizontal Time Dependent Positioning (HTDP) program interactively from this web page. A discussion of this issue can be found in the series of articles published in the Professional Surveyor (Snay and Soler 1999). Similar software and transformation procedures are available for many other developed countries.

The translations and rotations for all possible realizations of all three modern coordinate systems would result in dozens of sets of transformation values since each system has been almost continuously updated over the past decade. If the latest values were provided, they would be out of date by the time this manual is published. Therefore, the practitioner should consider contacting the following individuals/agencies for the latest and best information:

for WGS 84
Dr. Muneendra Kumar
National Imagery and Mapping Agency
4600 Sangamore Road
Bethesda, MD 20816-5003
kumarm@nima.mil

for NAD 83
Dr. Richard Snay
National Geodetic Survey
National Ocean Service
1315 East-West Highway
Silver Spring, MD 20910
rich@ngs.noaa.gov

for IERS
Dr. Claude Boucher
Central Bureau of the IERS – Observatoire de Paris
61, avenue de l'Observatoire
F-75014 Paris, France
boucher@ensg.ign.fr

3.3. Applications

Both datums and reference systems are used to determine new coordinates of points needed for boundary determination, navigation, construction, natural resource management and myriad other applications. Because datums have traditionally been divided into horizontal and vertical components, they are discussed accordingly.

3.3.1. Horizontal datums

Both local and global datums are used throughout the world. The typical procedure in surveying and geodesy is to occupy a monument with suitable equipment, such as a GPS receiver, and make observations. From these observations, coordinates of new points can be computed. There is an origin and an orientation associated with each datum but these are transparent to most practitioners. The coordinate values, azimuths to nearby points, origin, etc. can (usually) be obtained from the appropriate government agency in the country where the work is carried out.

In ocean areas, the horizontal network is simply *extended* for use in these areas. An implication of this extrapolation is that it generally leads to inaccurate horizontal positioning over these areas.

In their effort for safe marine and air navigation, the International Hydrographic Organization (IHO) in 1983 and International Civil Aviation Organization (ICAO) in 1989 have mandated the use of WGS 84 as a common geodetic system for global usage. Under this requirement, all nautical and aeronautical charts will be converted and civil airports worldwide surveyed in WGS 84.

As the use of the GPS in surveying and in automatic navigation became accepted for universal usage, the WGS 84 system was the most logical choice as a global geodetic system. In the interim, especially for marine navigation and nautical charts, local geodetic datums require transformation to WGS 84.

Due to the paucity of required common geodetic data, datum transformations could only be developed for a sub-set of the DIA's list. Table 3.2 lists one such sub-set (NIMA 1997).

3.3.2. Vertical datums

As mentioned earlier, the geoid is the ideal surface for referencing heights or elevations. However, from a practical point of view, MSL, as determined from tide gages, is usually utilized. The non-geodesist should be aware of the fact that elevations or heights are more complicated than they appear initially. For example, water does not necessarily run from a point that has a 'higher' GPS-determined-height to a 'lower' GPS-determined-height. This is because gravity affects the water such that this

Table 3.2 Geodetic datums/reference systems related to the World Geodetic System 1984 (through satellite ties)

Local geodetic datum	Associated ellipsoid
Adindan	Clarke 1880
Afgooye	Krassovsky 1940
Ain el Abd 1970	International 1924
American Samoa 1962	Clarke 1866
Anna 1 Astro 1965	Australian National
Antigua Island Astro 1943	Clarke 1880
Arc 1950	Clarke 1880
Arc 1960	Clarke 1880
Ascension Island 1958	International 1924
Astro Beacon "e" 1945	International 1924
Astro DOS 71/4	International 1924
Astro Tern Island (FRIG) 1961	International 1924
Astronomical Station 1952	International 1924
Australian Geodetic 1966	Australian National
Australian Geodetic 1984	Australian National
Ayabelle Lighthouse	Clarke 1880
Bellevue (IGN)	International 1924
Bermuda 1957	Clarke 1866
Bissau	International 1924
Bogota Observatory	International 1924
Campo Inchauspe	International 1924
Canton Astro 1966	International 1924
Cape	Clarke 1880
Cape Canaveral	Clarke 1866
Carthage	Clarke 1880
Chatham Island Astro 1971	International 1924
Chua Astro	International 1924
Coordinate System 1937 of Estonia	Bessel 1841
Corrego Alegre	International 1924
Dabola	Clarke 1880
Deception Island	Clarke 1880
Djakarta (Batavia)	Bessel 1841
DOS 1968	International 1924
Easter Island 1967	International 1924
European 1950	International 1924
European 1979	International 1924
Fort Thomas 1955	Clarke 1880
Gan 1970	International 1924
Geodetic Datum 1949	International 1924
Graciosa Base SW 1948	International 1924
Guam 1963	Clarke 1866
GUX 1 Astro	International 1924
Hjorsey 1955	International 1924
Hong Kong 1963	International 1924
Hu-Tzu-Shan	International 1924
Indian	Everest
Indian 1954	Everest
Indian 1960	Everest
Indian 1975	Everest
Indonesian 1974	Indonesian 1974
Ireland 1965	Modified airy

Table 3.2 (continued)

Local geodetic datum	Associated ellipsoid
ISTS 061 Astro 1968	International 1924
ISTS 073 Astro 1969	International 1924
Johnston Island 1961	International 1924
Kandawala	Everest
Kerguelen Island 1949	International 1924
Kertau 1948	Everest
Kusaie Astro 1951	International 1924
L. C. 5 Astro 1961	Clarke 1866
Leigon	Clarke 1880
Liberia 1964	Clarke 1880
Luzon	Clarke 1866
Mahe 1971	Clarke 1880
Massawa	Bessel 1841
Merchich	Clarke 1880
Midway Astro 1961	International 1924
Minna	Clarke 1880
Montserrat Island Astro 1958	Clarke 1880
M'Poraloko	Clarke 1880
Nahrwan	Clarke 1880
Naparima, BWI	International 1924
North American 1927	Clarke 1866
North American 1983	GRS 80
North Sahara 1959	Clarke 1880
Observatorio Meteorologico 1939	International 1924
Old Egyptian 1907	Helmert 1906
Old Hawaiian	Clarke 1866
Oman	Clarke 1880
Ordnance Survey of Great Britain 1936	Airy 1830
Pico de las Nieves	International 1924
Pitcairn Astro 1967	International 1924
Point 58	Clarke 1880
Pointe Noire 1948	Clarke 1880
Porto Santo 1936	International 1924
Provisional South American 1956	International 1924
Provisional South Chilean 1963	International 1924
Puerto Rico	Clarke 1866
Qatar National	International 1924
Qornoq	International 1924
Reunion	International 1924
Rome 1940	International 1924
S-42 (Pulkovo 1942)	Krassovsky 1940
Santo (DOS) 1965	International 1924
Sao Braz	International 1924
Sapper Hill 1943	International 1924
Schwarzeck	Bessel 1841
Selvagem Grande 1938	International 1924
Sierra Leone 1960	Clark 1880
S-JTSK	Bessel 1841
South American 1969	South American 1969
South Asia	Modified Fischer 1960
Timbalai 1948	Everest
Tokyo	Bessel 1841

Table 3.2 (continued)

Local geodetic datum	Associated ellipsoid
Tristan Astro 1968	International 1924
Viti Levu 1916	Clarke 1880
Voirol 1960	Clarke 1880
Wake-Eniwetok 1960	Hough 1960
Wake Island Astro 1952	International 1924
Zanderij	International 1924

seemingly strange phenomenon occurs. Therefore geodesists have developed easy-to-compute quantities that account for gravity effects.

The following terms are important to a reader who wants to pursue this further: Orthometric Height, Normal Orthometric Height and Quasi-geoid. The interested reader is referred to books on geodesy such as (Heiskanen and Moritz 1967).

The normal procedure in leveling is to occupy a monument (BM) and make observations with a level or other instrument. Given the elevation of the BM, the observed difference in elevation will permit the computation of the elevation of the new point. The datum enters into the picture because the elevation of the 'old' BM is referenced to the datum. However, in the vast majority of civil applications the difference in elevation is the needed quantity. For example, in determining the proper flow of liquid in a pipe, the difference in elevation, not the datum, is the critical quantity.

In coastal and ocean areas, the vertical datum, or a quantity related to it, is used as the water level on which the depths shown on nautical charts are based. Although a controversy still lingers concerning the use of Mean Low Water (MLW) and Mean Lower Low Water (MLLW) for a datum, it appears to have been resolved by the use of Lowest Astronomic Tide (LAT), (IHO 1998). The definitions and procedures needed to compute or model the tidal surfaces may vary between different agencies within the same country. One result of the varying definitions for the zero reference is that the bathymetric data from two different sources and ocean areas will not 'match' (Kumar 1996).

References

DIA, *Mapping, Charting, and Geodesy Support Systems, Data Elements and Related Features*, Technical Manual DIAM 65-5. Defense Intelligence Agency, Washington, DC, 1971.

Federal Register, 60(157), August 15, 1995.

Heiskanen, W. A. and Moritz, H., *Physical Geodesy*, W.H. Freeman and Co., San Francisco, CA, 1967.

IHO, Publication M-3, *Resolutions of the International Hydrographic Organization*, International Hydrographic Bureau, Monaco, 1998.

Kouba, J. and Popelar, J., *Modern Reference Frame for Precise Satellite Positioning and Navigation*, Proceedings of the International Symposium on Kinematic Systems in Geodesy, Geomatics and Navigation (FIS 94). Banff, Canada, August 1994, pp. 79–86.

Kumar, M., *Time-Invariant Bathymetry: A New Concept to Define and Survey Using GPS*, PACON International Proceedings, Hawaii, 1996.

Lemoine, F. G., Kenyon, S. C., Trimmer, R., Factor, J., Pavlis, N. K., Klosko, S. M., Chin, D. S., Torrence, M. H., Pavlis, E. C., Rapp, R. H. and Olson, T. R., *EGM96 The NASA GSFC and NIMA Joint Geopotential Model*. NASA Technical Memorandum, 1997.

McCarthy, D. D., 'IERS Conventions 1996', *IERS Technical Note*, 21, Paris, 1996.

Moritz, H., 'Geodetic Reference System 1980', *Bulletin Geodesique*, 54(3), Paris, 1980.

National Geodetic Survey, *Geodetic Glossary*, US Government Printing Office, Washington, DC, 1986.

NIMA, *Department of Defense World Geodetic System 1984, its Definition and Relationships with Local Geodetic Systems* (3rd ed.), National Imagery and Mapping Agency TR 8350.2, Bethesda, MD, 1997.

Snay, R. and Soler, T., *Professional Surveyor*, 19(10), and sequential series, 1999.

Torge, W., *Geodesy*, Walter de Gruyter, New York, 1980.

Zilkoski, D., Richards, J. and Young, G., 'Results of the general adjustment of the North American Vertical Datum of 1988', *Surveying and Land Information Systems*, 53(3), 1992.

Chapter 4

Coordinate transformations

Ayman Habib

4.1. Introduction

This chapter explains the concepts and the mechanics behind transforming coordinates between different systems. A right-handed coordinate system is always assumed. First, the concepts of two- and three- dimensional translation and rotation will be explained. Then, the seven-parameter transformation will be addressed along with some of its applications in geodetic activities. Finally, the general concepts and computational procedures for some map projections will be outlined.

4.2. Translation

The goal of this section is to establish the relationship between the coordinates of a given point and its coordinates in another coordinate systems that is shifted from the original system (i.e. they are parallel to each other with a shift in the origin).

4.2.1. Two-dimensional translation

In this section, two-dimensional coordinate systems are addressed. Assume that the two coordinate systems under consideration are (xy) and $(x'y')$, respectively. The second coordinate system $(x'y')$ is displaced through shifts x_T and y_T from the y- and x-axes of the first coordinate system (Figure 4.1). The mathematical relationship between the coordinates of a given point (P) with respect to these coordinate systems can be determined through visual inspection of Figure 4.1 which leads to the following equations:

$$x_p = x_T + x'_p$$

$$y_p = y_T + y'_p \qquad (4.1)$$

Where (x_p, y_p) are the coordinates of point P with respect to the xy- coordinate system; (x'_p, y'_p) are the coordinates of point P with respect to the $x'y'$- coordinate system.

4.2.2. Three-dimensional translation

Three-dimensional coordinate systems are dealt with in this section. Figure 4.2 illustrates the two coordinate systems (xyz) and $(x'y'z')$. The origin of the $(x'y'z')$ is shifted

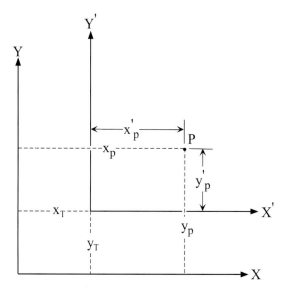

Figure 4.1 Two-dimensional translation.

with x_T, y_T and z_T relative to the origin of the first coordinate system (xyz). The coordinates of a certain point (P) with respect to these coordinate systems can be related to each other according to equation (4.2).

$$
\begin{bmatrix} x_p \\ y_p \\ z_p \end{bmatrix} = \begin{bmatrix} x_T \\ y_T \\ z_T \end{bmatrix} + \begin{bmatrix} x'_p \\ y'_p \\ z'_p \end{bmatrix}
\tag{4.2}
$$

Where (x_p, y_p, z_p) are the coordinates of point P with respect to the xyz- coordinate system and (x'_p, y'_p, z'_p) are the coordinates of point P with respect to the $x'y'z'$- coordinates system.

4.3. Rotation

The goal of this section is to establish the relationship between the coordinates of a given point with respect to two coordinate systems that are not parallel to each other and share the same origin (no shift is involved). One of the coordinate systems can be obtained from the other through a sequence of rotation angles.

4.3.1. Two-dimensional rotation

In this section, two-dimensional coordinate systems are considered. The coordinate system $(x'y')$, in Figure 4.3, can be obtained from the xy-coordinate system through a counter-clockwise rotation angle. The objective is to derive the mathematical relationship between the coordinates of a given point (P) with respect to these coordinate systems. Once again, by inspecting Figure 4.3 and using some trigonometric relation-

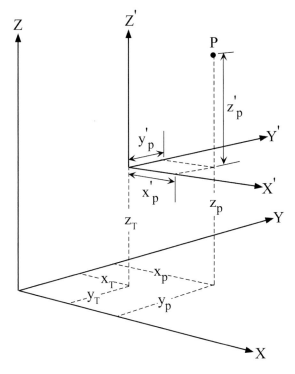

Figure 4.2 Three-dimensional translation.

ships, one can write equation (4.3):

$$\begin{bmatrix} x'_p \\ y'_p \end{bmatrix} = \begin{bmatrix} a + b \\ -c + d \end{bmatrix} = \begin{bmatrix} \cos\alpha x_p + \sin\alpha y_p \\ -\sin\alpha x_p + \cos\alpha y_p \end{bmatrix} \tag{4.3}$$

Where (x_p, y_p) are the coordinates of point P with respect to the xy- coordinate system; (x'_p, y'_p) are the coordinates of point P with respect to the $x'y'$- coordinate system.

Equation (4.3) can be rewritten in a matrix form as follows:

$$\begin{bmatrix} x'_p \\ y'_p \end{bmatrix} = \begin{bmatrix} \cos\alpha & \sin\alpha \\ -\sin\alpha & \cos\alpha \end{bmatrix} \begin{bmatrix} x_p \\ y_p \end{bmatrix} \tag{4.4}$$

A matrix is simply a notation representing a series of equations. For a more detailed explanation about rotation matrices and their properties, the reader can refer to Section 4.3.2.

4.3.2. Three-dimensional rotation

Now, three-dimensional coordinate systems are investigated. Assume that the two coordinate systems (xyz) and $(x'y'z')$ are under consideration. These two coordinate systems can be related to each other through a sequence of rotation angles (i.e. one

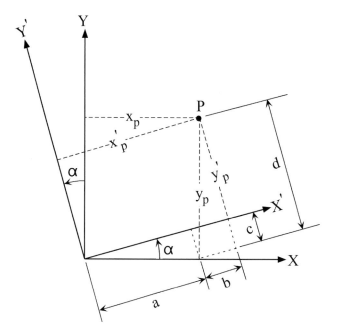

Figure 4.3 Two-dimensional rotation.

coordinate system can be obtained by rotating the other system around the x-, y- and z-axes, respectively). In Figure 4.4, the xyz- and the $x'y'z'$-coordinate systems can be made parallel by rotating the former system as follows:

- Primary rotation angle (ω) around the x-axis
- Secondary rotation angle (ϕ) around the y-axis
- Tertiary rotation angle (κ) around the z-axis

One should note that the rotation order could be arbitrarily chosen. The above-mentioned order is just an example of how to make the involved coordinate systems parallel. In this section, the relationship between the coordinates of a given point (P), with respect to these coordinate systems, as a function of the rotation angles is established.

(a) Primary rotation (ω) around the x-axis

Assume that the xyz-coordinate system has been rotated through an angle (ω) around the x-axis yielding the $x_\omega y_\omega z_\omega$-coordinate system (Figure 4.5). The mathematical relationship between the coordinates of a given point with respect to these coordinate systems can be seen in equation (4.5).

$$\begin{bmatrix} x_\omega \\ y_\omega \\ z_\omega \end{bmatrix} = \begin{bmatrix} 1 & 0 & 0 \\ 0 & \cos\omega & \sin\omega \\ 0 & -\sin\omega & \cos\omega \end{bmatrix} \begin{bmatrix} x \\ y \\ z \end{bmatrix} = R_\omega \begin{bmatrix} x \\ y \\ z \end{bmatrix} \tag{4.5}$$

(b) Secondary rotation (ϕ) around the y_ω-axis

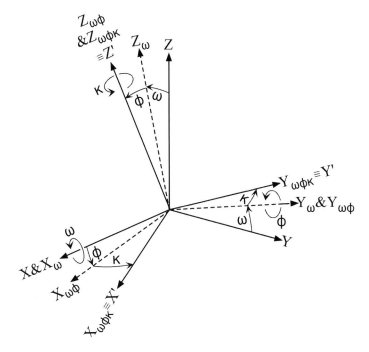

Figure 4.4 Three-dimensional rotation.

After the primary rotation (ω), assume that the $x_\omega y_\omega z_\omega$-coordinate system has been rotated with an angle (ϕ) around the $y\omega$-axis yielding the $x_{\omega\phi} y_{\omega\phi} z_{\omega\phi}$-coordinate system (Figure 4.6). The mathematical relationship between the coordinates of a given point with respect to these coordinate systems, as a function of the rotation angle ϕ can be seen in equation (4.6).

$$\begin{bmatrix} x_{\omega\phi} \\ y_{\omega\phi} \\ z_{\omega\phi} \end{bmatrix} = \begin{bmatrix} \cos\phi & 0 & -\sin\phi \\ 0 & 1 & 0 \\ \sin\phi & 0 & \cos\phi \end{bmatrix} \begin{bmatrix} x_\omega \\ y_\omega \\ z_\omega \end{bmatrix} = R_\phi \begin{bmatrix} x_\omega \\ y_\omega \\ z_\omega \end{bmatrix} \tag{4.6}$$

(c) Tertiary rotation (κ) around the $z_{\omega\phi}$-axis

Finally, assume that the $x_{\omega\phi} y_{\omega\phi} z_{\omega\phi}$-coordinate system has been rotated with an angle (κ) around the $z_{\omega\phi}$-axis yielding the $x_{\omega\phi\kappa} y_{\omega\phi\kappa} z_{\omega\phi\kappa}$-coordinate system (Figure 4.7). The mathematical relationship between the coordinates of a given point with respect to these coordinate systems, as a function of the rotation angle κ can be seen in equation (4.7).

$$\begin{bmatrix} x_{\omega\phi\kappa} \\ y_{\omega\phi\kappa} \\ z_{\omega\phi\kappa} \end{bmatrix} = \begin{bmatrix} \cos\kappa & \sin\kappa & 0 \\ -\sin\kappa & \cos\kappa & 0 \\ 0 & 0 & 1 \end{bmatrix} \begin{bmatrix} x_{\omega\phi} \\ y_{\omega\phi} \\ z_{\omega\phi} \end{bmatrix} = R_\kappa \begin{bmatrix} x_{\omega\phi} \\ y_{\omega\phi} \\ z_{\omega\phi} \end{bmatrix} \tag{4.7}$$

Assuming that the $x_{\omega\phi\kappa} y_{\omega\phi\kappa} z_{\omega\phi\kappa}$-coordinate system is parallel to the $x'y'z'$-coordinate system and by substituting equations (4.5) and (4.6) into equation (4.7), one gets the following equation:

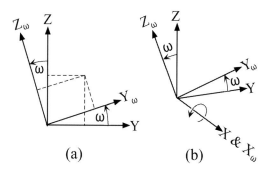

Figure 4.5 Primary rotation (ω) around the x-axis.

$$
\begin{bmatrix} x' \\ y' \\ z' \end{bmatrix} = R_\kappa R_\phi R_\omega \begin{bmatrix} x \\ y \\ z \end{bmatrix} = R(\omega, \phi, \kappa) \begin{bmatrix} x \\ y \\ z \end{bmatrix} \tag{4.8}
$$

The product of the three matrices ($R\kappa R_\phi R_\omega$) in equation (4.8) is referred to as the rotation matrix (R). It is a function of the rotation angles ω, ϕ and κ and gives the relationship between the coordinates of a given point with respect to the involved coordinated systems.

Remarks on the rotation matrix (R)

1 The rotation matrix is an orthogonal matrix. This means that its inverse will be equal to its transpose (i.e. $R^{-1} = R^T$). This stems from the fact that the magnitude (norm) of any column or row is unity (REA, 1997). Also, the dot product of any two rows or columns is zero.

2 The elements of the rotation matrix R will change as a result of changing the rotation order (i.e. $R_{\omega, \phi, \kappa} \neq R_{\phi, \omega, \kappa}$).

3 A positive rotation angle is one that is counter-clockwise when looking at the coordinate system with the positive direction of the rotation axis pointing towards us.

4.4. Three-dimensional similarity transformation (seven-parameter transformation)

In the previous two sections, the influence of rotation and translation between two coordinate systems was discussed. The scales along the axes of the involved coordinate systems were assumed to be equal. The impact of having uniform change in the scale will be considered in this section. Considering three rotation angles, one around each axis, and three translations, one in the direction of each axis, and a uniform scale change, a

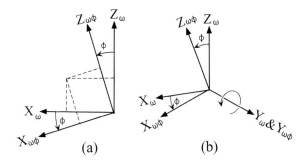

Figure 4.6 Secondary rotation (ϕ) around the y_ω-axis.

total of seven parameters must be involved in the transformation between the coordinate systems (Figure 4.8). This transformation is called a seven-parameter transformation, or sometimes a linear conformal transformation in three dimensions, or simply a three-dimensional similarity transformation. The mathematical model will be explained in the next section, followed by some geodetic and photogrammetric applications.

4.4.1. Mathematical model

If (x, y, z) and (x', y', z') are the rectangular coordinates of a given point before and after the transformation, the three-dimensional similarity transformation can be given by:

$$
\begin{bmatrix} x' \\ y' \\ z' \end{bmatrix} = \begin{bmatrix} x_T \\ y_T \\ z_T \end{bmatrix} + SR(\omega, \phi, \kappa) \begin{bmatrix} x \\ y \\ z \end{bmatrix} \tag{4.9}
$$

Where: S is the scale factor, (x_T, y_T, z_T) are the involved translations and $R(\omega, \phi, \kappa)$ is a three by three matrix which rotates the xyz -coordinate system to make it parallel to the $x'y'z'$ -system.

Equation (4.9) can be sequentially explained as follows:

- The xyz -coordinate system is made parallel to the $x'y'z'$ -system through a multiplication with the rotation matrix $R(\omega, \phi, \kappa)$.
- The uniform change in the scale is compensated for through the multiplication with the scale factor S.
- Finally the three translations x_T, y_T, z_T are applied in the x', y' and the z' directions, respectively.

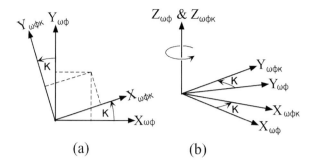

Figure 4.7 Tertiary rotation (κ) around the $z_{\omega\phi}$-axis.

4.4.2. Applications

To illustrate the importance of three-dimensional similarity transformation, its use in geodetic and photogrammetric applications is explained.

4.4.2.1. The collinearity equations

Photogrammetry can be defined as the art and science of deriving three-dimensional positions of objects from imagery (Kraus 1993). The main objective of photogrammetry is to inverse the process of photography. When the film inside the camera is exposed to light, light rays from the object space pass through the camera perspective center (the lens) until it hits the focal plane (film) producing images of the photographed objects. The main mathematical model that has been in use in the majority of photogrammetric applications is the collinearity equations. The collinearity equations mathematically describe the fact that the object point, the corresponding image point and the perspective center lie on a straight line, Moffitt and Mikhail (1980) (Figure 4.9).

The collinearity equations can be derived using the concept of the three-dimensional similarity transformation (Figure 4.10).

The reader should note that the focal plane is shown in the negative position (above the perspective center) in Figure 4.9 but it is shown in the positive position (below the perspective center) in Figure 4.10. Image coordinate measurements are accomplished on the positive, usually called the diapositive. The diapositive is used in all the photogrammetric mathematical derivations since it maintains directional relationship between points in object and image space. The directional relationship is not maintained between the object space and the negative (i.e. the negative is an inverted perspective representation of the object space).

Before deriving the collinearity equations, the following coordinate systems need to be defined (Figure 4.10):

- The ground coordinate system (XYZ): this can be a local coordinate system asso-

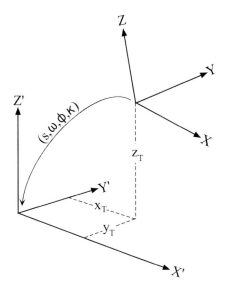

Figure 4.8 Three-dimensional similarity transformation.

ciated with the area being photographed (e.g. state plane coordinate system and height).

- The image coordinate system (*xyz*): this coordinate system is associated with the camera body. All the measurements in the image plane are referenced to this coordinate system.

The three angles ω, ϕ and κ describe the rotational relationship between the image and the ground coordinate systems. Now, let us define the following points:

- The origin of the ground coordinate system will be denoted by (*O*)
- The perspective center of the lens is denoted by (*P*)
- (*A*) and (*a*) will denote the object point and the corresponding image point, respectively

The vector (one dimensional array, REA, 97) connecting the origin of the ground coordinate system and the object point is given as:

$$\vec{V}_{OA} = \begin{bmatrix} X_A \\ Y_A \\ Z_A \end{bmatrix} \qquad (4.10)$$

One should note that this vector is referenced relative to the ground coordinate system. The vector connecting the origin of the ground coordinate system to the perspective center is defined by:

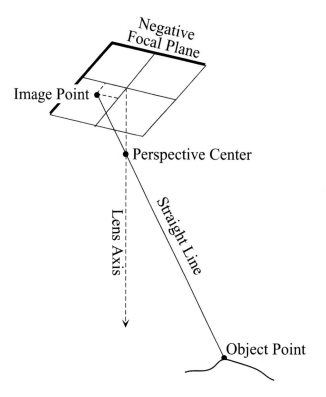

Figure 4.9 The principal of the collinearity condition.

$$\vec{V}_{OP} = \begin{bmatrix} X_P \\ Y_P \\ Z_P \end{bmatrix} \qquad (4.11)$$

Once again this vector is defined relative to the ground coordinate system. Finally, the vector connecting the perspective center to the image point is given by:

$$\vec{v}_{Pa} = \begin{bmatrix} x_a \\ y_a \\ -c \end{bmatrix} \qquad (4.12)$$

where (c) is the principal distance (the normal distance between the perspective center and the image plane). $(x_a, y_a, -c)$ are the coordinates of the image point (a) relative to the image coordinate system. The vector connecting the perspective center and the image point can be transformed into the ground coordinate system through a multiplication with the rotation matrix $R(\omega, \phi, \kappa)$ as follows:

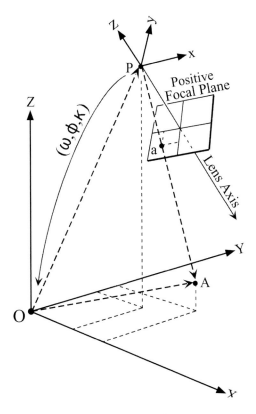

Figure 4.10 The concept of collinearity equations.

$$\vec{V}_{Pa} = R(\omega, \phi, \kappa) \begin{bmatrix} x_a \\ y_a \\ -c \end{bmatrix} \qquad (4.13)$$

The vector connecting the perspective center and the object point under consideration can be obtained by multiplying equation (4.13) with a scale factor, S (Figure 4.10).

$$\vec{V}_{PA} = SR(\omega, \phi, \kappa) \begin{bmatrix} x_a \\ y_a \\ -c \end{bmatrix} \qquad (4.14)$$

The three vectors in equations (4.10), (4.11) and (4.14) form a closed polygon. Therefore, the following equation can be written:

$$\vec{V}_{OA} = \vec{V}_{OP} + \vec{V}_{PA}$$

$$\begin{bmatrix} X_A \\ Y_A \\ Z_A \end{bmatrix} = \begin{bmatrix} X_P \\ Y_P \\ Z_P \end{bmatrix} + SR(\omega, \phi, \kappa) \begin{bmatrix} x_a \\ y_a \\ -c \end{bmatrix} \tag{4.15}$$

Note the similarity between equations (4.15) and (4.9). One should also note that these vectors are all referenced to the ground coordinate system. Rearranging the terms in equation (4.15) and using the orthogonality property of the rotation matrix (refer to Three-dimensional rotation), this equation can be rewritten as follows:

$$\begin{bmatrix} x_a \\ y_a \\ -c \end{bmatrix} = 1/SR^T(\omega, \phi, \kappa) \begin{bmatrix} X_A - X_P \\ Y_A - Y_P \\ Z_A - Z_P \end{bmatrix} \tag{4.16}$$

Equation (4.16) is one form of the collinearity equations. The classical and traditional form of the collinearity equations can be obtained by dividing the first two lines in equation (4.16) by the third line (thus, the scale factor S is eliminated), which leads to equation (4.17).

$$x_a = -c\frac{r_{11}(X_A - X_P) + r_{21}(Y_A - Y_P) + r_{31}(Z_A - Z_P)}{r_{13}(X_A - X_P) + r_{23}(Y_A - Y_P) + r_{33}(Z_A - Z_P)}$$

$$y_a = -c\frac{r_{12}(X_A - X_P) + r_{22}(Y_A - Y_P) + r_{32}(Z_A - Z_P)}{r_{13}(X_A - X_P) + r_{23}(Y_A - Y_P) + r_{33}(Z_A - Z_P)} \tag{4.17}$$

4.4.2.2. Datum to datum transformation

As mentioned in Chapter two, an ellipsoid of revolution, or simply a reference ellipsoid, is a reasonable approximation of the earth's surface. The shape (first two parameters) and the position of the reference ellipsoid in space are defined by the following parameters:

- The semi-major axis (a)
- The semi-minor axis (b), eccentricity (e) or flattening (f)
- The position of the center of the reference ellipsoid relative to the earth's center of gravity, and
- The rotational relationship between the ellipsoid axes and the natural coordinate system associated with the earth (as defined by the earth's axis of rotation and the equatorial plane).

The geodetic latitude (ϕ), the geodetic longitude (λ) and the ellipsoidal height (h) define the position of any point (P) in space. These quantities are specific for a certain reference ellipsoid. If, for any reason, the geodetic datum is changed – that is, the reference ellipsoid and its position – then the geodetic coordinates (ϕ, λ, h) will also change. This problem is simply a transformation of coordinates. In this section, the mathematical derivations of this type of coordinate transformation are not explained.

Rather, alternative approaches to solving this problem are described. For a detailed description, the reader can refer to Heiskanen and Moritz (1996).

First approach (direct approach):

In this approach, differential equations that relate changes in the geodetic coordinates of a given point to changes in the geodetic datum are formulated, see equation (4.18).

$$a\delta\phi = \sin\phi\cos\lambda\delta x_o + \sin\phi\sin\lambda\delta y_o - \cos\phi\delta z_o + 2a\sin\phi\cos\phi\delta f$$

$$a\cos\phi\delta\lambda = \sin\lambda\delta x_o - \cos\lambda\delta y_o$$

$$\delta h = -\cos\phi\cos\lambda\delta x_o - \cos\phi\sin\lambda\delta y_o - \sin\phi\delta z_o - \delta a + a\sin^2\phi\delta f \qquad (4.18)$$

where $(\delta\phi, \delta\lambda, \delta h)$ are the changes in the geodetic coordinates of a given point due to changes of $(\delta x_o, \delta y_o, \delta z_o, \delta a, \delta f)$ in the position, semi-major axis and the flattening of the reference ellipsoid, respectively. Note that in equation (4.18), the axes of the reference ellipsoids are assumed to be parallel (no rotation angles are involved).

Second approach (indirect approach):

In the previous approach, a mathematical function that directly relates the change in the geodetic coordinates to the changes in the parameters of the corresponding reference ellipsoid was shown. In the following approach, this transformation can be carried out sequentially as follows:

• Transform the geodetic coordinates into Cartesian geocentric coordinates using the shape parameters of the initial reference ellipsoid (refer to Section 2.6, equations (2.8)–(2.10)).
• Apply three-dimensional similarity transformation to the geocentric coordinates that were determined in the previous step. The similarity transformation should compensate for the change in the position and the orientation between the new and the original reference ellipsoids.
• Transform the new Cartesian geocentric coordinates into geodetic coordinates using the shape parameters of the new reference ellipsoid.

One should note that both direct and indirect approaches are expected to yield almost identical results. However, some approximations in the first approach are incorporated to simplify the transformation formulas. These approximations might lead to relatively less accurate results compared to the second approach.

4.5. Map projections and transformations

A map projection is a systematic representation of all or part of the surface of a round body, especially the earth, on a plane (Snyder 1987). The transformation from a round surface to a plane cannot be accomplished without distortion. If the map covers a large area of the earth's surface – e.g. a continent – distortions will be visually apparent. If the region covered by the map is small, distortions might be barely measurable using many projections, yet it can be serious with other projections.

The characteristics that are normally considered when choosing a map projection can be listed as follows:

- Area: Some map projection schemes are designed to be equal area. This means that equal areas on the map correspond to equal areas on the earth's surface.
- Shape: Map projections can be conformal. This means that relative local angles about every point on the map are shown correctly. One important result of conformality is that the local scale in every direction around any point is constant. Since local angles are correct, meridians intersect parallels at right angles on a conformal projection. One should note that no map can be both conformal and equal area.
- Scale: No map projection can show the scale correctly throughout the whole map. Rather, the map projection can be designed in such a way that guarantees uniform scale along one or more lines. Some map projections show true scale along every meridian.
- Direction: Conformal projection gives the relative local directions correctly at any given point. In this type of projection, the directions or azimuths of all points on the map are shown correctly with respect to the center.
- Special Characteristics: Some map projections offer special characteristics such as lines of constant directions are shown as straight lines. Some other projections show the shortest distance between two points as straight lines.

Developable surfaces are ones that can be transformed to a plane without distortion. There are three types of developable surfaces that can be used for map projection; namely the cylinder, the cone and the plane. All three surfaces are variations of the cone. A cylinder is a limiting form of a cone with an increasingly sharp apex. As the cone becomes flatter, its limit is a plane.

A cylinder, whose axis is coincident with the earth's polar axis, can be wrapped around the globe with its surface touching the equator. Or, it can intersect the globe (i.e. be secant to) at two parallels. In either case, the meridians may be projected onto the cylinder as equidistant straight lines perpendicular to the equator. Also, the parallels can be drawn as lines parallel to the equator. When the cylinder is cut along one of the meridians and unrolled, a cylindrical projection with straight meridians and straight parallels results. The Mercator projection is the best-known example of a cylindrical projection.

Similarly, a cone can be placed over the globe with its apex along the polar axis of the earth and its surface either touching the earth's surface at a particular parallel of latitude, or intersecting the earth's surface (i.e. secant to) at two parallels (Figure 2.9). In this case, the meridians can be projected onto the cone as equi-distant straight lines radiating from the apex. The parallels, on the other hand, can be drawn as lines around the circumference of the cone in planes perpendicular to the earth's axis. When the cone is cut along a meridian, unrolled and laid flat, the meridians remain straight lines and the parallels will appear as circular arcs centered at the apex. The angles between the meridians are shown smaller than the true angles. This type of projection is known as conic or conical map projection.

A plane tangent to the earth at one of its poles is the basis for polar azimuthal projections. The meridians are projected as straight lines radiating from a point. The meridians are spaced at their true angles instead of smaller angles of the conic projections. The parallels of latitude will appear as complete circles, centered at the pole.

The concepts outlined above might be modified in two ways and still produce cylindrical, conic or azimuthal projections.

1 As discussed earlier, the cylinder or cone may be secant to or cut the globe at two parallels (standard parallels) instead of being tangent at just one. This conceptually provides two standard parallels. Also, the plane may cut the globe at any parallel instead of touching a pole.
2 The axis of the cylindrical or cone can have a direction different from that of the earth's axis of rotation. Similarly, the plane may be tangent to any point other than the pole. This modification leads to oblique, transverse and equatorial projections. In this case, meridians and parallels of latitude are no longer straight lines or arcs of circles.

In the following sections, the projection formulas are listed in a manner suitable for computer computations. All the angular elements are assumed to be in radians.

4.5.1. The transverse Mercator projection

The main characteristics of the Transverse Mercator projection can be summarized as follows:

- It is a cylindrical (transverse) projection.
- It is a conformal projection.
- The central meridian, the meridians 90° away from the central meridians and the equator will appear as straight lines.
- Other meridians and parallels will appear as complex curves.
- The central meridian and the two nearly straight lines equidistant from and almost parallel to the central meridian will have true scale.
- This projection is mainly used for quadrangle maps at scales from 1:24,000 to 1:250,000.

4.5.1.1. Geodetic to xy-transverse Mercator (forward formulas)

Given the reciprocal of the flattening ($RF = 1/f$) and the semi-major axis (a) of the reference ellipsoid, together with the latitude and the longitude of the origin of the rectangular coordinates (ϕ_o, λ_o), and the scale factor along the central meridian, one can compute the following constants:

$$f = 1.0/RF \tag{4.19}$$

$$e^2 = 2f - f^2 \tag{4.20}$$

$$e'^2 = e^2/(1 - e^2) \tag{4.21}$$

$$p' = (1.0 - f)a \tag{4.22}$$

$$e_n = (a - p')/(a + p') \tag{4.23}$$

$$A = -1.5e_n + 9.0/16.0e_n^3 \tag{4.24}$$

$$B = 0.93750e_n^2 - 15.0/32.0e_n^4 \tag{4.25}$$

$$C = -35.0/48.0e_n^3 \tag{4.26}$$

$$U = 1.5e_n - 27.0/32.0e_n^3 \tag{4.27}$$

$$V = 1.31250e_n^2 - 55.0/32.0e_n^4 \tag{4.28}$$

$$W = 151.0/96.0e_n^3 \tag{4.29}$$

$$R = a(1.0 - e_n)(1.0 - e_n^2)(1.0 + 2.25e_n^2 + 225.0/64.0e_n^4) \tag{4.30}$$

$$M_o = \phi_o + A\sin(2.0\phi_o) + B\sin(4.0\phi_o) + C\sin(6.0\phi_o) \tag{4.31}$$

$$S_o = \text{Scale Factor } RM_o \tag{4.32}$$

For a given (ϕ, λ), one can compute the xy-transverse Mercator rectangular coordinates as follows:

$$M = \phi + A\sin(2.0\phi) + B\sin(4.0\phi) + C\sin(6.0\phi) \tag{4.33}$$

$$S = RM \text{ Scale Factor} \tag{4.34}$$

$$T_N = \sin\phi/\cos\phi \tag{4.35}$$

$$T_s = T_N^2 \tag{4.36}$$

$$ETS = e'^2\cos^2\phi \tag{4.37}$$

$$L = (\lambda - \lambda_o)\cos\phi \tag{4.38}$$

$$RN = (\text{Scale Factor } a)/\sqrt{1.0 - e^2\sin^2\phi} \tag{4.39}$$

$$A2 = 0.5RN \, T_N \tag{4.40}$$

$$A4 = (5.0 - T_s + ETS(9.0 + 4.0ETS))/12.0 \tag{4.41}$$

$$A6 = \{61.0 + T_s(T_s - 58.0) + ETS(270.0 - 330.0T_s)\}/360.0 \tag{4.42}$$

$$A1 = -RN \tag{4.43}$$

$$A3 = (1.0 - T_s + ETS)/6.0 \tag{4.44}$$

$$A5 = \{5.0 + T_s(T_s - 18.0) + ETS(14.0 - 58.0T_s)\}/120.0 \tag{4.45}$$

$$A7 = \{61.0 - 479.0T_s + 179.0T_s^2 - T_s^3\}/5040.0 \tag{4.46}$$

$$y = S - S_o + A2L^2[1.0 + L^2(A4 + A6L^2)] + \text{False Northing} \tag{4.47}$$

$$x = \text{False Easting} - A1L[1.0 + L^2\{A3 + L^2(A5 + A7L^2)\}] \tag{4.48}$$

Note that the false easting and false northing are used to avoid negative coordinates.

4.5.1.2. xy-Transverse Mercator into geodetic coordinate transformation (inverse formulas)

Given (x, y) rectangular coordinates, the corresponding geodetic coordinates need to be computed. First, compute the following constants:

$$f = 1.0/RF \tag{4.49}$$

$$e^2 = 2f - f^2 \tag{4.50}$$

$$e'^2 = e^2/(1 - e^2) \tag{4.51}$$

$$p' = (1.0 - f)a \tag{4.52}$$

$$e_n = (a - p')/(a + p') \tag{4.53}$$

$$C2 = -1.5e_n + 9.0/16.0e_n^3 \tag{4.54}$$

$$C4 = 15.0/16.0e_n^2 - 15.0/32.0e_n^4 \tag{4.55}$$

$$C6 = -35.0/48.0e_n^3 \tag{4.56}$$

$$C8 = 315.0/512.0e_n^4 \tag{4.57}$$

$$U0 = 2.0(C2 - 2.0C4 + 3.0C6 - 4.0C8) \tag{4.58}$$

$$U2 = 8.0(C4 - 4.0C6 + 10.0C8) \tag{4.59}$$

$$U4 = 32.0(C6 - 6.0C8) \tag{4.60}$$

$$U6 = 128.0C8 \tag{4.61}$$

$$C1 = 1.5e_n - 27.0/32.0e_n^3 \tag{4.62}$$

$$C3 = 21.0/16.0e_n^2 - 55.0/32.0e_n^4 \tag{4.63}$$

$$C5 = 151.0/96.0e_n^3 \tag{4.64}$$

$$C7 = 1097.0/512.0e_n^4 \tag{4.65}$$

$$V0 = 2.0(C1 - 2.0C3 + 3.0C5 - 4.0C7) \tag{4.66}$$

$$V2 = 8.0(C3 - 4.0C5 + 10.0C7) \tag{4.67}$$

$$V4 = 32.0(C5 - 6.0C7) \tag{4.68}$$

$$V6 = 128.0C7 \tag{4.69}$$

$$R = a(1.0 - e_n)(1.0 - e_n^2)(1.0 + 2.25e_n^2 + 225.0/64.0e_n^4) \tag{4.70}$$

$$M_o = \phi_o + \sin\phi_o\cos\phi_o(U0 + U2\cos^2\phi_o + U4\cos^4\phi_o + U6\cos^6\phi_o) \tag{4.71}$$

$$S_o = \text{Scale Factor } RM_o \tag{4.72}$$

Then, one can proceed to compute the corresponding geodetic coordinates as follows:

$$M = (y - \text{False Northing} + S_o)/(R \text{ Scale Factor}) \tag{4.73}$$

$$F = M + \sin M \cos M(V0 + V2\cos^2 M + V4\cos^4 M + V6\cos^6 M) \tag{4.74}$$

$$T_N = \sin F/\cos F \tag{4.75}$$

$$ETS = e'^2\cos^2 F \tag{4.76}$$

$$RN = a \text{ Scale Factor}/\sqrt{1 - e^2\sin^2 F} \tag{4.77}$$

$$Q = (x - \text{False Easting})/RN \tag{4.78}$$

$$B2 = -0.5T_N(1.0 + ETS) \tag{4.79}$$

$$B4 = -(5.0 + 3.0T_N^2 + ETS[1.0 - 9.0T_N^2] - 4.0ETS^2)/12.0 \tag{4.80}$$

$$B6 = (61.0 + 45.0T_N^2[2.0 + T_N^2] + ETS[46.0 - 252.0T_N^2 - 60T_N^4])/360.0 \tag{4.81}$$

$$B1 = 1.0 \tag{4.82}$$

$$B3 = -(1.0 + 2.0T_N^2 + ETS)/6.0 \tag{4.83}$$

$$B5 = (5.0 + T_N^2[28.0 + 24.0T_N^2] + ETS[6.0 + 8.0T_N^2])/120.0 \tag{4.84}$$

$$B7 = -(61.0 + 662.0T_N^2 + 1320.0T_N^4 + 720.0T_N^6)/5040.0 \tag{4.85}$$

$$\phi = F + B2Q^2(1.0 + Q^2[B4 + B6Q^2]) \tag{4.86}$$

$$L = B1Q(1.0 + Q^2[B3 + Q^2\{B5 + B7Q^2\}])$$ (4.87)

$$\lambda = L/\cos F + \lambda_o$$ (4.88)

4.5.2. The Lambert (conic) projection

The main characteristics of the Lambert projection can be summarized as follows:

- Lambert projection is conic.
- Lambert projection is conformal.
- Parallels are unequally spaced arcs of concentric circles, more closely spaced near the center of the map.
- Meridians are equally spaced radii of the same circles. Thus, the parallels will cut the meridians at a right angle.
- The scale will be true along the standard parallels.
- This projection is used for regions and countries with predominant east-west extent.
- The pole in the same hemisphere as the standard parallels will appear as a point. The other pole will be at infinity.

Now, the mathematical formulas to transform geodetic into xy-Lambert coordinates (forward equations) and the inverse formulas for transforming xy-Lambert into geodetic coordinates will be listed. In both cases, the following parameters that describe the reference datum and the location of the projection surface (the cone) are given:

- The semi major axis of the reference ellipsoid (a)
- The first eccentricity (e)
- The latitudes ϕ_1 and ϕ_2 of the standard parallels
- The latitude and the longitude ϕ_o and λ_o of the origin of the rectangular coordinates

Given these parameters, one can compute the following constants:

$$m = \frac{\cos\phi}{\sqrt{1.0 - e^2\sin^2\phi}}$$ (4.89)

substituting ϕ_1 and ϕ_2 yields m_1 and m_2

$$t = \frac{\tan(\pi/4 - \phi/2)}{\left(\dfrac{1.0 - e\sin\phi}{1.0 + e\sin\phi}\right)^{0.5e}}$$ (4.90)

ϕ_o, ϕ_1, ϕ_2 yields t_0, t_1 and t_2

$$n = \frac{\log m_1 - \log m_2}{\log t_1 - \log t_2}$$ (4.91)

$$F = \frac{m_1}{nt_1^n}$$ (4.92)

$$R_o = aFt_o^n \tag{4.93}$$

4.5.2.1. Geodetic to xy-Lambert coordinate transformation (forward equations)

Here, the geodetic coordinates (ϕ, λ) of any point are given. The goal is to compute the corresponding rectangular coordinates. The computational procedure can be listed as follows:

$$t = \frac{\tan(\pi/4 - \phi/2)}{\left(\dfrac{1.0 - e\sin\phi}{1.0 + e\sin\phi}\right)^{0.5e}} \tag{4.94}$$

$$R = a\,F\,t^n \tag{4.95}$$

$$\theta = n(\lambda - \lambda_o) \tag{4.96}$$

$$x = R\sin\theta + \text{False Easting} \tag{4.97}$$

$$y = R_o - R\cos\theta + \text{False Northing} \tag{4.98}$$

4.5.2.2. xy-Lambert to geodetic coordinate transformation (inverse equations)

Here, the rectangular coordinates are given for the purpose of computing the corresponding geodetic coordinates. The computation procedure should be as follows:

$$x_t = x - \text{False Easting} \tag{4.99}$$

$$y_t = y - \text{False Northing} \tag{4.100}$$

$$r = \frac{n}{|n|}\sqrt{x_t^2 + (R_o - y_t)^2} \tag{4.101}$$

$$\varTheta = \tan^{-1}\left(\frac{x_t}{R_o - y_t}\right) \tag{4.102}$$

$$T = \left(\frac{r}{aF}\right)^{1.0/n} \tag{4.103}$$

$$\phi = \pi/2.0 - 2.0\tan^{-1}\left(T\left(\frac{1.0 - e\sin\phi}{1.0 + e\sin\phi}\right)^{e/2.0}\right) \tag{4.104}$$

To compute the latitude ϕ in equation (4.104), one has to iterate starting from an initial value (e.g. $\phi = \pi/2.0 - 2.0\tan^{-1}(T)$). Finally, the longitude can be computed as follows:

$$\lambda = \varTheta/n + \lambda_o \tag{4.105}$$

4.5.3. The oblique Mercator projection

The main characteristics of the Oblique Mercator Projection can be summarized as follows:

- It is an oblique cylindrical projection
- It is a conformal projection
- The two meridians 180° apart are straight lines
- Other meridians and parallels are complex curves
- The scale will be true along the chosen central line
- Scale becomes infinite 90° from the central meridian
- It is mainly used for areas with greater extent in an oblique direction

Given the following parameters:

- semi-major axis (a) and the first eccentricity (e) of the reference ellipsoid
- The central scale factor
- The latitude and the longitude of the selected center of the map (ϕ_o, λ_c), and
- The azimuth (α_c) of the central line as it crosses the latitude (ϕ_o) measured east of north

One can start by computing the following constants:

$$B = \sqrt{1.0 + e^2\cos^4\phi_o/(1.0 - e^2)} \tag{4.106}$$

$$A = a\, B\ \text{Scale Factor}\sqrt{1.0 - e^2}/(1.0 - e^2\sin^2\phi_o) \tag{4.107}$$

$$t_o = \tan(\pi/4.0 - \phi_o/2.0)/\{(1.0 - e\sin\phi_o)/(1.0 + e\sin\phi_o)\}^{0.5e} \tag{4.108}$$

$$D = B\sqrt{(1.0 - e^2)}/\{\cos\phi_o\sqrt{1.0 - e^2\sin^2\phi_o}\} \tag{4.109}$$

$$F = D + \frac{\phi_o}{|\phi_o|}\sqrt{D^2 - 1.0} \tag{4.110}$$

$$E = Ft_o^B \tag{4.111}$$

$$G = 0.5(F - 1.0/F) \tag{4.112}$$

$$\gamma_o = \sin^{-1}\{\sin(\alpha_c)/D\} \tag{4.113}$$

$$\lambda_o = \lambda_c - \sin^{-1}\{G\tan(\gamma_o)\}/B \tag{4.114}$$

$$u_o = \frac{\phi_o}{|\phi_o|}A/B\tan^{-1}\{\sqrt{(D^2 - 1.0)}/\cos(\alpha_c)\} \tag{4.115}$$

$$v_o = 0.0 \tag{4.116}$$

4.5.3.1. Geodetic to xy-oblique Mercator coordinate transformation (forward equations)

Given the geodetic coordinates of one point (ϕ, λ), one can compute the corresponding rectangular coordinates as follows:

$$t = \tan(\pi/4.0 - \phi/2.0)/\{(1.0 - e\sin\phi)/(1.0 + e\sin\phi)\}^{0.5e} \tag{4.117}$$

$$Q = E/t^B \tag{4.118}$$

$$S = 0.5(Q - 1.0/Q) \tag{4.119}$$

$$T = 0.5(Q + 1.0/Q) \tag{4.120}$$

$$V = \sin(B\{\lambda - \lambda_o\}) \tag{4.121}$$

$$U = (-V\cos\gamma_o + S\sin\gamma_o)/T \tag{4.122}$$

$$v = A\log\left\{\frac{1.0 - U}{1.0 + U}\right\}/\{2.0B\} \tag{4.123}$$

$$u = A/B \tan^{-1}\{[S\cos\gamma_o + V\sin\gamma_o]/\cos[B(\lambda - \lambda_o)]\} \tag{4.124}$$

$$x = v\cos\alpha_c + u\sin\alpha_c + \text{False Easting} \tag{4.125}$$

$$y = u\cos\alpha_c - v\sin\alpha_c + \text{False Northing} \tag{4.126}$$

4.5.3.2. xy-Oblique Mercator to geodetic coordinate transformation (inverse equations)

Given the rectangular coordinates (x, y), the corresponding geodetic coordinates can be computed as follows:

$$x_r = x - \text{False Easting} \tag{4.127}$$

$$y_r = y - \text{False Northing} \tag{4.128}$$

$$v = x_r\cos\alpha_c - y_r\sin\alpha_c \tag{4.129}$$

$$u = y_r\cos\alpha_c + x_r\sin\alpha_c \tag{4.130}$$

$$Q' = 2.718281828^{-Bv/A} \tag{4.131}$$

$$S' = 0.5(Q' - 1.0/Q') \tag{4.132}$$

$$T' = 0.5(Q' + 1.0/Q') \tag{4.133}$$

$$V' = \sin(Bu/A) \tag{4.134}$$

$$U' = (V'\cos\gamma_o + S'\sin\gamma_o)/T' \tag{4.135}$$

$$t = [E/\sqrt{(1.0 + U')/(1.0 - U')}]^{1.0/B} \tag{4.136}$$

$$\phi = \pi/2.0 - 2.0\tan^{-1}(t[(1.0 - e\sin\phi)/(1.0 + e\sin\phi)]^{0.5e}) \tag{4.137}$$

One can iterate for ϕ using an initial estimate ($\phi = \pi/2.0 - 2.0\tan^{-1}t$). Finally, the longitude can be computed as follows:

$$\lambda = \lambda_o - \tan^{-1}\{(S'\cos\gamma_o - V'\sin\gamma_o)/\cos(Bu/A)\}/B \tag{4.138}$$

The above mentioned projections are employed in developing the state plane coordinate system in the US. The Lambert Conformal projection is used for states with dominant east-west extent. The Transverse Mercator projection is used for states with dominant north-south extent. The Oblique Mercator projection is used for states with oblique extent. The type of map projection and the associated constants (e.g. the standard parallels, the latitude and the longitude of the origin of the map, the scale factor, false easting, false northing, etc.) for each state can be seen in Snyder (1987). PC-Software for state plane coordinate transformations can be found in Moffitt and Bossler (1998).

References

Heiskanen, W. and Moritz, H. *Physical Geodesy* (Reprint), Institute of Physical Geodesy, Technical University, Graz, Austria, 1996.

Kraus, K., *Photogrammetry, Fundamentals and Standard Processes*, (vol. 1), Dummler, Bonn, 1993.

Moffitt, F. and Mikhail, E., *Photogrammetry* (3rd ed.), Harper & Row, New York, 1980.

Moffitt, F. and Bossler, J., *Surveying* (10th ed.), Addison Welsley Longman, Menlo Park, CA, 1998.

Snyder, J. P., *Map Projections – A Working Manual*, US Government Printing Office, Washington, DC, 1987.

REA (Staff of Research and Education Association), *Handbook of Mathematical, and Engineering Formulas, Tables, Functions, Graphs, Transforms*, Research and Education Association, Piscataway, NJ, 1997.

Chapter 5

Computer basics

Alan Saalfeld

What do you need to know or learn about computers to benefit from and successfully utilize the other sections of this manual? Surprisingly little! Computers have evolved from mysterious black boxes used by a few scientists and engineers into familiar home appliances that most teenagers use adeptly. Nevertheless, if you know nothing about computer systems, you will need to learn *something* simply because a computer is an absolutely indispensable tool in this day and age. In Global Positioning System (GPS), photogrammetric, remote sensing, surveying, and Geographical Information Systems (GIS) applications, you will work primarily with '*digital data*,' data sets that have been transformed into and delivered to you in computer-readable formats. *Computers are needed to read, transfer, store, retrieve, and process digital data.* If you do not yet own a computer, your only choice is not *whether* to get one, but rather deciding *which* system to acquire. If you do own a computer, you probably want to know if the system you own can handle the new functionality described in this manual or if you will need to upgrade, add to, or replace parts of your current system. In this chapter, we provide sufficient information about system characteristics and system choices to inform your decisions about acquiring, maintaining, and upgrading the computer system that is right for your needs. The presentation in this chapter will also try to highlight trends and developments in computer technology that should have the greatest influence on your choices for a system and for system components now and in the future.

5.1. Some guidelines for dealing with computer technology

First the good news: Finding a system to satisfy your needs will not be too difficult or too expensive because there are many suitable computer hardware and software products on the market at very reasonable prices. The daunting task for the uninitiated will be sorting through the seemingly endless permutations of choices to be made when you do not have a clear understanding of how one choice may impact another. We will offer common-sense guidelines to help focus your choices and inform your decisions. Here is an example of one such rule-of-thumb that is true for most business decisions (and is especially true when buying computer hardware and software):

All else being equal, always choose established products and companies that have been around for a while and that have a proven track record of reliability, service, and customer satisfaction.

It is important to remember that in the volatile, rapidly changing field of computer technology, 'a while' may mean 'a year or less,' so the fact that a company has a large customer base may also be relied upon to provide the inexperienced user with a sense of security and confidence in a still unfamiliar product. Companies like Microsoft, IBM, or ESRI have had their products tested over time by millions of users, and the products are constantly being fixed, improved, and fine-tuned by the industries' top software developers.

Our second guideline for acquiring a suitable computer system is aimed at improving chances that the components that need to work together actually do work together.

To avoid having to resolve conflicts due to incompatible parts and settings, buy assembled systems and subsystems whenever possible. This applies both to hardware and to software systems.

This guideline recommends that the novice computer user buy software suites and fully configured computer systems instead of buying individual customized computer programs or individual hardware components. Nowadays one may order a completely built computer system with components well chosen for smooth interaction and with software pre-installed and fully optimized. The vendor or manufacturer assumes responsibility for combining compatible hardware and software components and for setting various parameters and switches that allow the different subsystems to share resources and to share them efficiently.

5.2. Assembling the right computer system

In striving for compatibility, there is a logical order to putting the pieces together. We borrow our next 'common sense' advice on buying a computer from Dan Gookin (1998):

Five steps to buying a computer

1   Find out what you want your PC to do
2   Look for software to get that job done
3   Find the proper hardware to run the software
4   Shop for service and support
5   Buy it!

Dan Gookin

The key sequence of events here is (1) assess goals, (2) search for software that will achieve those goals, and (3) identify the hardware needed to run the appropriate software. The order is important. Buying the wrong hardware may leave the user with no appropriate software that runs on it. A logical corollary that needs be seriously considered by owners of established or 'legacy' information systems is that already owning the wrong hardware may actually become an impediment to successfully setting up a GIS/GPS/image processing/surveying/mapping system.

5.3. Computers and their operating systems

The way by which software applications and hardware communicate is through a software program called the operating system. Different computers require different operating systems. Personal Computers (PCs), sometimes called 'IBM clones' or 'x86 clones', are the most common computers, making up over 95 per cent of all computers

Table 5.1 Some computer/operating system configurations

Computer	PC		Mac	Sun	HP	IBM AIX
Some operating systems used by the computer:	DOS, Windows 3.x/95/98/NT, Windows 2000, LINUX		MacOS, System 7, System 8, System 9	Solaris, other UNIX	HP-IX, SCO UNIX, BSD UNIX	AIX, other UNIX

in the world. According to the Computer Industry Almanac (1998), in 1998 there were 129 million PCs in use throughout the US and 364.4 million PCs in use throughout the world. Over 120 million of the PCs in the US run one of the following operating systems: DOS, Windows 3.x, Windows 95, Windows 98, or Windows NT. To run on a particular computer, software applications must be tailored to the computer's operating system. Some applications run under DOS, but not under Windows NT, or *vice versa*. Applications written to run under Windows 98 will not run on a computer using a UNIX operating system. Different operating systems are like different languages. If the machine only speaks 'MacOS', then it does no good for software to try to talk to it in 'Windows 95'. PCs may also run LINUX, a PC-version of the UNIX operating system. (Note: UNIX is not a single operating system, but rather a family of similarly behaving systems.) Table 5.1 shows some of the most popular operating systems and the machines that they run on.

Not every software product is available for every operating system of every computer. We begin by summarizing available software solutions for each operating system. Several organizations and magazines publish lists of specialized software products along with the system requirements for using those products. The surveying trade journal, *Point of Beginning (POB)*, publishes annual reviews of GIS software, surveying software, and GPS post-processing software along with annual reviews of surveying hardware and tools. The complete reviews for the years 1998 and 1999 are online at *http://www.pobonline.com/surveys.htm*, and they provide excellent breakdowns of software capabilities and requirements. We summarize the software surveys of products and their operating systems in table format in Table 5.2. Since only PCs run DOS and Windows 95/98/NT operating systems, Table 5.2 clearly reveals that the vast majority of commercial GIS/GPS/surveying software products need PC hardware to run[1]. On the other hand, if your computer is not a PC, then your choice of software

Table 5.2 Numbers of software products reviewed in *POB* 1998-9 and classified by operating system

	No. of products	DOS	95/98/NT	UNIX	MacOS
1999 GIS software	60	6	54	11	2
1998 GPS software	30	3	30	0	0
1999 Surveying software	111	45	100	15	9
All software products	201	54	184	26	11
All software products (%)	100	27	91.5	13	5.5

[1] The percentages in Table 5.2 add to more than 100 per cent because some software products are written for more than one operating system.

products will be much more limited. These figures do not mean that you must choose a PC, but before you choose some other computer, you should make very sure that the rather limited selection of software products for that other computer's operating system will actually meet your data processing requirements.

5.4. Recent PC improvements

PCs have evolved dramatically in the past decade. They have become fully capable of storing and processing large spatial data sets. They have been fitted with a user-friendly Graphical User Interface (GUI) and with tools for programmers to work within that GUI and to program for it. Over the past few years, computer technology has evolved to a point where a moderately priced ($2,000–$3,000), properly configured PC has the hardware capability to easily handle almost every problem, process, data set, or computation arising in data processing for GIS, GPS applications, surveying, mapping, remote sensing, and photogrammetry. To illustrate the rapid improvement in computing power that has taken place over just the past 7 years, we have compiled Table 5.3 of advertised characteristics of systems selling for approximately $3,000 at six time points in that period.

To assure further comparability, we have chosen to compare the products of a single well-established manufacturer, Dell. Tables for other individual companies, such as Gateway, Compaq, IBM, Packard-Bell, and Micron, all of which advertised throughout the same time period, would be similar in their content and their progression. Some system configurations will facilitate specific types of operations, such as image processing, visualization, or graphic displays, and in section 5.8 we will highlight different optional features of computer systems that favor specific operations that arise in processing spatial data.

5.5. Trends in computer construction

It is important to understand how computer technology is evolving. Over time computers get smaller, faster, cheaper, friendlier (easier to use), and have greater data storage and data throughput capabilities. PCs currently dominate the computer market, and hardware market share drives software development: More software developers write programs for PCs and their operating systems because the developers can then sell more programs. As a result, of all types of software, PC software has the largest variety, the greatest dissemination, and the largest user base. Computer manufacturers, in turn, choose to build hardware that has all of this vast software pool just waiting for it, i.e. they build PCs. There is no reason for this economically driven symbiosis to slow down any time soon. In fact, some trends have been going on for more than three decades, and these trends have been lumped together succinctly into a maxim known as Moore's Law: *Computing power has increased exponentially (doubling every 18 months)*. Actually, this 'law' is nothing more than an empirically verified observation of change rates. This 'law,' as originally put forth by Moore, stated that the number of transistors on a chip doubles every 18 months. However, other related measures were also seen to double very regularly, including processor speed, hard drive size, number of instructions executed per second. Still other measures related to price were decreasing regularly, including cost of CPUs, cost of

Table 5.3 Evolution of the Dell $3,000 PC (1993–9) [a]

Month advertised	January 1993	January 1994	January 1996	February 1997	April 1998	November 1999
Processor type	i486 DX2	Pentium 60	Pentium 120	Pentium MMX	Pentium II	Pentium III
Processor speed (MHz)	50	60	120	200	300	733
System bus speed (MHz)	25	30	60	66	100	133
Hard drive size	170 MB	450 MB	1.6 GB	4.3 GB	6.4 GB Ultra	27.3 GB Ultra
HD speed/seektime	17 ms		11 ms	9.5 ms	5400 rpm/9.5 ms	7200 rpm/9 ms
RAM	4 MB	8 MB	16 MB EDO	32MB SDRAM	128 MB SDRAM	256 MB RDRAM
RAM speed	80 ns	70 ns	70 ns	66 MHz		400 MHz = 2.5 ns
Cache size (kB)/type	128/external	256/external	512/PipelineB	512/PipelineB	512/L2	512/L2
Graphics card		#9 GXE	9FX MGA	Matrox Millenium	NVidea	NVidea TNT
Graphics memory (MB)	1	1	2/DRAM	4/WRAM	4/AGP	32/4XAGP
CD-ROM	No	2X	6X EIDE	12X EIDE	2X DVD	8X DVD
Monitor screen (inches)	14	14	15	17	19	21
Modem	No	No	No	33.6 USR	56 USR	USR v.90
OS	DOS 5.0	Win 3.1/DOS 6	Windows 95	Windows 95	Windows 95	Windows 98SE
Software bundle	??	??	MS Office +	MS Office Pro	Home essentials	MS works 99
Slots/bus type	4 EISA, 2 ISA	3 ISA, 2 PCI	EISA, PCI	EISA, PCI	EISA, PCI	PCI, 4XAGP
Speakers	No	No	Altec ACS-31	Altec ACS-290	Altec ACS-295	Harmon Kardon
Advertised in	PC magazine	PC magazine	PC world	Byte	Byte	www.dell.com
Date	1/12/93	1/11/94	1/96	2/97	4/98	11/19/99
Advertised price ($)	2,999	2,999	2,949	2,984	2,999	3,074.90

[a] Data compiled by column from the following references: Dell Computer Ad (1993), Dell Computer Ad (1994), Dell Computer Ad (1996), Dell Computer Ad (1997), Dell Computer Ad (1998), Dell Computer Price Worksheet (1999).

Table 5.4 Illustrative improvements in the technology of a $3,000 base system

	January 1993	November 1999	Improvement
Processor speed (MHz)	50	733	14.6 ×
System bus speed (MHz)	12.5/25	133	10.6 × /5.3 ×
Hard drive size	170 MB	27.3 GB ultra	160 ×
Hard drive seek time (ms)	17	9	1.88 ×
RAM (MB)	4	256 RDRAM	64 ×
RAM speed	80 ns (12.5 MHz)	2.5 ns (400 MHz)	32 ×
Internal (on chip) cache size (kB)/type	8/internal	32/L1	4 ×
Cache size (kB)/type	128/external	512/L2	4 ×
Graphics memory (MB)	1	32/4 × AGP	32 ×
CD-ROM speed	1X	48 × (or 8 × DVD)	48 × (or 8 × DVD)
CD vs. DVD capacity	650 MB	up to 17 GB	up to 25 ×

memory chips, cost of hard drives, execution time for benchmark tests, and chip architecture separation distances. The most significant Consumer's Corollary to Moore's law, illustrated in Table 5.4, is that *performance per unit cost keeps increasing*.

The performance-per-unit-cost improvements have even caused some confusion in the vocabulary used to describe different types of computers. There used to be a clear performance distinction between workstations and PCs and a significant cost difference as well. Today, workstation and PC differences are narrowing as PCs gain speed and new features at a faster rate than workstations do. PCs are less than 1 year behind workstations in performance, some workstations are being priced as PCs, and some full-featured PCs are being advertised as workstations. Some traditional workstation manufacturers are even moving away from UNIX and their own proprietary computer chips to Microsoft's Windows NT operating system and Intel processors because of the large user base and software pool for Windows operating systems running on Intel hardware.

5.6. General software use considerations for all users

In this section we examine a typical computer system and what use different individuals might make of it. We have already seen that by far the most common computer is a PC with a Windows operating system. We also know from Table 5.2 that a PC with Windows is the system most likely to be found running GIS, GPS, or surveying applications software. Most PC systems sold in the US today are sold with some bundled easy-to-use software packages, commonly including an operating system such as Windows 98 or NT, a word-processing/office suite such as Microsoft Office 2000, and hardware (modems or network cards) and software (browsers) for accessing the Internet. Word processing programs are far and away the most used and most cost-effective applications. Nearly every business computer should be equipped with document preparation software of some kind. The Internet is an extensive network of linked computers sharing information resources and other resources of all kinds. Connecting to the Internet is easy. Internet use does not demand any computer sophistication. Libraries throughout the US offer free Internet access and assistance in learning to use the Internet at public computers at library sites. By the end of 1998, over 150 million people worldwide were using the Internet at least once a week, and over half of those people (76.5 million) lived in the US (*Computer Industry Almanac* 1998). By the end of 1999, the world total of frequent Internet users will have risen to 259 million worldwide, with 110 million in the US alone (*Computer Industry Almanac* 1999). The emergence of the World Wide Web (WWW, a graphics-oriented subdomain of the Internet), competition among numerous Internet service providers, and recent developments in browser technology have all combined to make the Internet easy and inexpensive for everyone to access and use.

Every computer user, from beginner to seasoned professional, should expect to use word-processing and web-browsing software in addition to any specialized application software. Next we look at how different users' computing experience affects how they utilize the computer.

5.7. Selection criteria for users with different experience levels

For the complete neophyte, ease of learning and ease of using the operating system, any bundled office software, and the Internet-service tools should be important criteria for choosing a system. Ease of use and ease of learning of the applications software tools must also be key considerations for the new computer user. An example of easy-to-use applications software is commercial GIS software in today's market. GIS software often includes tutorials, sample applications, help documentation, and even online courses to train users and to facilitate use of the system. Some of the more specialized software applications, such as surveying adjustment programs, do not come with such extremely user-friendly enhancements and complementary services. Economic considerations explain the greater development resources that can be and are invested in a high-demand system like a GIS versus the time and effort that go into a lower-demand system like a least-squares adjustment program. Each user should be aware that differences in software usability and the size of the potential user market are directly related to the time and money that a software developer can afford to invest to build the product. The Spartan no-frills look and feel of specialized software written only a few years ago is no longer the norm, however. Now there is easy-to-use software that builds friendly user interfaces made up of dialogue boxes, drop-down menus, file-browsers, and the like. In fact, today's software developers use component software to first build the friendly interfaces for their interactive products, and afterwards they add the 'guts:' the data processing and computational routines.

For the slightly more experienced computer user, familiarity with the operating system and bundled tools may also seem to be a big plus because it eliminates the need to learn something new. One must weigh the advantage of a familiar operating system against any shortage of specialized software for that operating system. For example, Macintosh operating systems are familiar to many in the US because of Macintosh's large presence in the school systems. Moreover, the Mac environment is arguably the most user-friendly and easiest to learn. Many Mac users find all other computers unfriendly by comparison and do not want to use other computers. Nevertheless, the 1998 *POB* GPS Processing Software Survey, which we summarized in Table 5.2, found no GPS processing software titles whatsoever that work with any

Macintosh computer operating systems. So if a loyal Mac user needs to process GPS data, they are out of luck unless they can write their own software. The latest Microsoft Windows environments, on the other hand, are almost as easy to learn, almost as easy to use, and they do support 91 per cent of the reviewed commercial software in GIS, GPS, and surveying, including 30 different programs for GPS processing. A computer user with limited computer experience, one who, for example, does not program, should concentrate on user-friendly, full-featured, self-contained software products, sometimes called turnkey systems. Turnkey systems consistently handle the many stages of processing spatial data to create displays or deliver numerical solutions. Geographic Information Systems, Image Processing Systems, and Computer Aided Design (CAD) Systems are examples of products that integrate diverse capabilities of data management for easy use by relatively unsophisticated computer users.

For the more experienced computer user, especially one who does program, additional opportunities for software use and development exist. The more experienced computer user will want to install software that includes some programming or scripting environment to be able to combine, modify, and possibly even create some of the basic mapping and GIS, GPS, and surveying routines. The programming languages C, C++, Java, and Visual Basic are currently the most popular programming and software development languages. All of the operating systems mentioned so far (and hence all of the computers) support some versions of the programming languages C, C++, and Java. Visual Basic, however, is a Microsoft product that only runs under Microsoft Windows operating systems. Visual Basic has become a favorite language among business programmers because it has many tools to build friendly interfaces quickly and robustly using point and click tools in a graphic environment. Microsoft also offers Visual J++ (Java) and Visual C++ in a coherent programming environment called Visual Studio. Visual programming languages permit prototyping and Rapid Application Development (RAD), two modern techniques for iteratively designing and building software systems quickly. Older programming languages such as Fortran and COBOL are still in use. They are still run on many older machines, especially for legacy business applications (COBOL) and non-commercial scientific applications (Fortran). Older programming languages are a bit like Latin–they can get a job done, if somewhat awkwardly, but they do not efficiently embody modern programming notions. Once again the lack of a broad user base affects the cost of software: Fortran and COBOL compilers for PCs cost much more than C or C++ compilers or Java or Visual Basic interpreters[2].

We summarize our recommended software components to use for applications described in this manual. These recommendations include operating system, business software, internet/browser software, GIS, CAD, and image processing turnkey systems, modern computer programming environments, and specialized software for GPS, photogrammetry, surveying applications and map integration as described in the cited *POB* Software Surveys (Table 5.5).

[2] Fortran, COBOL, C, and C++ are compiled programming languages. Java and BASIC are interpreted languages. Programs written in interpreted languages are converted to machine code one instruction at a time, and the resulting machine code is executed (run) before the next instruction is converted. Programs written in compiled languages are converted in their entirety to machine code before the entire resulting code is executed.

Table 5.5 Examples of software types and products for use with this manual

Software description	Examples of software packages
Operating system	Windows 98, Windows NT, Windows 2000
Basic office software	MS Office, or Corel Office Suite, or WordPerfect Office 2000
Browser software	Microsoft Internet Explorer or Netscape Navigator
Geographic information system	Arc/INFO under NT, or ArcView under Windows 98/NT/2000
CAD system	AutoCAD, MicroStation, or CorelCAD
Image processing system	Erdas IMAGINE, or ER Mapper
Programming environments	MS Visual Studio, Sun's Java SDK, C++ Builder
GPS, surveying software	See annual POB surveys for current details

Even though we have already seen that the PC is the only machine that offers support for the vast majority of the software we need, we have not looked closely at the different hardware options for PCs in terms of what best supports the software. We finish this chapter with a closer look at hardware components.

5.8. Computer hardware recommendations

We have already noted a number of facts that recommend the purchase and use of a PC for the applications described in this manual. Nevertheless, it is useful to summarize them for this final section on hardware recommendations. (1) The PC is the unrivaled winner for computer of choice throughout the US and the world and should continue to be so for some time. (2) The PC will run more GPS/GIS/surveying application software than any other type of machine. (3) Adding a high-end PC is quite inexpensive in terms of overall equipment investment in a business already using computers. (4) Via the Internet and possibly other local networks, a PC can communicate and share information with any existing networked platform in a business. These four points add up to the recommendation to buy a high-end PC no matter what system, if any, may be already in use. The following recommendations are for a PC configuration based on current technology and options available at this time. The features and the key functions of the components are described briefly to permit the reader to apply similar arguments to make wise choices in the future.

We will look briefly at the following system components: the microprocessor, the memory, the hard drive, the graphics card, the Internet connection, the monitor, and the CD or DVD drive option.

1   The microprocessor is the brain or Central Processing Unit (CPU) of the PC. Microprocessors are rated according to clock speed. The fastest currently available Intel microprocessors run at a speed of 733 MHz, which is 733 million clock cycles per second. Microprocessors with speeds of 800 MHz will soon be made available by Intel and also by AMD, Intel's major competitor. Microprocessors are also classified by the organization of the millions of transistors that make up the chip. Intel Pentium III processors and AMD Athlon processors, for example, have more transistors dedicated to floating point arithmetic operations than, say, Intel Celeron processors. Therefore, the Pentium III and Athlon processors are

much faster than Celeron processors at computationally intensive operations such as geometric computations for graphics. Software for GIS/GPS/surveying applications requires computationally intensive floating-point procedures. A Windows NT or Windows 2000 operating system allows more than one processor to share the computer workload. If a machine supports many users simultaneously, or if one user runs many applications simultaneously, then a multi-processor-capable operating system can distribute the work efficiently to the different processors. *Recommendation for the microprocessor:* get a fast Pentium III processor to handle computationally intensive operations; get a system with two processors only if it will have to run many tasks simultaneously; get an Athlon if cost difference between Pentium III and Athlon is an important savings to you.

2   Computer memory, Random Access Memory (RAM) is a volatile (i.e. fast-changing, temporary) data storage area very close to the CPU that can be accessed (to read or write data) very quickly, more than a hundred thousand times more quickly than a hard drive or a removable disk, for example. The mean seek time for RAM is 70 ns[3] (0.00000007 s) compared with a hard drive's mean seek time of 9 ms (0.009 s). The very low cost of SDRAM memory, at present $1–$2 per MB, argues for getting the largest memory possible if the system uses SDRAM and if the processing tasks are memory-intensive. Memory intensive tasks such as matrix inversion, create large amounts of intermediate data values through computations and not through disk accesses. If those data values are moved around in nanoseconds instead of milliseconds, the processing time may be reduced by five orders of magnitude. *Recommendation for RAM:* get at least 256 MB of SDRAM (or the new faster RDRAM, if you can afford it and if your system is new enough to support it).

3   The hard drive provides permanent storage for data sets. It also provides temporary storage for data being processed when the information being processed will not fit entirely into memory. Large hard drives nowadays consistently exceed 30 Gigabytes (GB), with several manufacturers selling drives of 37 GB or more for about $10 per GB. A 37 GB drive provides much more storage space than most applications could ever need. The Census Bureau's entire TIGER map database of every road segment, railroad, river, and other geographic features in the US only takes up 4 GB (compressed). EIDE Drive speeds have gotten slightly faster with spin rates increasing by 33 per cent to 7200 rpm, resulting in 9.0 ms mean access time. SCSI drives are currently a bit faster, but are also currently significantly more expensive than the non-SCSI drives that come standard with most PC packages. *Recommendation for the hard drive:* get a 37+ GB EIDE hard drive. Even if you do not need the space now, it is good to have room to expand. If you plan to do image processing, then your storage requirements can grow very fast.

4   Graphics cards have improved immensely in the past few years. Since graphics applications abound and SDRAM prices are so low, some new high-end PC systems calling themselves 'workstations' contain state-of-the-art graphics cards featuring 32 MB of SDRAM, the new Intel 82820 chip set for processing vectors

[3] Mean seek time for RAM has dropped dramatically recently from 70 to 10 to 2.5 ns for a new type of memory called RDRAM that operates on its own Rambus circuitry running at up to 400 MHz (half of the processor speed). New high-end machines are equipped with 128 MB or more of this fast memory. At $5–$6 per MB, the new RDRAM is currently considerably more expensive than SDRAM.

rapidly, and the new $4 \times$ AGP graphics bus that produces faster graphic through-put by employing higher data speeds (133 MHz) and wider data paths (64 bits). *Recommendation for the graphics card:* get a 32 MB SDRAM $4 \times$ AGP card that uses the new Intel graphics (82820) chip set or something better and newer.

5   Internet access via a fast network connection, cable, or fast modem, (in order of preference and speed) is increasingly important. Connectivity to the outside world is inexpensive and invaluable for any business. The vast pool of information and other resources made available by the Internet at low cost to all computer users is too good to pass up. *Recommendation for Internet access:* if you plan to use the Internet a lot, or if you plan to download or upload large data sets such as image files frequently, get a fast network connection (first choice) or cable connection (second choice) to the Internet if either is available at a reasonable price in your area. If neither is available, or if you do not expect to create heavy traffic on your Internet connection, then get a fast modem and subscribe to an Internet service that can utilize the full speed of your modem.

6   Monitors with large screens ($19''$, $21''$ or even larger) are a good investment if the display will play a critical role in the business operation. Operator-intensive activities, such as manual digitizing or visual validation of photogrammetric registration operations, can usually be done more accurately and with less eyestrain on larger, higher resolution screens. *Recommendation for a monitor:* if you will be working with mapping or GIS or detailed imagery, get a monitor with a large (at least $19''$) screen. Check to be sure that the monitor that you buy is not only compatible with your graphics card, but can also make good use of the features of your graphics card.

7   A DVD drive is a good investment for anyone working with extremely large data sets. The current access speed of state-of-the-art DVD drives is still eight times slower than a CD drive, but technology should soon allow the DVD drive to narrow the gap. A DVD drive can play both DVD-ROMs and CD-ROMS. A DVD disk can contain 17 GB of data, whereas a CD holds less than 0.7 GB. Spatial data sets, especially imagery, can and do exceed Gigabyte sizes. When the number of computers with DVD drives reaches a critical threshold, data providers will move to optional delivery of data in DVD format. Microsoft already makes its developer software and documentation available on a single DVD-ROM or on multiple CD-ROMs. *Recommendation for a DVD drive:* get a DVD drive that is at least $8 \times$ (eight times as fast as the original speed) with the next system that you buy, but do not go out immediately to buy a DVD drive to add to your system. First, there is not yet much data available on DVD-ROM, and second, the DVD drives will probably grow much faster long before data becomes so prolific that owning a DVD drive will be a necessity.

As we have illustrated in this section, key hardware components are many. Picking useful components in isolation is relatively easy to accomplish by following good sense guidelines. Making sure that the chosen components work well together is a much more difficult task. It is a task best left to the professional system builders who custom build systems from an extensive suite of compatible parts. A company such as Dell, Micron, or Gateway allows you to pick your processor, your memory and disk size and type, your graphics card, your monitor, and your choice of a DVD drive or a CD drive. Technicians at that company then put the compatible components together and guarantee that they will work correctly with each other. The final bit of advice for anyone about to buy a PC system echoes our earlier principles stated in section 5.1: *unless you are an experienced hardware installer, buy your system as fully configured as you can.* Do not plan to add components in 6 months if you know that you will need them in 6 months. Get them when you purchase the system. Even if parts get better in the 6 months that you wait, the parts may fail to be fully compatible with or fully utilized by original system parts. New type memory chips, for example, need updated chip sets and bus configurations that may in turn require a new motherboard. A new faster spinning second hard drive may be forced by the system controller to deliver data at the same slow rate as an already installed slower spinning drive. The potential for better performance or reduced price does not outweigh the potential frustration of having to pinpoint and try to fix a hardware/hardware or hardware/software compatibility mismatch.

References

Computer Industry Almanac Inc., Arlington Heights, IL, 1998. http://www.c-i-a.com/

Computer Industry Almanac Inc., Arlington Heights, IL, 1999. http://www.c-i-a.com/

Dell Computer Ad, *PC Magazine*, 12(1) inside back cover January 12, 1993.

Dell Computer Ad, *PC Magazine*, 13(1) inside back cover, January 11, 1994.

Dell Computer Ad, *PC World*, 14(1) inside back cover, January 1996.

Dell Computer Ad, *Byte*, 22(2) inside back cover, February 1997.

Dell Computer Ad, *Byte*, 23(4) inside back cover, April 1998.

Dell Computer Price Worksheet, *Dell Computer Website*, http://www.dell.com/

Gookin, D., *Buying a Computer for Dummies*, IDG Books, 1998.

GIS Software Survey, *Point of Beginning*, November 1999, pp. 34–44. http://www.pobonline.com/surveys.htm.

GPS Post-Processing Survey, *Point of Beginning*, August 1998, pp. 52–56. http://www.pobonline.com/surveys.htm.

Surveying Software Guide, *Point of Beginning*, June 1999, pp. 42–62. http://www.pobonline.com/surveys.htm.

Chapter 6

Basic electromagnetic radiation

Carolyn J. Merry

This chapter serves as an introduction to the basic concepts of electromagnetic radiation. Much of the material discussed in this chapter is taken from Avery and Berlin (1992), Jensen (2000), and Lillesand and Kiefer (1999). The reader is encouraged to consult these three textbooks for additional technical information.

Remotely sensed data of the earth are collected by special sensors that have the capability to record electromagnetic energy, acoustic energy (sound), or variations in force (gravity, magnetic). However, the focus of this chapter will be on the electromagnetic wavelengths that are recorded by aircraft or satellite sensors. Passive remote sensing records electromagnetic radiation that is reflected or emitted naturally from an object and normally depends on energy from the sun. In contrast, active remote sensing depends on electromagnetic energy that is generated by the sensor itself and then the sensor records the reflected energy from the object. Examples of active remote sensing are radar and sonar systems.

6.1. Electromagnetic energy

The world we live in is full of energy. For example, light, heat, electricity and sound are some of the forms of energy. One source of energy important to remote sensing is electromagnetic energy. The main source of electromagnetic energy received by the earth is light from the sun. Electromagnetic energy travels at the speed of light, which is about 299,792,458 m/s (about 186,000 miles/s). Electromagnetic energy is composed of two parts–an electrical wave and a magnetic wave that vibrate at right angles to each other (Figure 6.1). We measure the electromagnetic wave energy in terms of: wavelength, λ, which is the length from the top of one wave to the top of the next; amplitude, which is the height of the wave; and frequency, which is the number of waves that pass a specified point each second.

Light energy waves belong to a family called the electromagnetic spectrum. The electromagnetic spectrum includes visible rays, light that our eyes can detect, and invisible rays, such as gamma rays, X-rays, ultraviolet (UV) waves, infrared (IR) radiation, microwaves, and television and radio waves. As previously mentioned, electromagnetic energy (light waves) travels at the same speed–the speed of light–but their wavelengths are different and the light waves have different effects on earth materials. For certain wavelengths special instruments are used to record this electromagnetic energy in the form of images, which we call remote sensing imagery.

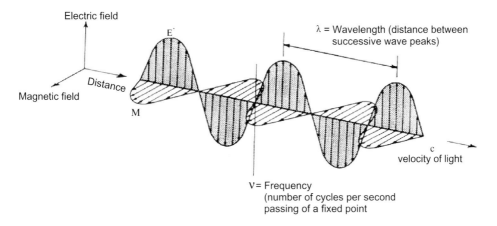

Figure 6.1 Electromagnetic wave (adapted from Lillesand and Kiefer 1999).

6.2. Electromagnetic spectrum

The electromagnetic spectrum defines the entire region of wavelengths. Normally, the spectrum is divided into arbitrary regions with a beginning and an ending wavelength. This interval is commonly referred to as a spectral band, channel or region (Figure 6.2, see plate 1). The spectral regions, which are approximate, are further defined in Table 6.1. The spectral regions are normally called UV, visible, near IR, middle IR, thermal IR and microwave. Below the UV region are the short wavelengths of gamma waves measured in nanometers (nm) and above the microwave region are the long wavelengths of radio waves measured in meters. However, these two wavelength regions are rarely used in terrestrial remote sensing because the earth's atmosphere absorbs the energy in these wavelength regions.

The UV region ranges from 0.03 to 0.4 micrometers (μm). The UV (literally meaning 'above the violet') region borders the violet portion of the visible wavelengths. These are the shortest wavelengths practical for remote sensing. The UV can be divided into the far UV (0.01–0.2 μm), middle UV (0.2–0.3 μm) and near UV (0.3–0.4 μm). The sun is the natural source of UV radiation. However, wavelengths shorter than 0.3 μm are unable to pass through the atmosphere and reach the earth's surface due to atmospheric absorption by the ozone layer. Only the 0.3–0.4 μm wavelength region is useful for terrestrial remote sensing.

The most common region of the electromagnetic spectrum used in remote sensing is the visible band, which spans from 0.4–0.7 μm (Figure 6.2, see plate 1). These limits correspond to the sensitivity of the human eye. Blue (0.4–0.5 μm), green (0.5–0.6 μm) and red (0.6–0.7 μm) represent the additive primary colors–colors that cannot be made from any other color. Combining the proper proportions of light that represent the three primary colors can produce all colors perceived by the human eye. Although sunlight seems to be uniform and homogeneous in color, sunlight is actually composed of various wavelengths of radiation, primarily in the UV, visible and near IR portions of the electromagnetic spectrum. The visible part of this radiation can be shown in its component colors when sunlight is passed through a prism, which bends the light through refraction according to wavelength (Figure 6.3).

Table 6.1 Spectral regions

Spectral region	Wavelength
Gamma rays	<0.03 nm
X-rays	0.03–30 nm
UV region:	0.03–0.4 μm
Far UV	0.01–0.2 μm
Middle UV	0.2–0.3 μm
Near UV (photographic UV band)	0.3–0.4 μm
Visible region:	0.4–0.7 μm
Visible blue	0.4–0.5 μm
Visible green	0.5–0.6 μm
Visible red	0.6–0.7 μm
Reflected near IR region:	0.7–3 μm
Photographic IR	0.7–1.3 μm
Middle IR	1.5–1.8 μm, 2.0–2.4 μm
Thermal IR (far IR)	3–5 μm, 8–14 μm (below ozone layer), 10.5–12.5 μm (above ozone layer)
Microwave	0.1–100 cm
Radar region:	0.1–100 cm
K	0.8–2.4 cm
X (3.0 cm)	2.4–3.8 cm
C (6 cm)	3.8–7.5 cm
S (8.0 cm, 12.6 cm)	7.5–15.0 cm
L (23.5 cm, 25.0 cm)	15.0–30.0 cm
P (68 cm)	30.0–100.0 cm
Radio	>100 cm

Most of the colors that we see are a result of the preferential reflection and absorption of wavelengths that make up white light. For example, chlorophyll in healthy vegetation selectively absorbs the blue and red wavelengths of white light to use in photosynthesis and reflects more of the green wavelengths. Thus, vegetation appears as green to our eyes. Snow is seen as white, since all wavelengths of visible light are scattered. Fresh basaltic lava appears black, as all wavelengths are absorbed, with no wavelengths of light being reflected back to the human eye.

The IR band includes wavelengths between the red light (0.7 μm) of the visible band and microwaves at 1000 μm. Infrared literally means 'below the red' because it is adjacent to red light. The reflected near IR region (0.7–1.3 μm) is used in black and white IR and color IR sensitive film. The middle IR includes energy at wavelengths ranging from 1.3 to 3 μm. Middle IR energy is detected using electro-optical sensors.

The thermal (far) IR band extends from 3 to 1000 μm. However, due to atmospheric attenuation, the wavelength regions of 3–5 μm and 8–14 μm are typically used for remote sensing studies. The thermal IR region is directly related to the sensation of heat. Heat energy, which is continuously emitted by the atmosphere and all features on the earth's surface, dominates the thermal IR band. Optical-mechanical scanners and special vidicon systems are typically used to record energy in this part of the electromagnetic spectrum.

The microwave region is from 1 mm to 1 m. The radio wavelengths are the longest waves. Microwaves can pass through clouds, precipitation, tree canopies and dry

Figure 6.3   White light passed through a prism will generate a spectrum of colors.

surficial deposits. There are two types of sensors that operate in the microwave region. Passive microwave systems detect natural microwave radiation that is emitted from the earth's surface. Radar (*ra*dio *de*tection *a*nd *r*anging) systems–active remote sensing–propagate artificial microwave radiation to the earth's surface and record the reflected component. Typical radar systems are listed in Table 6.1.

In summary, the spectral resolution of most remote sensing systems is described in terms of their bandwidths in the electromagnetic spectrum.

6.3. Energy transfer

Energy is the ability to do work and in the process of doing work, energy is moved from one place to another. The three ways that energy moves include conduction, convection and radiation.

Conduction is when a body transfers its kinetic energy to another body by physical contact. For example, when frying your eggs in a pan, heat is transferred from the electric coils on the stove to the frying pan. Another example is when you place your spoon in a hot cup of coffee. The spoon becomes hot because of the transfer of energy from the hot liquid to the spoon.

Convection is the transfer of kinetic energy from one place to another by physically moving the bodies. For example, convection occurs when warmer air rises in a room, setting up currents. Heating of the air near the ground during the early morning hours sets up convection currents in the atmosphere. Likewise, water in a lake will turn over during the fall, moving the cooling water on the surface to the bottom, mixing with the warmer water. Convection occurs in liquids and gases.

Radiation is the transfer of energy through a gas or a vacuum. This is of primary importance to remote sensing, since this is the only form of energy transfer that can take place. Energy transfer from the sun to the earth takes place in the vacuum of space.

There are two models that are used to describe electromagnetic energy–the wave model and the particle model.

6.3.1. Wave model

If one tries to see how light energy is moving, it is nearly impossible. Most physicists believe light energy travels like water, like wave motion. The light energy is carried along in very tiny ripples, which are much smaller than the actual waves in water. If you think of a cork in a pond, the waves make the cork move up and down, but the cork does not move in the direction of the water wave. In a similar manner the light energy waves are the same–each part of the wave is vibrating up and down at right angles like the cork, while the light energy moves along the wave.

Electromagnetic radiation is generated when an electrical charge is accelerated. The wavelength (λ) depends on the time that the charged particle is accelerated and is normally measured from peak to peak in micrometers (μm), nm or cm. The frequency (v) of the electromagnetic radiation depends on the number of accelerations per second and is measured as the number of wavelengths that pass a point for a given unit of time. For example, an electromagnetic wave that sends one crest every second is said to have a frequency of one cycle per second or 1 hertz (Hz).

The relationship between λ and v is (Boleman 1985):

$$c = \lambda v \tag{6.1}$$

where c is defined as the speed of light in a vacuum.

The frequency is then defined as:

$$v = c/\lambda \tag{6.2}$$

and the wavelength is defined as:

$$\lambda = c/v \tag{6.3}$$

From these two equations, the frequency is inversely proportional to wavelength, which means that the longer the wavelength, the lower the frequency or the shorter the wavelength, the higher the frequency. As electromagnetic radiation passes from one substance to another, the speed of light and wavelength will change while the frequency remains the same.

All earth objects–water, soil, rocks, and vegetation–above absolute zero (273°C or 0K) and the sun, emit electromagnetic energy. The sun is the initial source of electromagnetic energy that can be recorded by many remote sensing systems; radar and sonar systems are exceptions, as they generate their own source of electromagnetic energy.

A black body is defined as an object that totally absorbs and emits radiation at all wavelengths. For example, the sun is a 6,000K black body (Figure 6.4). The total emitted radiation from a black body (M_λ) is measured in watts per m^2 (W/m^2) and is proportional to the fourth power of its absolute temperature (T), measured in degrees Kelvin:

$$M_\lambda = \sigma T^4 \tag{6.4}$$

where: σ is the Stefan–Boltzmann constant of 5.6697×10^{-8} W/m^2/K^4.

This relationship is known as the Stefan–Boltzmann law. The law states that the amount of energy emitted by an earth object (or the sun) is a function of its temperature. The higher the temperature, the greater the amount of radiant energy emitted by the object.

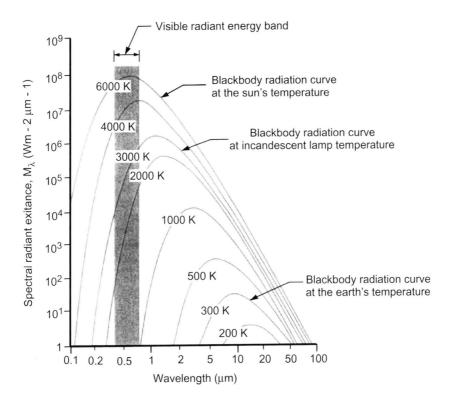

Figure 6.4 Spectral energy distribution from representative blackbodies (adapted from Lillesand and Kiefer 1999).

In addition to computing the total amount of energy, the dominant wavelength of this radiation can be calculated as:

$$\lambda_{max} = k/T \qquad\qquad (6.5)$$

where: k is a constant of 2,898 μmK and T is the absolute temperature, degrees K.

This is known as Wien's displacement law. For example, we can calculate the dominant wavelength of the radiation for the sun at 6,000K and the earth at 300K (27°C or 80°F) as:

$$\lambda_{max} = 2,898 \ \mu mK/6,000K$$

$$\lambda_{max} = 0.483 \ \mu m$$

and

$$\lambda_{max} = 2,898 \ \mu mK/300K$$

$$\lambda_{max} = 9.66 \ \mu m$$

Thus, the sun's maximum wavelength of radiation is 0.48 μm, located in the visible spectrum, which are the wavelengths sensitive to our eyes. The earth's peak wavelength at 9.66 μm is in the thermal region. As this relationship demonstrates, as the temperature of an object increases, its dominant wavelength (λ_{max}) shifts towards the shorter wavelengths of the electromagnetic spectrum.

6.3.2. Particle model

Electromagnetic energy can also be described as a particle model by considering that light travels from the sun as a stream of particles. Einstein found that when light interacts with matter the light has a different character and behaves as being composed of many individual bodies called photons that carry particle-like properties, such as energy and momentum. Physicists view electromagnetic energy interaction with matter by describing electromagnetic energy as discrete packets of energy or quanta.

An atom is made up of electrons – negatively-charged particles – that move around a positively-charged nucleus. The electron is kept in orbit around the nucleus by the interaction of the negative electron around the positive neutron. For the electron to move up to a higher level, work has to be performed, but energy needs to be available to move the electron up a level. Once an electron jumps up to a higher level, i.e. the electron becomes excited, then radiation is given off. This radiation is a single packet of electromagnetic radiation–a particle unit of light called a photon. Another word for this is a quantum, as the electron is making a quantum leap or a quantum jump in its orbital path around the nucleus.

Planck described the nature of radiation and proposed the quantum theory of electromagnetic radiation. A relationship between the frequency of radiation, as described by the wave model, and the quantum is (Boleman 1985):

$$Q = hv \tag{6.6}$$

where: Q is the energy of a quantum, in Joules (J); h is the Planck constant (6.626 × 10^{-34} J/s); v is the frequency of radiation.

If we take equation (6.3) and multiply it by 1 or h/h, then:

$$\lambda = hc/hv \tag{6.7}$$

By substituting Q for hv from equation (6.6), then the wavelength, λ, associated with a quantum of energy is:

$$\lambda = hc/Q$$

or rearranging:

$$Q = hc/\lambda$$

From this, the energy of a quantum is inversely proportional to its wavelength. In other words, the longer the wavelength, the lower its energy content. This inverse relationship is important to remote sensing. The relationship indicates that it is more difficult to detect longer wavelength energy being emitted, such as the thermal IR or microwave wavelengths, than those at shorter or visible wavelengths. In fact, remote sensing sensors typically 'look at' larger pixel sizes at thermal wavelengths or in the microwave region, so that there is enough electromagnetic energy to record. You experience this phenomenon also, by getting sunburn from UV rays when outdoors too long–there is a lot more energy at these shorter wavelengths.

6.4. Energy/matter interactions in the atmosphere

Electromagnetic energy travels from the sun to the earth. In the vacuum of space, not very much happens to electromagnetic energy. However, once the electromagnetic energy hits the earth's atmosphere, then the speed, the wavelength, the intensity, and the spectral distribution of energy may change. Four types of interactions are common–transmission, scattering, absorption and reflection.

6.4.1. Transmission

Transmission occurs when the incident radiation passes through the medium without any measurable attenuation. Essentially, the matter is transparent to the radiation. However, when electromagnetic energy encounters materials of different density, such as air or water, then refraction or the bending of light occurs when the light passes from one medium to another. Refraction of light occurs because the media are of different densities and the speed of electromagnetic energy will be different in each medium. The index of refraction, n, is a measure of the optical density of a substance. This index is the ratio of the speed of light in a vacuum, c, to the speed of light in a second medium, c_n:

$$n = c/c_n \tag{6.8}$$

The speed of light in a material can never reach the speed of light in a vacuum, therefore, the index of refraction is always greater than one. Light travels slower in water than air. For example, the index of refraction for the atmosphere is about 1.0002926 and for water it is about 1.33.

Snell's law can be used to describe refraction. This law states that for a given frequency of light, the product of the index of refraction and the sine of the angle between the ray and a line normal to the interface at the two media is constant:

$$n_1\sin\theta_1 = n_2\sin\theta_2 \tag{6.9}$$

The amount of refraction is a function of the angle θ made with the vertical, the distance involved (the greater the distance of the atmosphere, then the more the changes in density), and the density of the air involved (usually air is more dense near sea level than at higher elevations).

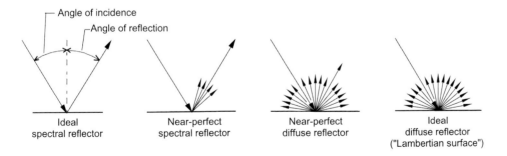

Figure 6.5 Specular vs diffuse reflectance (adapted from Lilleand and Kiefer 1999).

6.4.2. Scattering

Three types of scattering are important in remote sensing – Rayleigh, Mie and non-selective scattering. Particles in the atmosphere cause scattering of the electromagnetic energy. Scattering occurs when radiation is absorbed and then re-emitted by atoms or molecules. The direction of the scattering is impossible to predict.

Rayleigh scattering – molecular scattering – occurs when the effective diameter of the particle is many times smaller (<0.1) than the wavelength of the incoming electromagnetic energy. Usually air molecules, such as oxygen and nitrogen, cause scattering. Raleigh scattering occurs predominantly in the upper 4.5 km of the atmosphere. The amount of scattering is inversely related to the fourth power of the wavelength. For example, UV light at 0.3 μm is scattered approximately 16 times more ($(0.6/0.3)^4 = 16$) and blue light at 0.4 μm is scattered about five times more ($(0.6/0.4)^4 = 5.06$) than red light at 0.6 μm. Raleigh scattering causes the blue sky that we are used to seeing on clear, bright sunny days–the shorter violet and blue wavelengths are more efficiently and preferentially scattered than the longer orange and red wavelengths. In the morning or in the evening, the sun's light is passing through a longer path length. As a consequence, the red and orange wavelengths are hardly scattered at all, which makes for the red sunrises and red sunsets.

Mie scattering – non-molecular scattering – occurs in the lower 4.5 km of the atmosphere. The particles that are responsible for Mie scattering are roughly equal in size to the incoming wavelength. The actual size of particles ranges from 0.1 to 10 times (0.1 mm to several mm in diameter) the wavelength of the incident energy and are typically dust and other particles. Mie scattering is greater than Rayleigh scattering. Pollution particles–smoke and dust–in the air contribute even more to the red sunrises and sunsets.

Non-selective scattering takes place in the lowest portions of the atmosphere. The particles are roughly greater than 10 times the incoming wavelength. Since this type of scattering is non-selective, all wavelengths are scattered and appear as white light. Water droplets and ice crystals making up clouds and fog are examples of non-selective particle scattering, making clouds and fog to appear white.

The reason why scattering is important to remote sensing is that scattering can reduce the information content of imagery. Image contrast is lost, making it difficult to tell objects apart from one other. Filters on cameras have to be used to eliminate or filter out wavelengths that would normally cause scattering.

6.4.3. Absorption

Absorption is the process where radiant energy is absorbed and converted into other forms of energy. This can take place in the atmosphere or on the earth's surface. Absorption bands are a range of wavelengths in the electromagnetic spectrum, where the radiant energy is absorbed by a substance, such as water (H_2O), carbon dioxide (CO_2), oxygen (O_2), ozone (O_3) or nitrous oxide (N_2O). This is important for remote sensing because in these absorption bands, there is no energy available to be sensed by a remote sensing instrument. The visible spectrum – 0.4–0.7 μm – does not absorb all of the incident energy, but rather transmits the energy. This portion of the electromagnetic spectrum is known as an atmospheric window.

Chlorophyll in plants absorbs blue and red wavelengths for use in photosynthesis. This characteristic is used to map vegetation types. Also, water is an excellent absorber of energy for most wavelengths of light. Minerals important for geologic purposes also have unique absorption properties that are used for identification purposes.

6.4.4. Reflection

Reflection occurs when the electromagnetic radiation 'bounces off' an object. The incident radiation, the reflected radiation, and the vertical to the surface where the angles of incidence and reflection are measured, all lie in the same plane (Figure 6.5). Two types of reflecting surfaces are possible – specular and diffuse surfaces (Figure 6.5). Specular reflection occurs when the surface is essentially smooth and the angle of incidence is equal to the angle of reflection. The average surface profile height is several times smaller than the wavelength of the electromagnetic radiation striking the surface. For example, a calm water body will act like a near-perfect specular reflector.

If the surface has a large surface height relative to the wavelength of the incoming electromagnetic radiation, the reflected light rays will go in many different directions– this is known as diffuse radiation. Examples would include white powders, white paper or other similar materials. A perfectly diffuse surface occurs when the reflecting light rays leaving the surface are constant for any angle of reflectance. This is known as an ideal surface – a Lambertian surface (Schott 1997).

6.5. Energy/matter interactions with the terrain

An understanding and knowledge of energy-matter interactions of radiant energy is essential for accurate image interpretation and to extract biophysical information from remote sensing imagery. This understanding and knowledge is the key and focus of remote sensing research.

The amount of radiant energy absorbed, reflected or transmitted through an object per unit time is called the radiant flux (Φ) and is measured in Watts (W) (Figure 6.6). The radiation budget equation for incoming and outgoing radiant flux to a surface from any angle in a hemisphere can be stated as:

$$\Phi_{i\lambda} = r_\lambda + \tau_\lambda + \alpha_\lambda \tag{6.10}$$

where: $\Phi_{i\lambda}$ is the total amount of incoming radiant flux at a specific wavelength (λ); r_λ is the amount of energy reflected from the surface; τ_λ is the amount of energy transmitted through the surface; α_λ is the amount of energy absorbed by the surface.

Hemispherical reflectance (r_λ) is defined as the ratio of the radiant flux reflected from a surface to the radiant flux incident to it:

$$r_\lambda = \Phi_{\text{reflected}}/\Phi_{i\lambda} \tag{6.11}$$

Hemispherical transmittance (τ_λ) is defined as the ratio of the radiant flux transmitted through a surface to the radiant flux incident to it:

$$\tau_\lambda = \Phi_{\text{transmitted}}/\Phi_{i\lambda} \tag{6.12}$$

Hemispherical absorptance (α_λ) is defined by the relationship:

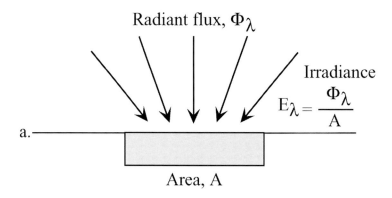

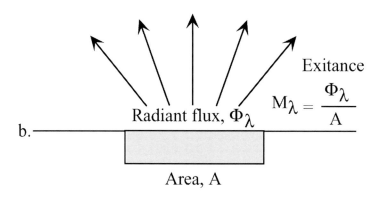

Figure 6.6 Radiant flux density (after Jensen 2000).

$$\alpha_\lambda = \Phi_{\text{absorped}}/\Phi_{i\lambda} \tag{6.13}$$

or

$$\alpha_\lambda = 1 - (r_\lambda - \tau_\lambda) \tag{6.14}$$

These three equations imply that radiant energy must be conserved. These radiometric quantities are useful to produce general statements about the spectral reflectance, absorptance and transmittance of earth terrain features. A percentage reflectance ($p_{r\lambda}$) can be obtained by taking equation (6.11), multiplying by 100, to get:

$$p_{r\lambda} = \Phi_{\text{reflected}}/\Phi_{i\lambda} \times 100 \tag{6.15}$$

This equation is used commonly in remote sensing research to describe the spectral reflectance characteristics of various terrain features. Figure 6.7 shows several typical spectral reflectance curves. These curves do not provide any information about the absorption or transmittance characteristics of the terrain features. However, since the remote sensing sensors – cameras and multispectral scanners – only record reflected

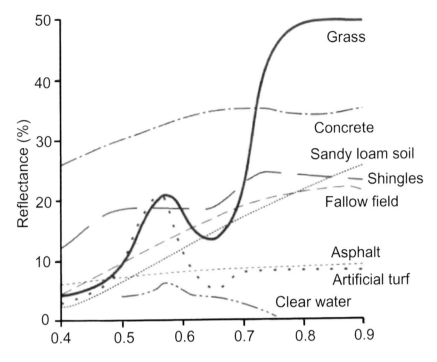

Figure 6.7 Typical spectral reflectance curves (after Jensen 2000).

energy, this information is still very valuable to the remote sensing interpreter and forms the basis for identifying and assessing objects on the ground.

These hemispherical reflectance, transmittance and absorptance radiometric quantities do not provide information on the exact amount of energy reaching an area on the ground from a given direction or about the exact amount and direction of energy leaving this same ground area. It is important to refine the radiometric measurement techniques so that more precise radiometric information can be extracted from remotely sensed data. Additional radiometric quantities will be introduced next.

Imagine an area 1×1 m^2 in size that is bathed in radiant flux (Φ) for specific wavelengths (Figure 6.6). The average radiant flux density – the irradiance, E_λ – will be the amount of radiant flux divided by the area of this flat surface in W/m^2:

$$E_\lambda = \Phi_\lambda/A \tag{6.16}$$

The amount of radiant flux leaving – the exitance, M_λ – will be the energy leaving per unit area, A, of this flat surface in W/m^2:

$$M_\lambda = \Phi_\lambda/A \tag{6.17}$$

Radiance is the most precise radiometric measurement used in remote sensing. Radiance for a given wavelength – L_λ – is the radiant flux leaving a surface in a given direction ($A\cos\theta$) and solid angle (Ω) per unit area:

$$L_\lambda = \theta/\Omega/A\cos\theta \tag{6.18}$$

Ideally, energy from the earth's surface will not become scattered in this solid angle field of view and 'contaminate' the radiant flux from the ground. However, most likely there will be scattering in the atmosphere and from nearby areas of the earth's surface that contribute to this energy in the solid angle field of view. As a result, atmospheric scattering, absorption, reflection and reflection – described earlier – will influence the radiant flux before the energy is recorded by the remote sensing system.

The discussion in this chapter has focused on the propagation of energy through the atmosphere and the interaction of this energy with terrain features. Detecting this electromagnetic energy is then performed either by using a photographic process – aerial photography – or by an electronic process – video cameras or multispectral scanners. Details of these two processes will be covered in Part 3.

In summary, an understanding of basic electromagnetic energy concepts is important when using remote sensing imagery. In this way the nature of interactions that take place from when the electromagnetic energy leaves the source, travels through the atmosphere, illuminates the terrain feature, and then is finally recorded by the remote sensing instrument can be understood. Additional information on remote sensing – sensors, processing the data, hardware and software, accuracy – are covered in much more detail in Part 3, Chapters 16–24.

References

Avery, T. E. and Berlin, G. L., *Fundamentals of Remote Sensing and Airphoto Interpretation* (5th ed.), MacMillan, New York, 1992.

Boleman, J., *Physics: an Introduction*, Prentice Hall, Englewood Cliffs, NJ, 1985.

Jensen, J. R., *Remote Sensing of the Environment: an Earth Resource Perspective*, Prentice Hall, Upper Saddle River, NJ, 2000.

Lillesand, T. M. and Kiefer, R. W., *Remote Sensing and Image Interpretation* (4th ed.), John Wiley & Sons, New York, 1999.

Schott, J. R., *Remote Sensing – the Image Chain Approach*, Oxford University Press, New York, 1997.

Part 2

Global Positioning System (GPS)

Chapter 7

Introducing the Global Positioning System

Chris Rizos

The NAVSTAR Global Positioning System (GPS) is a satellite-based radio-positioning and time-transfer system designed, financed, deployed, and operated by the US Department of Defense. GPS has also demonstrated a significant benefit to the civilian community, who are applying GPS to a rapidly expanding number of applications. The attractions of GPS are:

- Relatively high positioning accuracy, from meters down to the millimeter level.
- Capability of determining velocity and time, to an accuracy commensurate with position.
- No inter-station visibility is required for high precision positioning.
- Results are obtained with reference to a single, global datum.
- Signals are available to users anywhere on the earth: in the air, on the ground, or at sea.
- No user charges, requiring only relatively low-cost hardware.
- An all-weather system, available 24 hours a day.
- Position information is provided in three-dimensions.

Since its introduction to the civilian community in the early 1980s GPS has revolutionized geodesy, surveying and mapping. Indeed, the first users were geodetic surveyors, who applied GPS to the task of surveying the primary control networks that form the basis of all map data and digital databases. Today, around the world, GPS is the preferred technology for this geodetic application. However, as a result of progressive product innovations, the GPS technology is increasingly addressing the precise positioning needs of cadastral, engineering, environmental, planning and Geographical Information System (GIS) surveys, as well as a range of new machine, aircraft and ship location applications.

7.1. Background

Development work on GPS commenced within the US Department of Defense in 1973. The objective was to design and deploy an all-weather, 24 hour, global, satellite-based navigation system to support the positioning requirements of the US armed forces and its allies. For a background to the development of the GPS system the reader is referred to Parkinson (1994). GPS was intended to replace the large number of navigational systems already in use, and great importance was placed on the system's reliability and survivability. Therefore a number of stringent conditions had to be met:

- Suitable for all military platforms: aircraft (jet to helicopter), ships, land (vehicle-mounted to handheld) and space-based vehicles (missiles and satellites);
- Able to handle a wide variety of platform dynamics;
- A real-time positioning, velocity and time determination capability to an appropriate accuracy;
- The positioning results were to be available on a single, global, geodetic datum;
- The highest accuracy was to be restricted to the military user;
- Resistant to jamming (intentional and unintentional);
- Incorporating redundancy mechanisms to ensure the survivability of the system;
- A passive positioning system that did not require transmission of signals by the user;
- Able to provide the positioning service to an unlimited number of users;
- Use low-cost, low-power user hardware, and
- Was to be a replacement for the Transit satellite system, as well as other terrestrial navigation systems.

What was unforeseen by the system designers was the power of commercial product innovation, which has added significantly to the versatility of GPS, but in particular as a system for *precise positioning*. For example, GPS is able to support a variety of positioning and measurement modes in order to simultaneously satisfy a wide range of users; from those satisfied with navigational accuracy (of the order of 10 m), to those demanding very high (even sub-centimeter) positioning accuracy. *GPS has now so penetrated certain application areas so that it is difficult to imagine life without it!* Part 2 of this manual is not intended to be a comprehensive textbook on the GPS technology and its applications. Excellent general references to the engineering aspects of GPS are Kaplan (1996) and Parkinson and Spilker (1996). Texts dealing exclusively with the high precision GPS surveying techniques include Leick (1995), Hofmann-Wellenhof *et al.* (1998), and Teunissen and Kleusberg (1998).

A discussion of the GPS technology and applications starts with the identification of the three components (Figure 7.1).

- *The space segment:* the satellites and the transmitted signals.
- *The control segment:* the ground facilities carrying out the task of satellite tracking, orbit computations, telemetry and supervision necessary for routine operations.
- *The user segment:* the applications, equipment and computational techniques that are available to the users.

7.1.1. The space segment

The space segment consists of the constellation of spacecraft, and the signals that are broadcast by them, which allow users to determine position, velocity and time. The basic functions of the satellites are to:

- Receive and store data uploaded by the control segment

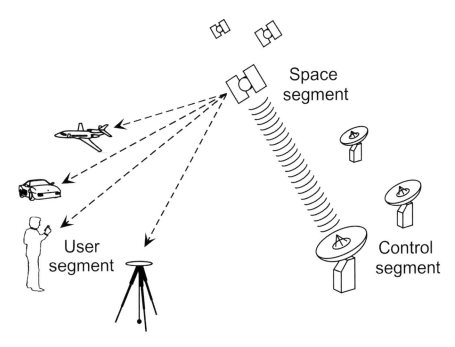

Figure 7.1 GPS system elements.

- Maintain accurate time by means of onboard atomic clocks, and
- Transmit information and signals to users on two L-band frequencies

Several constellations of GPS satellites have been deployed, and are planned. The first *experimental* satellite of the so-called 'Block I' constellation was launched in February 1978. The last of this 11 satellite series was launched in 1985. The *operational* constellation of GPS satellites, the 'Block II' and 'Block IIA' satellites, were launched from 1989 onwards. *Full Operational Capability* was declared on 17 July 1995 – the milestone reached when 24 'Block II/IIA' satellites were operating satisfactorily. There are 18 *replenishment* 'Block IIR' satellites, with the first launched in 1997. Currently 12 of these satellites are being redesigned as part of the 'GPS Modernization' program (see Section 15.4.1). The 'Block IIF' *follow-on* satellite series is still in the design phase, and the satellites are planned for launch from 2006 onwards with similar enhancements as the latter 'Block IIR' satellites, as well as having the ability to transmit a third frequency. The status of the current GPS satellite constellation, and such details as the launch and official commissioning date, the orbital plane and position within the plane, the satellite ID number(s), etc. can be obtained from several electronic GPS information sources on the Internet, for example the US Coast Guard Navigation Center (NAVCEN 2001).

At an altitude of approximately 20,200 km, a constellation of 24 functioning GPS satellites, located in six orbital planes inclined at about 63° to the equator (Figure 7.2), is sufficient to ensure that there will be *at least four satellites visible*, at any unobstructed site on the earth, at any time of the day. As the GPS satellites are in nearly circular orbits:

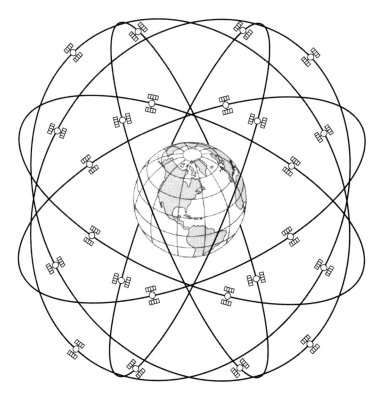

Figure 7.2 The GPS constellation 'birdcage' showing the 24 orbiting satellites.

- Their orbital period is approximately 11 h 58 min, so that each satellite makes two revolutions in one sidereal day (the period taken for the earth to complete one rotation about its axis with respect to the stars).
- At the end of a sidereal day the satellites are again over the same location on the earth.
- Reckoned in terms of a solar day (24 h in length), the satellites are in the same position in the sky about 4 min earlier each day.

Satellite visibility at any point on the earth, and for any time period, can be computed using 'mission planning' tools provided with standard GPS surveying software.

A GPS satellite may be above an observer's horizon for many hours, perhaps 6–7 h or more in the one pass. At various times of the day, and at various locations on the surface of the earth, the number of satellites and the length of time they are above an observer's horizon will vary. Although at certain times of the day there may be as many as 12 satellites visible simultaneously, there are nevertheless occasional periods of degraded satellite coverage (though naturally their frequency and duration will increase if satellites fail). 'Degraded satellite coverage' is typically defined in terms of the magnitude of the Dilution of Precision (DOP) factor, a measure of the quality of receiver-satellite geometry (see Section 9.3.3). The higher the DOP value, the poorer the satellite geometry.

Each GPS satellite transmits unique navigational signals centered on two L-band frequencies of the electromagnetic spectrum: L1 at 1575.42 MHz and L2 at 1227.60 MHz. (Two signals at different frequencies permit the ionospheric delay effect on the signal raypaths to be estimated – see Section 9.2.1.1, thus improving measurement accuracy.) At these two frequencies the signals are highly directional and can be reflected or blocked by solid objects. Clouds are easily penetrated, but the signals may be blocked by foliage (the extent of this is dependent on a number of factors, such as the type and density of the leaves and branches, and whether they are wet or dry, etc.). The satellite signal consists of the following components:

- Two L-band *carrier waves*
- *Ranging codes* modulated on the carrier waves
- Navigation message

The primary function of the ranging codes is to permit the *signal transit time* (from satellite to receiver) to be determined. The transit time when multiplied by the speed of light then gives a measure of the receiver-satellite 'range'. In reality the measurement process is more complex and the measurement is contaminated by a variety of biases and errors (Langley 1991b, 1993). The navigation message contains the satellite orbit (or ephemeris) information, satellite clock error parameters, and pertinent general system information necessary for real-time navigation to be performed. Although for positioning and timing the function of the GPS signal is quite straightforward, the stringent performance requirements of GPS are responsible for the complicated nature of the GPS signal structure. Table 7.1 summarizes the GPS system requirements and their corresponding implications on the signal characteristics (after Wells *et al.* 1986).

7.1.2. The control segment

The control segment consists of facilities necessary for satellite health monitoring, telemetry, tracking, command and control, and satellite orbit and clock error computations. There are currently five ground facility stations: Hawaii, Colorado Springs, Ascension Island, Diego Garcia and Kwajalein. All are operated by the US Department of Defense and perform the following functions:

- All five stations are *Monitor Stations*, equipped with GPS receivers to track the satellites. The resultant tracking data is sent to the Master Control Station (MCS).
- Colorado Springs is the *MCS* , where the tracking data are processed in order to compute the satellite ephemerides (or coordinates) and satellite clock error parameters. It is also the station that initiates all operations of the space segment, such as spacecraft maneuvering, signal encryption, satellite clock-keeping, etc.
- Three of the stations (Ascension Is., Diego Garcia, and Kwajalein) are *Upload Stations* through which data is telemetered to the satellites.

Each of the upload stations views all of the satellites at least once per day. All satellites are therefore in contact with an upload station several times a day, and new navigation messages as well as command telemetry can be transmitted to the GPS satellites on a regular basis. The computation of each satellite's ephemeris, and the determination of the each satellite's clock errors, are the most important tasks of

Table 7.1 GPS system requirements and the nature of GPS signal

System requirements	Implication on GPS signals
GPS has to be a multi-user system	• Signals can be simultaneously observed by unlimited numbers of users → Accomplished by one-way measurement to passive user equipment. • Signal has to have a relatively wide spatial coverage.
GPS has to provide real-time positioning and navigation capability for the users	• At a certain epoch, signals from several satellites have to be simultaneously observed by a single user → Each signal to have a unique code, so receiver can differentiate signals from different satellites. • Signal has to provide data for user to estimate range to the observed satellite in real-time → Signal has to enable time delay measurement by the user. • Signal has to provide the ephemeris data in real-time to the user → Ephemeris data is included in a broadcast message.
GPS has to serve both military and civilian users	• Signals have to provide two levels of accuracy for time delay measurements → Different codes for the military and civilian users. • Signal has to support the AS policy, in which the military code is encrypted to prevent unauthorized use.
GPS signal has to be resistant to jamming	• Requires a unique code structure. • Uses the 'spread spectrum' technique.
GPS can be used for precise positioning	• Provide range measurements at two frequencies, to compensate for ionospheric refraction effect. • Require carrier wave(s) with centimeter wavelength.

the control segment. The first is necessary because the GPS satellites function as 'orbiting control stations' and their coordinates must be known to a relatively high accuracy, while the latter permits a significant measurement bias to be reduced.

The product of the orbit computation process at the MCS is each satellite's *predicted ephemeris*, expressed in the reference system most appropriate for positioning: an *Earth-Centered-Earth-Fixed* (ECEF) reference system known as the World Geodetic System 1984 (WGS 84) (Chapter 3, Part 1). The accuracy with which the orbit is predicted is typically at the few meter level. The behavior of each GPS satellite clock is monitored against GPS Time, as maintained by an ensemble of atomic clocks at the MCS. The satellite clock *bias, drift* and *drift-rate* relative to GPS Time are explicitly determined at the same time as the estimation of the satellite ephemeris. The clock error behavior so determined is made available to all GPS users via clock error coefficients in a polynomial form broadcast in the navigation message (see Section 8.3.2). However, what is available to users is really a *prediction* of the clock behavior for some future time instant. Due to random deviations – even cesium and rubidium oscillators are not entirely predictable – the deterministic models of satellite clock error are only accurate to about 10 nanoseconds or so. This is not precise enough for range measurements that must satisfy the requirements of cm-level GPS positioning. Strategies have therefore to be implemented that will account for this *residual* range bias.

7.1.3. The user segment

This is the component of the GPS system with which users are most concerned – the space and control segments are largely transparent to the operations of the navigation function. Of interest is the range of GPS user applications, equipment, positioning strategies and data processing techniques that are now possible. The 'engine' of commercial GPS product development is, without doubt, the *user applications*. New applications are being continually identified, each with its unique requirements in terms of accuracy, reliability, operational constraints, user hardware, data processing algorithms, latency of the GPS results, and so on. As a result, GPS user equipment has undergone an extensive program of development that is continuing to this day. In this context, GPS *equipment* refers to the combination of hardware, software, and operational procedures or requirements. Chapter 9 discusses the various measurement models and data processing strategies, the hardware issues are introduced in Chapter 10, the various GPS techniques will be described in Chapter 11, while the field operations of relevance to precise GPS positioning will be dealt with in Chapters 12 and 13.

While military R&D has concentrated on achieving a high degree of miniaturization, modularization and reliability, the commercial equipment manufacturers have, in addition, sought to bring down costs and to develop features that enhance the capabilities of the positioning system. Civilian users have, from the earliest days of GPS availability, demanded increasing levels of performance, in particular higher accuracy, improved reliability and faster results. This is particularly true of the survey user seeking levels of accuracy several orders of magnitude higher than that required by the navigator. In some respects GPS user equipment development is being driven by the precise positioning applications – in much the same way that automotive technology often benefits from car racing. Another major influence on the development of GPS equipment has been the increasing variety of civilian applications. Although it is possible to categorize positioning applications according to many criteria, the most important from the perspective of geospatial applications are:

- Accuracy, which leads to a differentiation of the GPS user equipment and techniques into several sub-classes.
- Timeliness, whether the GPS results are required in real-time, or may be derived from post-mission data processing.
- Kinematics, distinguishing between static receiver positioning, and those applications in which the receiver is moving (or in the so-called 'kinematic' mode).

The different GPS positioning modes and data processing strategies are all essentially designed to account for biases (or systematic errors) in GPS measurements to different levels of accuracy (Section 7.3). In this regard there are two aspects of GPS that fundamentally influence the entire user segment: the user equipment, the data processing techniques, and the operational (field) procedures. They are:

1   The type of *measurement* that is used for the positioning solution (Sections 8.2 and 8.4). There is, on the one hand, the basic satellite-to-receiver 'range' measurements with a precision typically at the few meter level. However, for precise applications such as in surveying, carrier phase measurements must be used. These have precisions at the millimeter level, but require more complex data processing in order to realize cm-level positioning accuracy.

2   The *mode of positioning*, whether it is based on single-receiver techniques, or in terms of defining the position of one receiver relative to another that is located at a known position. Relative positioning is the standard mode of operation if accuracy better than a few meters is required, although the actual accuracy achieved depends on many factors.

7.1.3.1. Absolute positioning

In this mode of positioning the reference system must be rigorously defined and maintained, and total reliance is placed on the *integrity* of the coordinated points within the reference system. In general the coordinate origin of the coordinate system is the geocenter, and the axes of the system are defined in a conventional manner as in ECEF datums such as WGS 84 and ITRS (Section 7.2). Satellite single-point positioning is the process by which:

*   *Given* the position vector of the satellite being tracked (in the global system), and
*   *Given* a set of measurements from one or more ground tracking stations to the satellite (or satellites) being tracked;
*   *Determine* the position vector of the ground station(s).

Some space geodesy technologies can determine the absolute position of a ground station to a very high accuracy, as for example in the satellite laser ranging (SLR) technique. However, the coordinates of a GPS receiver *in an absolute sense* are determined to a much lower accuracy than the precision of the measurements themselves, because it is not possible to fully account for the effects of measurement biases. Combining data from two GPS receivers is an effective way of eliminating or mitigating the effects of unmodeled measurement biases.

7.1.3.2. Relative positioning

Conceptually, relative position is the difference between the two position vectors (in the *global system*), expressed in a local reference system with origin at one of the ground stations. Most of the error in absolute position are common to both sets of coordinates (due to the similar biases on all GPS measurements), and hence largely cancel from the baseline components. In this case the positioning accuracy approaches that of the basic measurement precision itself.

There are different ways in which such differential positioning can be implemented using GPS. Data processing techniques such as those implemented for GPS surveying are essentially concerned with the determination of the *baseline components* between simultaneously observing receivers (Figure 7.3).

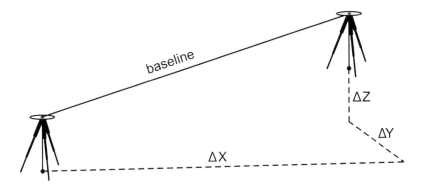

Figure 7.3 The baseline linking two simultaneously observing GPS receivers.

7.2. The issue of GPS datums

Chapter 3, Part 1, introduced the concept of datums and geodetic systems. In this section the modern geodetic reference systems are discussed from the viewpoint of GPS positioning.

7.2.1. WGS 84 system

The WGS 84 is defined and maintained by the US National Imagery and Mapping Agency (NIMA) as a *global geodetic datum* (NIMA 1997). It is the datum to which all GPS positioning information is referred by virtue of being the reference system of the broadcast GPS satellite ephemerides (Langley 1991a). The *realization* of the WGS 84 satellite datum is the catalogue of coordinates of over 1,500 geodetic stations (most of them active or past tracking stations) around the world. They fulfil the same function as national control benchmarks, that is, they provide the means by which a position can be related to a datum.

The relationship between WGS 84 (as well as other global datums) and local geodetic datums have been determined empirically (NIMA 1997), and transformation models of varying quality have been developed. Reference systems are periodically redefined, for various reasons, and the result is generally a small refinement in the datum definition, and a change in the numerical values of the coordinates of benchmarks. However, with dramatically improving tracking accuracy another phenomenon impacts on datum definition and its maintenance: *the motion of the tectonic plates across the earth's surface* (or 'continental drift'). This motion is measured in centimeters per year, with the fastest rates being over 10 cm/year. Nowadays this motion can be monitored and measured to sub-centimeter accuracy, on a global annual-average basis. In 1994 the GPS reference system underwent a subtle change to WGS 84(G730) to bring it into alignment with the same system as used by the International GPS Service to generate its precise GPS ephemerides. Another small change was made in 1996.

7.2.2. The international terrestrial reference frame

The WGS 84 system is the most widely used global reference system because it is the system in which the GPS satellite coordinates are expressed in the navigation message. Other satellite reference systems have been defined but these have mostly been for scientific purposes. However, since the mid 1980s geodesists have been using GPS to measure crustal motion, and to define more precise satellite datums. The latter were essentially by-products of the sophisticated data processing, which included the computation of the GPS satellite orbits. These surveys required coordinated tracking by GPS receivers spread over a wide region during the period of GPS survey 'campaigns'. Little interest was shown in these alternative datums until the network of tracking stations evolved into a *global one* that was maintained on a *permanent basis*, and the scientific community initiated a *project to define and maintain a datum at the highest level of accuracy*.

In 1991, the International Association of Geodesy decided to establish the International GPS Service for Geodynamics (IGS – nowadays the acronym 'IGS' stands for the 'International GPS Service') to promote and support activities such as the maintenance of a permanent network of GPS tracking stations, and the continuous computation of the satellite ephemerides and ground station coordinates. Both of these were preconditions to the definition and maintenance of a new satellite datum independently of the tracking network used to maintain the WGS 84 datum (and to provide the data for the computation of the GPS broadcast ephemerides). Routine activities commenced at the beginning of 1994 and the network now consists of about 50 core tracking stations located around the world, supplemented by more than 100 other stations (IGS 2001). The precise orbits of the GPS satellites (and other products) are available from the IGS via the Internet.

The definition of the reference system in which the coordinates of the IGS tracking stations are expressed and periodically re-determined is the responsibility of the International Earth Rotation Service (IERS). The reference system is known as the *International Terrestrial Reference System* (ITRS), and its definition and maintenance is dependent on a suitable combination of satellite laser ranging (SLR), very long baseline interferometry (VLBI) and GPS coordinate results (although increasingly it is the GPS system that is providing most of these data). Every other year a new combination of precise tracking results is performed, and the resulting new coordinates of SLR, VLBI and GPS tracking stations constitutes a new *International Terrestrial Reference Frame* (ITRF) or 'ITRF datum' which is referred to as 'ITRF yy', where 'yy' is the comput-ation year identifier. A further characteristic that sets the ITRS series of datums apart from WGS 84 is that the definition not only consists of the station coordinates, but also their *velocities* (due to continental and regional tectonic motion). Hence, it is possible to determine station coordinates within the datum, say ITRF 97, at some *epoch* such as the year 2000, by applying the velocity information and predicting the coordinates of the station at any time into the future (or the past). For example, the WGS 84(G730) reference system is identical to that of ITRF 91 at epoch 1994.0.

Such ITRS datums, initially dedicated to geodynamical applications requiring the highest possible precision, have been used increasingly as the fundamental basis for the redefinition of many nations geodetic datums. For example, the new Australian datum

– the Geocentric Datum of Australia 1994 – is defined as ITRF 92 at epoch 1994.0 (AUSLIG 2001). Of course countries are free to chose any of the ITRS datums (it is usually the latest), and define any epoch for their national datum (the year of GPS survey, or some reference date such as the year 2000). Only if both the ITRS datum (the designated ITRF yy) and the epoch are the same, can it be claimed that two countries have the same geodetic datum.

7.3. The performance of GPS

As far as users are concerned, there are a number of '*measures of performance*'. For example, how many observations are required to assure a certain level of accuracy is one measure that is important for survey-type applications. The less time required to collect observations, the more *productive* the GPS is, because productivity is closely related to the 'number of surveyed points per day'. Another measure of performance might be the maximum distance between two GPS receivers that would still assure a certain level of accuracy. However, the most common measure of performance is the positioning *accuracy*.

7.3.1. Factors influencing GPS accuracy

Biases and *errors* affect all GPS measurements. GPS biases may have the following characteristics:

1 Affect all measurements made by a receiver by an equal (or similar) amount.
2 Affect all measurements made to a particular satellite by an equal (or similar) amount.
3 Unique to a particular receiver-satellite observation.

7.3.1.1. Biases and errors

Their combined magnitude will affect the accuracy of the positioning results. Errors may be considered synonymous to internal instrument noise or *random errors*. Biases, on the other hand, may be defined as being those measurement errors that cause *true ranges* to be different from *measured ranges* by a 'systematic amount', such as, for example, all distances being measured either too short, or too long.

In the case of GPS, a very significant bias was *Selective Availability* (SA), a policy of the US government imposed on 25 March 1990, and finally revoked on the 1 May 2000. SA was a bias that caused all distances from a particular satellite, at an instant in time, to be in error by up to several tens of meters. The magnitude of the SA-induced bias varied from satellite-to-satellite, and over time, in an unpredictable manner. The policy *Anti-Spoofing* (AS), on the other hand, although not a signal bias, does affect positioning accuracy as it prevents civilian users access to the second GPS signal frequency. Measurements on two frequencies simultaneously is the best means by which the ionospheric refraction delay can be accounted for.

Biases must somehow be accounted for in the measurement model used for data processing if high accuracy is sought. There are several sources of bias with varying characteristics of magnitude, periodicity, satellite or receiver dependency, etc. Biases may have physical bases, such as the atmosphere effects on signal propagation, but may also enter at the data processing stage through imperfect knowledge of constants, for example any 'fixed' parameters such as the satellite orbit, station coordinates, etc. *Residual biases* may therefore arise from incorrect or incomplete observation modeling, and hence it is useful to assemble under the heading of 'errors' all random measurement process effects, as well as any unmodeled biases that remain after 'data reduction'.

7.3.1.2. Absolute and relative positioning

There are two GPS positioning modes which are fundamental to considerations of: (a) *bias propagation* into (and hence accuracy of) GPS results, and (b) the *datum* to which the GPS results refer. The first is *absolute or point positioning*, with respect to a well-defined coordinate system such as WGS 84 or the ITRS, and is often referred to as *Single-Point Positioning*. As the satellite coordinates are essential for the computation of user position, any error in these values (as well as the presence of other biases) will directly affect the quality of the position determination. The satellite-receiver geometry will also influence the error propagation into the GPS positioning results.

Higher accuracy is possible if the relative position of two GPS receivers, simultaneously tracking the same satellites, is computed. Because many errors will affect the absolute position of two or more GPS users to almost the same extent, these errors largely cancel when *differential or relative positioning* is carried out. This was particularly effective in overcoming the effect of SA-induced biases. There are different implementations of differential positioning procedures but all share the characteristic that the position of the GPS receiver of interest is derived *relative* to another fixed, or *reference*, receiver whose absolute coordinates are assumed to be known. One of these implementations, based on differencing the carrier phase data from the two receivers, is the standard mode for precise GPS techniques (Section 8.6.2).

7.3.1.3. Other factors influencing accuracy

Finally, GPS accuracy is also dependent on a host of other *operational, algorithmic and other factors*:

- Whether the user is moving or stationary. Clearly repeat observations at a static benchmark permit an improvement in precision due to the effect of averaging over time. A moving GPS receiver does not offer this possibility and the accuracy is dependent on single-epoch processing.
- Whether the results are required in real-time, or if post-processing of the data is possible. The luxury of post-processing the data permits more sophisticated modeling of the GPS data in order to improve the accuracy and reliability of the results.
- The level of measurement noise has a considerable influence on the precision attainable with GPS. Low measurement noise would be expected to result in comparatively high accuracy. Hence carrier phase measurements are the basis for high accuracy techniques (Section 8.4), while pseudorange measurements are used for comparatively low accuracy applications (Section 8.2).

- The degree of redundancy in the solution as provided by extra measurements, which may be a function of the number of tracked satellites as well as the number of observables (e.g. carrier phase and pseudorange data on L1 and L2).
- The algorithm type may also impact on GPS accuracy (although this is largely influenced by the observable being processed and the mode of positioning). In the case of carrier phase-based positioning, to ensure cm-level accuracy it is crucial that a so-called 'ambiguity-fixed' solution be obtained (Section 9.4.3).
- 'Data enhancements' and 'solution aiding' techniques may be employed. For example, the use of carrier phase-smoothed pseudorange data, external data such as from inertial navigation systems (and other such devices), additional constraints, etc.

7.3.2. Accuracy versus positioning mode

Figure 7.4 illustrates the different positioning accuracy associated with the different GPS positioning modes (accuracy is quoted as two-sigma values, i.e. 95 per cent confidence level). The following comments may be made with respect to this diagram:

1 The top half refers to Single-Point Positioning (SPP), the lower half to the relative positioning mode.
2 The basic SPP services provided by the US Department of Defense are the Standard Positioning Service (SPS) and the Precise Positioning Service (PPS), both intended

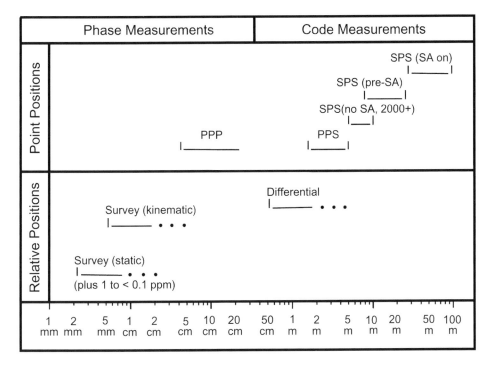

Figure 7.4 GPS accuracy and positioning modes.

for single-epoch positioning.

3   There is a wide range of horizontal SPS and PPS accuracy possible due to a variety of factors:

  a   100 m level accuracy SPS positioning with SA on, *as a result of an artificial degradation of the system*.

  b   5–15 m level accuracy of SPS positioning without SA, representing the current 'natural' accuracy ceiling when using basic navigation-type GPS receivers, because of the difficulty in accounting for the ionospheric bias in the single-frequency C/A-code measurements.

  c   10–50 per cent improvement is possible using dual-frequency GPS receivers.

  d   2–10 m level accuracy PPS positioning, using dual-frequency P-code pseudo-range measurements.

  e   Dual-frequency GPS, coupled with the high accuracy satellite clock and ephemeris data provided by the IGS, can deliver a 50 per cent improvement in basic SPS accuracy.

4   Surprisingly, the averaging of SPS results for up to 60 min at a single benchmark does not significantly improve positioning accuracy, with recent studies indicating an improvement of the order of 10–20 per cent compared to single-epoch solutions.

5   The carrier phase-based procedures are typically only applied in the relative positioning mode for most engineering, surveying and geodetic applications, and *relative* position accuracy is usually expressed in terms of *parts per million* ('ppm' – e.g. 1 cm error in 10 km).

6   Carrier phase-based positioning may be in the single-epoch mode (as is necessary for kinematic positioning), or takes advantage of the receiver being static in order to collect data over an *observation session*.

7   *Precise Point Positioning* is possible using carrier phase data, with accuracies better than a decimeter possible if the observation session is several hours in length (Section 14.2.3).

8   The accuracy of carrier phase-based positioning techniques is a function of baseline length, number of observations, length of observation session, whether ambiguities have been fixed to their integer values or not, and others.

9   In all cases, the vertical accuracy is about 2–3 times worse than the horizontal positioning accuracy.

The 'resolution of the carrier phase ambiguities' is central to precise carrier phase-based positioning in many surveying and engineering applications and requires the determination of the exact number of integer wavelengths in the carrier measurement of satellite-to-receiver distance (Section 9.4.2).

It should be emphasized that GPS was originally designed to provide accuracies of the order of a *dekameter* (ten meters) or so in the SPP mode, and is optimized for real-time operations. All other developments to improve this basic accuracy capability must be viewed in this context. As a general axiom of GPS positioning, the higher the accuracy sought, the more effort (in time, instrumentation and processing sophistication) that is required.

7.4. High precision GPS positioning

GPS is having a profound impact on society. It is estimated that the worldwide market for GPS receiver equipment in 2000 was about US$10 billion, but the annual market for services may be several times this value! Market surveys suggest that the greatest growth is expected to be consumer markets such as in-vehicle applications, integration of GPS and cellular phones, and portable GPS for outdoor recreation and similar activities. These are expected to ultimately account for more than 60 per cent of the GPS market. The penetration of GPS into many applications (and in particular into consumer devices) helps make the processes and products of geospatial information technology more and more a part of the mainstream 'information society'. However, in the following chapters the focus will be on the surveying and mapping disciplines, and how GPS is now an indispensable tool for geospatial professionals.

7.4.1. GPS in support of geospatial applications

In this manual the authors have adopted a very broad definition of 'GPS surveying', encompassing all applications where coordinate information is sought in support of mapping or geospatial applications. In general such applications:

- Are of *comparatively high accuracy*. This is, of course, a subjective judgement, but in general 'high accuracy' implies a level of coordinate precision much higher than originally intended of GPS. As GPS is a navigation system designed to deliver dekameter-level SPP accuracy, the accuracy threshold for *surveying* may be arbitrarily set at the sub-meter level, while *mapping* accuracy's may be satisfied by Differential GPS (DGPS) techniques that can deliver accuracy's at the few meter level. *In this manual 'GPS surveying' will be considered synonymous with carrier phase-based positioning.*
- Require the use of *unique observation procedures, measurement technologies and data analysis*. In fact, the development of distinctive field procedures, specialized instrumentation and sophisticated software is the hallmark of 'GPS surveying'.
- Do *not require positioning information 'urgently'*. 'Navigation', on the other hand, is concerned with the safe passage of vehicles, ships and aircraft, and hence demands location information in *real-time*.
- In general permits *post-processing of data* to obtain the highest accuracy possible.
- Has as its raison d'entre, the *production of a map*, or the establishment of a *network of coordinated points* which support the traditional tasks of the surveying discipline, as well as new applications such as GIS.

In the case of *land surveying applications*, the characteristics of GPS satellite surveying are:

1 The points being coordinated are generally *stationary*.
2 Depending on the accuracy sought, GPS *data are collected over some 'observation session'*, ranging in length from a few seconds to several hours, or more.
3 Restricted to the relative positioning mode of operation.
4 In general (depending on the accuracy sought) the measurements used for the data reduction are those made on the satellites *L-band carrier waves*.

5 Generally associated with the *traditional surveying and mapping functions*, but accomplished using GPS techniques in less time, to a higher accuracy (for little extra effort) and with greater efficiency.

A convenient approach is to adopt a geospatial applications classification on the basis of accuracy requirements. Four classes can be identified on this basis:

Scientific Surveys (category A):	better than 1 ppm
Geodetic Surveys (category B):	1–10 ppm
General Surveying (category C):	lower than 10 ppm
Mapping/Geolocation (category D):	better than 2 m

Category A primarily consist of those surveys undertaken in support of precise engineering, deformation analysis, and geodynamic applications. Category B includes geodetic surveys undertaken for the establishment, densification and maintenance of control networks. Category C primarily encompasses lower accuracy surveys, primarily to support engineering and cadastral applications, geophysical prospecting, etc. Category D includes all other general purpose 'geolocation' surveys intended to coordinate objects or features for map production and GIS data capture (Chapter 25, Part 4). Users in the latter two categories form the majority of the GPS user community. Category A and B users may provide the 'technology-pull' impetus for the development of new instrumentation and processing strategies, which may ultimately be adopted by the category C and D users. Note, this classification scheme is entirely *arbitrary*, and does not relate to any specification of 'order' or 'class' of survey as may be defined by national or state survey agencies.

7.4.2. Using GPS in the field

With respect to category D users (using the pseudorange-based techniques), the planning issues, as well as the field and office procedures, are not as stringent as for the GPS surveying users. Hence most of the attention will be focused on carrier phase-based techniques. Some comments to the *operational aspects* of GPS surveying (categories A, B, C above):

1 Survey planning considerations are derived from:

    a The nature and aim of the survey project – *as for conventional surveys.*

    b The unique characteristics of GPS, and in particular no requirement for receiver intervisibility – *a simplification in survey design.*

    c The number of points to be surveyed, the resources at the surveyor's disposal, and the strategy to be used for propagating the survey – *a logistical challenge.*

    d Prudent survey practice, requiring redundant and check measurements to be incorporated into the network design.

2 Field operations are characterized by requirements for:

    a Clear skyview.

    b Setup of antennas over ground marks.

    c Simultaneous operation of two or more GPS receivers.

    d   Common data collection over some observation session (if in static mode).

    e   Deployment of GPS hardware to new stations.

3   Field validation of data collected, in order to:

    a   Verify sufficient common data collected at all sites operating simultaneously.

    b   Verify quality of data to ensure that acceptable results will be obtained.

    c   Where data dropout is high, or a station has not collected sufficient data, reoccupation may be necessary.

4   Office calculations:

    a   To obtain GPS solutions for single sessions or baselines.

    b   To combine the baseline results into a network solution.

    c   To incorporate external information (e.g. local control station coordinates), and hence modify the GPS-only network solution.

    d   To transform the GPS results (if necessary) to the local geodetic datum, and to derive orthometric heights.

    e   To verify the accuracy and reliability of the GPS survey.

The GPS project planning issues are discussed in Chapter 12. Chapter 13 deals with field operations.

7.4.3. GPS competitiveness

GPS needs to be competitive against conventional terrestrial techniques of surveying. Several criteria for judging the utility of GPS can be identified:

1   Cost benefit: *issues such as the capital cost of equipment, ongoing operational costs, data processing costs, development, training and maintenance costs.* This can best be measured according to productivity. The direct cost of a GPS survey (not including equipment and training costs) can be estimated during the planning phase. It needs to be established whether competing technologies offer lower costs.

2   Ease of Use: *issues such as servicing, timeliness of results, expertise of users.* Experience indicates that the primary factors affecting servicing are those of distance to the servicing agents, their technical expertise, and their customer service. To ensure high quality results in a reasonable time (real-time operations may not be required) it is important that all personnel be adequately trained.

3   Accuracy: *obviously related to the class of application.* The level of accuracy sought will directly influence many other factors such as: type of instrumentation, technique to be used, sophistication of software, cost of survey, field operations, etc.

4   External factors such as availability of satellite ephemerides and other performance constraints such as superior GPS networks to connect into, base station operation, etc.

GPS *complements* the traditional EDM-theodolite techniques for routine surveying activities. Indeed the traditional techniques are likely to continue playing the dominant role for some time to come, unless the conditions for survey are ideal from a GPS point of view. In that case, the 'Real-Time Kinematic' (RTK) technique will be favored (Section 11.2.2). For mapping of features, for GIS-type applications, GPS is the ideal low-to-moderate accuracy point coordination tool. Finally, for high precision (geodetic) positioning, particularly over long distances, GPS is without peer (Teunissen and Kleusberg 1998).

References

AUSLIG, *Australian Surveying and Land Information Group web page*, 2001. http://www.auslig.gov.au/ausgda/gdastrat.htm

Hofmann-Wellenhof, B., Lichtenegger, H. and Collins, J., *Global Positioning System: Theory and Practice* (4th ed.), Springer-Verlag, 1998, 389 pp.

IGS, *International GPS Service web page*, 2001. http://igscb.jpl.nasa.gov.

Kaplan, E. (ed.), *Understanding GPS: Principles and Applications*. Artech House, 1996, 570 pp.

Langley, R. B., 'The orbits of GPS satellites', *GPS World*, 2(3), 1991a, 50–53.

Langley, R. B., 'Time, clocks, and GPS', *GPS World*, 2(10), 1991b, 38–42.

Langley, R. B., 'The GPS observables', *GPS World*, 4(4), 1993, 52–59.

Leick, A., *GPS Satellite Surveying* (2nd ed.), John Wiley & Sons, 1995, 560 pp.

NAVCEN, *US Coast Guard Navigation Center web page*, 2001. http://www.navcen.uscg.gov/gps/.

NIMA, *World Geodetic System 1984 (WGS 84) – Its Definition and Relationships with Local Geodetic Systems* (3rd ed.), National Imagery and Mapping Agency, Technical Report, 1997, 120 pp.

Parkinson, B. W., 'GPS eyewitness: the early years', *GPS World*, 5(9), 1994, 32–45.

Parkinson, B. W. and Spilker, J. J. (eds.), *Global Positioning System: Theory and Applications*, vol. I (694 pp.) and vol. II (632 pp.), American Institute of Aeronautics and Astronautics, Inc., 1996.

Teunissen, P. J. and Kleusberg, A. (eds.), *GPS for Geodesy* (2nd ed.), Springer, Berlin, 1998.

Wells, D. E., Beck, N., Delikaragolou, D., Kleusberg, A., Krakiwsky, E. J., Lachapelle, G., Langley, R. B., Nakiboglu, M., Schwarz, K. P., Tranquilla, J. M. and Vanicek, P., *Guide to GPS Positioning*, Canadian GPS Associates, Fredericton, NB, Canada, 1986, 600 pp.

Chapter 8

Fundamentals of GPS signals and data

Hasanuddin Abidin

8.1. GPS signal characteristics

8.1.1. Signal structure

GPS satellites transmit signals at two frequencies, designated L1 and L2, on which three binary modulations are impressed: the C/A-code, the P(Y)-code, and the broadcast (or navigation) message, as depicted in Figure 8.1. L1 is the principal GPS carrier signal with a frequency of 1575.42 MHz, and is modulated with the P(Y)-code, the C/A-code, and the navigation message. The second signal, L2, is transmitted at a frequency of 1227.60 MHz and modulated only with the P(Y)-code and the navigation message. This second signal was primarily established to provide a means of estimating the ionospheric delays to GPS measurements. (Each GPS satellite also transmits a L3 signal at 1381.05 MHz, associated with its dual role as a nuclear burst detection satellite, as well as an S-band telemetry signal, however these will not be discussed further.)

The 'precision' P(Y)-code has a bit rate of 10.23 MHz; while the 'coarse acquisition' C/A-code and navigation message have bit rates of 1.023 MHz and 50 Hz, respec-

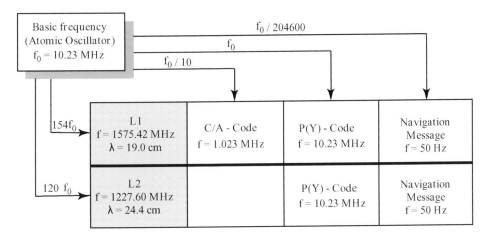

Figure 8.1 Current structure of the GPS L1 and L2 satellite signals.

tively. The P(Y)-code is an encryption of the published P-code by a code sequence referred to as the 'W-code' (resulting in the so-called 'Y-code'), and accessible only by the US Department of Defense (and other authorized) users. This encryption of the P-code was imposed on 31 January 1994 for all satellites under the so-called *Anti-Spoofing* (AS) policy. The characteristics of the signal modulations are summarized in Table 8.1. A detailed explanation of the characteristics of the GPS signals can be found in such engineering texts as Parkinson and Spilker (1996) and Kaplan (1996).

All signals transmitted by GPS satellites are right-hand polarized, coherently derived from a basic frequency of 10.23 MHz (Figure 8.1), and are transmitted within a bandwidth of about 20.46 MHz (at both center frequencies) by a shaped-beam antenna array on the nadir-facing side of the satellite. Each satellite transmits a unique C/A-code and a unique 1-week long segment of the P(Y)-code. The transmitted power levels are 23.8 and 19.7 dBW for the P(Y)-code signals on L1 and L2, respectively, and 28.8 dBW for the C/A-code signal (Langley 1998). (*Decibel*, abbreviated as dB, is a unit for the logarithmic measure of the relative power of a signal, hence a 3 dB increase in the strength of a signal corresponds to a doubling of the power level. dBW indicates actual power of a signal compared to a reference of 1 Watt.) The strength of the electromagnetic wave decreases during propagation primarily as a function of transmitter–receiver distance. Thus, the GPS signal arriving at the receiver antenna is relatively weak, with roughly the same strength as the signal from geostationary TV satellites. However, the carefully designed GPS signal structure allows the use of small antennas, as opposed to large TV dishes.

Since both the signal transmitting satellite and the receiver move with respect to each other, the signal arriving at the antenna rotating with the earth is *Doppler-shifted*. Consequently the received frequency differs from the transmitted frequency by an amount which is proportional to the radial velocity of the satellite with respect to the receiver. Signals from different satellites therefore arrive at the receiver antenna with different Doppler-shifted frequencies. For example, assuming a stationary receiver in the satellite orbit plane, the maximum frequency shift of about ± 6 kHz due to Doppler effect would be at the epoch of local horizon crossing (maximum satellite–receiver distance about 25,783 km), where the radial velocity is a maximum (about ± 0.9 km/s). Naturally there is no Doppler frequency shift at the point of closest approach (the zenith, at an approximate distance of 20,183 km).

8.1.2. Signal coverage

The GPS signal is transmitted by the satellite towards the earth in the form of a *signal beam*, as illustrated in Figure 8.2 (Parkinson and Spilker 1996). The figure shows that the signal covers not only the earth's surface (where the signal can be used for positioning and navigation) but above it as well, as long as the user is within the main beam of the GPS signal, but outside the earth's shadow. This is useful for a variety of space applications.

In order to achieve high accuracy, positioning by GPS has to be performed in the *differential* mode (as discussed in the Section 7.1.3). In this mode, two GPS receivers observe the same satellites, at the same time, and the *relative* position between the two receivers is estimated. If one considers the spatial coverage of the GPS signal from a satellite as depicted in Figure 8.2, then the distances between two points on the earth's

Table 8.1 Characteristics of the GPS signal modulations.

Component	C/A-code	P(Y)-code	Navigation message
Bit generation rate	1.023 Mbps (10^6 bits/s)	10.23 Mbps	50 bps
Bit length	≈293 m	≈29.3 m	≈5950 km
Repetition rate	1 ms	1 week	N/A
Code type	37 unique codes (pseudo-random gold codes)	37 one-week segments (PRN code)	N/A
Carrier frequency	L1	L1, L2	L1, L2
Expected minimum signal strength at the user receiver, referenced to 0 dBic antenna[a]	−160 dBW	−163 dBW (on L1) −166 dBW (on L2)	N/A
Fundamental characteristics	Easy to acquire due to its short period and low cross-correlation between different PRN codes; thus easy to rapidly distinguish among signals arriving simultaneously from multiple satellites. Acquired before P(Y)-code so as to provide the timing information necessary to acquire the more complex P(Y)-code	More accurate and jam resistant than C/A-code. Not available to civilian users under AS. Resistant to mutual interference when signals are received simultaneously from multiple satellites. Difficult to acquire, i.e. receiver correlator must be timed to within roughly one P-code chip to allow correlation	Provides satellite health status, time, ephemeris data, various correction terms and hand-over-word which tells receiver where to start the search for the P(Y)-code. Total transmission time is 30 s

[a] dBic describes antenna gain in dB with respect to a circularly-polarized isotropic radiator (a hypothetical ideal reference antenna).

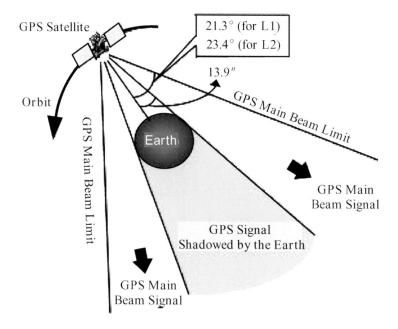

Figure 8.2 Spatial coverage of the GPS signal beam.

surface that can still view the same satellite, as a function of the observation *mask angle* is indicated in Figure 8.3. For example, even with an elevation mask angle of 15° (the local elevation angle below which the satellite signals will not be tracked), two receivers on the earth's surface separated as far apart as about 13,500 km can still view the same GPS satellite. However, both receivers 'see' the signal very near the horizon and in general, since the signal will be obstructed by either topography or objects around the antenna, receiver separations will be much less than those shown in Figure 8.3. For most surveying/mapping applications, maximum receiver separations are of the order of tens of kilometers (and generally less).

8.2. Codes and pseudorange measurements

8.2.1. Pseudo-random noise codes

There are two Pseudo-Random Noise (PRN) codes, which are modulated on the signals transmitted by a GPS satellite: the P(Y)-code and the C/A-code. The two main functions of these codes are: (a) to provide time delay measurements so the user can determine the distance from the receiver's antenna to the observed satellite (either code could be used, but the P(Y)-code provides a more precise range estimate than the C/A-code), and (b) to help the receiver in differentiating the incoming signals from different satellites. These codes are sequences of binary values (zeros and ones), and although the sequence *appears to be random*, each code has a unique structure generated by a mathematical algorithm (Figure 8.4). One version of the code is generated within the satellite, and the identical code sequence is replicated within the

GPS Orbit

GPS Satellite

≈ 28 °

≈ 20,000 km

d

α

Earth

R

Mask Angle	α	Arc Length (d)
0 °	152 °	16889 km
5 °	142 °	15778 km
10 °	132 °	14667 km
15 °	122 °	13556 km
Earth radius: R ≈ 6378 km		

GPS Antenna

Mask Angle

Mask angle = minimum elevation of observed satellites.

Figure 8.3   Visibility of the same GPS satellite from two points on the earth's surface.

receiver. Two such identical codes will only be aligned at *zero lag* (i.e. when the sequence of one code is time-shifted to the instant when all the 0s and 1s in that code match the sequence in the other code). Because the codes are generated by either the satellite clock, or the receiver clock, they are in fact a means of representing the time defined by the respective clock. If the two clocks are synchronized to the same time system, then the clock times can be compared and the difference is a measure of the time taken for a signal to travel from satellite to receiver. From this the distance can be estimated (Section 8.2.2).

As indicated in Table 8.1, each C/A-code is a unique sequence of 1023 binary numbers which repeats itself every 1 ms. Each binary bit of the C/A-code is generated at the rate of 1.023 MHz and has a 'duration' or 'length' of approximately 1 μs (or about 293 m in units of length).

In contrast to the C/A-code, the P(Y)-code consists of a much longer binary sequence of 2.3547×10^{14} binary numbers, and its pattern will not repeat itself until after 266 days. The P(Y)-code is generated at a rate ten times faster than the

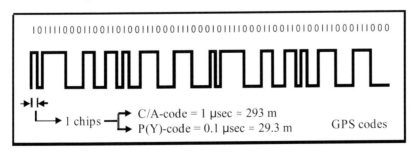

101111000110011010011100011100010111100011001101001110001110 00

1 chips

C/A-code = 1 μsec ≈ 293 m
P(Y)-code = 0.1 μsec ≈ 29.3 m

GPS codes

Figure 8.4   Example of a sequence of binary numbers of a PRN code.

C/A-code, i.e. 10.23 MHz. This means each bit has a duration of approximately 0.1 μs (or a 'length' of about 29 m). Each GPS satellite is assigned a unique 1 week segment of the P(Y)-code, and this segment is set back to zero each week at midnight (0 h UT) from Saturday to Sunday.

8.2.2. Determining satellite–receiver range

By acquiring the P(Y)-code, or the C/A-code, the observer can measure the distance or *range* to the satellite. The basic principle for obtaining this range is the so-called 'code-correlation' technique whereby the incoming code from the satellite is correlated with a replica of the corresponding code generated inside the receiver, as depicted in Figure 8.5. Both codes are generated using the same mathematical algorithm. The time shift (dt) required to align the two codes is, in principle, the time required by the signal carrying the code to travel from the satellite to the receiver. Multiplying dt with the speed of light results in an estimate of the range. This range is referred to as a *pseudorange*, because it is still biased by the time offset (or mis-synchronization) between the satellite clock and receiver clock used to measure the time delay.

In general, the precision of a pseudorange measurement is about 1 per cent of its code length (or *resolution*). The nominal precision of the P(Y)-code pseudorange is therefore about 0.3 m, and for the C/A-code pseudorange it is about 3 m. Besides being more precise, the P(Y)-code pseudorange measurement is more resistant to the effects of multipath and jamming/interference. Moreover, since the P(Y)-code is modulated on both the L1 and L2 signals, the user can obtain pseudorange measurements at both frequencies, that is the P(Y)-L1 pseudorange and the P(Y)-L2 pseudorange, so that by combining these two measurements it is possible to derive a new pseudorange that is not affected by the ionospheric delay (Section 9.2.1.1). However, due to the implementation of the AS policy only authorized users can gain access to the P(Y)-code directly to make pseudorange measurements using the code-correlation technique. Civilian receivers must employ different signal processing techniques (Section 10.2.4, Table 10.1) to make dual-frequency measurements. Apart from the receivers used for many GPS surveying applications, where dual-frequency measurements are a prerequisite for obtaining fast centimeter-level accuracy coordinates, most civilian GPS receivers intended for navigation applications only observe the C/A-code pseudorange, and hence are referred to as *single-frequency navigation receivers*.

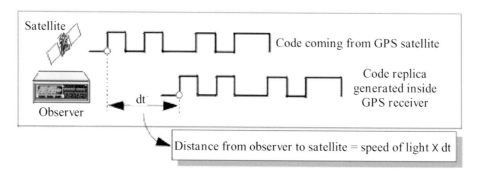

Figure 8.5 The principle of obtaining a range measurement using PRN codes.

8.3. Broadcast navigation message

Besides the ranging codes, GPS signals are also modulated with the *navigation message*. This contains information such as the satellite's orbital data (the so-called broadcast ephemeris), satellite almanac data, satellite clock correction parameters, satellite health and constellation status, ionospheric model parameters for single-frequency users, and the offset between the GPS and UTC (*Universal Time Coordinated*) time systems. The content of the navigation message is continuously updated by the GPS control segment (Section 7.1.2), and broadcast to the users by the GPS satellites.

8.3.1. Broadcast ephemeris

The most important data contained within the navigation message is the *broadcast ephemeris*. This ephemeris is in the form of Keplerian orbital elements and their perturbations, as listed in Table 8.2 (Wells *et al.* 1986; Seeber 1993). These parameters are illustrated in Figure 8.6. The coordinates of the satellite in the WGS 84 datum (Section 7.2.1) at every observation epoch can be computed using the algorithm given in Seeber (1993).

Table 8.2 Content of GPS clock and broadcast ephemeris messages

Time parameters	
t_{oe}	Reference time for the ephemeris parameters (s)
t_{oc}	Reference time for the clock parameters (s)
a_0, a_1, a_2	Polynomial coefficients for satellite clock correction, i.e. representing the bias (s), drift (s/s), and drift-rate (s/s^2) components
IOD	Issue of data (arbitrary identification number)
Satellite orbit parameters	
$\sqrt{a}$	Square root of the semi-major axis (m$^{1/2}$)
e	Eccentricity of the orbit (dimensionless)
i_o	Inclination of the orbit at t_{oe} (semicircles)
Ω_o	Longitude of the ascending node at t_{oe} (semicircles)
ω	Argument of perigee (semicircles)
M_o	Mean anomaly at t_{oe} (semicircles)
Orbital perturbation parameters	
Δn	Mean motion difference from computed value (semicircles/s)
$\dot{\Omega}$	Rate of change of right ascension (semicircles/s)
idot	Rate of change of inclination (semicircles/s)
C_{us} and C_{uc}	Amplitude of the sine and cosine harmonic correction terms to the argument of latitude (rad)
C_{is} and C_{ic}	Amplitude of the sine and cosine harmonic correction terms to the inclination angle (rad)
C_{rs} and C_{rc}	Amplitude of the sine and cosine harmonic correction terms to the orbit radius (m)

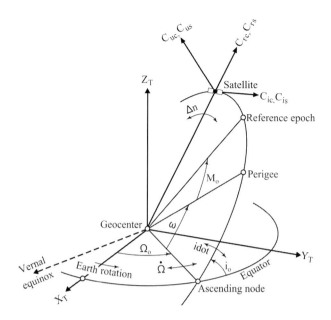

Figure 8.6 Geometric visualization of the GPS broadcast ephemeris parameters.

8.3.2. Broadcast satellite clock error model

The satellite clock error at any time t can be computed from the following model:

$$dt^j = \alpha_0 + \alpha_1(t - t_{oc}) + \alpha_2(t - t_{oc})^2 \tag{8.1}$$

where α_0, α_1, and α_2 are the clock offset, the clock drift-rate, and half the clock drift acceleration at the reference clock time t_{oc} (*time of clock*). Equation (8.1) provides a good prediction for the satellite clock behavior because the GPS satellites use high quality atomic clocks (cesium and rubidium oscillators) which have a stability of about 1 part in 10^{13} over a period of 1 day. Since Selective Availability (SA) was turned off the satellite clock correction computed from equation (8.1) has a *residual* error of the order of a few meters. (However, when SA was switched on during the 1990s this error was of the order of 20–25 m.)

8.4. Carrier waves and carrier phase measurements

The main function of the GPS carrier waves, L1 and L2, is to 'carry' the PRN codes and navigation message to the receiver. The codes and navigation messages are modulated onto the carrier waves using the bi-phase shift key modulation technique, as illustrated in Figure 8.7 (e.g. Hofmann-Wellenhof *et al.* 1998). When the data (code and navigation message) value for a binary bit is 0 and it changes to a value of 1 (or visa versa, 1 to 0) then the carrier phase is shifted by 180°. When there is no change in the value of the adjacent bits, then there is no change in the phase.

Phase measurements made on the carrier waves can also be used to derive very precise range measurements to the satellites, which are sometimes referred to also

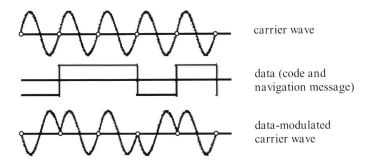

carrier wave

data (code and
navigation message)

data-modulated
carrier wave

Figure 8.7 Bi-phase shift key modulation of the GPS carrier wave.

as 'phase ranges' or 'carrier ranges', or simply *carrier measurements*. For precise GPS applications such as for surveying and geodesy, the carrier measurements must be used in place of pseudoranges.

8.4.1. Ambiguous carrier phase

The carrier phase measurement is the difference between the reference phase signal (which has the form of a sine wave) generated by the receiver and the phase of the incoming GPS signal (after stripping away the code and navigation message modulations), and is in fact the *beat signal*. The measurement of phase by a GPS receiver is complicated by the fact that the receiver cannot measure directly the complete range from the receiver to the satellite. At the initial epoch of signal acquisition (t_0) only the *fraction* of the cycle of the beat phase can be measured by the receiver. (One cycle corresponds to one *wavelength* of the sine wave that is the carrier signal without modulation, which for the L1 carrier wave is about 19 cm and for L2 is approximately 24 cm.) The remaining *integer* number of cycles to the satellite cannot be measured directly. This unknown integer number of cycles is usually termed the 'initial cycle (phase) ambiguity' (N). Subsequently, the receiver only counts the number of integer cycles that are being tracked. As long as there is no loss-of-lock of tracking of the satellite, the value of N remains a *constant*.

The phase measurement at a certain epoch t_i, i.e. $\varphi(t_i)$, is therefore equal to the fractional phase (Fr) at that epoch augmented with the number of full cycles 'counted' (Int) since t_0:

$$\varphi(t_i) = \mathrm{Fr}(\varphi(t_i)) + \mathrm{Int}(\varphi; t_0, t_i) + N(t_0) \tag{8.2}$$

The above equation indicates that the GPS carrier phase measurement (when expressed in length units) is not a true or absolute range from receiver to the satellite as in the case of the pseudorange, but is an *ambiguous* range. There is an unobserved part of the range caused by the initial cycle ambiguity of the phase, as illustrated in Figure 8.8. In order to convert this ambiguous range into a true range, the cycle ambiguity N has to be estimated. If the value of the cycle ambiguity can be correctly estimated, then the resultant *carrier range* will be transformed into a very precise range measurement (at the few millimeter level), and can be used for high precision position-

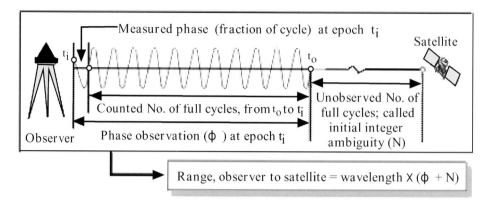

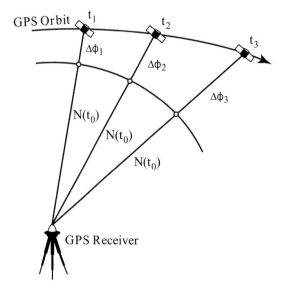

Figure 8.8 Phase (carrier) range determination using GPS phase measurement.

ing. However, the carrier range has the opposte ionospheric effect to the regular pseudorange (cf. equations (8.4) and (8.5)). Correctly estimating the value of the integer cycle ambiguity is not an easy task. This computation is referred to as *ambiguity resolution* (Section 9.4.2).

8.4.2. Unambiguous carrier range

From equation (8.2) it can be seen that the first two terms on the right hand side will change because the satellite is moving. If the two terms are expressed as:

$$\Delta\varphi_i = \text{Fr}(\varphi(t_i)) + \text{Int}(\varphi; t_0, t_i) \tag{8.3}$$

then the geometrical interpretation of this change in carrier phase (range) with time is

Figure 8.9 Geometrical interpretation of carrier range and cycle ambiguity.

illustrated in Figure 8.9. Note that this representation assumes that there is continuous lock on the signal. If loss-of-lock on the signal results, due to any cause, then the second term on the right hand side of equation (8.3) (the integer 'count') is corrupted and a *cycle slip* has occurred. Another interpretation of the cycle slip is that the ambiguity term in equation (8.2) is not the same constant as at initial signal lock-on. The value of N has changed (or 'jumped') after the epoch of the cycle slip by a quantity which is equal to the integer number of cycles in the cycle slip. Cycle slip 'repair' is a crucial operation in carrier phase data processing, as it ensures that the ambiguity term is a constant for the entire observation session. Remember, the value of N must be determined and, if possible, 'resolved' to its likeliest integer value if the ambiguous carrier phase measurements made during an entire observation session are to be converted to precise carrier range observables.

8.5. GPS signal propagation

In its propagation from the satellite to the observer's antenna the signal has to travel through several layers of the atmosphere, as illustrated in Figure 8.10. Due to the refraction inside the tropospheric and ionospheric layers, the speed and direction of the signal will be affected, and in turn the measured pseudorange and carrier phase will be *biased*. In addition, signal propagation may be influenced by other phenomena such as intentional or accidental interference due to multipath and jamming.

8.5.1. Ionospheric and tropospheric delay

The ionosphere will affect the speed, frequency, direction, and polarization of GPS signals, as well as cause amplitude and phase scintillation (Parkinson and Spilker 1996). By far the largest effect will be on the speed of the signals, which will directly

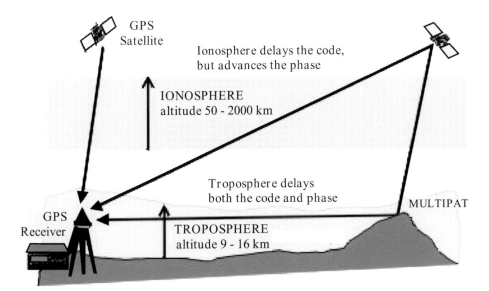

Figure 8.10 Propagation effects on the GPS signal.

affect the carrier range and pseudorange measurements to the satellite. The ionosphere will *delay* the code measurement (the pseudorange) and *advance* the signal phase (the carrier range), by the same amount. The thickness of the ionosphere is several hundreds of kilometers, and is generally considered to start from about 50 km above the surface of the earth.

The slant range error caused by the ionosphere can be up to about 100 m, and its magnitude is dependent on factors such as the time of day (the ionosphere is 'quieter' at night), latitude of the observer, time of the year, period within the 11 year sunspot cycle (the highest ionospheric activity for the current cycle was in 2001), elevation angle to the satellite, and frequency of the signal. The estimated maximum rate-of-change of ionospheric propagation delay is about 19 cm/s, which corresponds to about 1 cycle on L1 (Parkinson and Spilker 1996). Very rapidly changing ionospheric conditions can cause losses of lock, especially on the L2 frequency under AS. When dual-frequency observations are available, a linear combination of measurements made on L1 and L2 can be constructed that eliminates the first-order ionospheric delay (Section 9.2.1.1). In the case of single-frequency users, the broadcast navigation message contains parameters of an approximate model for the ionospheric delay, and although it is preferable to having no model, perhaps only about 50 per cent of the true delay can be represented by such a model.

The troposphere will also affect the speed and direction of the GPS signals, as well as cause some attenuation of the signal. As with the ionosphere, the largest effect will be on the speed of the signal and, in this case, the troposphere will *delay* both the code and phase measurements by the same amount. The tropospheric delay effects are of the order of 2–25 m, a minimum in the vertical (or zenith) direction and becoming larger as the elevation angle of the signal reduces down to the horizon. Both the wet and dry components of the troposphere contribute to the delay, with the former being the most difficult to estimate. The troposphere is only about 10 km thick, and hence signals reaching high altitude points will have less troposphere to travel through, and hence will experience less tropospheric delay.

8.5.2. Multipath effects

In addition, the GPS signal can experience *multipath*, a phenomenon in which a signal arrives at the antenna via two or more paths, that is, a direct signal and one or more reflected signals (Figure 8.10). The signals will interfere with each other and the resultant signal will cause an error in the pseudorange and carrier phase measurements. The range errors caused by multipath will vary depending on the relative geometry of the satellites, reflecting surfaces and the GPS antenna; and also the properties and dynamics of the reflecting surfaces. In the case of the carrier phase, the multipath effect can be up to about 5 cm, that is, 0.25 cycles of wavelength (Hofmann-Wellenhof *et al.* 1998); while for the pseudoranges it can be many tens of meters. Multipath is considered one of the limiting factors to improvements in GPS carrier phase-based positioning 'productivity'. That is, the presence of multipath disturbance impacts on the reliability of ambiguity resolution, particularly when observation sessions are very short (as is the case of 'rapid-static' and kinematic 'on-the-fly' GPS surveying techniques – Section 11.1). Hence strategies must be implemented that deal with such eventualities, such as making extra (redundant) surveys of the same points.

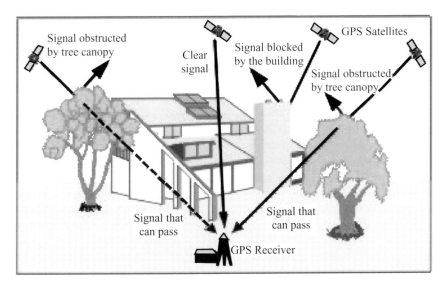

Figure 8.11   Obstructions and interference to the GPS signal.

It should also be noted here that in order to be used for positioning, the signal transmitted by the satellites has to reach the GPS antenna. If the satellite signal is blocked by the topography, foliage or other structures, then measurements cannot be made (Figure 8.11). GPS positioning therefore cannot be performed in tunnels, indoors or under water. In areas with many high-rise buildings and tall trees, that can obstruct the signal reception from the satellites (and cause a lot of cycle slips), it can also be expected that positioning with GPS will be a less optimal technique than in relatively open areas. There are a few strategies that can be used to overcome the signal obstruction problem. The obvious one is to choose a location for the antenna site that has the clearest visibility to the sky. In such cases where the observation site has been fixed beforehand, then elevating the antenna above the obstructing objects, for example above the tree canopy, can help. If it is permitted, cutting the trees around the observation side is another option. Finally, the siting of antennas near sources of microwave radiation, such as microwave communication towers, radio and TV broadcasting stations, or cell phone towers, is to be avoided wherever possible so as to minimize the chances of accidental jamming of the GPS signals.

8.6. Differencing of GPS data

The effects of the above propagation biases and errors, along with others such as satellite ephemeris and clock errors, will contaminate derived parameters such as position, velocity, and time information. Hence accounting for such effects is always necessary in order to ensure high quality results. The most effective strategy for eliminating or mitigating measurement biases is to take advantage of the fact that observations are made to several GPS satellites, from two or more receivers (in relative positioning mode), at the same time. *Differencing* GPS measurements is the means by which complex observation models (which must explicitly include all the different

GPS signal biases and errors) can be *simplified* in order to yield high accuracy results, even in the presence of many measurement biases. The reason is that many biases are common to several measurements, and hence can effectively cancel in combinations of different data. For example, in the case of the satellite clock bias, all measurements made to a particular satellite, at the same time, by any number of receivers, will be 'contaminated' by this bias. That is, all measurements are either too 'short' or too 'long' by the same amount. In the case of the tropospheric delay, all measurements made by receivers to the same satellite will be very *nearly* affected by the same bias, depending on how close the receivers are (for short baselines essentially the same tropospheric conditions will affect both sets of measurements at an epoch, to the same satellite, but as the receiver separation increases this correlation decreases).

The pseudorange and carrier phase data are related to the various parameters of interest and to the measurement biases through the following mathematical relations (Leick 1995; Hofmann-Wellenhof *et al.* 1998):

$$P_i^j = \rho_i^j + \text{trop}_i^j + \text{ion}_{qi}^j + c(\text{recclk}_i - \text{satclk}^j) + \text{mlpr}_i^j + \varepsilon\text{pr}_i^j \qquad (8.4)$$

$$L_i^j = \rho_i^j + \text{trop}_i^j - \text{ion}_{qi}^j + c(\text{recclk}_i - \text{satclk}^j) + \lambda_q N_{qi}^j + \text{mlph}_i^j + \varepsilon\text{ph}_i^j \qquad (8.5)$$

where the subscript refers to the receiver identifier i and signal frequency q, the superscript refers to the satellite identifier j, and all units are in meters unless otherwise stated: P_i^j is the pseudorange at frequency f_q ($q = 1, 2$), L_i^j is the carrier phase at frequency f_q ($q = 1, 2$), ρ is the geometric range between the receiver and the satellite, trop_i^j is the range bias caused by tropospheric delay, ion_{qi}^j is the range bias caused by ionospheric delay at frequency f_q, recclk_i is the receiver clock bias (in seconds), satclk^j is the satellite clock bias (in seconds), c is the speed of electromagnetic radiation in a vacuum (in m/s), $\text{mlpr}_i^j, \text{mlph}_i^j$ are the multipath disturbance on the P_i^j and L_i^j observables, N_1, N_2 are the cycle ambiguities of the L1 and L2 signals (integer cycles, each with wavelength λ_q), and $\varepsilon\text{pr}_i^j, \varepsilon\text{ph}_i^j$ is the measurement noise for the P_i^j and L_i^j observables.

Note, apart from the ambiguity term, and constants such as c and λ, all other terms vary with time, hence equations (8.4) and (8.5) are valid only for a certain measurement *epoch*. The geometric range is the fundamental quantity, related to receiver i and satellite j Cartesian coordinates (x,y,z and X,Y,Z, respectively) by the well known expression, at some instant in time t:

$$\rho_i^j(t) = \sqrt{[X^j(t) - x_i]^2 + [Y^j(t) - y_i]^2 + [Z^j(t) - z_i]^2} \qquad (8.6)$$

The pseudorange and carrier phase measurements can be differenced within the data processing software to create new range data combinations with specific properties. Differencing of GPS data can be performed in several modes, and depending on the way in which the differencing is carried out, *One-Way* (OW), *Single-Difference* (SD), *Double-Difference* (DD), and *Triple-Difference* (TD) observables can be constructed. OW data is the basic range data from one receiver to one satellite at one frequency, as represented by equations (8.4) and (8.5). Different combinations of OW pseudorange and carrier phase measurements to the same satellite, from the same receiver, are possible. Some of these combinations have beneficial characteristics for cycle slip editing, ambiguity resolution, or coordinate parameter estimation, making dual-

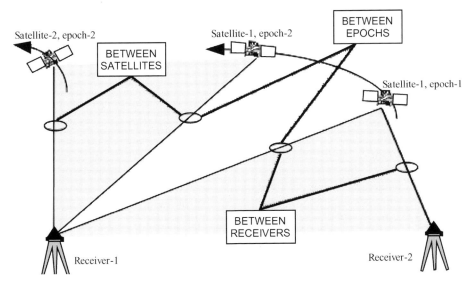

Figure 8.12 Data differencing modes.

frequency instrumentation desirable for all high accuracy applications. Chapter 9 presents the basic mathematical observation models for pseudorange and carrier phase used in GPS data processing.

GPS data differencing can be carried out between different receivers, different satellites, or different epochs, as illustrated in Figure 8.12. It should be noted that between-receiver and between-satellite differencing are performed among the measurements made at the same observation epoch, and requires that all GPS receivers deployed in the field should track the same satellites during the same time period, in a synchronized fashion (Section 12.3.3).

There are several consequences of the GPS data differencing process:

- differencing can eliminate or reduce the effects of many biases, and
- differencing will reduce the quantity of data, however
- differencing will introduce mathematical correlations among the data, and
- differencing will increase the noise level of the differenced data.

Commercial data processing software combines measurements from a pair of GPS receivers, involving different satellites, both L1 and L2 frequencies (if dual-frequency instrumentation is used) and different data types (carrier phase and precise pseudorange data in the case of state-of-the-art receivers), by the process of data differencing to construct the 'observables' that are the input for the baseline estimation algorithm.

8.6.1. Single-differenced data

Single-Differenced (SD) data is the difference between two OW measurements. Depending on the manner of differencing, there are three types of SD observables, namely *between-receiver* SD, *between-satellite* SD, and *between-epoch* SD. The measurements may be carrier phase or pseudorange, on either of the two frequencies.

Table 8.3 Characteristics of errors and biases in SD data

Errors/biases	SD data		
	Between-receiver (Δ)	*Between-satellite (∇)*	*Between-epoch (δ)*
Satellite Clock	Eliminated	–	–
Ephemeris	Reduced (depending on distance between receivers)	–	–
Receiver clock	–	Eliminated	–
Ionospheric	Reduced (depending on distance between receivers)	Reduced (depending on angular separation between satellites)	Reduced (depending on time interval between epochs)
Tropospheric	Reduced (depending on distance between receivers)	Reduced (depending on angular separation between satellites)	Reduced (depending on time interval between epochs)
Cycle ambiguity	–	–	Eliminated (if no cycle slips)
Noise	Increased by $\sqrt{2}$	Increased by $\sqrt{2}$	Increased by $\sqrt{2}$

Each type of SD differencing will have different effects, as summarized in Table 8.3. Between-receiver differencing of measurements involving the same satellite signal will eliminate the effects of satellite clock errors, and will reduce the effects of tropospheric and ionospheric biases, depending on the distance between the receivers. Between-satellite differencing will eliminate the effects of receiver clock errors. In the case of between-epoch differencing, the cycle ambiguity of the phase observation will be eliminated when there is no cycle 'slip' between the two epochs, and depending on the time interval between epochs. The reader is referred to such texts as Hofmann-Wellenhof *et al.* (1998) and Leick (1995) for the mathematical details.

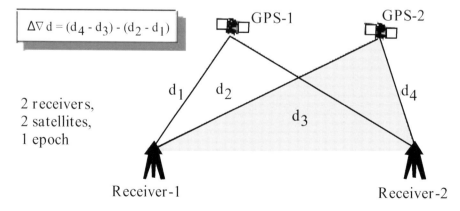

$$\Delta\nabla d = (d_4 - d_3) - (d_2 - d_1)$$

GPS-1 GPS-2

2 receivers, 2 satellites, 1 epoch

d_1 d_2 d_3 d_4

Receiver-1 Receiver-2

Figure 8.13 Receiver–satellite double-difference.

Table 8.4 Characteristics of errors and biases in DD data

Errors/biases	DD data		
	Receiver–satellite ($\Delta\nabla$)	Receiver–epoch ($\Delta\delta$)	Satellite–epoch ($\nabla\delta$)
Satellite clock	Eliminated	Eliminated	–
Ephemeris	Reduced (depending on distance between receivers)	Reduced (depending on distance between receivers)	–
Receiver clock	Eliminated	–	Eliminated
Ionospheric	Reduced (depending on distance between receivers)	Reduced (depending on distance between receivers)	Reduced (depending on time interval between epochs and the distance between satellites)
Tropospheric	Reduced (depending on distance between receivers)	Reduced (depending on distance between receivers)	Reduced (depending on time interval between epochs and the distance between satellites)
Cycle ambiguity	–	Eliminated (if no cycle slips)	Eliminated (if no cycle slips)
Noise	Increased by 2	Increased by 2	Increased by 2

8.6.2. Double-differenced data

Double-Differenced (DD) data is the differencing result of two SD observables, or four OW measurements. Depending on the types of SD data being differenced, three DD observables can be identified: *receiver–satellite* DD, *satellite-epoch* DD, and *receiver-epoch* DD. The geometry of the receiver–satellite DD observable is depicted in Figure 8.13, and the characteristics of the differencing process are summarized in Table 8.4.

Manipulating equation (8.5) leads to the following model of the standard carrier phase DD for either the L1 or L2 carrier phase observation, where all clock errors have been eliminated and the tropospheric and ionospheric biases are assumed to have been significantly reduced (superscripts are satellites j and l, subscripts are receivers i and m):

$$\lambda.(\varphi_i^j - \varphi_m^j - \varphi_i^l + \varphi_m^l) = (\rho_i^j - \rho_m^j - \rho_i^l + \rho_m^l) + (\text{trop}_i^j - \text{trop}_m^j - \text{trop}_i^l + \text{trop}_m^l) -$$

$$(\text{ion}_i^j - \text{ion}_m^j - \text{ion}_i^l + \text{ion}_m^l) + \lambda(N_i^j - N_m^j - N_i^l + N_m^l) +$$

$$(\text{mlph}_i^j - \text{mlph}_m^j - \text{mlph}_i^l + \text{mlph}_m^l) + (\varepsilon\text{ph}_i^j - \varepsilon\text{ph}_m^j - \varepsilon\text{ph}_i^l + \varepsilon\text{ph}_m^l) \qquad (8.7)$$

Which, after adopting the commonly used double-differencing symbology, and lumping the multipath and noise terms together, leads to the following expression:

$$\lambda\Delta\nabla\varphi_{im}^{jl} = \Delta\nabla\rho_{im}^{jl} + \Delta\nabla\text{trop}_{im}^{jl} - \Delta\nabla\text{ion}_{im}^{jl} + \lambda\Delta\nabla N_{im}^{jl} + \Delta\nabla\varepsilon_{im}^{jl} \qquad (8.8)$$

The above receiver–satellite DD phase combination is the standard observable used in carrier phase-based GPS positioning for survey applications, but with a further simplification through the deletion of the double-differenced atmospheric bias terms (equation (9.3)). However, the cycle ambiguity term remains an integer value, and still needs to be estimated in the case of the GPS surveying techniques described in Section 11.1. For a particular measurement epoch, different DD observables can be constructed with different pairs of satellites, as well as involving different frequency (L1 and L2) and data type (pseudorange and carrier phase) combinations.

8.6.3. Triple-differenced data

Triple-Difference (TD) data is the difference between two DD, or four SD, or eight OW data. Regardless of the differencing sequences between DD data, only one TD observable is obtained, that is the receiver-satellite-epoch TD. In GPS surveying, the TD phase data is usually used for automatic cycle slip detection and repair, and for determining the approximate baseline vector to be used as a priori values in the baseline estimation process.

References

Hofmann-Wellenhof, B., Lichtenegger, H. and Collins, J., *Global Positioning System: Theory and Practice* (4th ed.), Springer-Verlag, Berlin, 1998, 389 pp.

Kaplan, E. (ed.), *Understanding GPS: Principles and Applications*, Artech House, 1996, 570 pp.

Langley, R. B., 'A primer on GPS antennas', *GPS World*, 9(7), 1998, 50–54.

Leick, A., *GPS Satellite Surveying* (2nd ed.), John Wiley & Sons, New York, 1995, 560 pp.

Parkinson, B. W. and Spilker, J. J. (eds.), *Global Positioning System: Theory and Applications*, vols. I (694 pp.) & II (632 pp.), American Institute of Aeronautics and Astronautics, Inc., 1996.

Seeber, G., *Satellite Geodesy, Foundations, Methods, and Applications*. Walter de Gruyter, Berlin, 1993, 531 pp.

Wells, D. E., Beck, N., Delikaragolou, D., Kleusberg, A., Krakiwsky, E. J., Lachapelle, G., Langley, R. B., Nakiboglu, M., Schwarz, K. P., Tranquilla, J. M. and Vanicek, P., *Guide to GPS Positioning*, Canadian GPS Associates, Fredericton, NB, Canada, 1986, 600 pp.

Chapter 9

GPS mathematical models for single point and baseline solutions

George Dedes and Chris Rizos

This chapter briefly describes the mathematical models for the GPS measurements for the two most common position estimation techniques: (a) Single-Point Positioning (SPP) using pseudorange data, and (b) baseline determination using double-differenced carrier phase data. For a detailed treatment of these models, and the computational procedures used in GPS data processing, the reader is referred to such texts as Leick (1995), Hofmann-Wellenhof *et al.* (1998) and Strang and Borre (1997). The last of these references is particularly useful for those interested in the algorithms themselves as it supplies sample MATLAB code for SPP and baseline determination.

9.1. GPS mathematical models

9.1.1. Pseudorange mathematical model

A GPS receiver uses the transmitted Pseudo-Random Noise (PRN) codes to measure the point of maximum correlation between the incoming satellite PRN code and the receiver-generated replica of the same C/A-code (Section 8.2.1). This process produces an observation of the (code-derived) range or, equivalently, the *signal transmit time* t^j of the received signal at time t_i. Through this process the propagation delay $(t_i - t^j)$ is measured, which is scaled by the speed of electromagnetic radiation in a vacuum (c), to generate the pseudorange observable (Langley 1993):

$$P_i^j(t_i) = c.(t_i - t^j) \tag{9.1}$$

The reason this observable is referred to as a *pseudorange* is due to the fact that the receiver and satellite clock errors affect the measurement of propagation delay, and therefore the receiver–satellite range, in a systematic manner. (The satellite clock controls the signal generation, and the receiver clock controls the code's replica generation). In addition, the pseudoranges are also affected by the ionospheric and tropospheric delays, multipath, and receiver noise, and can be modeled for either the L1 or L2 observable by equations (8.4) and (8.6). However, a simplified mathematical model is generally adopted for pseudorange-based SPP:

$$P_i^j(t_i) = \sqrt{[X^j(t^j) - x_i(t_i)]^2 + [Y^j(t^j) - y_i(t_i)]^2 + [Z^j(t^j) - z_i(t_i)]^2} + c.\text{recclk}_i \tag{9.2}$$

All terms have been previously defined in Section 8.6.

9.1.2. Double-differenced carrier phase mathematical model

Although code-range measurement technology has advanced substantially, the most precise observable will always be the carrier phase measurement φ_i^j. This observable is equal to the *difference* between the phase φ_i of the receiver-generated carrier wave at signal reception time, and the phase φ^j of the satellite-generated carrier phase at transmission time. Ideally the carrier phase observable would be equal to the total number of full carrier wavelengths plus the fractional cycle between the transmitting satellite antenna and the receiver antenna at any instant. However, the GPS receiver cannot distinguish one carrier cycle from another and the best that it can do is to measure the fractional phase when it locks onto the satellite, and then to keep track of the phase changes thereafter. As a result, the initial phase is *ambiguous* by a quantity N_i^j, that is an unknown integer number of cycles (equation (8.2) and Figure 8.9).

The basic observable for carrier phase-based GPS positioning is *not* the one-way phase, but rather it is the double-differenced observable involving simultaneous measurements to two satellites and two receivers as represented by equation (8.8). The principle advantage of such an observable is the elimination of the receiver and satellite clock errors. Assuming the double-differenced atmospheric biases are negligible, and only considering the parameters of interest, the following simplified observation model is obtained:

$$\lambda . \Delta \nabla \varphi_{im}^{jl} = \Delta \nabla \rho_{im}^{jl} + \lambda . \Delta \nabla N_{im}^{jl} \tag{9.3}$$

There are now four coordinate triplets involved: two coordinate sets for the receiver i and m, as well as two sets of coordinates for satellites j and l. *What if the ionospheric bias is not negligible?* One option is to form the 'ionosphere-free' observable from dual-frequency measurements (Section 9.2.1.1).

9.2. Preparing for data processing

There are several issues regarding GPS data processing including:

- The degree of 'pre-processing' of the data.
- The parameter estimation technique used.
- The data combinations used in processing.
- The parameterization of the observations.
- The quality control procedures used to evaluate the solutions.
- The options and capabilities of the data processing software.

The reader is referred to, for example, the texts of Leick (1995) and Hofmann-Wellenhof *et al.* (1998) for details. In this chapter only brief remarks will be made with regard to some of the abovementioned issues, with particular attention being paid to 'ambiguity resolution' for carrier phase data processing.

9.2.1. Combinations of dual-frequency GPS data

GPS data (either pseudorange or carrier phase) at the L1 and L2 frequencies can be linearly combined in different ways.

9.2.1.1. The ionosphere-free combinations

A first-order approximation for the ionospheric delay (in units of meters) at frequency q is (Leick 1995):

$$\frac{ion_q}{c} \approx \frac{(1.35 \times 10^{-7}).STEC}{f_q^2} \tag{9.4}$$

where f_q is the frequency (in Hz), c is the speed of light, and STEC is the slant total electron content of a column of ionosphere condensed onto a disc (in units of free electrons per square meter). Note that the higher the frequency, the smaller the ionospheric delay, and therefore the delay on the L1 signal is less than on the L2 signal. The relationship between the ionospheric delays on the two GPS frequencies can be expressed as:

$$f_1^2.ion_1 = f_2^2.ion_2 \tag{9.5}$$

Hence the L2 ionospheric delay is approximately 1.647 times that on L1 ($1.647 \approx f_1^2/f_2^2$).

Equation (8.5) can be simplified (and expressed in units of cycles of the f_q frequency, noting that $\lambda = c/f$) by dropping some terms and leaving off the subscripts and superscripts:

$$\varphi_q = \frac{f_q}{c}.(\rho - ion_q) + N_q \tag{9.6}$$

Constructing equation (9.6) for both the L1 and L2 measurements, then multiplying each equation by the associated frequency, and finally subtracting the two equations yields the following relation:

$$f_1.\varphi_1 - f_2.\varphi_2 = \frac{f_1^2 - f_2^2}{c}.\rho - \frac{1}{c}(f_1^2.ion_1 - f_2^2.ion_2) + f_1.N_1 - f_2.N_2 \tag{9.7}$$

where the second term on the right hand side of equation (9.7) is zero, due to the relation at equation (9.5). In order to combine the L1 and L2 phase observations, which are in units of cycles (of different wavelengths for L1 and L2), they have to be converted to the same units, for example, scaling by the L1 frequency:

$$\frac{f_1(f_1.\varphi_1 - f_2.\varphi_2)}{f_1^2 - f_2^2} = \frac{f_1}{c}.\rho + \frac{f_1(f_1.N_1 - f_2.N_2)}{f_1^2 - f_2^2} \tag{9.8}$$

yields the following expressions for the *ionosphere-free* L1 phase measurement:

$$\varphi_{1ion-free} = \alpha_1.\varphi_1 + \alpha_2.\varphi_2$$

$$\varphi_{1ion-free} = \frac{f_1}{c}.\rho + \alpha_1.N_1 + \alpha_2.N_2 \tag{9.9}$$

where $\alpha_1 = f_1^2/(f_1^2 - f_2^2) \approx 2.546$ and $\alpha_2 = -f_1f_2/(f_1^2 - f_2^2) \approx -1.984$. Alternatively, if equation (9.7) were scaled by the L2 frequency, the ionosphere-free L2 phase measurement would have been obtained, having the same form as equation (9.9) except that α_1 would be replaced by β_1, and α_2 replaced by β_2, where $\beta_1 = f_1f_2/(f_1^2 - f_2^2) \approx 1.984$ and $\beta_2 = -f_2^2/(f_1^2 - f_2^2) \approx -1.546$. The ambiguity term for the ionosphere-free combination is no longer an integer, being $N_{1ion-free} \approx 2.546N_1 - 1.984N_2$ (when expressed in units of L1 cycles) or $N_{2ion-free} \approx 1.984N_1 - 1.546N_2$ (when expressed in units of L2 cycles). By a similar process an expression for the ionosphere-free pseudo-range combination can be obtained:

$$P_{ion-free} = \frac{f_1^2.P_1 - f_2^2.P_2}{f_1^2 - f_2^2} \tag{9.10}$$

9.2.1.2. Linear combinations of GPS data: general form

Other linear combinations of GPS phase measurements are useful for ambiguity resolution (Hofmann-Wellenhof *et al.* 1998). A general form for the linear combination of GPS data at different frequencies can be developed (with its cycle ambiguity still an integer):

$$\varphi_{n,m} = n.\varphi_1 + m.\varphi_2 \tag{9.11}$$

where n and m are integers, and φ_1 and φ_2 are the phase measurements made on the L1 and L2 carrier waves. The ambiguity of the combined observable $N_{n,m}$ is related to the cycle ambiguity of the L1 and L2 signals, N_1 and N_2, through the following equation:

$$N_{n,m} = n.N_1 + m.N_2 \tag{9.12}$$

For example, the *widelane* combination is defined by $n = 1$ and $m = -1$, and the effective wavelength of this linear combination is approximately 86 cm $\approx c/(f_1-f_2)$. The processing of this observable (and the resolution of its ambiguity) is useful for long baseline determination.

9.2.2. The least-squares solution

It is assumed that equations (9.2) and (9.3) express the relationship between the 'true' values of the observations and the underlying parameters. The mathematical basis of least-squares solutions requires the definition of two models: (a) the functional model, and (b) the stochastic model. The former is based on the observation models represented by, for example, equations (9.2) or (9.3). The latter is tantamount to describing the accuracy of the measurement and its statistical properties.

Because least-squares estimation assumes a linear mathematical model, and both the pseudorange and carrier phase observables are *non-linear* with respect to both the receiver and satellite position vectors (see equation (8.6)), then equations (9.2) and (9.3) must be *linearized* before proceeding. This is done by using approximate values of the unknown receiver coordinate parameters, so that the estimated parameters become the *corrections* to these initial values, not the values themselves. The least-squares estimation problem is usually described in terms of matrices (represented here by symbols that are underlined), for which the general solution is (Strang and Borre 1997):

$$\underline{\Delta x} = (\underline{A}^T \underline{\Sigma}^{-1} \underline{A})^{-1} \underline{A}^T \underline{\Sigma}^{-1} \underline{\Delta P} \tag{9.13}$$

where $\underline{\Delta x}$ is the matrix containing the estimable parameters, $\underline{A}$ is the 'design matrix' containing the partial derivatives of the processed observable with respect to the estimable parameters, $\underline{\Delta P}$ is the matrix representing the residual quantities: observations minus the 'calculated observations' (generated using the approximate values of the estimable parameters), $\underline{\Sigma}$ is the 'covariance matrix', and is also the inverse of the so-called 'weight matrix of the observations', $\underline{A}^T \underline{\Sigma}^{-1} \underline{A}$ is the 'normal equation matrix', and the 'covariance matrix of the solution' is $\underline{\Sigma}_{\Delta x} = (\underline{A}^T \underline{\Sigma}^{-1} \underline{A})^{-1}$, which is the inverse of the normal equation matrix.

'Quality' information can be derived from the elements of the covariance matrix of the solution, such as the standard deviations of the parameters, as well as being useful for further statistical testing. Assuming the observations are all independent (and hence the weight matrix is diagonal), each with the same standard deviation σ, then a simplified expression for this matrix is:

$$\underline{\Sigma}_{\Delta x} = \sigma^2 (\underline{A}^T \underline{A})^{-1} \tag{9.14}$$

In general the solution to a non-linear problem must be iterated by updating the approximate values of the estimable parameters using the current solution for $\underline{\Delta x}$. Then the process is repeated until a convergent solution is obtained (generally within 2–3 iterations). If the functional and stochastic models are correct, then the observation residuals in the $\underline{\Delta P}$ matrix should be small and randomly distributed.

9.3. Single-point positioning (SPP)

Although geodesists, surveyors and GIS professionals typically use two (or more) GPS receivers in the relative positioning mode, the primary objective of the GPS system is to allow positioning anywhere, under any weather conditions, 24 hours a day, with a single receiver. This mode of positioning is nevertheless important for positioning features for inventory (i.e. GIS) purposes. Single receiver positioning is generally referred to as 'single-point positioning' or 'absolute positioning', and can be performed using pseudoranges alone, or (under certain circumstances) carrier phase measurements, or both phase and pseudoranges. Only the former will be discussed here.

9.3.1. Pseudorange-based positioning

The following comments may be made to the solution procedure based on equation (9.2) (Langley 1991):

- There are four parameters that must be estimated: the 3-D coordinates of the receiver and the receiver clock error, requiring simultaneous observations from receiver i to four satellites ($j = 1...4$) for a unique solution.
- If pseudorange measurements to more than four satellites are made, a least-squares solution is employed to derive the optimal solution.
- An estimate of the satellite clock error is computed using parameters contained within the satellite navigation message (equation (8.1)), and applied as a *correction* to the pseudorange observation P, hence no satellite clock parameters need to be estimated. Since Selective Availability (SA) was switched off on 1 May 2000, this clock error estimate is accurate to a few meters of equivalent range.

- Most GPS receivers have inbuilt models for the ionospheric and tropospheric bias. If dual-frequency pseudorange observations are available, a new observable which is a combination of L1 and L2 measurements can be generated that is 'ionosphere-free' (equation (9.10)).
- Any errors in the known coordinates of the satellite (X,Y,Z) will impact directly on the accuracy of the solution. Hence ephemeris errors, together with the satellite clock error model, are one of the limiting factors to SPP accuracy.
- The *reference datum* in which the receiver coordinates are expressed is that of the known (fixed in equation (8.6)) satellite coordinates which, if derived from the broadcast ephemeris, is WGS 84 (Section 7.2.1).
- Such a solution can be obtained instantaneously (as more than four satellites are visible at the same time if the skyview is clear). Hence the user's GPS receiver may be moving, and each independent solution is a 'snapshot' of the receiver's location at a particular epoch. However, SPP solutions can also be computed *post-mission*, in which case more precise models of the satellite clock error and the satellite ephemeris may be used (see Section 14.2.3).

The factors influencing the final positioning accuracy may be gauged from equations (9.13) and (9.14), that is:

1. the measurement noise (or random) error, represented by σ,
2. the magnitude of any unmodeled biases, such as satellite ephemeris and clock, multipath and atmospheric biases, that may also be absorbed into a larger value for σ, and
3. the factors that govern the magnitude of the elements of the design matrix $\underline{A}$.

A combination of (1) and (2) is referred to in GPS parlance as the 'user equivalent range error', while the factor at (3) is defined in terms of a quantity known as 'dilution of precision'. Both of these are discussed in subsequent sections.

9.3.2. User equivalent range error

If the standard deviation σ of the pseudoranges represents both the measurement and the modeling errors, and it is assumed that these errors are all independent, then the quantity σ is equal to the square root of the sum of the variances of all these errors. If these errors include the internal receiver noise, the residual satellite clock error (remaining after the broadcast model is applied), ephemeris errors, atmospheric errors, and multipath, then the resulting standard deviation σ is known as the User Equivalent Range Error (UERE).

For the Standard Positioning Service (SPS – see Section 7.3.2), the UERE was of the order of 25 m with SA on. However, with SA switched off the SPS UERE value has dropped to below the 5 m level, and is now dominated by the ionospheric and multipath errors. For the Precise Positioning Service (i.e. the service available to military users), using dual-frequency receivers, the UERE value is of the order of 1–2 m.

9.3.3. Dilution of Precision

The most important result for a GPS user, besides knowing the receiver's 3-D position, is the accuracy of the estimated position. *How much have the pseudorange measurement and modeling errors affected the estimated position?* To answer this question it is necessary to study the covariance matrix of the solution $\Sigma_{\Delta x}$. If it is assumed that all the pseudoranges have the same standard deviation, then equation (9.14) is the expression defining the relationship between the solution accuracy and the measurement precision.

The 4×4 covariance matrix $\underline{\Sigma}_{\Delta x} = \sigma^2 (\underline{A}^T \underline{A})^{-1}$ has ten independent (non-identical) elements and is a combination of the satellite geometry (i.e. the matrix $\underline{A}$), and the quality of the pseudorange measurements (including both random and systematic components). In order to reduce this information to just one number, which depends only on the satellite geometry, one can take the square root of the trace of the matrix, divided by the standard deviation of each pseudorange measurement (Strang and Borre 1997, Hofmann-Wellenhof *et al.* 1998). If UERE is substituted for σ, the resulting quantity is known as the Geometric Dilution of Precision (GDOP):

$$\text{GDOP} = \frac{\sqrt{\text{trace}(\Sigma_s)}}{\text{UERE}} = \frac{\sqrt{\sigma_x^2 + \sigma_y^2 + \sigma_z^2 + \sigma_{\text{clk}}^2}}{\text{UERE}} = \frac{\sqrt{\sigma_e^2 + \sigma_n^2 + \sigma_u^2 + \sigma_{\text{clk}}^2}}{\text{UERE}} \quad (9.15)$$

where σ_x, σ_y, σ_z and σ_{clk} are the standard deviations of the x, y, z components and receiver clock solution (taken from the diagonal elements of the covariance matrix), respectively, and σ_e, σ_n, σ_u are the standard deviations of the estimated position along the east, north and up directions. The GDOP *factor* is a dimensionless quantity that is independent of the measurement errors, and dependent only on receiver–satellite geometry. An estimate of the overall solution error therefore is:

$$\sigma_s = \sqrt{\text{trace}(\Sigma_s)} = \text{UERE.GDOP} = \sqrt{\sigma_x^2 + \sigma_y^2 + \sigma_z^2 + \sigma_{\text{clk}}^2}$$

$$= \sqrt{\sigma_e^2 + \sigma_n^2 + \sigma_u^2 + \sigma_{\text{clk}}^2} \quad (9.16)$$

Note that the pseudorange standard deviation is *multiplied* by the GDOP scaling factor, and because this scaling factor is greater than one it *amplifies* the pseudorange error, or 'dilutes' the precision of the position determination. Many times it is desirable to examine the quality of the 3-D position itself, or that of its horizontal and vertical components separately. In such cases only the corresponding elements of the covariance matrix need be considered. For example, using only the first three diagonal elements $\sigma_x^2, \sigma_y^2, \sigma_z^2$ the Position Dilution of Precision (PDOP) can be obtained:

$$\text{PDOP} = \frac{\sqrt{\sigma_x^2 + \sigma_y^2 + \sigma_z^2}}{\text{UERE}} = \frac{\sqrt{\sigma_e^2 + \sigma_n^2 + \sigma_u^2}}{\text{UERE}} \quad (9.17)$$

The last term in equations (9.15)–(9.17) is included because the trace of the covariance matrix is unaffected when transforming the x, y, z coordinates to the local e, n, u coordinate system. Hence it is also possible to determine the Horizontal Dilution of Precision (HDOP) and the Vertical Dilution of Precision (VDOP):

$$\text{HDOP} = \frac{\sqrt{\sigma_e^2 + \sigma_n^2}}{\text{UERE}} \qquad (9.18)$$

$$\text{VDOP} = \frac{\sqrt{\sigma_u^2}}{\text{UERE}} = \frac{\sigma_u}{\text{UERE}} \qquad (9.19)$$

Typically more satellites yield smaller DOP values, and hence more accurate position solutions. Also, in general, the HDOP values are between one and two and the VDOP values are larger than the HDOP values, indicating that the vertical positions are less accurate than the horizontal positions (Section 7.3.2).

Although DOP values are of particular interest for SPP applications, they also have a role in indicating the quality of a position determined using certain carrier phase-based techniques. These are the techniques based on very short static periods such as *stop-and-go*, and the *kinematic* procedures (Section 11.1). Such techniques are sensitive to receiver–satellite geometry, and DOP values are a convenient way of expressing this relationship (Section 12.3.5). 'Mission planning' software can be used to compute the DOP values at a certain location, for a specific instant in time, or for a span of time. Because the orbits of the GPS satellites repeat in such a way that they rise and set each day (but about 4 min earlier from one day to the next), then the DOP values will vary on a predictable daily basis.

9.4. Baseline determination

The objective of relative positioning is to estimate the 3-D *baseline* vector between the known (generally assumed stationary) and the unknown point or points (Figure 7.3). Although different processing schemes are possible, using either pseudorange or carrier phase data, only the carrier phase-based techniques capable of cm-level accuracy will be discussed here. Many of the systematic errors in carrier phase measurements are completely (or nearly so) eliminated by taking Double-Differenced (DD) combinations of the simultaneous phase observations recorded by receivers at the known and unknown points (Figure 8.13). Several commercial software packages use Equation (9.3) to model the DD observables. However, when it is compared with the model in equation (8.8), it is clear that all systematic biases due to the clocks, multipath and atmospheric signal delay are assumed to have canceled. The degree of cancellation of the atmospheric biases is dependent on the baseline length, since these common errors tend to decorrelate with increasing receiver separation. Clock biases can indeed be assumed to have canceled completely. Multipath is more problematic, and when present acts to distort (or 'bias') the solution. Hence all efforts should be made to ensure that the observation environment is 'multipath-free', or that the receiver's electronics and antenna are able to mitigate as much of the effect as possible.

9.4.1. Using double-differenced carrier phase observables

There are a number of comments that may be made regarding the observable model in equation (9.3):

- There are two classes of estimable parameters: the 3-D receiver coordinates and the double-differenced ambiguities.

- The *integer* nature of the ambiguities is preserved following double-differencing.
- The ambiguities are assumed to be *constants* for as long as measurements are made, hence *all* data contributes to the estimation of these parameters. If, for whatever reason, the signal tracking to a satellite is broken, then the ambiguities involving that satellite before the break are *not equal* to the ambiguities after the break. That is, a *'cycle slip'* has occurred.
- One set of coordinates in the observation model are held fixed in the solution, while the other receiver's coordinates are estimated, requiring that the pair of GPS receivers be deployed with one at a *reference* station of known location and the second receiver at the station whose coordinates are to be determined.
- In principle, the second receiver can be in motion, requiring a new set of coordinates to be derived at each epoch. This mode of positioning is referred to as the *kinematic technique* (Section 11.1.3).
- The entire computational process may be carried out in *post-mission* mode, after the recorded measurements from both receivers are brought together within the baseline determination software, or in *real-time*, if the reference station measurements are transmitted to the second receiver (see Section 11.2.2, Real-Time Kinematic (RTIC)).

When a least-squares solution is implemented to estimate the coordinate and ambiguity parameters, the main challenges are:

1 To reliably estimate both the coordinate and ambiguity parameters.
2 To ensure the errors and biases in the observable model remain small (compared to the wavelength of the carrier signal – approximately 19 cm for L1 and 24 cm for L2).
3 To take advantage of the integer nature of the ambiguities.

In general, (1) requires that observations are collected by the pair of GPS receivers over some *observation session* whose length varies according to a number of factors, principally the distance between the receivers (or the baseline length), the accuracy required, and the GPS technique used (Section 12.3). Factor (2) typically requires that the baseline length be kept below a few tens of kilometers, and that no cycle slips are present in the DD observables. (If cycle slips do occur, then these cycle slips should be 'repaired' or new ambiguity parameters should be introduced in the mathematical model. "Repairing" cycle slips allows the modeling of the ambiguity parameters as constants for the entire observation period.) Factor (3) is crucial to high precision carrier phase-based GPS positioning techniques, and is discussed in the subsequent sections of this chapter.

9.4.2. Resolving ambiguities

What is ambiguity resolution? It is an improvement of the accuracy of carrier phase-based GPS positioning by means of a mathematical process whereby the ambiguous carrier phase measurements (Section 8.4.1) are converted to unambiguous 'carrier range' measurements (Section 8.4.2). In other words, the *integer* valued ambiguity term in equations (8.2), (8.5), (8.7), (8.8) and (9.3) is determined, and applied as a correction to the measured carrier phase, thereby creating a precise pseudorange observable (from equation (9.3)):

$$\Delta \nabla L_{im}^{jl} = \Delta \nabla \rho_{im}^{jl} \qquad (9.20)$$

where $\Delta \nabla L_{im}^{jl}$ is the DD carrier range.

The starting point for the Ambiguity Resolution (AR) process is the real-valued estimates of the DD ambiguities $\Delta \nabla N_{im}^{jl}$ derived from a least-squares solution using the observation model in equation (9.3). If these estimates are 'reliable' then AR can be attempted with some degree of success. In the early days of GPS positioning, 1 h or more of carrier phase measurements were collected to ensure the 'separability' of the position and ambiguity parameters within the least-squares solution. As the observation session lengthened, not only did the accuracy of the ambiguities increase but also the coordinate parameters. In the limit, there would be no need to resolve the ambiguities because the coordinates would have been determined to cm-level accuracy (see Section 12.1.1). Only with the drive to shorten observation sessions did it become crucial that AR be undertaken so that there was a *jump* in the coordinate accuracy when using the observation model in equation (9.20) instead of equation (9.3). This accuracy jump is illustrated in Figure 9.1 and discussed in Section 9.4.3.

There are several steps in the baseline determination process:

1 Define the a priori values of the ambiguity parameters, generally from the least-squares solution using model equation (9.3).
2 Use a *search* algorithm to identify likely integer values for the ambiguities.
3 Employ a *decision-making* algorithm to select the 'best' set of integer values.
4 Apply some *validation* tests to check whether these integer ambiguity values are indeed correct.

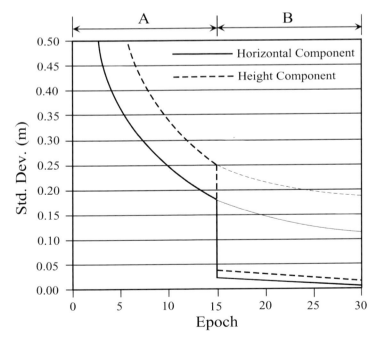

Figure 9.1   The evolution in quality of a baseline solution as ambiguities are resolved.

5   If the ambiguity values are reliable then convert the carrier phase data to carrier range, and use another least-squares solution with model equation (9.20).

Considerable R&D has been invested in developing AR procedures for the 'high productivity' GPS surveying techniques described in Section 11.1 (Han and Rizos 1997). The quest has been for techniques that ensure reliable AR even with relatively small amounts of data. AR *reliability* (steps (3)–(5) above) is a function of:

- baseline length (typically the receivers are not separated by more than about 20–30 km),
- the number of satellites (the more the better),
- whether satellites rise or set during the session (continuous tracking of satellites across the entire observation session is preferred),
- the satellite–receiver geometry, as characterized by DOP values,
- the degree of multipath disturbance to the measurements (the less the better),
- whether observations are made on both carrier frequencies (it is much easier to resolve ambiguities when dual-frequency observations are available), and
- the length of the observation session (the longer the better).

Although the above remarks imply that ambiguity resolution is a process which can only be applied to static GPS carrier phase data, during the last decade the AR algorithms have been significantly refined and the receiver hardware improved to such a point that AR can be carried out using just a few tens of seconds of tracking, even if the receiver is in motion. However, such performance requires: (a) dual-frequency carrier phase and precise pseudorange data, (b) high quality GPS measurements, (c) tracking to six or more satellites, and (d) the baseline length to be of the order of ten kilometers or less.

9.4.3. Comments on the carrier phase-based solutions

There are two types of DD carrier phase data solutions:

1   A so-called 'ambiguity-free' solution (also referred to as an 'ambiguity-float' or 'bias-float' solution), in which both the coordinates (one end of a baseline) and the ambiguity parameters are estimated as real-valued quantities. This is based on the observation model defined by equation (9.3).

2   A so-called 'ambiguity-fixed' solution (also sometimes referred to as a 'bias-fixed' solution), in which only the coordinate parameters are estimated. This is based on the observation model defined by equation (9.20), and is therefore a very strong (and accurate) solution. However, if the AR is unreliable, and one or more of the ambiguity values is incorrect by one or more integers (for L1 each integer corresponds to an error of about 19 cm in the derived carrier range), then the subsequent ambiguity-fixed solution is *biased* by several decimeters!

The significant impact that AR makes to carrier phase-based GPS positioning is best understood with reference to Figure 9.1, where the solution quality (in terms of the standard deviation of the coordinate components) is plotted against measurement epoch. With increasing epoch number extra carrier phase observations have been included in the least-squares solution, leading to a decrease in the standard deviations (or in other words, an increase in accuracy).

There are several further comments that can be made with respect to Figure 9.1:

- In region A the precision (and accuracy) of the coordinates steadily improves as more data is collected.
- As soon as sufficient data is available to resolve the ambiguities (at epoch 15) a dramatic improvement in the coordinate parameter precision is evident. This is the rationale behind the rapid-static, kinematic initialization, or on-the-fly AR GPS techniques; that is, collect just enough data to ensure that an ambiguity-fixed solution is obtained.
- In region B, when the unambiguous carrier range data are processed, there is no significant improvement in the quality of the coordinate solution, and in effect there is no justification for continuing to collect data past epoch 15. This is the scenario encountered in stop-and-go GPS positioning (Section 11.1.2), where just a few epochs of measurements are sufficient to obtain cm-level accuracy, or in the case of kinematic positioning (Section 11.1.3) there would be just independent epochs along the receiver trajectory.
- If enough data is collected over an observation session, the precision of the ambiguity-free solution will steadily improve, converging to that obtained from an ambiguity-fixed solution. This means that AR is not required in order to obtain high accuracy coordinate results. In other words, AR is optional and if there is any indication that the AR is incorrect that solution may be discarded in preference to the ambiguity-free solution (see Section 12.1.1 vis-à-vis Section 12.1.2).
- In conventional static GPS surveying the data is post-processed, and it is therefore not known *a priori* at what point (or even if) sufficient data has been collected to ensure an ambiguity-fixed solution is obtained. Hence *conservative* observation session lengths are recommended (Section 12.3.3). Furthermore, accuracy is assured because *all* the measurements (including those taken before epoch 15) contribute to the ambiguity-fixed solution.
- The benefit of real-time carrier phase-based techniques (Section 11.2.2) is that only sufficient data need be collected to derive an ambiguity-fixed solution, and then just sufficient additional epochs of carrier range data to determine the coordinates. Of course, in the kinematic mode this would be just one measurement epoch.

Ambiguity resolution is probably the most uniquely identifiable characteristic of high precision GPS positioning. Furthermore, there have been dramatic improvements to the *productivity* of carrier phase-based GPS surveying techniques because the 'time-to-AR' has been significantly reduced: from an hour of more during the 1980s, to several tens of minutes in the early 1990s, to the current observation session length of a few tens of seconds. There are, however, several trends that indicate continued improvement in AR performance can be expected during the next decade (Sections 15.1, 15.3 and 15.4), primarily through the relaxation of some of the constraints to very rapid AR. In particular the distance between receivers will lengthen several fold, and the length of the observation session will decrease to the point where single-epoch AR will be feasible. The latter, when implemented in real-time, will make carrier phase-based relative positioning as robust and easy to use as current pseudorange-based differential GPS techniques (Section 11.2.1).

References

Han, S. and Rizos, C., 'Comparing GPS ambiguity resolution techniques', *GPS World*, 8(10), 1997, 54–61.

Hofmann-Wellenhof, B., Lichtenegger, H. and Collins, J., *Global Positioning System: Theory and Practice* (4th ed.), Springer-Verlag, Berlin, 1998, 389 pp.

Langley, R. B., 'The mathematics of GPS', *GPS World*, 2(7), 1991, 45–50.

Langley, R. B., 'The GPS observables', *GPS World*, 4(4), 1993, 52–59.

Leick, A., *GPS Satellite Surveying* (2nd ed.), John Wiley & Sons, New York, 1995, 560 pp.

Strang, G. and Borre, K., *Linear Algebra, Geodesy, and GPS*, Wellesley-Cambridge Press, 1997, 624 pp.

Chapter 10

GPS instrumentation issues

Dorota Grejner-Brzezinska

This chapter briefly introduces the important issues related to GPS instrumentation and tries to address questions such as:

- What are the components of a GPS receiver?
- What should I consider when purchasing a GPS receiver?
- Why do GPS receivers for surveying applications cost so much more than for GIS/mapping?
- Will my receiver suffer from interference or multipath, and can the right choice of antenna help in this regard?
- How is a dual-frequency receiver different from a single-frequency receiver?
- What does real-time positioning imply for the user?
- What are some of the memory considerations for GPS receivers?
- What are the power requirements for GPS receivers?
- Must I use a choke ring antenna for the highest performance?
- Can the same GPS receiver be used for all or most types of positioning applications?
- What configurations are now possible for user GPS equipment?

It is beyond the scope of this chapter to provide the reader with a full list of GPS products, their characteristics, and their manufacturers, especially with new receivers being continually released onto the market. One source of information is the GPS Mapping/Positioning Receiver Survey published in the journal *Point of Beginning* (POB 2000). The most complete source of such information is the *GPS World* magazine, in particular the *GPS World Buyers Guide* (usually in the June edition), the *GPS World Receiver Survey* (usually in the January edition), and the *GPS World Showcase* (special editions in August and December each year), highlighting technology and product innovations.

GPS receivers belong to the GPS User Segment which covers the entire spectrum of applications, equipment and computational techniques that are available to the systems' users (Section 7.1.3). Over the past two decades civilian as well as military GPS instrumentation has evolved rapidly, with one of the most dramatic trends being the simultaneous reduction in hardware costs and increase in positioning performance. Even though military receivers and many civilian application-specific instruments have evolved in different directions, one can ask: *are all GPS receivers essentially equivalent, apart from functionality and user software?* In general, the answer is *yes* – all GPS receivers consist of essentially the same functional blocks,

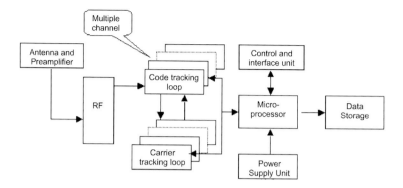

Figure 10.1 Conceptual architecture of a GPS receiver.

even if their implementation differs from one *class* of receivers to another. By far, the majority of receivers manufactured today are of the C/A-code single-frequency variety, suitable for most civilian navigation and general positioning applications. However, for high precision geodesy, and most surveying applications, dual-frequency hardware is standard.

The primary components of a generic GPS receiver are illustrated schematically in Figure 10.1 (Langley 1991). A GPS receiver system must carry out the following tasks:

- Select the satellites to be tracked.
- Search and acquire each of the selected GPS satellite signals.
- Recover the broadcast navigation data for every satellite.
- Make pseudorange, range-rate and/or carrier phase measurements.
- Compute position, velocity and time information.
- Optionally record the data for post-processing.
- Optionally transmit the data to another receiver via a radio modem for real-time differential solutions.
- Optionally accept user commands, and display results, via a portable control/display unit or a PC.

10.1. Antenna and preamplifier

An *antenna* is defined as a device concerned with the transition between a guided wave and a free-space wave, and vice versa (Johnson 1993). On transmission, an antenna accepts energy from a transmission line and radiates it into space, and on reception it collects energy from an incident wave and sends it down a transmission line. The GPS receiving antenna detects an electromagnetic signal arriving from a satellite, and after bandpass filtering, which provides adequate filter selectivity to attenuate adjacent channel interference, and initial preamplification, transfers the signal to the Radio Frequency (RF) section for further processing. A typical GPS antenna is omnidirectional, having essentially a non-directional gain pattern in azimuth, though the pattern does change somewhat with elevation angle.

In general, antennas can be designed as L1-only or dual-frequency, when both L1 and L2 carrier signals are tracked. The most common GPS antenna types currently available are monopole (asymmetric antennas) and dipole (symmetric antennas) configurations, quadrifilar helices, spiral helices, and microstrips. A helical antenna consists of a single conductor (or multiple conductors) wound into a shape with circular or elliptical cross section. Generally helical antennas are wound with a single conductor, however, a helix can be designed with bifilar, quadrifilar or multifilar windings (Johnson 1993). A quadrifilar helix antenna is frequently used with handheld GPS receivers as it does not need a ground plane. The microstrip antenna, also sometimes called a 'patch' antenna, is usually constructed of one or more elements that are photoetched on one side of a double-coated printed-circuit board. It can be circular or rectangular in shape, made up of one or more patches of metal, separated from a ground plane by a dielectric sheet, and may have single- or dual-frequency capability (Langley 1998b). Microstrip antennas are the most common, primarily because of relatively easy implementation and miniaturization, ruggedness and general suitability for kinematic applications.

GPS antennas are often protected against possible damage by a plastic housing (radome) designed to minimize attenuation of the signal (Langley 1995b, 1998a). For precise geodetic applications, an additional ground plane or choke ring antenna can be used (Figure 10.2 shows some typical high precision antennas, see plate 2). The *ground plane* is a horizontally-oriented metallic disk, centered at the GPS antenna's base, which shields the antenna from any signals arriving from below the antenna. The *choke ring* assembly is essentially a ground plane containing a series of concentric circular troughs one-quarter wavelength deep, designed to eliminate multipath surface waves (Section 8.5.2).

10.1.1. Preamplifier section

Most GPS antennas are combined with a low-noise preamplifier (sometimes supported by additional filters either before or after the preamplifier). An antenna when combined with a preamplifier (usually housed in the base of the antenna, between the output of the antenna and the feeder line to the receiver) is called an *active* antenna, as opposed to a *passive* antenna that does not have such a pre-amplifier. The pre-amplifier boosts the signal level before feeding it to the receiver's RF section. However, caution must be exercised when using an antenna with a preamplifier not supported by the receiver manufacturer. Preamplifier's noise and gain must be within the receiver's acceptable range, and the voltage and current supplied by the receiver to the preamplifier must be compatible with the antenna's characteristics. A similar warning applies when using antenna 'splitters' to connect two GPS receivers to a 'zero baseline' (i.e. a single antenna). Total gain of active antennas ranges from 20 to 50 dB, while power consumption is between 12 and 32 mA at 5 V DC. An amplified antenna can help prevail over problems such as loss-of-lock due to antenna dynamics, or partially obstructed skyview due to foliage cover, where a passive antenna may totally fail.

10.1.2. Cables and connectors

An important issue is the proper selection of the antenna cable, as the signal attenuation depends on the type and length of the cable used. The signal power can be significantly lowered if the coupling between the antenna and the receiver is faulty (e.g. due to different *impedance* of the connecting cable and the antenna, or simply by having dirty connectors), or the cable itself not being of the correct rating (Langley 1998b). (Impedance is essentially the ratio of the voltage to the current, measured in Ohms (Ω).) Flexible RG/U coaxial transmission cables with 50 Ω impedance, consisting of a solid or stranded inner conductor, a plastic-dielectric support, and a braided outer conductor, are most commonly used to connect the GPS antenna to the receiver. Since signal loss increases linearly with cable length (and is slightly higher for high frequencies), for long cable runs it is recommended that a low-loss cable such as Belden 9913, Belden 8214, RG-58C, RG-62 or RG-62A, be used. Another way to minimize the signal attenuation on long cables is to place an additional preamplifier between the antenna and the cable. For short runs, however, cheaper RG-174 or Belden 9201 cables would be sufficient.

A variety of cable connector types are used for GPS equipment. The most commonly used (in both male and female varieties) are: BNC, F, MCX, N-type, lemo, OSX, SMA, SMB and TNC (Langley 1998b). A full list of coaxial cables and frequently used connectors is given in Johnson (1993) and Storm (2001). Antenna cables are dispensed with entirely in cases where the antenna and receiver electronics are integrated within one unit, as in the so-called 'smart' antennas depicted in Figure 10.3 (see plate 3).

10.1.3. Antenna physical characteristics

In typical GPS antennas the physical (or geometric) center of the antenna usually does not coincide with the phase center (or electrical center) of the antenna where the microwave measurements are made. Moreover, the phase centers for L1 and L2 generally do not coincide, and different types of antennas have different locations of their phase centers. In addition, the location of the phase center can vary with azimuth and elevation of the satellites, and the intensity of the incoming signal. This effect should not, in general, exceed 1–2 cm, and for modern microstrip antennas it reaches only a few millimeters. Nevertheless, in precise static GPS positioning applications it is good practice to always align the carefully leveled antennas in the same direction (e.g. local magnetic north), which results in cancellation in both length and orientation of the offset between the physical and phase centers when the same type of antenna is used at both ends of a baseline.

Since the GPS signal arrives at the phase center (L1 or L2), yet for surveying and mapping applications the coordinates of the ground mark are sought, the observations have to be mathematically reduced to the ground mark location, using the *antenna height*. The GPS antenna is usually mounted on a tribrach attached to a tripod, pillar or range pole, and oriented along the local plumb line directly above the ground mark to be surveyed. The antenna height is the vertical distance between the ground mark and the antenna phase center. (Sometimes the slant height is measured to the bottom of the ground plane or the base of the choke rings, and the vertical height is computed using manufacturer-provided information about phase center location above the reference plane.)

10.1.4. Antenna electrical characteristics

The measure of antenna gain over a range of elevations and azimuths is known as the *gain pattern*, where *gain* refers to an antenna's ability to successfully receive a weak signal, or its ability to concentrate in a particular direction the power accepted by the antenna. The gain pattern, similar to the phase center location, changes with the direction of the incoming signal. Since the antenna design must satisfy several, and at times conflicting, requirements, uniform gain in all directions is not necessarily desirable for a GPS antenna. For example, low elevation signals usually have to be filtered out, since too much gain at low elevation will contribute to high multipath and atmospheric signal attenuation. At the same time, too much filtering at lower elevations would limit the minimum elevation at which satellites can be observed, resulting in weaker geometry.

Antenna *bandwidth* is defined as the frequency band over which the antenna's performance (described in terms of gain pattern, polarization, input impedance, etc.) is 'sufficiently good' (Langley 1998b). Antennas may be narrowband or wideband. Since GPS is a narrowband system (within a bandwidth of about 20 MHz, Section 8.1.1), most GPS antennas are also of the narrowband type. On the other hand, bandwidth has to be large enough to assure the antenna's proper functioning over the intended range of Doppler-shifted frequencies. For example, antennas accepting only the central lobe of the C/A-code might have a very narrow bandwidth of ±2 MHz, while antennas accepting more than the central lobe of C/A-code, or tracking the P(Y)-code, have a wider bandwidth of ±10–20 MHz. Dual-frequency L1/L2 microstrip antennas are usually designed as two-patch antennas, one patch for each frequency, each with a bandwidth of ±10–20 MHz. The bandwidth of microstrip antennas is proportional to the thickness of the substrate used. Interested readers are referred to Johnson (1993), Weill (1997), Langley (1998a,b), and references cited therein, for more information about antennas.

10.2. RF front-end section and the signal processing block

The signal processing functions of a receiver are the 'heart' of a GPS receiver, performing the following functions (Parkinson and Spilker 1996):

- Precorrelation sampling and filtering, and Automatic Gain Control (AGC).
- Signal splitting into multiple signal processing channels.
- Doppler frequency shift removal.
- Generation of the reference PRN codes.
- Satellite signal acquisition.
- Code and carrier tracking from multiple satellites.
- System data demodulation from the satellite signals.
- Making pseudorange measurements from the PRN codes.
- Making carrier frequency measurements from the satellite signals.
- Extracting signal-to-noise information from the satellite signals.
- Estimating the relationship to GPS system time.

The basic components of the RF section are a precision quartz crystal oscillator used to generate the 10.23 MHz reference frequency, multipliers to obtain higher frequencies, filters to eliminate unwanted frequencies, and signal mixers. The RF section receives the signal from the antenna, and translates the arriving (Doppler-shifted) frequency to a lower one, called the *beat* or *intermediate frequency (IF)*, by mixing the incoming signal with a pure sinusoidal one generated by the receiver's oscillator. As a result, the modulation of the IF remains the same, only the trans-formed carrier frequency becomes the difference between the original signal and the one generated locally. The IF signal produced by the RF section is subsequently processed by the signal tracking loops.

10.2.1. Multi-channel architecture

An important characteristic of the RF section is the number of tracking channels, and hence the number of satellites that can be tracked simultaneously. Older GPS receivers had a limited number of channels (as few as one), which required sequencing through satellites in order to make enough measurements for three-dimensional positioning. The current generation of GPS receivers is based on dedicated channel architecture, where every channel tracks one satellite, either on the L1 or L2 frequency. Since the accuracy, reliability and speed of obtaining position results increases with the number of satellites used in the solution, a good quality receiver should have enough channels to track all visible satellites at any given time. Typically geodetic/survey-grade GPS systems are able to track up to 12 satellites (total of 24 dual-frequency channels), which is especially important for real-time applications, or when being used as a reference station receiver.

10.2.2. Tracking loops

The *tracking loop* is a mechanism that enables a receiver to track a signal that is changing in frequency and in time. In simple terms, as long as there is signal 'lock-on', a tracking loop performs a comparison between the incoming signal and the local one generated within the receiver, and 'adjusts' the internal signal to 'match' the external one. There are essentially two types of tracking loops used in GPS receivers: *delay-lock*(or code-tracking) loop (DLL) and the *phase-lock*(or carrier) tracking loop (PLL). Dual-frequency receivers have separate channels (and thus separate tracking loops) for both frequencies. A tracking loop must be tunable within a frequency range (i.e. must have sufficient bandwidth) to accommodate the residual frequency offset of the modulated signal caused by Doppler shifts (maximum $\pm$ 6 kHz), residual user clock drift, bias frequency offset, and data modulation.

10.2.3. Signal acquisition with DLL and PLL

The DLL applies a 'code-correlation' technique to align a PRN code (either P(Y) or C/A) from the incoming signal with its replica generated within the receiver (Section 8.2). The alignment is achieved by shifting the internal signal in time. In principle, the amount of 'shift' that is applied corresponds to the time required for the incoming signal to travel from the satellite to the receiver. This time interval multiplied by the speed of light generates the range (more correctly the *pseudorange*) between the phase centers of the receiver and the GPS satellite antennas. Once the DLL is locked, the PRN code is removed from the signal, by first mixing it with the locally generated signal, and secondly by applying filtering to narrow the resulting bandwidth down to about 100 Hz. Through this process the GPS receiver reaches the necessary signal-to-noise ratio (SNR) value to compensate for the gain limitations of a physically small antenna (Langley 1991). At this point the navigation message data can be extracted from the IF signal by the PLL, which performs the alignment of the phase of the beat frequency signal with the phase of the locally generated carrier replica. Once the PLL is locked to the signal it will continue tracking the variations in the phase of the incoming carrier as the satellite-receiver range changes. The carrier beat phase observable (in cycles) is obtained by counting the elapsed cycles and by measuring the fractional phase of the locked local oscillator signal (Section 8.4.1).

10.2.4. Signal tracking in the presence of anti-spoofing

The code-correlation tracking procedure described above works well when the receiver can generate the replica PRN code. However, under the policy of *Anti-Spoofing* (AS), which is the current mode of GPS operation, the P-code is replaced with the encrypted Y-code. Hence different tracking techniques must be applied in civilian receivers to make measurements on the L2 frequency (remember, there is no C/A-code on L2!). Thus, AS is the reason why L2 measurements are harder to make than L1 measurements. To overcome the P-code encryption, various codeless squaring, cross-correlation and quasi-codeless tracking techniques have been developed. Only a small number of GPS receiver manufacturers have patented (or licensed) rights to produce and market dual-frequency instrumentation. This is one of the main reasons why dual-frequency instrumentation is much more expensive than single-frequency (L1) instruments.

It should be mentioned, however, that all these methods of recovering L2 carrier in the presence of AS show SNR degradation with respect to the code-correlation technique applicable under no-AS conditions (Table 10.1). Consequently, the weaker signal (indicated by the smaller SNR) recovered under AS is more susceptible to jamming (or interference) as well as to ionospheric scintillation effects. The so-called 'Z-tracking technique' is superior to all other techniques in this regard (Ashjaee 1993). When the second (and third) civilian frequencies are made available on the new GPS satellites slated for launch from about 2003 onwards, standard code-correlating techniques will be used to make dual- (and triple-) frequency measurements (Section 15.4.1).

Table 10.1 L2 signal recovery under AS (Ashjaee and Lorenz 1992)

Component	Squaring technique	Cross-correlation	Code-correlation plus squaring	Z-tracking
LI C/A code	No	Yes	Yes	Yes
Y2 code	No	Y2-YI	Yes	Yes
L2 wavelength	Half	Full	Half	Full
SNR (dB)	−16	−13	−3	0
SNR degradation with respect to code correlation (no AS) (dB)	−30	−27	−17	−14

10.3. Other instrumental components

10.3.1. Microprocessor

GPS receivers perform numerous operations in real time, such as acquiring and tracking of the satellite signals, decoding the broadcast message, timekeeping, range measurement, multipath and interference mitigation, etc. All of these operations are coordinated and controlled by a *microprocessor*. Other functions that might be performed by the microprocessor are data filtering to reduce the noise, position estimation, datum conversion, communication with the user via the control and display unit, and managing the data flow through the receiver's communication port. With respect to the latter, an industry standard output message format that is used on almost all GPS receivers is that known as NMEA 0183 (Langley 1995a). The National Marine Electronics Association, a professional trade association serving all segments of the marine marketplace, developed a uniform interface standard for digital data exchange between different marine electronic devices, navigation equipment and communication equipment. NMEA 0183 data messages use plain text in ASCII (American Standard Code for Information Interchange) format, which makes it comparatively easy to connect a GPS receiver to a PC or other device in order to pass coordinate information to other devices and users.

10.3.2. Control and display unit

The control device, usually designed as a keypad/display unit, is used to input commands from the user for selecting different data acquisition options, and other important parameters such as cut-off elevation angle, data recording interval, antenna height, etc. as well as to display information about the receiver's operations, navigation results, observed GPS satellite constellation, etc. Most receivers intended for surveying and mapping have command/display capabilities with extensive menus and prompting instructions, even offering on-line help. GPS receivers designed for integration into navigation systems usually do not have their own interface units; in such cases the communication with the receiver is facilitated through an external PC via messages in the NMEA format (Langley 1995a). Among the currently approved 60 or so NMEA sentence types, nine are specific to GPS receivers: ALM–GPS almanac data, GBS–GPS satellite fault detection, GGA–GPS fix data, GRS–GPS range residuals, GSA–GPS DOP (dilution of precision) and active satellites, GST–GPS pseudorange noise statistics, GSV–GPS satellite in view, MSK–MSK (minimum-shift keying) receiver/GPS receiver interface, MSS–MSK receiver signal status. Manufacturers can also define product-specific NMEA messages.

10.3.3. Data storage

Many GPS receivers, but primarily the ones intended for surveying and mapping applications, provide an option for *storing* the measurements and the navigation message for data post-processing. The storage media most frequently used in GPS receivers are removable memory cards or some form of solid state (RAM) memory. Manufacturers usually offer different memory capacity options, and in the case of extended storage requirements, extra memory can usually be added. The storage capacity needs of a GPS receiver are dependent on the length of the data acquisition session, the type of observable(s) to be recorded, the number of tracking channels, and the data recording interval (which can be a measurement every 120 s for geodesy and surveying, to anything up to 20 times a second for special kinematic applications!). A complete set of dual-frequency measurements requires about 100 bytes per epoch; multiplying this by the number of channels (maximum number of satellites to be observed) and the total number of epochs of observation, gives an estimate of the total memory needed. For example, recording data every 10 s for 10 h with 10 satellites in view will require $360 \times 10 \times 10 \times 100 = 3.6$ Mbytes of data storage.

Receivers storing data internally must have an RS-232 or some other kind of communication port to allow data transfer between the receiver and a PC. For example, transferring data to a PC via a serial port takes about 100 s/Mbyte (thus 16 Mbytes of data takes a little less than half an hour). Some modern receivers use fast parallel ports, which significantly reduce the data transfer time. An alternative is to use a PCMCIA card plugged into the receiver and download via a PC (with PCMCIA card reader).

10.3.4. Power supply

Current GPS instrumentation is comparatively energy efficient and typically uses low voltage DC power provided by either internal rechargeable batteries, or external batteries (see Table 10.2), allowing for many hours of continuous operation. Typical power consumption for geodetic/survey-grade GPS receivers ranges from 2 to 9 W. External power can also be supplied to a GPS receiver via an AC–DC converter. The lower the power consumption of the receiver, the more field operations can be carried out with a single battery. Fieldwork problems due to loss of power supply is considered an 'ultimate sin', that nevertheless occurs more frequently than it should!

10.3.5. Radio links

Both Differential GPS (DGPS) and Real-Time Kinematic (RTK) systems require a device to facilitate the reception of messages broadcast by one or more GPS base stations (Section 11.2). Each DGPS service provider requires their own specific communication link (Section 14.1). In general, the appropriate radio/coms equipment can be obtained from third party suppliers, or the service providers themselves. In the case of RTK systems, the radio link is sold as part of the GPS receiver kit (see Table 11.3 for typical radio links).

10.4. Choosing a GPS receiver

The 2000 Receiver Survey in the *GPS World* magazine lists 495 receivers from 58 manufacturers (*GPS World*, January 2001). *Why are there so many GPS receivers on the market?* The answer is: because there are so many different applications of GPS, and new uses spring up every day. Starting from the lower end of the accuracy requirements, handheld GPS receivers operate at the single-point accuracy level (<10 m without Selective Availability), while receivers used for mapping and GIS data collec-

Table 10.2 Batteries commonly used for GPS equipment (source: http://www.cadex.com)

Characteristic	NiCd	Lithium Ion	Sealed Lead Acid (SLA)
Energy density (W/kg)	40–60	100	30
Cycle life	1500	Up to 1000; 2-year life time	200–300
Fast charge time (h)	1–1.5	3–4	8–16
Overcharge tolerance	Moderate	Very low	High
Self-discharge per month (%)	20	10	5
Nominal cell voltage (V)	1.2	3.6	2
Maintenance requirement	30–60 days	Not required	3–6 months
Size/weight	Small for portable equipment	Small	Typically large
Typical cost in US$	50 (7.5 V)	100 (7.2 V)	25 (6 V)
Other characteristics	• Recommended slow charge of a new battery for 24 h • Performs best if fully discharged periodically Sheer endurance • High current capabilities • The only commercial battery that accepts charge at extremely low temperatures	• Does not need periodic full discharges • Lightest of all batteries • May cause thermal run-away • Usually sold in a pack, complete with a protection circuit • Non-rechargeable version available • Most expensive	• Good storage life but can never be recharged to its full potential • Should be stored in charged state • Most economic for high-power applications where size and weight is not a consideration

tion typically require a positioning accuracy in the range of sub-meter to a few meters. Both of these classes of GPS receiver are single-frequency units, designed to operate in real-time. The highest accuracy specifications are naturally related to surveying/geo-detic-type applications. In this case, a dual-frequency receiver, collecting data for post-processing, is usually specified. RTK techniques are increasingly used when the appropriate reference-to-rover receiver data communications link is available.

How to define a set of selection criteria? The first task is simply to list the functionalities and characteristics that distinguish one receiver from another. The second task is to assign some scale of values to these, to reflect their 'relative level of desirability'. Both of these are daunting tasks! An incomplete list of functionalities might include: type of measurements, number of channels, single- or dual-frequency, type of data storage, antenna type, nature of the instrument 'packaging', functionality of software, pricing, service and maintenance contract, range of applications, multipath and interference rejection qualities.

Those responsible for the selection of GPS equipment for a specific application may find the assignment a challenge if brochures and product specifications provided by the manufacturers are the only information sources consulted. The buyer has to correctly interpret the specifications, and should be able to ask the right questions in order to gain information regarding the conditions (and constraints) under which the specified performance is achieved. Confusion over terminology is also quite common, making it often difficult to compare 'apples and oranges'. For information on GPS receiver performance the reader is referred to Van Dierendonck (1995), Langley (1997a,b), Kaplan (1996), and Parkinson and Spilker (1996). In the following sections several issues that might aid the selection of the 'right GPS receiver' will be discussed.

10.4.1. The antenna for 'The Job'

The size and physical characteristics, which can be different depending on the application requirements, represent constraints on antenna choice. Hence, in the case of top-of-the-line GPS receivers, the manufacturers offer several different antenna options. The 2000 Antenna Survey in the *GPS World* magazine lists 218 antennas from 38 manufacturers (*GPS World*, February 2001).

Is there such a thing as an 'ideal GPS antenna', or are there always some trade-offs to be made? Unfortunately there is no *ideal* antenna suitable for *all* GPS applications. Each user must make a decision based on application-specific criteria, keeping in mind that each antenna type has a character of its own, and even those that appear to be physically similar might have significantly different performance characteristics. This issue becomes of crucial importance when the highest accuracy is required. For example, an antenna that provides a low elevation angle gain cut-off might not be the best to meet the requirements of low phase center variation with zenith angle, and vice versa. Since most multipath signals arrive from directions near the horizon, shaping the pattern to have low gain in these directions may reject such signals. Such a design for a high profile antenna with a low gain towards the horizon is appropriate for a ground reference station. On the other hand, a high profile antenna is not desirable for airborne use where it is necessary to receive signals even while the aircraft undergoes severe pitching and rolling, and where sharp nulls in the gain pattern might cause loss of signal tracking.

10.4.1.1. Static positioning

Surveying/geodetic antennas for static positioning are usually much larger, and have a ground plane or choke rings added, compared to compact light-weight antennas intended for kinematic applications. It is worth mentioning here that for some antennas, such as in the case of microstrip antennas, a ground plane not only contributes to multipath mitigation, but also increases the antenna's zenith gain (to a limit, the larger the ground plane, the higher the antenna's gain in the zenith direction). Choke ring antennas can almost entirely eliminate the multipath problem for the surface waves and the signals reflected from the ground. (The multipath from objects above the antenna still represents a significant challenge for antenna designers.) The performance of the choke ring antenna is usually better for the L2 frequency than for L1, the reason being that the choke ring can be most easily optimized only for one frequency. If the choke ring is designed for L1, it will have no effect on L2 signals, whereas a choke ring antenna designed for L2 multipath suppression still has some benefits for L1. Further antenna developments are occurring and one manufacturer now claims mitigation of multipath on both frequencies. Another manufacturer claims to have developed a low-weight antenna with similar performance to a choke ring antenna.

10.4.1.2. Kinematic applications

Airborne antennas, for example, should have the capability of acquiring signals from below the antenna horizon. In addition, bulky antennas are generally inappropriate for aircraft, and hence light-weight small microstrip antennas are preferred for such applications. The primary criteria for antennas used for kinematic surveying and mapping are small size, light weight and multipath rejection (and for high-productivity survey techniques, dual-frequency antennas must be used). Such antennas can be fitted atop a pole that can be used to position the antenna accurately over a ground mark, or placed on an antenna pole that is carried in a backpack. Hence the small 'frisbee' style antenna is favored (Figures 10.5 and 10.7). For some RTK configurations the GPS antenna is integrated with the radio link antenna in the one pole-mounted housing. Furthermore, in some instrument models the antenna(s) and receiver are integrated within the one unit (Section 10.4.4.1).

10.4.1.3. Recreational and DGPS-capable receivers

The majority of handheld GPS receivers have a very small, single-frequency antenna embedded within the receiver. The antennas for handheld receivers usually comprise quadrifilar helices, consisting of two bifilar helical loops, orthogonally-oriented on a common axis, which do not require a ground plane (as opposed to the microstrip types used for surveying-type antennas). There is a trend nowadays for antennas intended to receive DGPS messages (e.g. from L-band geostationary satellite transmissions, or mobile telephony services, etc.) to be integrated within the same housing used for the GPS signal antenna.

10.4.2. GPS performance

What is meant by GPS performance? It is not simply a question of one receiver brand or model being more 'accurate' than another. In general, given a *specified accuracy* the receiver performance will match that accuracy standard only if a range of criteria are satisfied 'in the real world'. For example, relative positioning accuracy will typically be a function of the number of satellites being tracked, the measurement type that is being processed, the length of data acquisition time, operational mode (such as whether the receiver is moving or stationary), quality of observation conditions, degree of signal obstructions, baseline length, and environmental effects, and consequently the *actual* performance will vary as these conditions change.

10.4.2.1. Defining performance parameters

It should be noted that the commercial data sheets, stating the achievable positioning accuracy, might be misleading because the claimed performance of the receiver is typically achieved under ideal observing conditions. Testing a GPS receiver on either a pre-established baseline network, or using a 'zero baseline', is one way to gauge the degree to which the *ideal* performance degrades when real operating conditions are experienced. In general, the high-productivity carrier phase-based positioning techniques such as rapid-static, stop-and-go and kinematic (Section 11.1) will be the most sensitive to deviations of actual observing conditions from ideal ones.

Often more important than the accuracy of the results under certain observing conditions are other measures of 'good performance'. The following is an incomplete list (and does not distinguish between survey-grade receivers intended for post-processed modes of positioning, and relatively basic navigation receivers): susceptibility to fade or interference, time-to-first-fix, reacquisition time after signal loss, measurement data rate, power consumption, multipath rejection, and so on. Clearly, it is not a simple matter to unambiguously define 'good performance'.

10.4.2.2. Dual-frequency or single-frequency GPS receivers

Dual-frequency instruments can make twice the number of measurements than single-frequency instruments. Furthermore, the pseudorange and carrier phase measurements on the L1 and L2 frequencies can be linearly combined to generate new observables with specific desired qualities. The special combination of $2.546 \times$ L1 $- 1.546 \times$ L2 creates an observable that has had the ionospheric delay eliminated, and is known as the 'ionosphere-free' observable (Section 9.2.1.1). (L1 and L2 can be either the pseudorange or carrier phase measurements.) Another important dual-frequency carrier phase combination is $\phi_1 - \phi_2$, which is known as the 'wide-lane' combination because the effective wavelength is 86.2 cm (compared with 19 and 24 cm for the L1 and L2 carrier waves, respectively). In summary, dual-frequency instrumentation is insisted upon if:

1   High precision baseline determination over distances of 20–30 km or more are sought.
2   High-productivity GPS surveying techniques are to be used.

Dual-frequency measurements are required for the former because they permit the biases due to residual ionospheric delay (remaining after differencing the simultaneous measurements from two receivers to the same satellite) to be eliminated. In the case of the latter group of techniques, rapid ambiguity resolution can only be done reliably if combinations of the L1 and L2 measurements are possible. Hence, for these reasons dual-frequency phase-tracking instrumentation may be considered to represent the 'top-of-the-line' GPS hardware.

10.4.2.3. Radio interference

GPS receivers can track the satellite signals under favorable conditions, but what distinguishes one receiver from another is typically how well the tracking can be performed (and the resultant quality of measurements) under less than ideal environmental conditions. Several kinds of interference can disrupt a GPS receiver's operation: in-band emissions, nearby-band emissions, harmonics, and jamming (Parkinson and Spilker 1996). As more and more sources of microwave transmissions appear every day, signal interference has become a major concern to the GSP user community (Johannesse 1997). Under less than ideal conditions radio interference can, at the very least, reduce the GPS signal's apparent strength (the SNR), and consequently the accuracy, or at worse even block the signal entirely. Medium-level interference would cause frequent losses-of-lock or cycle slips, and might render the data virtually useless.

10.4.2.4. Multipath interference

Multipath is a special form of signal interference. The GPS signal is generated using a spread spectrum technique that inherently rejects interference caused by long-delay multipath (Section 8.5.2). However, short-delay multipath still causes problems, especially for kinematic applications and static positioning using very short observation spans, thus special techniques have been developed to reject multipath errors in GPS measurements.

10.4.3. Hardware component quality

Often GPS manufacturers will try to (favorably) differentiate their equipment from that of other manufacturers on the basis of the quality of the hardware components.

10.4.3.1. Receiver clock

How good is the GPS receiver oscillator (clock)? Is this important to the user? The differential mode of positioning using carrier phase data requires that the measurements undergo 'double-differencing' during the baseline estimation process (Section 8.6.2), where the error in range attributable to the receiver oscillator is eliminated. There are indeed differences in the quality of the receiver oscillator from one manufacturer to another, however for the vast majority of surveying applications there is no noticeable degradation in the quality of relative positioning results that can be attributed to this source alone.

10.4.3.2. Receiver noise

Since a GPS receiver, as in the case of any measuring device, is not 'perfect', measurements will always be made with a certain level of 'noise'. The most basic kind of noise is so-called *thermal noise*, produced by the random movement of the electrons in any material that has a temperature above 0 degrees Kelvin (Langley 1998a,b). The commonly used measure of the received signal strength is the signal-to-noise-ratio, *SNR*. The larger the SNR value, the stronger the signal. In the case of RF and IF, the most commonly used measure of the signal's strength is the carrier-to-noise-power-density ratio, C/N_o, defined as a ratio of the power level of the signal carrier to the noise power in a 1 Hz bandwidth (Van Dierendonck, 1995; Langley 1998b).

10.4.3.3. Receiver measurement precision

C/N_o is considered a primary parameter describing the GPS receiver performance as its value determines the precision of the pseudorange and carrier phase measurements. Typical values of C/N_o for modern survey-grade GPS receivers (L1 C/A-code) range between 45 and 50 dB-Hz. For example, for C/N_o equal to 45 dB-Hz, and signal bandwidth of 0.8 Hz, the RMS tracking error due to receiver thermal noise for the C/A-code is 1.04 m. For GPS receivers with narrow correlators (Van Dierendonck *et al.* 1992) with the same bandwidth and C/N_o, the RMS is only 0.39 cm. Hence, there is indeed a difference in the noise level (and therefore the resulting pseudorange measurement precision) between so-called 'narrow correlator' receivers and standard GPS receivers. The RMS tracking error due to noise for a carrier-tracking loop with C/N_o of 45 dB-Hz and signal bandwidth of 2 Hz is only about 0.2 mm for the L1 frequency (Braasch 1994; Langley 1998a). While similar receiver types from the different manufacturers would be expected to generate L1 carrier phase measurements with very similar precision, the quality of measurements of L2 carrier phase is a function of the signal processing technique used (Table 10.1).

10.4.3.4. Rejection of radio interference

The most common interference protection used in GPS receivers are: (a) null-steering antennas, also known as a 'controlled radiation pattern' antennas (which unfortunately are rather bulky and expensive, and hence inappropriate for many applications), (b) narrow front-end filters and narrowed, aided tracking loops (such on-receiver techniques are generally limited in the kinds of interference they can reject), and (c) a recently developed Interference Suppression Unit (ISU). An ISU consists of a simple patch antenna and electronic unit that plugs directly into the GPS receiver antenna port, and various tests have proven its effectiveness for interference suppression (Braasch and Snyder 1998). The interested reader is referred to Ward (1994) for a review of interference monitoring, mitigation, and analysis techniques.

10.4.3.5. Multipath rejection techniques

Existing multipath rejection technology is intended to improve the C/A-code-based pseudorange measurement, and can potentially increase the single-point positioning

accuracy by about 50 per cent. In general, signal processing techniques can reject the multipath signal only if the multipath distance (the difference between the direct and the indirect paths – Figure 8.11) is roughly more than about 10 m. In a typical geodetic/surveying application, however, the antenna is about 2 m above the ground, thus the multipath distance reaches at most 4m, and consequently signal processing techniques cannot fully mitigate the effects of reflected signals on pseudorange measurements. In the case of carrier phase measurements the problem is significantly reduced, being a disturbance of the order of several centimeters. This is, nevertheless, sufficient to impact on the accuracy of techniques that rely on sophisticated 'on-the-fly' ambiguity resolution algorithms to achieve high accuracy with very small quantities of carrier phase data (Section 9.4.2). For more discussion on multipath and its influence on GPS receivers the reader is referred to Van Dierendonck *et al.* (1992), Van Nee (1993), Braasch (1994), Weill (1997), and Parkinson and Spilker (1996).

10.4.4. The receiver 'form-factor'

GPS receivers come in all 'shapes and sizes'. In the case of off-the-shelf products (as opposed to 'developer's kits' consisting of boardsets or microchips), the GPS 'package' is generally designed to suit the application. Hence low-cost recreational receivers consist of a small unit in which all components are integrated into an attractive handheld device. In the case of the top-of-the-line GPS receivers intended for surveying and mapping, several different configurations can be purchased from the manu-facturers, or the products are built in a modular fashion so that the various components may be connected together (via cables or direct connectors) in different ways (Figures 10.4–10.7).

10.4.4.1. Smart antennas/integrated receivers

For applications such as tracking weather buoys, time-tagging of seismic or other events, navigation, or precise timing and synchronization of wireless voice and data networks, so-called 'smart antennas' are often used. These are essentially self-contained, shielded and sealed units that house a GPS receiver, an antenna, a power supply, processor and other supporting circuitry in a single ruggedized enclosure, which mounts like an antenna (Figure 10.3, see plate 3). These units are typically single-frequency instruments, making pseudorange measurements and are intended for automatic, continuous operation. Once power is applied a smart antenna starts immediately to produce position, course, speed and time information through a serial interface port. A second serial port enabling real-time differential GPS operation, or for outputting an accurate timing pulse synchronized to the sub-microsecond level, may also be provided. Perhaps their defining characteristic is that they have a very rudimentary control/display capability consisting of little more than an on-off switch and/or an indicator light (or lights). Even these may be missing in some models.

Smart antennas have also evolved into comprehensive survey packages. For exam-ple, the *Locus Survey System* from Magellan Corporation/Ashtech Precision Products, or fully integrated GPS receivers such as the *Odyssey* or *Regency* from Topcon Positioning Systems, as well as several Trimble and Leica products. An integrated receiver houses all electrical circuitry, with all its associated features and the data

storage medium, and may also include a radio modem, removable battery module, and several serial interfaces. Smart antennas/integrated receivers such as these provide alternative easy-to-use, cable-free, light-weight, low-power options for receivers intended for use as high-productivity GPS surveying tools (Figure 10.6, see plate 6).

10.4.4.2. GIS/mapping receivers

In contrast to the 'smart antenna'-type instruments, these receivers are distinguished by having both a LCD display/command unit through which instructions and user-entered data is input, and a DGPS signal decoder. In some products the unit is almost indistinguishable from a basic handheld GPS receiver, with all circuitry integrated into one unit. In contrast, some systems consist of a data logger device that functions as the display/control unit, possibly equipped with a miniature QWERTY keyboard (or other data input devices, such as a barcode reader) to allow feature attribute data to be entered. A laser range finder, with magnetic compass and inclinometer, may be attached to the GPS system in order to 'shoot' to features in the field-of-view of the GPS point. In this way from one GPS point (which can be surveyed easily because it has clear skyview), many landmarks and features can be coordinated, even if they are located where GPS signal shading would have made them difficult to map using GPS alone.

The antenna as well as the receiver box may be in one module, or consist of individual components. For maximum flexibility the DGPS signal decoder may be a separate interchangeable module (e.g. to allow for the use of a variety of DGPS signal input options – Section 14.1). In such configurations all components apart from the data-logger device are usually stored in a backpack (which also carries the batteries and the antenna pole), and therefore the system has a similar appearance to a 'poleless' RTK instrument (Figure 10.7, see plate 7).

10.4.4.3. GPS surveying receivers

These are typically the most expensive class of GPS receiver. Hence they should be able to be used in a variety of positioning modes, such as standard DGPS positioning (e.g. for GIS/mapping); static baseline determination; mounting on aircraft, land or marine platforms for kinematic positioning; and man-portable units for high-productivity GPS surveying. Although there are receivers optimized for special high precision applications, most top-of-the-line GPS receivers intended for surveying are capable of being used in a variety of modes.

In cases where the point being coordinated is a ground mark, dual- and single-frequency carrier phase-tracking GPS receivers may be mounted on tripods (Figure 10.4, see plate 4), control pillars, or special range poles (Figures 10.5 and 10.6, see plates 5 and 6). The pole-mounted configuration is generally the most appropriate one for high-productivity survey techniques. Real-time as well as the post-mission data processing operation can be supported. The antenna may be integrated within the same housing as the GPS receiver (Figure 10.6, see plate 6), or separate from it. The latter does allow for more flexibility in that in some circumstances the GPS antenna may be located some distance from where the receiver (and power supply) is housed. Furthermore, such a configuration permits different antennas to be used for certain

applications (including choke ring antennas), as well as being more convenient when mounted on a moving platform such as a vehicle or aircraft. The backpack-mounted configuration (Figure 10.7, see plate 7) may be the most convenient for RTK operations, or when the system is used for GIS/mapping, as discussed earlier.

10.4.4.4. Special high accuracy applications

The applications of carrier phase-tracking GPS receivers were traditionally those associated with geodesy and surveying, that is, static positioning. In such a positioning mode, the receiver was tripod or pillar-mounted, and was expected to collect data for an observation session that might range from an hour to several days in length. High-productivity techniques have challenged this standard static position mode. Nevertheless, there is an increasing need for GPS receivers to be set up *permanently* as static reference receivers, at continuously operating base stations, that are part of DGPS service networks (Section 14.2.1), or the International GPS Service, or a variety of specialized networks addressing such applications as geodynamics, deformation monitoring, and atmospheric sensing. In such cases the form-factor is not critical. Typically, the antenna is of the choke ring variety, the receiver is attached to a PC and/or communications link, and the power supply is assured (and uninterruptable).

Increasingly carrier phase-tracking GPS receivers are being used for kinematic applications such as machinery guidance and control. The RTK operating mode is used, and the most crucial design characteristic is that the receiver box must be robust, as it has to operate reliably in extreme environments. The receiver will typically be operating in hot (or very cold) and dusty (or wet) conditions, attached to a vibrating vehicle. Hence, all components (antenna, receiver, radio link, cables and connectors) must be ruggedized to a significant degree. Such receivers are unlikely to be used for the man-portable high-productivity surveying applications referred to earlier.

References

Ashjaee, J., 'An analysis of Y-code tracking techniques and associated technologies', *Geodetical Info Magazine*, 7(7), 1993, 26–30.

Ashjaee, J. and Lorenz, R., 'Precision GPS Surveying after Y-code', *Proc. 5th Int. Tech. Meeting of the Satellite Division of the US Institute of Navigation*, 1992, pp. 657–659.

Braasch, M. S., 'Isolation of GPS multipath and receiver tracking errors', *Navigation*, 41(4), 1994, 415–434.

Braasch, M. S. and Snyder, C. A., 'Running interference: testing a suppression unit', *GPS World*, 9(3), 1998, 50–54.

Johannesse, R., 'Interference: sources and symptoms', *GPS World*, 8(11), 1997, 44–48.

Johnson, R. C. (ed.), *Antenna Engineering Handbook* (3rd ed.), McGraw-Hill, New York, 1993, 1392 pp.

Kaplan, E. (ed.), *Understanding GPS: Principles and Applications*, Artech House, 1996, 570 pp.

Langley, R. B., 'The GPS receiver: an introduction', *GPS World*, 2(1), 1991, 50–53.

Langley, R. B., 'NMEA 0183: a GPS receiver interface standard', *GPS World*, 6(7), 1995a, 54–57.

Langley, R. B., 'A GPS glossary', *GPS World*, 6(10), 1995b, 61–63.

Langley, R. B., 'The GPS error budget', *GPS World*, 8(3), 1997a, 51–56.

Langley, R. B., 'The GPS receiver system noise', *GPS World*, 8(6), 1997b, 40–45.

Langley, R. B., 'Propagation of the GPS signals and GPS receivers and the observables', *GPS for Geodesy* (2nd ed.), Teunissen, P. J. and Kleusberg, A., (eds.), Springer, Berlin, 1998a, pp. 112–185.

Langley, R. B., 'A primer on GPS antennas', *GPS World*, 9(7), 1998b, 50–54.

Parkinson, B. W. and Spilker, J. J. (eds.), *Global Positioning System: Theory and Applications*, vols. I (694 pp.) and II (632 pp.), American Institute of Aeronautics and Astronautics, Inc., 1996.

POB, 'GPS mapping/positioning receiver survey, Point of Beginning, *POB* 25(12), 2000, 56–58.

Storm Products, 2001, web page, http://www.stormproducts.com/.

Van Dierendonck, A. J., 'Understanding GPS receiver terminology: a tutorial', *GPS World*, 6(1), 1995, 34–44.

Van Dierendonck, A. J., Fenton, A. and Ford, T., 'Theory and performance of narrow correlator spacing in a GPS receiver', *Navigation*, 39(3), 1992, 265–283.

Van Nee, R., 'Spread spectrum code and carrier synchronization errors caused by multipath and interference', *IEEE Transactions on Aerospace and Electronic Systems*, 29(4), 1993, 1359–1365.

Ward, P. W., 'GPS receiver RF interference monitoring, mitigation, and analysis techniques', *Navigation*, 41(4), 1994, 367–391.

Weill, L. R., 'Conquering multipath: the GPS accuracy battle', *GPS World*, 8(4), 1997, 59–66.

Chapter 11

Making sense of the **GPS** techniques

Chris Rizos

In Section 7.1.3, the user segment was defined as the entire spectrum of applications equipment, operational procedures and computational techniques that provide the users with the position results. There are a wide variety of Global Positioning System (GPS) applications, which is matched by a similar diversity of user equipment and techniques. Nevertheless, the most fundamental classification system for GPS techniques is based on the type of observable that is tracked (see Sections 8.2 and 8.4): (a) civilian navigation/positioning receivers using the C/A-code on the L1 frequency, (b) military navigation/positioning receivers using the satellite P(Y)-codes on both L-band frequencies, (c) single-frequency (L1) carrier phase tracking receivers, and (d) dual-frequency carrier phase tracking receivers. When these classes of hardware are used in the appropriate manner for relative positioning, the accuracy that is achieved ranges from a few meters in the case of standard pseudorange-based techniques, to the sub-centimeter level in the case of carrier phase-based techniques. Although Single-Point Positioning (SPP) accuracy of 5–10 m is now possible, in this manual it will be assumed that for most geospatial applications such as surveying and mapping/ Geographical Information Systems (GIS), only the relative positioning techniques are of relevance.

The following classes of relative positioning techniques can therefore be identified:

1 *Static and kinematic GPS surveying techniques:* high precision techniques based on post-processing of carrier phase measurements.
2 *Differential GPS (DGPS):* instantaneous low-to-moderate accuracy positioning and mapping technique based on pseudorange measurements.
3 *Real-Time Kinematic (RTK) techniques:* versatile high precision techniques that use carrier phase measurements in an instantaneous positioning mode.

The post-processed GPS surveying techniques are discussed in Section 11.1, with particular emphasis on the new 'high productivity' techniques, which are today widely utilized for many surveying/mapping tasks. Chapter 12 deals with the project planning issues that are relevant for both the post-processed and real-time techniques. Chapter 13 discusses GPS projects from the applications and field operations perspective.

The DGPS and RTK techniques, because they are able to deliver results in real-time, are very powerful GPS positioning technologies. They are discussed in Section 11.2, and the communication link issues identified in Section 11.3. In the case of these real-time techniques there is a clear distinction between the functions of the receiver at the reference station on the one hand, and the user equipment on the other. Such a distinction is less obvious in the case of post-processed techniques, where all the

equipment tends to be owned and/or operated by the surveyor/contractor. In such a case whether one receiver or the other is designated 'reference' or 'user' is largely immaterial (especially in the case of static survey techniques). This notion of 'owner-ship' introduces a new element in any consideration of user techniques. Namely, some techniques can be seen as requiring the provision of a *service*, while other techniques are completely under the control of the surveyor/contractor. In Sections 14.1 and 14.2, the former is discussed in relation to trends in the provision of real-time (and other) services to GPS users. This is in contrast to the user procedures that must be specified for the latter class of techniques, in the form of guidelines or recommended practices, as outlined in Section 12.2.

11.1. High productivity GPS surveying techniques

There are essentially two types of conventional static GPS surveying techniques:

1   *Ultra precise, long baseline GPS techniques*: accuracy from a few parts per million to several parts per billion, characterized by top-of-the-line GPS receivers and antennas, many hours (and even days) of observations, and data processing using sophisticated 'scientific' software.
2   *Medium-to-short baseline GPS survey techniques*: accuracy at the few parts per million level for baselines typically <50 km, to support control network appli-cations, with data processing being carried out by commercial software packages.

The main weaknesses of such procedures are that the observation times are comparatively long, the results are obtained well after the field survey, and the field procedures are rigid. During the late 1980s considerable attention was paid to these issues, as they were considered to be unnecessarily restrictive for precise GPS technol-ogy. That is, if antennas could be moving during a GPS survey, then new applications for the GPS technology could be addressed. If the length of time required to collect phase data for a reliable solution could be shortened, then GPS survey productivity would improve and the technology would be attractive for many more surveying applications. If the results could be obtained immediately after the measurements have been made, then GPS could be used for time-critical missions such as engineering stakeout, etc. RTK is discussed in Section 11.2.2, and hence in this section the focus will be on the *high productivity* carrier phase-based techniques for which post-proces-sing of the data is the norm.

In the last decade or so new GPS surveying methods have been developed with the two *liberating* characteristics of: (a) static antenna setups no longer having to be insisted upon, and (b) long observation sessions no longer essential in order to achieve survey level accuracy. These modern GPS surveying techniques are given a variety of names by the different GPS manufacturers, but the following generic terminology will be used in this manual:

- *Rapid-static* positioning technique
- *Stop-and-go* technique
- *Kinematic* positioning technique

All require the use of specialized hardware and software, as well as new field procedures. GPS receivers capable of executing these types of surveys can also be

used for conventional static GPS surveying. Although the field procedures are different from conventional static GPS surveying, the principles of planning, quality control and network processing are very similar.

Each of the techniques represents a technological solution to the problem of obtaining *high productivity* (coordinating as many points in as short a period of time as possible) and/or versatility (e.g. the ability to obtain results even while the receiver is in motion and/or in real-time) without sacrificing very much in terms of *accuracy* and *reliability*. However, none of these techniques is as accurate or reliable as conventional static GPS surveying, and because each of these techniques has its distinctive advantages (Section 11.1.4) they are all offered as positioning options within modern GPS products. Hence, the surveyor/contractor has access to a diverse 'toolkit' of carrier phase-based GPS techniques from which they must make an appropriate selection (Sections 12.1 and 12.2).

11.1.1. Rapid-static GPS surveying

Also often referred to as fast-static or quick-static. The following characteristics distinguish rapid-static techniques from other methods of static GPS surveying:

- *Observation time requirements*: these are significantly shorter than for conventional static GPS surveying, and are a function of baseline length, whether dual-frequency instruments are used, number of satellites being tracked and satellite geometry. Typically the receivers need only occupy a baseline for a period of 10–30 minutes (the lower value corresponding to baselines < 5 km and tracking six or more satellites; the upper value being for longer baselines up to 20 km, and/or where tracking is to only four satellites).
- *Hardware requirements*: in most systems only dual-frequency phase measurements are sufficient, while in other system configurations dual-frequency pseudorange measurements are *also* required. It is rare to 'mix' different brands of receivers and software, compared to static GPS surveying techniques.
- *Specialized software*: the basis of this technique is the ability of the software to *resolve* the ambiguities using a very short observation period. There is a variety of software, with different characteristics and levels of sophistication, but the fundamental requirement is a *fast ambiguity resolution* (AR) capability. Dual-frequency data is essential for fast AR (Section 9.4.2).

The field procedures are much like those for conventional static GPS surveying except that: (a) the station occupation times are shorter, (b) the baselines should be comparatively short, (c) the satellite geometry favorable, and (d) signal disturbances such as multipath should be kept to a minimum. It is not possible to define exactly how much data needs to be collected in order to produce quality baselines every time, that is, 'ambiguity-fixed' solutions (Section 9.4.3). Equipment user manuals typically give guidelines in this regard. Some receivers also provide an audio and/or visual *indication* when enough data has been collected in the field (but this cannot be confirmed until the data is downloaded and processing is completed). If the real-time positioning mode is employed (Section 11.2.2) then the 'data quantity gamble' for rapid-static GPS surveying can be overcome.

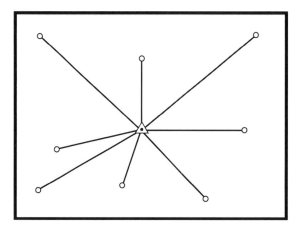

Figure 11.1   Field procedure for the rapid-static surveying technique.

The rapid-static technique is well suited for short range applications such as control densification and engineering surveys, or any job where many points need to be surveyed (Figure 11.1). Unlike the *kinematic* and *stop-and-go* techniques, there is no need to maintain lock on the satellites when moving from one point to another.

11.1.2. 'Stop-and-go' GPS surveying

This is a true *kinematic* technique because the user's receiver continues to track satellites while it is in motion. It is known as the stop-and-go (or semi-kinematic) technique because the coordinates of the receiver are only of interest when it is stationary (the 'stop' part), but the receiver continues to function while it is being moved (the 'go' part) from one stationary setup to the next. There are in fact three stages to the operation:

1   *The initial ambiguity resolution:* carried out before the stop-and-go survey commences. The determination of the ambiguities by the software can be carried out using any method, but in general it is one of the following:

  • A conventional static (or rapid-static) GPS survey determines the baseline from the reference station receiver to the first of the points occupied by the user's receiver. An 'ambiguity-fixed' solution provides an estimate of the integer values of the ambiguities that are then used in subsequent positioning.
  • Setup both receivers over a known baseline, usually surveyed previously by GPS, and derive the values of the ambiguities in this way.
  • Employ a procedure known as 'antenna swap'. Two tripods are setup a few meters apart, each with an antenna on them (the exact baseline length need not be known). Each receiver collects data for a few minutes (tracking the same satellites). The antennas are then carefully lifted from the tripods and swapped, that is, the receiver one antenna is placed where the receiver two antenna had been, and visa versa. After a few more minutes the antennas are

swapped again. The software is able to resolve the ambiguities over this very short baseline.

- The most versatile technique is to resolve the ambiguities 'on-the-fly' (that is, while the receiver is tracking satellites but the receiver/antenna is moving). The impact on precise GPS positioning of On-The-Fly Ambiguity Resolution (OTF-AR) is discussed in Section 11.1.4.

2   *The receiver in motion*: once the ambiguities have been resolved, the survey can begin. The user's receiver is moved from point to point, collecting just a minute or so of data. *It is vital that the antenna continues to track the satellites*. In this way the resolved ambiguities are valid for all future phase observations, in effect converting all carrier phase data to unambiguous 'carrier-range' data (by applying the integer ambiguities as data corrections). As soon as the signals are disrupted (causing a cycle slip) then the ambiguities have to be reinitialized (or recomputed). Bringing the receiver back to the last surveyed point, and redetermining the ambiguities can most easily be done using the 'known baseline' method (see above).

3   *The stationary receiver*: the 'carrier-range' data is then processed in the double-differenced mode to determine the coordinates of the user receiver relative to the static reference receiver (Section 9.4.1). *The trajectory of the antenna is not of interest, only the stationary points which are visited by the receiver.*

The technique is well suited when many points close together have to be surveyed, and the terrain poses no significant problems in terms of signal disruption. The accuracy attainable is about the same as for the rapid-static technique. The software has to sort out the recorded data for the different points, and to differentiate the *kinematic* or 'go' data (not of interest) from the *static* or 'stop' data (of interest). (The technique can also be implemented in real-time if a communications link is provided to transmit the data from the reference receiver to the roving receiver.) The survey is typically carried out in the manner illustrated in Figure 11.2.

One *negative* characteristic of this technique is the requirement that signal lock *must*

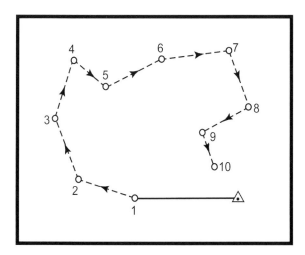

Figure 11.2   Field procedure for the stop-and-go surveying technique.

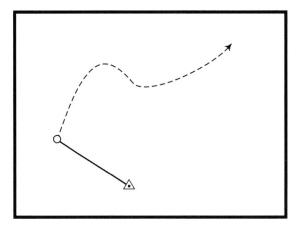

Figure 11.3 The kinematic GPS surveying technique.

be maintained on the satellites by the user receiver as it moves from point to point. This may require special antenna mounts on vehicles if the survey is carried out over a large area. Mobile configurations are shown in Figures 10.5–10.7.

11.1.3. Kinematic GPS surveying

This is a generalization of the stop-and-go technique. Instead of only coordinating the stationary points and disregarding the trajectory of the roving antenna as it moves from point-to-point, the intention of *kinematic* surveying is to determine the position of the antenna *while it is in motion* (Figure 11.3). In many other respects the technique is similar to the stop-and-go technique. That is, the ambiguities must be resolved *before* starting the survey, and the ambiguities must be reinitialized *during* the survey when a cycle slip occurs. However, for many applications, such as the positioning of an aircraft or a ship, it is impractical to reinitialize the ambiguities if the roving antenna has to return to a stationary control point. Today the 'kinematic' GPS surveying technique is undergoing rapid improvement and OTF-AR is a routine procedure, making kinematic surveying techniques ideal for road centerline surveys, topographic surveys, hydrographic surveys, airborne applications, and many more.

11.1.4. Choosing the appropriate technique

Each of these high productivity GPS surveying techniques has its strengths and weaknesses, however, all are less accurate than the conventional GPS surveying technique. This should not be too great a drawback as it is often not necessary that relative accuracy of 1 ppm (parts per million – equivalent to 1 cm error in 10 km) be insisted upon. Often a combination of conventional static and GPS techniques such as the ones described above makes for an ideal solution to a surveying problem. One of the challenges for the GPS surveyor is to select the best combination of techniques for the terrain, distance and logistical constraints that they face. This is discussed in Section 12.3.

The development of OTF-AR algorithms is a dramatic step forward because static ambiguity reinitialization is no longer necessary for any positioning technique. The ambiguities will be resolved *while the antenna is moving* to the next stationary survey point. If a point X has been surveyed (that is, a few minutes of 'carrier-range' tracking data have been collected), and as the antenna is moved from point X to point Y, an obstruction blocks the signals and causes cycle slips to occur, then the antenna does not have to go back to point X. New ambiguities can be resolved 'on-the-fly' as the antenna moves from X to Y. However, there must be a sufficient period of uninterrupted tracking for this to take place. During this 'dead time' (top part of Figure 11.4) cm-level positioning is not possible if OTF-AR is implemented in real-time, but *is* possible with the post-processing mode because the data before AR has been completed can be 'backwards' corrected and then used to generate ambiguity-fixed baseline results.

Although the time-to-AR varies (and is influenced by the baseline length, satellite geometry, and several other factors), the required period of continuous carrier phase measurement is *of the order of a few tens of seconds to several minutes*. In an extremely unfavorable scenario, there may be so many signal obstructions that there is

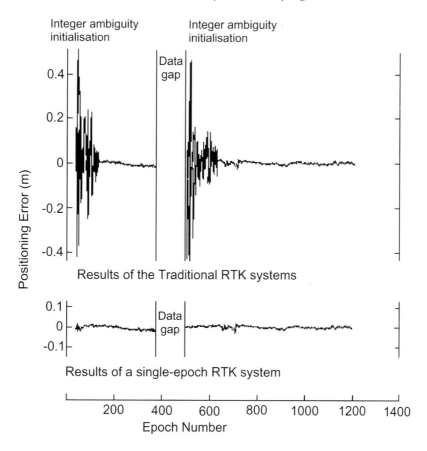

Figure 11.4 'Dead time' indicated for OTF-AR after a break in GPS signals, compared with the 'ideal' case of single-epoch AR that does not suffer from this problem.

12.1.7. Differential GPS (DGPS) techniques

DGPS is the pseudorange-based relative positioning technique that delivers accuracy in the range of sub-meter to a few meters in real-time (Section 11.2.1). This accuracy is mainly a function of distance to the reference station, but if this distance is less than about 100 km the major impact on accuracy is satellite geometry, the quality of the measurements and multipath effects on the pseudorange measurements. Invariably DGPS positioning is carried out by a user using an established DGPS service, so that the establishment of a reference station by the user is not required (Section 14.1). *Wide Area* DGPS (WADGPS) is an enhancement of the basic DGPS technique in that a network of reference stations located across a region (typically set up many hundreds of kilometers apart) is used to generate the DGPS corrections that are then transmitted to the user (Section 11.3.3). Such a scheme allows for the modeling of distance-dependent biases such as atmospheric refraction and satellite orbit errors, making possible meter-level accuracy even if the nearest WADGPS reference station is located hundreds of kilometers away. The Wide Area Augmentation System (WAAS) is capable of delivering WADGPS service to suitably equipped users (Section 14.1.1.2).

12.2. Standards and specifications

The specifications for any GPS survey project are usually set by the client and it is the role of the surveyor/contractor to indicate the cost of achieving those specifications, while assuring the client that surveys will be undertaken to 'best practice' standards. However, no international set of standards for the many variants of GPS surveys exists. Many survey authorities at both the national and state level have developed *standards and specifications* (S&S), some in a general form that cover all types of surveys (not only GPS), and others that are for GPS alone at the applications-specific level (e.g. for cadastral, or engineering surveys). Schinkle (1998) outlines a set of guidelines on how the New York State Department of Transport undertakes GPS engineering control surveys. This may be used as a 'template' to assist in the design of guidelines appropriate for the reader's own application, or industry.

However, the pace at which GPS technology has changed over the past 10 years means that much of the S&S documentation may be somewhat dated. Some documents, such as the Australian SP1 specifications (ICSM 2000) and the New Zealand specifications (DOSLI 1994, 1996a) have undergone considerable revision. Other specifications, such as Craymer *et al.* (1990), GSD (1992), FGCC (1984, 1989) and CGCC (1993) are essentially pre-RTK documents, that have lost some of their relevance in the face of rapid technological developments.[2] GSD (1992) proposes a novel solution to this problem. Rather than a description of specific procedures to be followed, it is suggested that an *acceptance test* or *validation survey* be conducted to evaluate the performance of the proposed survey system for a particular type of production survey. Such an approach does not appear to have been adopted in other countries.

The published guidelines for GPS surveys (some of which are referred to above) may be the only *official* S&S documents available to guide the surveyor (the GPS user

[2] The authors understand that both the US and Canadian S&S are undergoing revision at present.

manuals do not deal with national or state survey practice, and are therefore of limited use in this regard). Nevertheless, the surveyor/contractor should be familiar with all relevant S&S as the client requirements may be couched in the language of such documents. In addition, quality assurance practice demands that GPS field and office procedures are consistent, and the use of S&S ensures such consistency.

12.2.1. Accuracy standards

GPS 'standards and specifications' provide the basis by which 'levels of service' may be defined. *However, different recommended practices are associated with different accuracy standards.* The various national geodetic control authorities typically use different terminology for the *classes* of GPS survey, and have varying numerical accuracy limits used to define the categories. Sections 13.1 and 13.2 describe the US accuracy standards, and some typical applications.

In general these categories are distinguished by some relative accuracy measure. For example, in Australia, relative error is defined for the various categories of survey by the specification of the maximum allowable 'base error' and 'line-length error' (at the 95 per cent confidence level) for the semi-major axis of the relative error ellipse (or ellipsoid): $e = a + b.L$, where L is the inter-receiver distance in kilometers, the quantities e and a are in millimeters, and b is expressed in parts per million. The 'base error' component (a) incorporates effects such as error in centering over a mark, antenna height measurement and antenna phase center variation, while the 'line-length error' component (b) represents the various distance-dependent biases that impact on GPS baseline determination. On the other hand, in the US the different survey categories are currently being redefined in terms of coordinate accuracy, e.g. 1, 2, 5, 10, and 20 cm (FGDC 2001).

Furthermore, the accuracy standards are typically different for the horizontal position component and for the vertical component, with the added complication that *ellipsoidal heights* may be treated differently to the *orthometric heights* (NGS 1997). In general, however, the pre-existing accuracy standards are maintained, but augmented by several higher accuracy standards that are only applicable for GPS surveys. Many countries are recognizing that there are now essentially *two* classifications for accuracy:

- An *internal* one based on the minimally constrained adjustment of the GPS-only survey.
- An *external* one based on a constrained adjustment, where the coordinates of existing control stations of the GPS network are held fixed (or constrained to some degree) to their published values.

In Australia and New Zealand, the former corresponds to a survey's *class*, while the latter defines its *order* (DOSLI 1996b; ICSM 2000). In the case of the US, the former refers to the classification of the survey according to a *local accuracy* measure, while the latter is the classification according to a *network accuracy* measure.

For each category of survey accuracy the S&S generally recommend different practices for:

- Network design

- Instrumentation and survey technique
- Field procedures and monumentation
- Office reduction procedures
- Calibration and result validation procedures

Note that different countries not only may have a different classification system for accuracy, but also different recommended practices for each class of GPS survey.

12.2.2. Quality management

Though 'quality' is nowadays a much abused buzzword, there is nevertheless an unprecedented interest in measuring, assuring, verifying and improving the quality of products and services. The language of *Quality Management* (QM) is replete with terms such as *Quality Control* (QC) and *Quality Assurance* (QA). Although the two terms are often considered synonymous, a good working definition is that: (a) *QA* refers to the set of recommended practices and procedures that maximize the chances that the product or service will satisfy the client's requirements, at an agreed to cost, while (b) *QC* refers to the procedures used to monitor and verify the level of quality achieved, and if it is inadequate, to detect the source of the problem and, if possible, remedy it.

QM is a management *system* which provides a framework for a consistent approach to managing *all* aspects of an organization's operations. The intention is to 'get it right first time, every time', rather than the historical approach of 'let's check if we got it right – if not let's fix it up'. However, perhaps the defining characteristic of QM is that it promotes an organizational *culture* in which all quality procedures are subject to continual scrutiny and improvement, supported by a management system which encourages identified improvements to be put into practice. Although 'rules-of-thumb' (and commonsense) generally suffice when it comes to GPS project planning and execution, many national and state survey authorities have also developed guidelines and recommended practices for GPS positioning.

The success of a survey in meeting the client's specifications is more assured if QC procedures are incorporated from the very start. When GPS is used as a survey tool, *quality control* procedures and measures are little different to those employed during conventional surveying. The basic principles of incorporating *redundancy*, *repeat or check measurements*, design of *closed loop configurations* and careful *network adjustment* are all applicable to GPS surveys as well. Many of these have been specified in the appropriate S&S documents as a function of the level of accuracy sought (Section 12.2.1).

All projects can assume to have been initiated and planned by the 'campaign manager' or 'project manager', usually an experienced surveyor/contractor, who is responsible for carrying out (or at the very least supervising):

- The survey planning, the technical as well as the administrative tasks such as contacting the relevant authorities to gain station access permission; booking accommodation and transport; investigation of facilities such as communications, power supplies (for computers and receivers); etc.
- The pre-mission tasks such as logistical planning; testing and validating equipment; reconnaissance; site preparation; briefing of staff; etc.

- The day-to-day management of the field parties, including preparation of contingency plans in the event of instrument failures, etc. and implementing changes to the observing schedule.
- The supervision of the field processing of GPS data to check data validity; perform a preliminary network adjustment as the survey progresses; recommend alterations to the schedule in order to re-observe stations; observe new points; etc.
- Overseeing the data processing, quality control and final network adjustment; report writing; and ensuring that the product delivered to the client is according to the agreed specifications.
- Archiving of data and data records.

12.3. Selecting the appropriate technique

One of the most important planning issues is selecting the *appropriate* GPS survey technique from the range listed in Table 12.1. Factors such as capital cost of the equipment, occupation time-per-point and mode of processing, will affect the economics of any particular survey project. The final factor, vulnerability to poor sky visibility, may rule out certain observation techniques in environments where GPS signals are obstructed. Note that many of these considerations are in fact addressed in S&S documents and in the GPS manufacturer's user manuals – another reason why the surveyor/contractor should be familiar with the recommended practices for each GPS technique.

12.3.1. Distance from reference station

A primary concern is the distance of the user receiver from the reference station. Ideally, baseline lengths should be as short as possible. As the inter-receiver distance increases, more sophisticated (and expensive) techniques must be adopted. For example, DGPS is the primary tool for mapping-type surveys out to distances of 100 km, but for longer distances the WADGPS technique should be used. In the case of carrier phase-based positioning techniques, RTK-GPS at the 1–2 cm level is only viable for baseline lengths less than 10 km (and success at this range depends very much on the reliability of the radio link, as well as a host of other factors). On the other hand, static GPS survey techniques are the most robust of the high precision techniques, especially if the baseline lengths are longer than several tens of kilometres.

S&S generally give guidelines as to which techniques can be used for different length of baseline and different accuracy levels (Section 12.2.1). As a general rule, to achieve high accuracy over greater distances requires longer observation spans, with distances beyond about 20 km being particularly problematic using commercial GPS data processing software.

12.3.2. Accuracy versus cost

Ideally, the client should specify the accuracy requirement for a survey project. Table 12.1 lists the viable techniques to achieve requisite accuracies, although it is emphasized that care must be taken when using a GPS technique close to the limits of the specified accuracy specifications. (S&S tend to be conservative, and give guidelines for use of particular GPS techniques that have relatively wide safety margins.) For example, while most DGPS systems can easily guarantee 5 m accuracy, attaining 0.5 m or better will require carrier phase-tracking GPS hardware and the appropriate software. It is also important to recognize that this stated accuracy assumes good sky visibility and satellite geometry. For all techniques, achievable confidence limits can degrade in environments which could be classed as 'GPS unfriendly', such as obstructed sky view and unfavorable multipath environments.

Problems can arise when a client, with very little expertise in positioning or GPS performance, does not actually know his or her own requirements. Current GPS technology offers obvious choices in the 0.5–5 m accuracy range (DGPS, WADGPS) and less than 0.05 m accuracy range (RTK, static, rapid-static). However, the 0.05–0.5 m accuracy 'gap' poses special problems. This anomaly is due to the vast difference in raw measurement precision of the pseudorange vis-à-vis carrier phase, and the consequent differences in processing procedures. Carrier phase-based techniques 'blow-out' in accuracy to the sub-meter level only when the ambiguities cannot be resolved using the data that has been collected. In such cases, an 'ambiguity-float' solution may be sub-optimal (in that it *fails* to deliver cm-level accuracy), but could still be acceptable for certain types of medium accuracy mapping surveys. The best strategy may be to only consider GPS techniques which can be relied on to supply results that are *more* accurate than the client's specifications. For example, use carrier phase-based techniques for any specification calling for an accuracy higher than 0.5 m. However, if cm-level accuracy must be *assured*, the survey will require special care (and could be significantly more expensive).

The main factors influencing the cost of a survey are: capital cost of equipment, number of points to be surveyed, time taken to complete the necessary fieldwork, time to process data and evaluate results. In general, the *faster and more accurate* a survey, the *more expensive the hardware is but with lower operational costs* in terms of manpower and field expenses. Capital costs of GPS hardware required for different techniques are indicated only in a relative sense in Table 12.1. Often, the choice may be between hiring equipment for the project and purchasing equipment for a range of projects.

12.3.3. Occupation time-per-point

Generally, the quality of a computed coordinate increases with longer occupation time. However, because of its impact on 'productivity', the length of time a GPS receiver must physically occupy a single point will significantly impact on the total cost of the survey. Therefore, observing a large number of points, each for long periods, is the most expensive GPS option. Only WADGPS and DGPS techniques offer truly continuous point 'pick-up', but at comparatively low accuracy. Single- and dual-frequency kinematic and static surveys offer continuous point 'pick-up', so long as the receiver maintains signal lock while moving from point to point. The necessity for initialization and re-initialization, to resolve the carrier phase ambiguities, can significantly increase the time taken to complete these types of survey, even with modern OTF-AR algorithms. RTK-GPS is the most accurate real-time fast pick-up technique, but RTK typically requires a minimum of a few seconds at each point (not including time for OTF-AR if it is required due to loss-of-lock of the signal as the receiver is moved from point to point). RTK can be run in continuous mode but with a lessened reliability.

Of the precise static survey options, *rapid-static GPS* is the most commonly used technique for measuring baselines less than 20 km in length. The rapid-static technique is more reliable than RTK in terms of quality control, but considerably less productive as the observation period at a point may be 10 min or more. For high accuracy surveys beyond 20 km from the reference station, conventional static GPS techniques still are preferred, with station occupation times varying from 30 min to several hours depending on the accuracy required and the application. Recommended station occupation times are given in the GPS manufacturer's manuals, but be aware that S&S guidelines typically give more conservative (i.e. longer) times. It must be emphasized that both receivers defining the baseline to be surveyed must track the same satellites for the same period of time, as it is only the common data epochs that can be processed. Furthermore, both receivers must be configured to make measurements at the same 'data rate', e.g. every 15–120 s (the longer interval for static techniques, the shorter interval for 'stop-and-go' GPS), to ensure that simultaneous measurements are made by both receivers. Rates as fast as 1 s, or even faster, are used for many kinematic applications.

12.3.4. Processing mode

Techniques that rely on real-time data processing imply that the receiver in the field computes the position and associated quality information. Post-processing techniques rely on a receiver logging raw data, which is then downloaded into a data processing software package. Both modes have their advantages and disadvantages.

Post-processing allows more flexibility in terms of data 'manipulation' and control over the processing options, hence the results tend to be more reliable. Sophisticated data post-processing software can ensure greater quality control, as there is the opportunity to 'snoop' forwards and backwards through the data looking for outliers and low quality data. Real-time systems, on the other hand, have access only to prior data (they have no knowledge of future data), and a limited amount of processor time in which to deliver the solution. Therefore, internal real-time quality indicators must be estimated based on relatively scant supply of data. Furthermore, in post-processing, the user may view data at leisure and identify and edit 'bad data' manually, rather than relying on an automated data (pre-)processor. This approach can be time consuming, and also relies on the analyst being experienced in GPS data processing. However, the main attraction of real-time processing is the speed at which points can be surveyed and the coordinates determined (some applications such as engineering stake-out require the coordinates straight away). In addition, coordinate data and feature attributes from standard mapping surveys can be logged in real-time and, post-survey, can be downloaded directly into GIS or data visualization software.

12.3.5. Dilution of Precision (DOP)

As a rule, the shorter the occupation time-per-point, the more critical the number of satellites and the satellite geometry will be on the quality of the coordinate solution hence careful attention needs to be made to this issue. The sensitivity of a coordinate solution to poor geometry (i.e. high DOP – Section 9.3.3) decreases as more data are collected. Therefore, conventional static surveys are least sensitive to low numbers of visible satellites and/or poor DOP, while real-time GPS techniques such as DGPS, WADGPS and RTK-GPS are more sensitive. RTK-GPS surveys, requiring a minimum of five or six satellites for ambiguity initialization (or re-initialization if lock is lost) are the most vulnerable to poor geometry. S&S may specify maximum DOP values and the minimum number of satellites for particular GPS techniques or applications.

For a full 24 satellite constellation, the (DoD 1993) document states that, with a elevation masking angle of 5°, six or more satellites will be visible for more than 99.9 per cent of the time. Typically, seven, eight or nine satellites are available for 90 per cent of the time. Hence even the removal of one satellite does not compromise a situation in which at least five satellites are in view almost continuously and, most of the time, at least two or more combinations of four satellites will support a Position DOP (PDOP) of six or less. However, for most survey work, a 15° elevation mask is specified, and the loss of improved satellite geometry caused by raising the elevation mask angle from 5° to 15° is offset by removing noisier data (contaminated by atmospheric and multipath effects from those low-elevation satellites).

12.3.6. Poor sky visibility

Sky visibility is a major influence on the appropriateness of the chosen survey-type, and Table 12.1 gives an indication of how vulnerable each measurement technique is to poor observing conditions. All 'high productivity' positioning methods (rapid-static, stop-and-go, kinematic) are less reliable in sub-optimal observing conditions, such as encountered in urban and wooded areas, than static survey techniques. The reason is that visibility (and multipath interference problems) can, to a significant extent, be overcome by occupying the points for longer time periods. To ensure maximum flexibility and adaptiveness to changing visibility conditions, consider the use of a combination of different survey techniques on the project.

Is a 'clear sky' scenario realistic? The planning phase of any GPS survey should include an assessment of the environment in which the survey is to be undertaken through site reconnaissance (Section 12.5.7). If many points are to be surveyed, obviously the quality of sky visibility can vary greatly from point to point. It should always be kept in mind that the stark differences between a constellation viewed in clear sky, and that which can be tracked in an urban area, can greatly reduce the efficiency of GPS in such environments. Commercial mission planning software can be used to identify times when the geometry of the visible satellite constellation is poor *if the true pattern of satellite signal masking is known a priori at each point to be surveyed*, so that these times of degraded coverage can be avoided.

12.3.7. Multipath disturbance

Multipath can severely degrade the accuracy specifications indicated in Table 12.1. Multipath is often associated with poor sky visibility (tall buildings and other obstructions typically being sources of multipath), but can also be severe if an antenna is mounted on an aircraft, on vessels or vehicles. To some extent (but not completely), multipath can be averaged out by longer time periods of static observation. Therefore, fast positioning techniques are most sensitive to multipath. However, multipath can cause significant degradation for *all* types of GPS positioning, and some effort to find relatively multipath-free sites should be made in the planning phase of the project.

12.4. The issue of quality control in project design

Because many QC practices require extra work, the final network-cum-project design is usually a *compromise* between technical requirements and economics, worked out within the framework of explicit recommended practices or S&S for GPS surveys (or at the very least, prudent practices) that ensure the job gets done to the appropriate standards of accuracy and reliability (Section 12.2.2). In other words, sufficient redundant measurements should be made to ensure adequate QC, but not such an increased survey effort that would significantly raise the cost of the survey. There is, however, some confusion as to what is meant by 'quality control' in the context of GPS surveys.

12.4.1. Internal quality control indicators

Internal QC indicators are those provided to users by the firmware in GPS receivers or by GPS post-data processing software. The user will usually have little influence over how these indicators are generated, or may even be unaware of how the indicators are computed, and will therefore have to accept them at face value. Examples of internal quality indicators are: a receiver stating that ambiguities have been fixed in a RTK-GPS survey; a receiver observing for a set amount of time in a rapid-static survey and then indicating to the user that a sufficient amount of data have been collected to estimate a reliable solution; the standard deviations of the final baseline vector from a post-processed static survey; and so on.

In these examples, software must rely on internal statistical analysis tools to assess the quality of the data, and hence the quality of the final solution. But there are limits to what these tools can indicate. For example, if an incorrect height of antenna in a RTK-GPS survey is recorded, the processing software will not be able to identify such a problem, even though all coordinates computed during the survey will be incorrect. The software operates using a specified set of statistical parameters which, in general, work quite well. However, situations do arise when such parameters may no longer be valid, because they do not represent the prevailing conditions particularly well. In such cases, internal quality indicators may not detect whether GPS surveys are producing spurious results. This is especially the case when validating whether the resolved ambiguities are in fact correct. Final coordinate standard deviation values generated by the baseline determination software are an internal precision measure (how well the GPS data fits a pre-specified mathematical model), rather than an accuracy measure (how well the final solution matches the true coordinates). It is assumed all systematic errors have been removed and certain a priori assumptions about the quality of the raw observations, and in kinematic cases, the dynamics of the receiver, have to be made. Therefore, internal quality indicators cannot to be assumed to be an indication of the true accuracy of a position solution. However, these quality indicators are not necessarily incorrect. Indeed, they are excellent for indicating problems with the data, but their reliability is inadequate for assessing the quality of a GPS survey.

12.4.2. External quality control indicators

External QC indicators are a form of *independent* check on GPS solutions. They are in fact the only valid indicators for ensuring survey specifications have been met, and therefore should be factored into the GPS survey planning process. Most S&S documents will set out guidelines for external QC indicators, and define the tolerances that must be met if the GPS survey is to be accepted as having satisfied some specified level of accuracy. The main procedures for testing GPS baseline solutions are: (a) repeatability, (b) polygon misclose computation, and (c) comparison with known control station coordinates. In practice, the applicability of each of these procedures depends mainly on the type of GPS technique chosen for a particular project and how the receivers are deployed in the field to carry out the survey. This is discussed below in the context of two basic types of GPS survey, one for establishing control points, and the other a form of survey for fast pick-up of features.

12.4.2.1. Quality control for monumented surveys

Any type of static survey will invariably involve placing an antenna over a monumented point. The standard survey practice of observing a redundant network of baselines connecting unknown points to a series of known control stations should be observed. External QC indicators are then derived from: (a) comparison of individual baseline results which have been measured more than once, (b) linking several independent baselines into a closed polygon and determining the extent of the misclose (in three dimensions), and (c) a three-dimensional least-squares adjustment. The latter not only permits the adjusted coordinates of some points to be compared to their known values (if care is taken to ensure that the datum of the GPS survey and that of the known control stations is the same), but also permits the horizontal and vertical error ellipses to be analyzed. Often S&S documents will define the maximum allowable size of error ellipses for a certain class (or order – or whatever the terminology used) of survey. When planning a static control survey, the survey of a 'reasonable' number of redundant (i.e. extra) and repeat baselines must be observed. Although this is usually specified in S&S documentation, a rule-of-thumb is, for example, to ensure that 10 per cent of baselines be resurveyed and that each point be connected to a minimum of three independent baselines.

12.4.2.2. Quality control for fast pick-up surveys

The nature of fast pick-up surveys tends to be somewhat different to the monumented surveys referred to above. Although some of these survey points may be monumented (or, at the very least, located at revisitable locations), the majority of points are likely to be unmonumented. This poses a challenging QC problem as standard least squares adjustment techniques cannot be used at most points. If a point is unmonumented, it is impossible to return to the precise location of the initial survey, and therefore it is not possible to make secondary (redundant) observations. Reasonable QC techniques for fast pick-up surveys include: detecting outlier results by using a priori knowledge of the geometry of the surface or feature being surveyed; forming polygons whose misclose can be checked; using a second, independent positioning technique (e.g. a second reference station); ensuring that the observation conditions are good (e.g. maintaining PDOP < 6 and track at least five satellites).

12.5. Project planning tasks

Planning can be considered as comprising the following tasks:

1 *Project Design*: involves project layout and network design, and is driven primarily by accuracy and station location/density requirements (defined by the client), productivity/economic considerations (of concern to all parties involved), and S&S (promoted by the state or national survey authority).
2 *Observation Schedule*: giving consideration to such factors as the number of GPS receivers, the GPS technique used, occupation time per point, number of points to be surveyed per day, requirement for multiple point occupancy, etc.
3 *Instrumentation and Personnel*: instrumentation appropriate for the GPS technique being used, which, if not available in-house, could be hired. In addition, trained personnel are required in order to carry out the fieldwork and to process the data in a professional manner.
4 *Logistical Considerations*: issues such as transportation and receiver deployment strategies (appropriate to ensure observation schedule can be adhered to), special site requirements (e.g. power availability, station intervisibility, etc.), and factors related to network design and QC such as multiple point occupancy, etc.
5 *Reconnaissance*: which may or may not be necessary, depending upon how critical the GPS points are to be to the overall project, whether permanent marks will be established, the GPS technique that will be used, etc.

Note, it is not always necessary to have elaborate project plans. For example, if the intention is merely to coordinate a large number of points from a single reference station, then one receiver needs to be setup at a point of known location while the other visits each point in turn, but perhaps incorporating the appropriate QC procedures (Section 12.4.2.2).

The earlier discussion in this chapter was concerned with the complex relationship between the GPS survey technique that is appropriate for a project, S&S to be considered (and crucial issues that arise), and quality control procedures that should be incorporated within the planning process. Additional planning issues that must be addressed include:

- The receiver deployment strategy to be used
- The placement of survey marks
- The datum for the survey and the occupation of existing control stations
- The transformation of the height information
- Site reconnaissance
- Sky visibility
- Training, and appropriate field and office procedures.

Some of these will be considered also in Chapter 13, where the execution of typical GPS surveys in the US will be discussed from the practical point of view, and in the context of good survey practice.

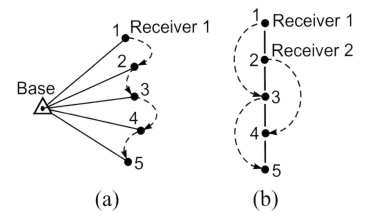

Figure 12.1 Receiver deployment strategies, (a) 'radiation' mode, (b) 'traverse' mode.

12.5.1. Receiver deployment

There are a number of possible receiver deployment schemes that can be used. Each has its advantages and disadvantages with respect to logistical considerations such as cost, time, manpower, etc. Generally, some reference station (or stations) is (are) occupied for all or some of the project, and the other receivers move between static sessions to predefined points, or points may be occupied on a random basis during RTK-GPS or 'stop-and-go' surveys. A combination of the 'reference station' or 'radiation' mode, and the 'leap-frog' or 'traverse' mode, as shown in Figure 12.1, is usually used. Static or rapid-static surveys typically use the 'leap-frog' mode of deployment, while the RTK-GPS, kinematic and stop-and-go techniques are best used in the 'radiation' mode.

There is a danger in the *overuse* of the 'radiation' mode of survey because, unless all the points are visited more than once, all *radiated* baselines are in reality 'no check' baselines. It is tempting to consider a two-reference station configuration as having a natural *built-in* redundancy (there are two baselines coming into each point, one from each of the reference stations). However, although the point is no longer fixed by a single 'no check' baseline radiation, the additional baseline from the second reference station *may* not be considered truly independent as, for example, the same error in height of antenna or station misidentification can still be made.

12.5.2. Placement of marks

One of the significant advantages of the GPS survey technique over conventional surveying techniques is that surveyed points may be placed *where they are required*, irrespective of whether intervisibility between stations is preserved. Generally, the GPS points would be expected to be clustered around the project *focus*, for example, a road, dam, powerline corridor, etc. This is in contrast to a traditional geodetic control network that was generally evenly spaced, and the control stations located in prominent locations such as at the tops of hills to ensure that they were visible from afar. Nevertheless, some intervisibility of stations may be necessary to permit the use of

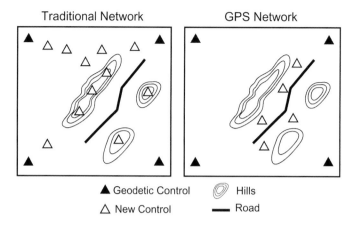

Figure 12.2   Terrain need not influence point selection.

conventional terrestrial survey techniques as well. In addition, extra survey stations that 'carry in' the control from the nearest geodetic control stations to the project area are not usually necessary for GPS work. *Hence even spacing of stations and selection of stations on the basis of terrain are no longer important considerations* (Figure 12.2). Monumentation guidelines are generally provided by the national or state geodetic agency.

12.5.3. Datum issues

Of critical importance at the planning stage is to ensure that the datum to which the results of the survey are related is the one required by the client. The first stage is to ensure that coordinates of all control stations being used in the survey are known and, importantly, that the datum they are associated with is also known. It may be necessary to execute a final transformation to change the surveyed coordinates to the required datum and coordinate type (e.g. E, N, H or lat, long, height) using the appropriate transformation parameters (Chapter 4). The number and distribution of known control stations, and the accuracy with which the coordinates of the known stations are required is strongly dependent on the use to which the known stations will be put. Guidelines can usually be found in the relevant GPS S&S. For most purposes a minimum of three or four control stations located around the perimeter of the project area is sufficient.

12.5.4. Survey existing control stations

Some reasons for surveying both existing (i.e. known) control stations and new points are:

- Required by the relevant GPS survey S&S
- For the determination of local transformation parameters between the GPS datum and the local control datum

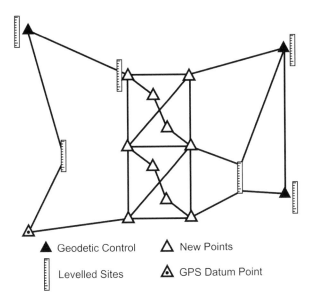

▲ Geodetic Control △ New Points

▌ Levelled Sites △ GPS Datum Point

Figure 12.3 A GPS network linking existing control stations, new sites and the GPS datum
station.

- For quality control purposes
- In order to determine the geoid-ellipsoid separation, and
- In order to connect new GPS points into the surrounding control network.

However, a minimum of one known station must be used as the datum station in the
GPS survey. The control station coordinates may be in the GPS satellite datum (Section
7.2) or a local geodetic system (Section 3.1). Furthermore, only the horizontal components may be known, or their leveled height, or both, as indicated in Figure 12.3.

12.5.5. Height reduction

It is often necessary to reduce the GPS ellipsoidal heights to orthometric heights
(Section 13.2.4). This reduction requires that geoid information for the local area
be available, or at the very least there are some stations which are surveyed in the
project area which also have leveled (or orthometric) heights. If several stations have
leveled heights, and these points are surveyed using GPS to determine their ellipsoidal
heights as well (relative to the datum station), then the difference in the two heights at
these stations is a local estimate of the geoid-ellipsoid separation, or geoid height
(Gnipp 2000).

12.5.6. Site reconnaissance

Reconnaissance issues for GPS surveying include:

- *Satellite availability*: satellite selection, satellite health, observation window, etc.

- *Satellite visibility*: checking on-site obstructions.
- Clearly identifying the groundmark over which the GPS antenna is to be set up.
- Identifying, if necessary, eccentric stations to be occupied if the primary ground-mark cannot be used, and other azimuth stations if required.
- Station access: critical for minimizing non-productive travel times and unscheduled delays in getting on-site.
- Site conditions: on-site power? multipath environment?

Information on station access should be clearly stated in words as well as described in some graphical form. This is critical for minimizing down-time due to difficulties in finding stations, or if access involves caretakers (e.g. to visit the roofs of buildings), etc. As with details of the station access and point description, the area around the site should be studied carefully. Depending on the objective of the project, and its accuracy requirements, this task may be very elaborate and include, for example:

- Investigating the provision of on-site power.
- Testing soil stability and defining the appropriate antenna mount (tripod, pillar, etc.).
- Noting the presence of any potential multipath causing structures.
- Noting any UHF, TV radio, microwave, cellular phone or radar transmitters as they could affect a receiver's operation.
- Establishing permanent monumentation – using previous marks helps avoid this.
- Establishing nearby azimuth marks.
- Clearing the area of possible obstructions caused by trees or shrubbery.
- Taking photographs of the surrounding area, including any tree cover.

In some countries, account may have to be taken of the season. For example, tree cover may be much thicker during summer than at other times of the year.

12.5.7. Assessing sky visibility

To schedule a GPS survey the following factors need to be taken into account:

- Satellites are not normally tracked below an elevation of 15°–20° due to large atmospheric refraction errors at low elevation angles.
- There is a 24 hour observation window for GPS.
- The length of observation session depends on the technique being used.
- The satellites' positions in the sky are predictable. They can be computed and output in a convenient graphical form, and taken out into the field during reconnaissance.
- Satellite availability is best visualized in the form of a *skyplot*, which is a graph of satellite tracks on a zenithal projection centered at the GPS ground station. The satellite azimuth and elevation is shown as a function of time and can be computed using standard GPS 'mission planning' software.

How to best use skyplots during reconnaissance? The obstructions at a site can be plotted on a zenith plot (or a photograph of the zenith could be taken with a camera fitted with a fish-eye lens!) and compared with the skyplot. Skyplots annotated with possible obstructions are also useful during actual data tracking because if the data quality is found to be poor, the possible source of interference can be identified and an appropriate note made to assist subsequent data processing. *How may the obstructions be mapped during reconnaissance?* The basic tools are a compass and a clinometer, and every field party should be equipped with them. However, if the site location has not yet been fixed with certainty, as may be the case with sites in urban environments, the reconnaissance assumes greater importance. As there may be many candidate locations for each site, some shortcut techniques should be considered (Santerre and Boulianne 1995). For example, keep several graphs handy that relate distance to and height of a potential obstruction, to the elevation angle that it subtends. Efficient reconnaissance techniques can be developed if the heights of buildings and trees are known, and distances are easily measured using, for example, a handheld laser range finder. However, the pattern of signal obstructions is very sensitive to the height of the antenna. By increasing the antenna height in urban areas using some form of telescopic pole, the extent of signal obstruction may be greatly reduced.

12.5.8. Training and appropriate procedures

Although many of the procedures followed for GPS surveys are similar to those for conventional surveys, it must nevertheless be emphasized that field and office staff must be well trained and familiar with the various procedures that are to be followed. Under QM principles (Section 12.2.2) procedures should be clearly documented, including those concerned with:

- Field documentation such as logging sheets, instructions, supporting maps, etc.
- Equipment handling procedures such as GPS receiver setup and demount, antenna height measurement, battery care, etc.
- On-site procedures such as monitoring of GPS data collection, eccentric station surveys, monumentation, ties to conventional surveys, etc.
- Data download and GPS data handling procedures.
- Computational procedures, including baseline and network processing, height and datum transformation, and report writing.
- Equipment testing as required by S&S, as well as normal maintenance and care.

The 'testing' of GPS equipment or the 'qualification' of GPS surveyors are just some of the proposals that have been suggested (Section 13.4.4). However, this manual is not a 'cookbook' for executing GPS surveys. The intention is to draw attention to the various factors that must be considered when planning and executing GPS projects. The next chapter will illustrate how many of the issues raised in the earlier sections are addressed by US surveyors in typical GPS survey and mapping applications.

References

CGCC, *Specifications (Proposed) for High-Production GPS Surveying Techniques*, California Geodetic Control Commission, 1993, 16 pp.

Craymer, M., Wells, D. E., Vanicek, P. and Devlin, R. L., 'Specifications for urban GPS surveys', *Surveying and Land Information Systems*, 50(4), 1990, 251–259.

DoD, *Global Positioning System Standard Positioning Service Signal Specification*. US Department of Defense, 1993, 46 pp. + appendices.

DOSLI, *Good GPS Survey Practice Guidelines*. Department of Survey and Land Information, New Zealand, 1994, 27 pp. + appendices.

DOSLI, *New Zealand Geodetic Survey Specifications: GPS Surveys*. Department of Survey and Land Information, New Zealand, 1996a, 16 pp.

DOSLI, *New Zealand Standards of Accuracy for Geodetic Surveys*. Department of Survey and Land Information, New Zealand, 1996b, 16 pp.

Gnipp, J., 'A short, practical guide for the (American) land surveyor: vertical GPS survey standards', *GIM International*, 14(5) 2000, 36–37.

GSD, *Guidelines and Specifications for GPS Surveys*. Geodetic Survey Division, Geomatics Canada, Natural Resources Canada, release 2.1, 1992, 63 pp., available at http://www.geod.nrcan.gc.ca/products/html-public/GSDpublications/GPS_guidelines/English/Guidelines.pdf.

FGCC, *Standards and Specifications for Geodetic Control Networks*. Federal Geodetic Control Committee (now the Federal Geodetic Control Subcommittee of the Federal Geographic Data Committee), National Ocean and Atmospheric Administration (NOAA), 1984, 31 pp.

FGCC, *Geometric Geodetic Accuracy Standards and Specifications for Using GPS Relative Positioning Techniques*. Federal Geodetic Control Committee (now the Federal Geodetic Control Subcommittee of the Federal Geographic Data Committee), National Ocean and Atmospheric Administration (NOAA), version 5, 1989, 48 pp.

FGDC, *Geospatial Positioning Accuracy Standards, Part 2: Standards for Geodetic Networks*. Federal Geographic Data Committee, National Ocean and Atmospheric Administration (NOAA), 2001, available at http://www.fgdc.gov/standards/status/sub1_2.html.

ICSM, *Standards and Practices for Control Surveys*, Australian Inter-Governmental Committee on Surveying and Mapping, version 1.4, special publication no. 1, 2000, available at http://www.anzlic.org.au/icsm/publications/sp1/sp1.htm.

NGS, *Guidelines for Establishing GPS-Derived Ellipsoid Heights (Standards: 2 cm and 5 cm)*. National Geodetic Survey Tech. Manual NGS-58, available at http://www.ngs.noaa.gov/PUBS_LIB/NGS-58.html 1997.

Santerre, R. and Boulianne, M., 'New tools for urban GPS surveyors', *GPS World*, 6(2), 1995, 49–54.

Schinkle, K., 'A GPS how-to: conducting highway surveys the NYSDOT way', *GPS World*, 9(2), 1998, 34–40.

Chapter 13

Carrying out a **GPS** surveying/ mapping task

Mike Stanoikovich and Chris Rizos

GPS works differently from the techniques based on optical instrumentation that surveyors have used for decades. Those who do not understand the limitations of the equipment, or are unaware of the field and office procedures that must be followed, can misuse Global Positioning System (GPS). If used incorrectly GPS can mean 'Good Positioning Sometimes', yielding erroneous coordinates that could prove very costly to both the surveyor and the client. Chapters 11 and 12 have dealt at length with the GPS techniques available to the surveyor, with particular emphasis given to issues that influence the *quality* of outcomes. Hence, in addition to discussing the characteristics and limitations of the various GPS techniques, the authors have also dealt with such topics as GPS Standards and Specifications (S&S), the concepts of quality control, and best practice guidelines for planning and executing successful GPS surveys. In this chapter the lessons of earlier chapters are applied to some typical GPS surveys. Although the examples are drawn from the US, many of the principles have general relevance to the practice in other countries.

A GPS surveying project usually begins with a client approaching a geospatial professional to generate a set of coordinates or to prepare a map using the GPS technology. In the US, most states require that GPS measurements used to provide geographic coordinates for surveys, mapping, or for Geographical Information Systems (GIS) databases must be performed under the direct supervision of a licensed land surveyor in the state where the work is being performed. Four GPS survey/ mapping tasks are examined in this chapter:

- Using GPS to establish geodetic control networks
- Using GPS for photogrammetric mapping and engineering design projects
- Using GPS to conduct a utility infrastructure inventory; and
- Using GPS to survey heights in the local vertical datum

Before proceeding it is necessary to consider in a little more detail (than given in Section 12.2) the GPS S&S for surveys in the US.

13.1. US Standards and Specifications for **GPS** relative positioning techniques

National S&S exist for static GPS (FGCC 1989). However, because of the age of these S&S, the US currently lacks national guidelines for, e.g. kinematic or real-time carrier phase-based GPS techniques. As a result it is the responsibility of professional surveyors to develop 'internal' guidelines that satisfy the positional requirements of their projects. Any S&S (or internal guidelines where no official S&S exist) must be considered side-by-side with the various accuracy standards for US surveys. Until recently the survey accuracy standards were defined by the document (FGCC 1984), in terms of the relative accuracy of network points (and therefore dependent on the distance between neighboring control points). There is currently a transition under-way to a new set of standards based on coordinate accuracy, i.e. 1, 2, 5, 10, 20 cm (FGDC 2001). The US National Geodetic Survey (NGS) is the federal agency respon-sible for establishing and maintaining the National Spatial Reference System (NSRS) – the national database of geodetic control stations (NSRS 2001). The *core* group of stations that make up the High-Accuracy Reference Network (HARN) was surveyed by NGS and other state, local, and private agencies, using GPS techniques. The NGS is also responsible for defining the standards for many other aspects of surveying, for example with regards to establishing permanent monumentation (NGS 1978).

In many states, the particular state's Department of Transportation (DOT) has adopted its own set of surveying standards. Many of these include specifications for GPS positioning (Schinkle 1998). Before performing work within a state in the US, it is important to contact the DOT to obtain relevant surveying and GPS standards. Some states' DOT, for example, do not accept elevations produced by GPS and require that elevations be established using standard differential leveling techniques. While most municipalities in the US have established surveying standards for cadastral surveys, most have not defined GPS specifications. Therefore, it is the surveyor's responsibility to set GPS quality standards that meet the positional requirements of a project. Gener-ally, by default, the national specifications for GPS positioning techniques are applic-able.

In Section 7.4.1 four *generic* survey accuracy categories were defined in order to illustrate how different surveying and mapping applications could be addressed using the GPS technology. The various national geodetic authorities typically use a variety of terminology for the *classes* of GPS survey, and have different numerical accuracy limits to define the categories (Section 12.2.1). Below is a brief overview of the US survey categories and the associated surveying/mapping applications to which they relate. Many state, county, and local government agencies refer to these specifications in their *Request for Proposals* (RFP).

13.1.1. Highest accuracy geodetic/scientific surveys

These survey categories were known as Order A (1:1,000,000 accuracy) and Order B (1:500,000 accuracy) in the older S&S (FGCC 1984), and generally correspond to the 1 cm and 2 cm categories in the new standards (FGDC 2001). Such surveys are intended mostly for establishing monitoring stations to measure tectonic plate move-ment, and for various scientific purposes. In addition, such surveys are used to coor-dinate the HARN stations at the federal and state level. Surveyors establishing these networks must follow the S&S published by (FGCC 1989).[1] These S&S are very strict, requiring, for example, dual-frequency receivers, long static observation sessions, the measurement of redundant baselines, and strict guidelines for monitoring weather

---

[1] To be soon superseded with new specifications (personal correspondence by author with NGS).

conditions and other factors. Obviously this method of GPS surveying is the most expensive technique and should be used only when such very precise positioning accuracy is required. See Section 13.2.1 for further details concerning such surveys.

13.1.2. First-order surveys

Such surveys are used mostly for establishing local area control networks, which in the old classification required a relative accuracy of the order of 1:100,000. These can be used as a method of *densification* of the HARN by state and local agencies. In addition many city and county governments use this type of control network as the foundation for creating a GIS. Both static (Sections 12.1.1 and 12.1.3) and rapid-static (Section 12.1.2) techniques can be used to establish such control networks. Static surveys can be performed using either single- or dual-frequency receivers by following all (FGCC 1989) procedures. Rapid-static techniques must be performed using dual-frequency receivers.

13.1.3. Control surveys

These surveys are used primarily to establish photogrammetric control, for engineering projects, engineering design, and construction layout projects, with a relative accuracy requirement ranging from 1:50,000 to 1:10,000 (old standards), or one or more decimeter coordinate accuracy (new standards) in relation to the NSRS. This type of survey network is typically used as the reference network for any subsequent stop-and-go (Section 11.1.2), kinematic (Section 11.1.3) or Real Time Kinematic (RTK) (Section 11.2.2) GPS surveys. Although there are no explicit national S&S for kinematic/RTK-GPS positioning, many RFPs now contain language directing the surveyor/contractor to use these survey techniques for such surveys. Care must therefore be taken to explain to the client that such procedures are *high productivity* techniques that have a larger risk associated with them than the static or rapid-static techniques. It may be necessary to use appropriate quality control procedures during the survey (Section 12.4).

13.1.4. Orthometric heights

NGS S&S for obtaining orthometric heights exist for staff-based leveling techniques using either conventional geodetic levels or the digital electronic barcode level:

*   Geodetic Leveling: National Ocean and Atmospheric Administration (NOAA) Manual NOS NGS 3 (August 1981); and
*   Interim Federal Geodetic Control Subcommittee (FGCS) Specifications and Procedures to Incorporate Electronic Digital/Bar – Code Leveling Systems, Version 4.0 (July 1994).

No national specifications exist for obtaining orthometric heights using GPS techniques. However, guidelines have been developed for GPS-derived ellipsoidal heights (NGS 1997). It is recommended that surveyors who want to determine orthometric heights using GPS should use the geoid model *GEOID99* to obtain separation values between the ellipsoid and geoid at a points of interest (NGS 2001a). Though this geoid model is claimed to be accurate to a few centimeters, it is important to correct for the local datum shift by surveying several benchmarks that have orthometric height values, following the procedure described in Section 13.2.4. Many countries have now computed such geoid models to assist those wishing to derive orthometric heights from GPS surveys (e.g. in Australia the geoid model is referred to as *AUSGeoid98* – AUSLIG 2001). Such a technique cannot, of course, be used in those countries where a precise gravimetrically-determined geoid model is not available. An alternative to using a geoid model is to determine the offset between the GPS height datum and the leveled height datum using the *geometric* technique described in Section 13.2.4.

13.2. Surveying/mapping applications: guidelines for GPS

13.2.1. Application 1: using GPS to establish geodetic control networks

In 1987 the NGS began establishing the HARN for the US states and territories, with a station spacing of about 75–125 km (the spacing varies from state to state). At present there are over 3,700 HARN stations, all having horizontal and vertical (orthometric and ellipsoidal) values to the highest accuracy standard (FGCC 1984). However, while these stations are important for the nation's geodetic framework, they are impractical for everyday survey applications because of the long distances between stations.

In many states, the NGS, in coordination with the state DOT or the state geodetic agency, has *densified* the HARN network to produce station spacing of the order of 25–30 km. Many of the densification projects are performed to either the old Order A or Order B specifications (or 1–2 cm according to the new standards). The coordinates generated by such surveys are submitted to the NGS for incorporation in the NSRS, a process generally referred to as '*bluebooking*'. When a GPS project is 'bluebooked' (NGS 2001b), the stations and data associated with them are maintained by the NGS, and included in any future re-adjustment of the national network (see Section 13.4.3).

Over the past decade, many county and municipal governments have implemented GIS to manage their geospatial information. In most cases, a GPS control network has been established to provide the coordinate foundation. Although creating a geodetic control framework is a significant cost, it is relatively insignificant compared to the cost of creating a complete GIS database. However, the GPS survey techniques used to establish this framework should be accurate enough to adequately support current and future applications of GIS. If the GIS will be a long-term geospatial information management system, then the geodetic control network should be designed to support such an effort. If the GIS will be updated and maintained using GPS technology, the control network may have to be in a fairly dense grid, with station spacing as high as 5 km.

Static GPS surveying is the only method approved by the NGS for establishing GPS control networks, and specifications such as (FGCC 1989; and updates) must be strictly adhered to. Surveyors who fail to comply completely with such guidelines will have their results excluded from the NSRS. Typically what is defined is:

1   The GPS receiver and antenna hardware to be used.
2   Guidelines concerning maximum and minimum inter-receiver distances, ancillary measurements, type of monumentation, length of observation sessions, testing and validation procedures to be followed, etc.

3   The GPS network design principles to be followed, including the number of repeat baselines to be measured, the extra (redundant) baselines to be measured, the number and type of control stations to be connected to, and so on.
4   The data processing software and options that must be used.
5   The documentation and reporting guidelines to be followed.

The design of a static GPS survey for the establishment of a control network may be represented by Figure 12.3, where it can be seen that there are many *redundant* baselines that have been measured. The final result will be a set of coordinates for the stations generated either by a least-squares adjustment of the individual baselines, or by scientific GPS software that can process the data from the entire campaign in a single rigorous adjustment. The latter is the most accurate data processing strategy, while the former, although less rigorous, is the standard methodology using commercial software (Section 13.4.2).

13.2.2. Application 2: GPS for photogrammetric mapping and engineering projects

Photogrammetric mapping projects are often performed to support a city- or county-wide GIS. As photogrammetry in general still uses ground control, and the project requirements must dictate the number of horizontal and vertical control points needed to support the project and the GPS technique to be used. The accuracy requirements in turn are mainly dependent on the mapping scale of the project, the required contour interval and the national mapping accuracy standards. In general, agencies require that the horizontal and vertical accuracy standards meet those of the minor control surveys mentioned in Section 13.1.3. Where it is not specified which S&S are to be followed for the GPS survey, the project contract specifications may require that the positional accuracy that meets, for example, the US National Map Accuracy Standards (USGS 1947), or equivalent standards in other countries.

Planimetric mapping with contours for engineering design applications requires a greater density of horizontal and vertical points. This is most economically done by using pre-established control network stations established using static or rapid-static techniques, and to then densify this network using the post-processed kinematic, or RTK-GPS methods in the 'stop-and-go' mode of operation (Section 11.1.2).[2] If dual-frequency GPS receivers are used, ambiguity resolution is possible 'on-the-fly' (OTF) (Section 11.1.4), so that, for example, ambiguities can be resolved while the surveyor is moving from point-to-point, even if satellite lock is lost or interrupted during travel. In this mode, the surveyor travels from point-to-point collecting enough data at the photogrammetric or engineering marks to achieve the results desired. The following comments may be made with regard to these surveys:

*   The NSRS horizontal and vertical control should be used as the basis of the control project. This relates the local control network to the national reference system.
*   Control stations (or benchmarks in the case of vertical surveys) should be evenly

---

[2] With efficient OTF-AR, the 'go' part of the 'stop-and-go' survey need not have continuous satellite lock.

distributed across the region. Good practice is to strive to have control points in a minimum of three of the four compass quadrants.

- Static and rapid-static networks should be used to establish the local control network upon which the kinematic survey (post-processed or real-time) is based. Where S&S are not clear on such matters as observation session length (as a function of baseline length and number of observed satellites), manufacturers' recommendations are usually sufficient.

- It is good practice when performing rapid-static surveys for a minimum of 50 per cent of the points to be observed two times. In the case of kinematic surveys, perhaps 100 per cent should be observed twice. The use of fixed height tripods, or bipods, is recommended in order to minimize the possibility of antenna height errors.

- In the case of post-processed kinematic surveys, it is recommended that several reference stations be used, and to compute the baselines independently, as indicated in Figure 13.1. Then compare the coordinate values calculated from the different reference stations to check the repeatability of these measurements.

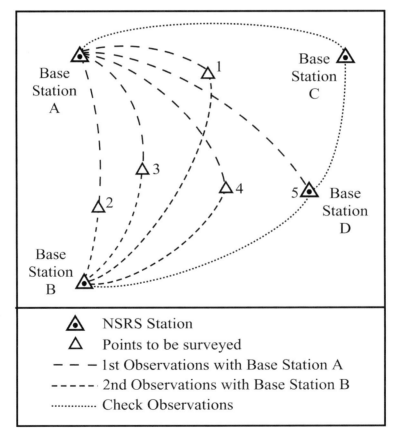

Figure 13.1 Performing a post-processed kinematic GPS survey for establishing photogrammetric/engineering control using two reference stations (for checking).

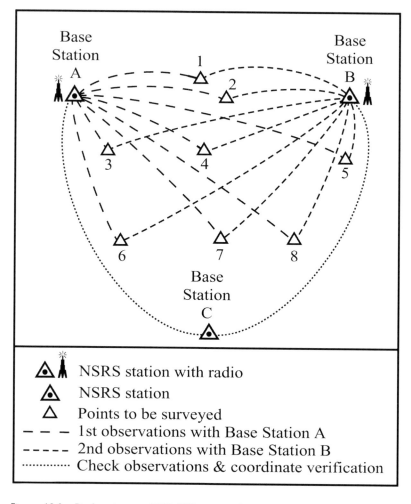

Figure 13.2 Performing an RTK-GPS survey for establishing photogrammetric/engineering control using two reference stations (for checking).

- RTK-GPS is best suited for surveying a large number of points in a relatively small area (Section 11.2.2). It allows the surveyor to obtain coordinate values, and to analyze the measurement quality, immediately in the field. When linked to a pen-based computer and mobile Computer Aided Design (CAD) package, RTK-GPS allows the surveyor to 'navigate inside' the mobile CAD system and create a finished mapping product while in the field.
- In order to 'Quality Control' RTK-GPS surveys, coordinates should be established using one reference station and then observed a second time using a second RTK reference, as illustrated in Figure 13.2. Although using two simultaneous RTK reference stations is an acceptable check (if the RTK-GPS equipment has the facility to switch between the two transmitting stations), if possible the surveyor should observe the same station several hours later to minimize the chances of introducing systematic errors.

13.2.3. Application 3: using GPS to conduct a utility infrastructure inventory

Differential GPS (DGPS) is ideally suited for determining positions with an accuracy in the range of sub-meter to a few meters in real-time (Section 11.2.1), using very little data – typically the station occupation time is less than 30 seconds. The technique is robust, with no need to solve for any ambiguities. This means that satellite lock can be lost frequently without the need to re-initialize, making it easier to coordinate points near tall buildings and under heavy tree canopies. However, use of GPS in this environment will be very susceptible to multipath. It is recommended the user take into account the possible degradation before certifying positional quality. Either the Local Area DGPS (LADGPS) or the Wide Area DGPS (WADGPS) technique may be used (Section 11.3.3), and there are several commercial and free-to-air services available to users (Section 14.1). Although it is possible to use DGPS techniques in the post-processed mode, the majority of DGPS applications are addressed in real-time using an appropriate communications data link to carry the DGPS correction messages to users (see Sections 11.3.1 and 11.3.2). While DGPS can yield sub-meter coordinate results, the height information obtained is much less accurate (perhaps only of the order of a few meters, or worse).

The DGPS positioning technique is frequently used for asset and utility inventories, when the client wants to establish the coordinates of certain structures or assets, and tie their locations to GIS data related to those structures or assets. The main applications for such surveys are *asset management* and *utility maintenance* purposes. DGPS is very cost-effective when compared with rapid-static GPS, post-processed kinematic GPS, or RTK-GPS. In particular no reference stations need to be established by the surveyor/contractor. Besides saving time and being less logistically challenging, DGPS systems are less expensive than carrier phase-based equipment. In addition, the equipment is smaller and the GPS receiver can be linked to other mapping devices such as pen-based computers to create a system ideally suited for the rapid GIS-type mapping. Other devices can be added to this mobile mapping system, including voice recognition software, laser measuring devices, barcode scanners, digital cameras, and video.

The same system used to map the infrastructure may be used later to perform maintenance and work order management. This can be conducted either at a central docking station or through wireless technology. Instructions on location and work to be performed are sent directly from the central office to maintenance crews in the field. The maintenance crew, using DGPS linked to a mobile computer, can navigate to the location, access the work order, and perform the necessary work. Once the work has been completed, the maintenance crew can then transmit the information back to the central office, where the GIS database is updated, and the crew can then navigate to the next assignment.

13.2.4. Application 4: using GPS to determine heights

Static and rapid-static GPS techniques are capable of establishing the relative *ellipsoidal height* between two simultaneously tracking GPS receivers to a few centimeters accuracy (see S&S such as those defined in NGS 1997). The accuracy of height difference determination using kinematic or stop-and-go techniques is, however, a little worse. For many engineering and mapping applications the height quantity required is the *orthometric height*. The relationship between orthometric height and ellipsoidal height is represented by the simple formula (Hofmann-Wellenhof *et al.* 1998):

$$H = h - N$$

where H is the orthometric height, h is the ellipsoidal height, and N is the geoid height (or the geoid-ellipsoid separation). In the GPS case, it is the relative height quantities that are of interest, that is:

$$\Delta H = \Delta h - \Delta N$$

where Δ represents the *difference* in the respective height quantities between the reference station and the survey point.

The quantity ΔN can be derived from two sources:

1   Using a geoid model such as *GEOID99* in the US (NGS 2001a), or *AUSGeoid98* in Australia (AUSLIG 2001).
2   Estimating the difference from surveys of several points which have heights in both systems, and then interpolating to other surveyed points.

The former is comparatively easy to implement in those areas where a high quality gravimetric geoid model is available. Even under such circumstances the accuracy of the geoid height difference may only be of the order of a 3–5 cm over baseline lengths of the order of a few tens of kilometers (and worse in mountainous areas, or areas where there is a deficit of high quality gravity data). Hence the accuracy of the orthometric height difference may only be 5–10 cm at best. The latter method has several advantages in that it does not require an explicit knowledge of the geoid-ellipsoid separation, but derives that information – that is, the *difference* between the ellipsoid height datum of the GPS network and the Mean Sea Level (MSL) datum – from those GPS surveyed stations that have heights in both systems. That is, the GPS heights are related to the adopted ellipsoidal height of the datum station (or stations) held fixed in the baseline solution (Section 9.4.1), but the MSL heights have been independently determined using standard leveling techniques. If there is a systematic offset between these two datums, this *geometric* technique will be able to accommodate them, while the gravimetric technique will *not* be able to do so.

The implementation of the geometric technique for correcting GPS-derived ellipsoidal height differences to heights on the local MSL datum is as follows:

1   A number of benchmarks with leveled (MSL) heights are identified and included in the set of points that are surveyed using GPS. *Ideally these points should be well distributed across the area, as well as having some of them around the perimeter of the survey area.*
2   As a result of the GPS survey, the ellipsoidal heights of the above mentioned benchmarks are determined. *The two sets of heights permit the offset between the MSL and the ellipsoidal height datums to be determined.* If the MSL/leveling datum is coincident with the geoid, then the offset is identical to the geoid height. On the other hand, if the leveling datum is not the geoid, then the offset cannot be interpreted as the geoid height but does fulfil the function of a correction: from the GPS height datum to the local leveling datum.
3   The offset will typically *vary* from benchmark-to-benchmark, but the pattern of

offsets can be used to predict the height correction at other points where only the GPS height has been determined. This 'pattern' can be realized in the form of a simple model, for example as a plane of best-fit through the benchmark corrections. *The height correction is then interpolated at any other point in the survey area using this model.*

This geometric technique is a *pragmatic* solution to the problem of how to convert heights on the GPS datum to heights consistent with the local leveling datum, and has general applicability, unlike the use of a prior computed geoid height model such as *GEOID99*.

13.3. On-site procedures

The objective of well designed procedures is to ensure that the survey is carried out according to plan (Chapter 12), in order that good quality data is collected, from which coordinates results are generated to the required level of accuracy and with the appropriate QC measures. Obviously the higher the survey accuracy requirements, the more elaborate are the field and office procedures. In this section brief comments are made with regards to issues impacting on the efficient execution of GPS fieldwork. Detailed instructions and checklists for such may be developed by surveyors themselves, using official S&S as guidelines.

13.3.1. Antenna setup and height measurement

Antenna setup and height measurement errors are, unfortunately, very common. The following are some procedures that minimize the chances of making such errors:

- The antenna normally bears a *direction indicator* that should be oriented in the same direction at all sites using a compass. This ensures that any antenna center offset (as measured from the mechanical center to the electrical phase center – Section 10.1.3) will propagate into the baseline solution (groundmark-to-groundmark) in a systematic manner.
- The same antenna, receiver and cabling should be maintained together in a 'kit'.
- Because of the high precision of GPS surveys, the *centering of the antennas is important*. If centering is poor, the accuracy of the overall survey will suffer; hence plumbobs should be avoided. Tribrachs with built-in optical plummets should be regularly calibrated.
- The antenna assembly should be mounted on a standard survey tribrach with an optical plummet, on a good quality survey tripod.
- Setting up on a geodetic pillar is, of course, reasonably effortless and to be preferred.
- Care must be taken with *antenna height measurement*.

As the latter is probably the most critical of all antenna setup operations some further comments must be made. The height of the antenna above the station marker, measured to the standard reference point on the antenna housing, should be measured to the nearest millimeter. As this is a common source of error, the measurement should be checked (e.g. by independent measurement by another person). Although different antenna types have different recommendations for height measurement, all antenna height measurements must be carefully noted, preferably with a diagram. In the case of kinematic or stop-and-go surveys the antenna is usually mounted on a pole or bipod with constant height. Hence the chances of making a height measurement error are low (though not zero!).

13.3.2. Field log sheets

In the case of RTK-GPS and DGPS, all data is electronically recorded for later download. The following is therefore mostly relevant for post-processed techniques. A field log sheet should be used, on which pertinent information concerning the site being occupied and the data collection process itself is entered. These sheets may be very comprehensive (running to many pages) in the case of the highest accuracy GPS surveys, or simply a few lines on a standardized booking sheet for rapid mapping techniques such as stop-and-go, RTK-GPS or DGPS. A log sheet for a *static* GPS survey would typically contain some or all of the following information:

- Date and time, field crew details, etc.
- Station name and number (including aliases, site codes, etc.).
- Session number, or other campaign indicator.
- Serial numbers of receiver, antenna, data logger, memory card, etc.
- Start and end time of observations (actual and planned).
- Satellites observed during session (actual and planned).
- Antenna height (several measurements), and eccentric station offsets (if used).
- Weather (general remarks), and meteorological observations (if measured).
- Receiver operation parameters such as data recording rate, type of observations being made, elevation mask angle used, data format used, etc.
- Any receiver, battery, operator or tracking problems that were noticed.

Examples of field log sheets are available from the GPS manufacturers or from national S&S documents, or may be designed by the agency or organization carrying out the survey.

13.3.3. Eccentric station survey

Unlike conventional ground surveys, GPS techniques require a comparatively clear sky view above the elevation mask angle (typically set at 15°–20°). Sometimes the ground-mark that is to be surveyed does not satisfy this condition, perhaps because it is a previously monumented station. In such circumstances an *eccentric station* may be surveyed by GPS, and the necessary site measurements made in order to reduce the baseline components to the required groundmark.

13.3.4. Checklists

From the *Quality Management* (QM) point of view (Section 12.2.2) it is good practice to develop a checklist of on-site procedures – referred to as Standard Operating Procedures (SOPs) – that must be followed (some of which would require appropriate entries in the field log sheets):

- GPS receiver initialization procedures
- Setup and orientation of the antenna
- Correct cable connection of antenna-to-receiver, receiver-to-battery, receiver-to-radio, etc.
- Double-checking of centering and antenna height measurement
- Receiver startup procedure (e.g. entry of site number, height of antenna, etc.)
- Survey of eccentric station
- Temperature, pressure and humidity measurements (if required)
- Monitoring receiver operation and data recording (see below)
- Field log sheet entries (see above)
- Photographs or sketches of point occupancy
- Procedures at completion of session (e.g. communications, data transfer to PC, etc.)
- Instructions in event of receiver problems (i.e. contingency plans, etc.)

A GPS receiver can display a lot of information, hence a checklist with regards to the monitoring of receiver operation is useful (especially if the field crew is unfamiliar with the hardware). This checklist could include monitoring: battery status; memory capacity remaining; satellites being tracked; real-time navigation position solution; satellite health messages; date and time; elevation and azimuth of satellites; signal-to-noise ratios; antenna connection indicator; tracking channel status; and amount of data being recorded. The 'ultimate GPS fieldwork sins' are summarized in Table 13.1. The development of SOPs is one way to guard against poor fieldwork.

13.3.5. Field office procedures

It cannot be over emphasized that data should be processed as soon as possible after the observation session in order to assure the quality of the survey at an early stage. As a prerequisite therefore, all data should be systematically catalogued and archived between observation sessions (if there is time), or at the end of the working day at the very latest. Many problems can be identified at this stage. The following are some typical field office procedures:

- Data handling tasks – transfer of data from receiver to PC
- Data verification, backup and archiving in field office
- Preliminary computation of baseline(s)
- Preliminary QC checks, such as the inspection of repeated baselines, loop closures, and evaluation of (incomplete) minimally constrained network
- Management of field crews – develop contingency plans for repeated observation sessions
- Oversee calibration and testing of field equipment
- Preparation of campaign report, and maintain reporting to head office and/or the client

Without data safely downloaded from the GPS receiver, the survey work should never be considered complete. Therefore: (1) download data from the receiver 'ASAP' (GPS receivers have many hours of internal memory, so daily download is a reasonable routine); (2) delete files from receiver memory when data download procedure has been verified; (3) download to PC hard disk, then to floppy disks, and then make

Table 13.1 GPS fieldwork blunders

Common problems	Remedies/advice
Power loss causing termination of survey	Always have backup power supplies
Cable problems affecting operations	Keep them in good condition
Incorrect receiver operation	Field staff must be trained, and follow appropriate checklists
Data collection not coordinated	Good teamwork, and have well designed logistical procedures and checklists
Loss of data after survey	Follow systematic data management procedures
Setup on wrong station	Reconnaissance and good on-site procedures
Antenna height measurement error	Check, and recheck readings

backup copies; (4) store backup disks separately; (5) label and write-protect floppy disks (be ruthlessly systematic in following a disk labeling convention); (6) cross-reference field log sheets to data files; and (7) verify data download (e.g. check number and size of files).

13.4. Tying up 'loose ends'

There are many additional tasks that might be considered to be 'part-and-parcel' of GPS surveying. Some of them are briefly discussed here. The reader is referred to GPS textbooks (Hofmann-Wellenhof *et al.* 1998), and to manufacturers' user manuals for further details.

13.4.1. RINEX files

RINEX (*Receiver INdependent EXchange*) is a data file format devised in 1989 for geodetic applications requiring the international exchange of GPS data sets, gathered during global campaigns, by different brands of receivers (Gurtner 1994). It has now been adopted as the standard exchange and archive format for GPS surveying and precise navigation applications as well. RINEX Version 2.2, which is able to handle kinematic GPS data, is now in common use. The RINEX format has the following characteristics:

- ASCII format, with a maximum of 80 characters per record
- Phase data recorded in cycles of the L1 or L2 carrier frequency, pseudorange data in meters
- All receiver-dependent calibrations are assumed to have been applied to the data
- Time-tag is the time of the observation in the receiver clock time frame
- Separate measurement, navigation message, and meteorological data file formats

Most GPS measurement systems will *output* RINEX formatted data files. Many commercial GPS software will also permit the *input* of RINEX data files for processing. This has the following implications: (a) data from one brand of receiver may be processed within another commercial software package, and (b) data from surveys using a mixed set of GPS receivers can be processed within one software package. In reality, however, the processing of data collected by a certain brand of GPS receiver using software not explicitly designed to process this data can be problematical. This is especially the case for rapid survey techniques that demand OTF ambiguity resolution. The reason is that the RINEX format does *not* include all the receiver-generated data a software package may need to optimally process small amounts of data.

13.4.2. Network adjustment

In general, a GPS survey project involves the use of a small number of receivers to coordinate a large number of points. The area of operations may span distances of merely a few kilometers (as on an engineering site), to several hundred kilometers, or even thousands of kilometers in the case of scientific surveys. A typical GPS survey, such as for mapping or control densification, involves distances of the order of several tens of kilometers. The survey may be carried out using conventional static GPS survey techniques, or high productivity techniques such as stop-and-go, RTK-GPS, etc. For moderate to low accuracy GPS surveys (e.g. as for the applications described in Sections 13.2.2 and 13.2.3):

- Points are either surveyed as 'no check' baselines from a single reference station (Section 12.5.1), or are surveyed from two or more reference stations.
- The datum for the surveyed points is the reference station(s)'s coordinate(s) which are assumed to be known in the '*GPS framework*', that is, either WGS 84 or one of the International Terrestrial Reference System (ITRS) reference frames (Section 7.2).
- The coordinates of the surveyed points are easily derived from one or more baseline solutions *without* the need for a network adjustment. (Where there are two coordinates derived from two independent baselines, using the GPS survey mode depicted in Figures 13.1 or 13.2, the mean is taken if the difference – which is a QC measure – is not larger than some specified threshold value.)
- If the coordinates are to be given in a local geodetic datum that is *not* geocentric, then an 'official' transformation model is generally applied.

In the case of surveys for setting up a control network *linked* to the national geodetic framework (i.e. the NSRS in the US), or which is a *densification* of the national geodetic framework (Section 13.2.1), a network adjustment is generally necessary. The following comments may be made with respect to this *secondary* operation:[3]

- There must be a sufficient number of *redundant* baselines to warrant a network adjustment.
- Redundant baseline surveys are usually mandated in national S&S.
- The various baselines may be connected to each other, and to the surrounding geodetic control network in a complex manner (e.g. see Figure 12.3).
- Several existing geodetic control stations are usually surveyed, and hence are 'linked into' the GPS survey (Section 12.5.4).
- The network adjustment may require only one control station's coordinates to be held fixed (in a so-called *minimally constrained adjustment*), or several of them held fixed, or nearly so (which is usually the case for densification projects).

---

[3] The 'primary' adjustment is the baseline processing step.

- The network adjustment requires all the baseline solutions, and their associated variance-covariance matrices, as *input* – the output is an adjusted set of coordinates for the network, in the datum defined by the fixed control station(s), and the resulting quality information.
- Network adjustment software is nowadays provided as part of the commercial GPS data processing software package.
- Executing a network adjustment requires skill on the part of the analyst, and national S&S sometimes define the manner in which the adjustment is to be carried out and the QC that must be conducted on the final network solution.

13.4.3. NGS 'bluebook' submission

In the US, a GPS survey established as a base for a state, county, or municipal GIS, or for general control purposes, should be tied to the NSRS through a process known as *'bluebooking'.* The NGS will only 'bluebook' projects that meet or exceed the appropriate FGCS specifications, and that were *pre-approved* by the NGS for inclusion. The following steps must be followed before any GPS data is acquired (NGS 2001b):

1 Contact the NGS adviser in the state where the GPS work is being performed (if no state adviser exists, contact the NGS headquarters).
2 Inform the NGS of the project's intent, and ask for any recommendations or suggestions.
3 Submit a preliminary control network station map showing all existing NGS control.
4 Submit a station observation plan for approval.
5 Perform GPS observations.
6 Submit the final report (computed coordinates, etc.).

After the final report has been submitted the data is checked by NGS for completeness, whether it adheres to FGCS specifications, and for the quality of the computed coordinates. Once the coordinates have been accepted, the NGS will include the station and coordinate information in the NSRS database. This data is available to the public, in either hard copy or digital format. In addition, these coordinates will be automatically included in any future spatial re-adjustment of the US national geodetic network. The 'bluebooking' process may take up to 1 year from the planning stage to the publication of the final coordinates by the NGS.

13.4.4. The testing of GPS

Many national S&S documents have a section concerning the issue of GPS 'testing' (see, e.g. FGCC 1989; GSD 1992; DOSLI 1996; ICSM 2000). The 'test' procedure may be intended to validate a particular GPS technique, or an individual piece of receiver equipment. Sometimes it is the surveyor that is being 'tested' or 'qualified' (GSD 1992). In addition, cadastral surveys using GPS receive special attention due to the need to satisfy regulations for *'legal traceability'.* The dilemma facing individuals and government certifying agencies is deciding on just what aspect of the GPS technology to actually test, and then to decide on how to go about doing it. Consider the following possibilities:

1 Testing of the GPS 'system' as a whole – *primarily the space and control segments.*
2 Testing a class or brand of GPS instrument, involving the combination of hardware and software – *the user segment.*
3 Testing a particular GPS instrument unit – on a regular basis, or in response to a suspected problem.
4 Testing the skill of personnel – *in field and data processing procedures.*

However, certain difficulties shared by GPS testing procedures should be acknowledged:

- The magnitude of GPS biases is a function of time – how can a test be conclusive if it is carried out only once?
- Some GPS biases are a function of geographic location – how can a test be considered conclusive if it is carried out in only one location?
- The propagation of most GPS biases into the baseline solution is a complex combination of factors, such as time, location, baseline length and satellite geometry – *should all baseline lengths be tested?*
- The quality of GPS baselines is not just a function of errors and biases, but also the length of observation session and the type of carrier phase solution – *what operational procedures for data collection and baseline processing should be insisted upon?*
- What should be the outcome of the testing – a lifetime 'stamp-of-approval', or an annual 'certification'?

GPS 'tests', or checking procedures, can be grouped under four categories:

1 *Accreditation Tests:* These may test new GPS systems, e.g. in order to 'accredit' them for use in cadastral surveys. They may also be proposed as a form of 'competency' or 'qualification' test for surveyor/contractors. These would typically be once-off tests, and would only be undertaken if new technology or procedures were to be considered for use.
2 *Certification Tests:* These are the tests would be deemed to be necessary at some regular interval, e.g. on an annual basis, in a manner analogous to Electronic Distance Measurement calibration procedures.
3 *Validation Tests:* These are test procedures which may be insisted upon at regular intervals.
4 *QA Practices and Verification:* These are not true 'tests', but are quality assurance procedures built into the recommended field and office practices within S&S documentation.

The de facto 'accreditation' test for new GPS surveying systems has, since 1983, been the FGCS network survey test (FGCS 2001). The test involves the survey of a network of stations in Washington DC with baselines varying in length from less than 1 km to over 100 km. To date over 30 GPS systems have been 'tested'. 'Certification' or 'calibration' tests are those that, like the FGCS test, involve the survey of mini-network of stations and the comparison of the results against the known coordinates. The difference is that instead of there being a specially designed network, any portion of the HARN or NSRS network can be used, and they can be performed on a regular basis. 'Validation' tests are those self-administered tests that the surveyor can apply on a daily basis, or more frequently than the 'calibration tests'. Examples of such tests include: (1) zero baseline tests, (2) re-measuring certain baselines, and (3) checking the coordinates of several control points of the NSRS network.

A 'zero baseline' test can be used to study the precision of the receiver measurements (and hence its correct operation), as well as the data processing software. The experimental setup, as the name implies, involves connecting two GPS receivers to the same antenna. The antenna 'splitter' can be purchased from specialist electronics shops. Obviously this test cannot be easily applied to integrated antenna/receiver systems (Section 10.4.4.1). When two receivers share the same antenna, biases such as those which are satellite (clock and ephemeris) and atmospheric path (troposphere and ionosphere) dependent, as well as errors such as multipath cancel during data processing. The quality of the resulting 'zero baseline' is therefore a function of random observation error. An important advantage of this test is that it is comparatively simple to administer – no specialized software or groundtruth data is required, and the location of the antenna is immaterial.

Re-measuring a baseline that was previously observed using GPS is a practice similar to that required of cadastral surveyors. This is not an onerous requirement, and therefore has merit. In addition, this double observed baseline provides some measure of redundancy for any subsequent least-squares network adjustment (Section 13.4.2). It therefore has a double benefit. Guidelines would have to be established regarding the selection of baseline, what baseline differences can be tolerated before the surveyor needs to be concerned, etc. In addition, the practice of measuring between established control marks is also familiar to surveyors, who are often required to observe between existing control points in order to establish datum and starting bearings.

Different S&S may place a different emphasis on one type of test over another. Furthermore, such testing may only be insisted upon for certain category of surveys, such as cadastral surveys. It is therefore crucial that the surveyor/contractor be familiar with any requirements for such test procedures.

References

AUSLIG, *AUSGeoid98. Australian Surveying and Land Information Group web page*, 2001, http://www.auslig.gov.au/geodesy/ausgeoid/geoid.htm.

DOSLI, *New Zealand Geodetic Survey Specifications: GPS Surveys*, Department of Survey and Land Information, New Zealand, 1996, 16 pp.

FGCC, *Standards and Specifications for Geodetic Control Networks*, Federal Geodetic Control Committee (now the Federal Geodetic Control Subcommittee of the Federal Geographic Data Committee), National Ocean and Atmospheric Administration (NOAA), 1984, 31 pp.

FGCC, *Geometric Geodetic Accuracy Standards and Specifications for Using GPS Relative Positioning Techniques*, Federal Geodetic Control Committee (now the Federal Geodetic Control Subcommittee of the Federal Geographic Data Committee), NOAA, version 5, dated 11 May 1988 and reprinted with corrections, 1989, 48 pp.

FGCS, *Federal Geodetic Control Subcommittee Instruments Working Group web page*, 2001, http://www.ngs.noaa.gov/FGCS/instruments.

FGDC. *Geospatial Positioning Acuracy Standards, Part 2: Standards for Geodetic Networks*, Federal Geographic Data Commitee, National Ocean and Atmospheric Administration (NOAA), 2001, available at http://www.fgdc.gov/standards/status/sub1_2.html.

GSD, *Guidelines and Specifications for GPS Surveys*. Geodetic Survey Division, Geomatics Canada, Natural Resources Canada, release 2.1, 1992, 63 pp., available at http://www.geod.nrcan.gc.ca/products/html-public/GSDpublications/GPS_guidelines/English/Guidelines.pdf.

Gurtner, W., 'RINEX: The receiver-independent exchange format'. *GPS World*, 5(7), 1994, 48–52.

Hofmann-Wellenhof, B., Lichtenegger, H. and Collins, J., *Global Positioning System: Theory and Practice* (4th ed.), Springer-Verlag, 1998, 389 pp.

ICSM, *Standards and Practices for Control Surveys*. Australian Inter-Governmental Committee on Surveying and Mapping, version 1.4, special publication no.1, 2000, available at http://www.anzlic.org.au/icsm/publications/sp1/sp1.htm.

NGS, *Geodetic Benchmarks*. NOAA Manual NOS NGS 1, 1978, 50 pp.

NGS, *Guidelines for Establishing GPS-Derived Ellipsoid Heights (Standards: 2 cm and 5 cm)*. National Geodetic Survey Tech. Manual NGS-58, 1997, available at http://www.ngs.noaa.gov/PUBS_LIB/NGS-58.html.

NGS, *Geoid99 web page*, 2001a, http://www.ngs.noaa.gov/GEOID/

NGS, *National Geodetic Survey Bluebook web page*, 2001b, http://www.ngs.noaa.gov/FGCS/BlueBook.

NSRS, *National Spatial Reference System web page*, 2001, http://mapindex.nos.noaa.gov/mapfinderhtml3/surround/geodetic/geodetic.html.

Schinkle, K., 'A GPS how-to: conducting highway surveys the NYSDOT way'. *GPS World*, 9(2), 1998, 34–40.

USGS, *National Map Accuracy Standards*. US Geological Survey 1947, available at http://rmmcweb.cr.usgs.gov/public/nmpstds/acrodocs/nmas/NMAS647.PDF.

Chapter 14

Servicing the GPS user

Gérard Lachapelle, Sam Ryan and Chris Rizos

GPS users may now use a range of services to assist them in designing, executing and analyzing Global Positioning System surveys. The main services are those offered to support *real-time operations*. However, apart from these services, GPS users can take advantage of data and products generated by *infrastructure* that has been deployed and operated as a result of initiatives taken at the national and international level. In addition, there are a wide variety of information, data and product services available via the Internet. These include GPS manufacturers and agents of their products, government agencies, national professional associations, academic institutions, international organizations, and many others. Almost all can be accessed via the World Wide Web (WWW).

14.1. Real-time DGPS services

Differential GPS (DGPS) is a form of GPS *augmentation*. Augmentations are used to enhance accuracy, integrity, reliability and availability. Accuracy may be further subdivided into three categories, namely predictable, relative and repeatable accuracy. Although the removal of Selective Availability (SA) in May 2000 renders DGPS unnecessary for *some* applications, DGPS is still a requirement for any application where an accuracy consistently better than 5 m is required, and where reliability and the other parameters matter. This is because the total effect of atmospheric and ephemeris biases on Single-Point Positioning (SPP) is still larger than 5 m in most cases. Since these errors are spatially (and temporally) correlated, DGPS is effective in reducing them.

Differential services imply *real-time* services, although it is understood that postprocessing can also be used to enhance post-mission accuracy. However, rarely is a DGPS service established only to address surveying/mapping applications, and often the system requirements are those defined for precise navigation. Consequently, many users operating a wide range of GPS receiver equipment must be able to receive and process DGPS transmissions (Section 11.2.1). The Radio Technical Committee for Maritime Services (RTCM) Special Committee 104 has drafted a standard format for the correction messages necessary to ensure an *open* real-time DGPS system (RTCM 1998). According to RTCM recommendations, the pseudorange correction message transmission consists of a sequence of different message types (Table 11.2).

Assuming the necessary infrastructure for a DGPS service is in place (reference receivers, communication links, and transmission protocols), the most important system parameters are:

- *Predictable accuracy*: The accuracy of a system's position solution with respect to a charted or mapped point. Both the position solution and the chart/map must be based on the same geodetic datum.
- *Repeatable accuracy*: The accuracy with which a user can return to a position whose coordinate has been measured at a previous time with the same system.
- *Relative accuracy*: The accuracy with which a user can measure position relative to that of another user of the same navigation system at the same time.
- *Availability*: The percentage of time that the services of the system are usable.
- *Coverage*: The surface area in which the signals are adequate to permit the user to determine position to a specified level of accuracy.
- *Reliability*: The probability that the system will perform its function within defined performance limits for a specified period of time under given operating conditions.
- *Integrity*: The ability of the system to provide timely warnings to users when the system should not be used.

The needs and requirements for DGPS services, in relation to various air, marine and land navigation applications are defined, for example, in the US Federal Radionavigation Plan (DoD/DoT 2000), and the reader is referred to this publication for background information. This manual, on the other hand, will focus on aspects of the available DGPS services that can support geospatial applications.

DGPS services can be found in many countries, primarily because the majority of system developers and service providers are global companies operating across different geographic regions. Some are commercial systems, and others broadcast messages 'free-to-air'. Several of them are mature systems, while others will expand as additional base stations are established and new technologies and user markets are developed. In general such services may be characterized according to the following:

- Whether LADGPS or WADGPS implementation.
- The type of communications link used.
- Whether the service addresses a specific group of users, or is a general service.
- The nature of the organization providing the service, is it a government agency, an academic institution or a private company?
- Whether the service is freely available, or whether it is operated as a commercial activity.
- Whether it is restricted to RTCM pseudorange correction broadcasts, or the transmission of carrier phase data, or both.
- Whether the system supports post-processed DGPS by archiving the reference station data.
- Whether the service uses a single reference station, or is part of a network of DGPS stations.
- The sophistication of the quality control measures that are in place.

14.1.1. Satellite-based augmentation DGPS systems

In the case of commercial Wide Area DGPS (WADGPS) services, *satellite augmentation* implies that communication satellites are used to provide the data link between user and service provider. However, future satellite-based augmentation systems will have additional navigation signals transmitted by satellites other than those of the GPS constellation. Both of these classes of services are briefly reviewed.

14.1.1.1. Commercial WADGPS services

There are currently two global WADGPS services, one offered by the Dutch multinational company *Fugro* and the other by the French company *Thales*. The Fugro *OmniSTAR* service relies on a large number of ground-based reference stations around the world (OmniSTAR 2001). The operations of sub-networks in different regions are coordinated from a small number of central base stations. The reference stations within a particular sub-network send their data by landline (or other means) to the base station, where the correction data is generated and passed to an upload station that sends this data to a geostationary satellite. The satellite acts as a 'bent pipe', and transmits these corrections to users. The procedure is identical across the world except that different geostationary satellites may be used. In the early days of the service the Inmarsat satellites were used (for which a user required a gimbal-mounted directional antenna). Nowadays L-band mobile satellite communication systems are used in many parts of the world, which permit users to receive the message transmission via small modem devices with short 'whip' antennas.

The OmniSTAR DGPS service is a strictly commercial operation and offers several levels of service to users on land. In its simplest configuration it operates as a Local Area DGPS (LADGPS) (RTCM messages from one reference station), which can deliver a few meters accuracy up to a few hundred kilometers from the station, with accuracy degradation being a function of the distance from the nearest reference station. A WADGPS option uses a proprietary Fugro format to combine the correction messages from several reference stations, delivering sub-meter accuracy (Section 11.3.3). Users need a special decoder attached to their receiver in order to convert the transmitted message into RTCM format before it is passed to the serial port of their GPS receiver. The advantage of the service is that it is available anywhere in the world (onshore and offshore). It is, however, comparatively expensive on a day-rate basis.

As with the *Fugro* system, *Thales* (formerly *Racal*) has an extensive network of DGPS stations around the world. In every other respect the two systems are direct competitors. Thales's DGPS service, known as *Landstar*, is a commercial operation that also offers users both a LADGPS and WADGPS option (Thales 2001). As with OmniSTAR, this is a comparatively expensive service with user charges several times higher than LADGPS services based on ground-based augmentation.

14.1.1.2. Aeronautical wide area augmentation system (WAAS)

This system, which includes a satellite-based and a ground-based component, is being deployed by the US Federal Aviation Administration (FAA) to support aviation navigation for the en route through to precision approach phases of flight (FAA 2001a). Its components are shown in Figure 14.1. The satellite-based component consists of several (up to four) geostationary satellites that will have two major functions, namely the transmission of GPS differential corrections to users and the provision of a ranging capability to improve availability, reliability and integrity over the US.

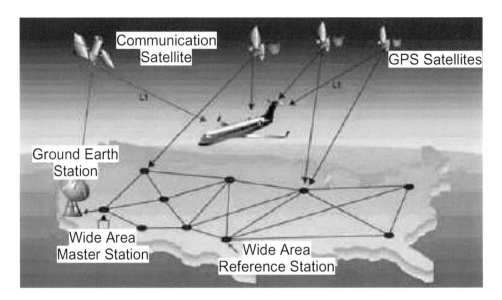

Figure 14.1 US WAAS concept.

The geo-stationary satellite transmission will be on L1 using a C/A-code for ranging and a uniquely structured message to broadcast differential corrections and integrity information to users. Reception of the differential corrections from the geostationary satellites is subject to the line-of-sight condition, similar to the GPS signal.

The ground-based segment includes some 25 reference stations, two processing sites, Wide-Area Master Stations (WMS), two Ground Earth Stations (GES) to uplink the GPS differential corrections to the geostationary satellites and timing information to the geostationary satellites, and software to calculate differential corrections as well as to perform various integrity functions. The specified accuracy of the system is 7.6 m, 95 per cent of the time (FAA 2001a). However, the actual accuracy is expected to be significantly better than this. The reliability of WAAS is near 100 per cent, and the integrity, which is extremely important for aviation, is very high.

GPS receivers will have to be 'WAAS-ready' in order to receive and use the WAAS message. However, WAAS is intended to be a public service and anyone equipped with the appropriate receiver will have access to it. This means that for geospatial applications requiring sub-meter level positioning accuracy WAAS may become a viable alternative to the commercial WADGPS services.

There are plans to extend the WAAS into the southern part of Canada in the decade ahead. Other countries or continents are also deploying similar augmentation systems. The European Geostationary Navigation Overlay System (EGNOS) and the Japanese Multi-Function Satellite-Based Satellite Augmentation System (MSAS) are the other WAAS-like systems under development.

14.1.2. Ground-based DGPS services

These invariably are of the LADGPS type (Section 11.3.3). Although a variety of services exist, many of them locally organized, there is one service that has the unique characteristics of: (a) having been originally established to serve only one industry, (b) is internationally recognized, and (c) the messages are transmitted free-to-air.

14.1.2.1. Public marine DGPS services

This DGPS service was intended only for maritime applications, yet is compatible with international systems being established around the world in order that marine users are able to acquire and use transmitted RTCM messages wherever they go. This service utilizes the marine beacon frequencies of 285–325 kHz (IALA 1999). The advantages of this frequency band are its availability for navigation use and the fact that ground waves propagate long distances (several hundreds of kilometers). Over land the range is less and depends on soil conductivity, however, it is still of the order of a few hundreds of kilometers, which is well over the radio horizon.

The planned and operational stations (as of the start of 2001) are shown in Figure 14.2, the DGPS stations are shown in dark grey with light grey coverage contours. Marine radiobeacon DGPS systems are a worldwide endeavor with national services being set up and maintained by the maritime authority of each country. Both the International Marine Organization (IMO) and the International Association of Lighthouse Authorities (IALA) have produced specifications and guidelines for marine DGPS systems (IALA 1999). In the US and Canada, the US Coast Guard (USCG) and Canadian Coast Guard (CCG) are responsible for these services.

The USCG and the CCG declared their DGPS systems to be fully operational on 12 March 1999 and 25 May 2000, respectively. The initial objectives of this Maritime

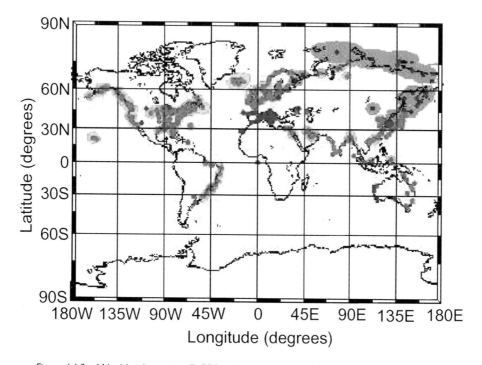

Figure 14.2   Worldwide marine DGPS reference stations (planned and operational).

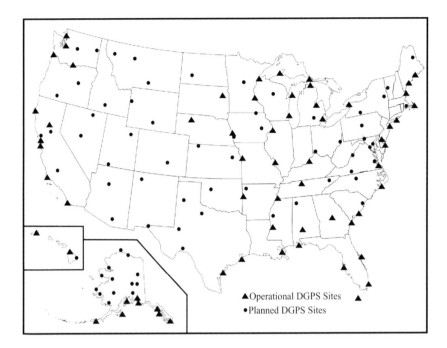

Figure 14.3   The US Nationwide DGPS reference station network.

DGPS Service were to support the harbor entrance and approach phase of navigation, vessel traffic control, as an aid to navigation, and exclusive economic zone ocean mapping. Navigation in inland waterways was later added as another objective. The service, being very reliable and available in the coverage areas, has been well received by users well beyond the maritime community, particularly by farmers using DGPS-guided vehicles for 'precision agriculture'. As a result, it is being extended to cover the entire continental US under the *Nationwide DGPS (NDGPS) Service* project (USCG 2001). When completed in around 2002 the network will number 125–135 DGPS reference stations (Figure 14.3).

14.1.2.2. DGPS via the radio data service (RDS)

In many developed countries, commercial companies, mostly located in capital cities or where there is a large (generally niche) local market for DGPS services, are providing a terrestrial LDGPS service via FM radio station broadcasts. The encrypted RTCM messages are modulated on the sideband RDS signal (Langley 1993). (The Radio Data System – RDS – is a protocol for encrypting digital data on FM sidelobe signals developed by the European Broadcasting Union.) This service takes good advantage of existing radio service infrastructure and allows for the use of very small low-cost FM receivers no larger than a pager to receive and decode the RTCM messages. The DGPS range is limited to that of the FM signal reception range. However, with the expansion of free-to-air DGPS services on the one hand (see above), and the removal of SA leading to improved SPP on the other hand, the market for such services for land vehicle applications has contracted. Nevertheless, research is underway to use RDS (or a similar technology) for city-based Real-Time Kinematic (RTK) services.

14.1.2.3. US local area augmentation system (LAAS)

The US FAA is deploying this system principally to support precision approach and landing Category II/III applications (FAA 2001b). The system will be deployed around selected airports and will service a limited area around these airports. A local ground system will generate differential corrections from the GPS, WAAS geostationary satellites and airport pseudolites (when used). A VHF transmitter will then be used to broadcast the correction and integrity messages to local users. However, this system, due to its localized applications and other restrictions, will not be of interest to users outside the aviation community.

14.2. Services and data from continuously-operating GPS networks

Modern high productivity GPS surveying techniques such as *rapid-static* and *stop-and-go*, as well as the *kinematic* positioning techniques, are best carried out with a static reference receiver operating at a site with known coordinates and the user's receiver moving from point to point (Section 12.5.1). In such a scenario the base station must operate *continuously* without reference to the times when the user receiver(s) is(are) turned on and collecting data. *How to improve the efficiency of high precision GPS positioning by avoiding the need for every user to have to establish their own base station for a particular project?*

There is already a trend in many countries for the establishment of Continuously-Operating Reference Station networks (CORS) to support a range of applications. These are also sometimes referred to as *Active Control Stations* or *GPS Fiducial Networks*. These networks are generally operated by government organizations as a free service (though there are some receivers run by private companies and from whom data may be purchased). By using CORS data, users are, in principle, able to perform high precision positioning using a single GPS receiver, hence *halving equipment costs*. Both pseudorange-based DGPS and carrier phase-based positioning relative to the CORS network stations is possible. Although such CORS networks can be easily enhanced to offer real-time DGPS/RTK services, the majority are simply collecting measurements and archiving the data files for post-processing. There is little doubt that ultimately many CORS networks will evolve into *multi-functional infrastructure* to support the full range of user positioning needs in their area of influence.

14.2.1. CORS networks

CORS networks have mainly been established in support of local geodynamic applications. An example is the Geographical Survey Institute's (GSI) GEONET in Japan, established in the aftermath of the catastrophic 1995 Kobe earthquake. The GEONET consists of almost one thousand GPS receivers deployed across the country, linked to the main archive center in GSI Tsukuba headquarters (GSI 2001). Another example is the Southern California Integrated GPS Network (SCIGN), comprising over one hundred GPS reference stations jointly operated by several academic and government

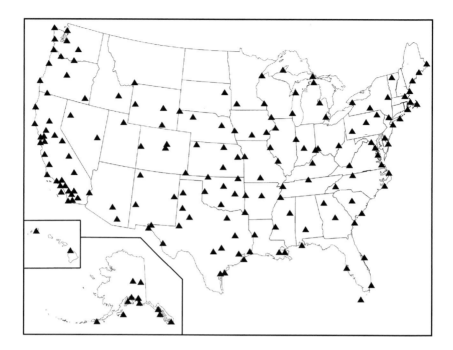

Figure 14.4   Stations of the US CORS network (NGPS 2000).

organizations (SCIGN 2001). CORS network stations have also been established in many countries as part of the International GPS Service, and contribute to the definition and maintenance of the International Terrestrial Reference System (ITRS) (Section 7.2.2).

Recently CORS networks have also been established in many countries simply to support high precision geospatial applications. It is not possible to list here even a small sample of the national CORS networks currently operational (or planned), however, two examples will be used to highlight some of the issues concerning such networks. The first is the US CORS Network (Figures 14.4 and 14.5), a cooperative program coordinated by the National Geodetic Survey (NGS) (NGS 2001). This network consists of almost 200 reference stations, operated by a variety of government, academic, commercial and private organizations, and includes some reference stations of the US Nationwide DGPS network (Figure 14.3). In contrast to this continental scale network, the second example is the Hong Kong Lands Department's CORS network (Rizos 2000) (Figure 14.6), which is intended to service an area of approximately 1,000 km^2.

Such CORS network stations typically log dual-frequency GPS data at 15 or 30 s intervals (though there is now a trend towards 1 s data logging), and make the available the data to users as RINEX files (Section 13.4.1). They may also provide data sets for data processing services (Section 14.2.3).

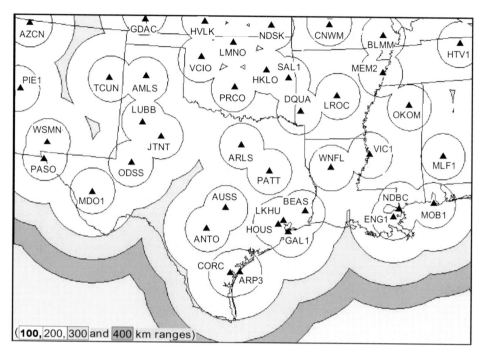

Figure 14.5   CORS network stations in the southern states, note the radii marked around the stations.

14.2.2. Implications for GPS surveying in the post-mission mode

In this section the discussion is concerned with how CORS networks can support carrier phase-based static and kinematic GPS positioning in the post-mission mode of data processing. As already mentioned, increasingly the justification for the establishment of CORS networks is to support precise GPS positioning applications. The simplest means of implementing such a service is to allow the user to download the necessary data files via the Internet. The data, when combined with that collected by their own GPS receiver, can be processed within standard baseline determination software.

Although such a scheme appears to offer advantages, present implementations require that data from the *nearest* GPS base station be downloaded to the user's computer before processing can be carried out. However, in practice, current GPS carrier phase-based baseline estimation techniques that use 'on-the-fly' Ambiguity Resolution (OTF-AR) algorithms require the reference receiver to be less than 10 km distant from the user receiver. (The further away the base station the longer the observation time for a baseline determination, and hence the less productive GPS surveying techniques become.) But the establishment of a network of GPS reference receivers at a density to support all (or most) GPS surveys is currently not feasible. For example, in the case of the GEONET in Japan, despite the large number of GPS receivers deployed across the country, the average receiver spacing is 20–30 km. Many of the US CORS stations are more than 100 km apart (Figure 14.5). Hence,

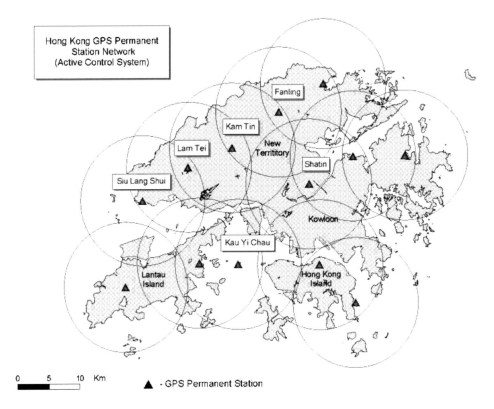

Figure 14.6   The Hong Kong GPS permanent station network (courtesy of Survey and Mapping Office, Lands Department of Hong Kong).

unless the user's receiver was operating nearby to a CORS network station, single (reference-to-user receiver) baseline determination techniques such as those implemented within commercial GPS data processing software can not be used.

In the case of CORS networks intended to service small countries or individual cities, such as the Hong Kong network, users are fortunate to always have at least one reference station within a radius of 10 km, hence there is no constraint to using CORS data for high productivity GPS surveys. In fact, the Hong Kong network is so designed that the vast majority of users can use data from *two* reference stations (Figure 14.6), and be able to determine two independent baselines connecting the user receiver to the CORS network. In this way the extra baseline can be used as a quality control measure (Section 12.4.2.2). One simple solution to this 'distance to reference station' dilemma in the case of CORS networks spread across large regions is to increase the density of reference stations in the major cities or where user demand is highest. Techniques that take advantage of *network-based processing strategies* (in place of single baseline processing) will overcome the 'distance to reference station' constraint (Section 15.3.2).

14.2.3. GPS data processing services

One of the more innovative implementations of CORS-based processing is *not* to require the surveyor to download *any* GPS reference receiver data. Instead, the user records their GPS data and then submits (or uploads) the data file to a central server via an Internet web page (this can even be done from the field via 'smart' cell phones and Internet-ready mobile devices). After the processing is complete the user is informed by email where the result file can be downloaded. Such a service has the additional advantage in that it obviates the need for surveyors to maintain their own PC-based data processing software. Furthermore, this service is ideally suited for the more complex network-based processing that uses *all* the CORS network data surrounding the survey area. Two types of services are currently available:

1 User's data is incorporated with data from a set of CORS network stations and processed using sophisticated 'scientific' GPS processing software. Such processing is equivalent to that applied for conventional static GPS surveying except that the network-based processing ensures that high accuracy is possible even if the distance of the user receiver from the nearest CORS network station is hundreds of kilometers.

2 User's data is processed in the *SPP* mode (Witchayangkoon and Segantine 1999), except that precise satellite clock and orbital information is used in place of those provided in the navigation message (Section 8.3). Although the CORS data is not used directly in such a processing scheme, IGS CORS data is used to derive the precise clock and ephemeris information.

There are several CORS-based processing services currently available, and more are expected to come online in the next few years. Table 14.1 describes three such services, two US-based and one Australian. These services only support static GPS positioning.

14.3. Information, data and product services for geospatial professionals

There is a broad range of information sources concerning GPS services, data and products, as well as surveying/mapping in general. The Internet is now the primary source of such information.

14.3.1. National land, mapping or geodetic agencies

These provide a wealth of information, data and services concerning GPS status, mapping products, datum information, geoid height models, transformation parameters, GPS data and ephemeris information, standards & specifications (S & S), and valuable links to government and non-government WWW sites. Examples from North America and Australia include:

* US National Geodetic Survey, http://www.ngs.noaa.gov/
* Geomatics Canada, http://www.geocan.nrcan.gc.ca/org/indexe.html
* Australian Surveying and Land Information Group (AUSLIG), http://www.auslig.gov.au/
* Land Information New Zealand, http://www.linz.govt.nz/

Table 14.1 Comparison of online GPS processing services

Services		AUSLIG[a]	JPL[b]	Scripps[c]
Baseline determination?		Yes	No	Yes
Precise point positioning?		No	Yes	No
Multiple data submission in one go?		Up to seven RINEX files from one or multiple sites closer than 5000 km	Single RINEX file	Single RINEX file
Data spans multiple days?		OK	No	Not specified
Input of GPS antenna Height?		Yes	No	Yes
Data submitted	Dual- or single-frequency?	Dual	Dual	Dual
	Static or kinematic?	Static	Static	Static
	Minimum data span?	2 h	Not specified	Not specified
	Data sampling rate?	30 s	Not specified	120 s max
	Data format?	RINEX format (filename: IGS naming convention)	RINEX format (filename: IGS naming convention)	RINEX format (filename: IGS naming convention)
	Data compression?	UNIX compressed, GZipped and/or Hatanaka compressed	UNIX compressed	UNIX compressed
	Data submission?	Anonymous ftp and uploading	Anonymous ftp only	Anonymous ftp and uploading
Resources used		IGS products and data from the nearest GPS stations	FLINN orbit	Reference stations can be selected
Processing times		Single site with one day data: 15 min	5–10 min	15 min

Table 14.1 (continued)

Services		AUSLIG[a]	JPL[b]	Scripps[c]
Results delivered	ITRS coordinates	ITRF97	ITRF96 and WGS84	ITRF97 and WGS84
	Local coordinates	Geocentric Datum of Australia (GDA94)	No	No
	Format	One PDF file, one email, no map	8 text files, one email, no map	One email, one map
	Quality	24 h data: 1 cm horizontal and 1–2 cm vertical, 2 h data: 2 cm horizontal and 5 cm vertical	24 h data: a few mm horizontal and cm vertical, 2 h data: 12 cm, 3 h data: 5 cm	Not given
Age of data		1 January 1998 to current date (predicted orbit may be used)	July 1995 to two weeks before the current date	Not given
Software used		MicroCosm	GIPSY	GAMIT

[a] AUSLIG GPS web processing service: Information: http://www.auslig.gov.au/geodesy/sgc/wwwgps/wwwfaq.htm, Interface: http://www.auslig.gov.au/cgi-bin/gps.cgi. Results: ftp://ftp.auslig.gov.au/sgac/wwwgps/gps######.pdf (###### 5 digit job number)

[b] JPL GPS web processing service (Auto-GIPSY): Information: http://www.unavco.ucar.edu/processing/gipsy/auto_gipsy_info.html, Results: ftp://sideshow.jpl.nasa.gov/pub/ag/############ (############ job designator)

[c] Scripps GPS web processing service: Information: http://lox.ucsd.edu/~matthijs/Pythagoras/HelpPythagoras.html, Interface: http://lox.ucsd.edu/cgi-bin/Pythagoras.cgi, Uploading site: ftp://jon.ucsd.edu/pub/rinex

14.3.2. Academic institutions

Many geography, geomatics, geospatial and surveying departments in North America offer short courses in pertinent subjects. Lists of departments and their specialization's can be found on many web sites. Good starting points are the web pages of the national geodetic agencies (see above).

14.3.3. International associations

There are a number of international associations, some scientific, others discipline-based, that are sources of information on geospatial matters that are of global relevance. Examples are:

- International Association of Geodesy, http://www.gfy.ku.dk/~iag/
- International Geographic Union, http://www.igu-net.org/
- International Cartographic Association, http://www.icaci.org/
- International Hydrographic Organization, http://iho.shom.fr/
- International Federation of Surveyors, http://www.fig.net/figtree/
- International Society for Photogrammetry and Remote Sensing, http://www.isprs.org/
- International GPS Service, http://igscb.jpl.nasa.gov/

14.3.4. National associations and institutions

These bodies generally deal with professional matters such as training, ethical standards, and so on, but also host conferences, publish journals and convene working groups on a variety of topics. There are, in addition, associations or groups organized at the state or province level, or for particular class of geospatial professional or industry (e.g. hydrographic surveyors, government employees, etc.), or common interest groups (e.g. GIS-based resource mapping, etc.). The main national bodies in North America are:

- American Congress on Surveying and Mapping, http://www.acsm.net/
- Canadian Institute of Geomatics, http://www.cig-acsg.ca/

14.3.5. Private organizations

As far as GPS is concerned these are mainly the GPS manufacturers, but include specialist consulting companies. It is not possible to list these organizations in this manual, but the reader will find many such lists under 'web links' at various sites. However, several of the more comprehensive and up to date sites are:

- http://www.navtechgps.com/links.asp
- http://gauss.gge.unb.ca/GPS.INTERNET.SERVICES.HTML
- http://www.gmat.unsw.edu.au/snap/gps/gps_links.htm

References

Department of Defense/Department of Transport, *U.S. Federal Radionavigation Plan 1999*, 2000, 196 pp., available from http://www.navcen.uscg.gov/pubs/frp1999/default.htm.

FAA, *Federal Aviation Authority web page*, 2001a, http://gps.faa.gov/Programs/WAAS/waas.htm.

FAA, *Federal Aviation Authority web page*, 2001b, http://gps.faa.gov/Programs/LAAS/laas.htm.

GSI, *Geographical Survey Institute GEONET web page*, 2001, http://www.gsi.go.jp/ENGLISH/index.html.

IALA, *IALA Guidelines for the Performance and Monitoring of a DGNSS Service in the Band 283.5–325 kHz*, International Association of Lighthouse Authorities, March, 1999.

Langley, R. B., 'Communication links for DGPS', *GPS World*, 4 (5), 1993, 47–51.

OmniSTAR, *Fugro web page*, 2001, http://www.omnistar.com/.

NGS, *National Geodetic Survey CORS web page*, 2001, http://www.ngs.noaa.gov/CORS/cors-data.html.

Radio Technical Committee for Maritime Services (RTCM), *RTCM Recommended Standards for Differential GNSS (Global Navigation Satellite Systems) Service, version 2.2*, 1998, RTCM Paper 11-98/SC104-STD, developed by RTCM Special Committee No. 104.

Rizos, C., 'GPS surveying technology – why doesn't every surveyor own a kit?' *Surveying World*, 8(4), 2000, 26–29.

SCIGN, *Southern California Integrated GPS Network web page*, 2001, http://www.scign.org/.

Thales, *web page*, 2001, http://www.thales-landstar.com/.

USCG, *US Coast Guard National Differential GPS Project web page*, 2001, http://www.navcen.uscg.gov/dgps/ndgps/default.htm.

Witchayangkoon, B. and Segantine, P., 'Testing JPL's PPP service', *GPS Solutions*, 3(1), 1999, 73–76.

Chapter 15

Where do we go from here?

Chris Rizos

Over the last two decades GPS has evolved from an exclusively US military system to a *global utility* that benefits users around the world in many different applications. GPS has clearly ushered in a new era in which any person, no matter where they are, has access to a low-cost positioning technology that does not require special skills to operate. This is certain to lead to many new GPS applications and *ubiquitous* positioning will transform society in a profound way. Although speculating on this is beyond the scope of this manual, readers should nevertheless appreciate that it is not always possible to isolate discipline-specific GPS technology and applications issues from those trends and developments that are emerging *at the general societal level*.

Surveying and geodesy were the first civilian applications of GPS. Since the early 1980s high accuracy GPS positioning instrumentation, software and techniques have been progressively refined to the point that 'GPS surveying' is challenging traditional surveying procedures for many applications. In Part 2 of this manual the authors have attempted to draw attention to those developments that are of particular significance for geospatial applications. *Is there room for further improvement in GPS? Will the GPS techniques ultimately supersede all other technologies?* Predicting the future is always a risky enterprise! However, as stated in Section 7.4.3, GPS is likely to *complement* the traditional EDM-theodolite techniques for routine surveying activities for many years to come. Nevertheless there are many applications, including some forms of mapping and low-to-moderate precision positioning, for which GPS (and follow-on satellite-based systems) will be the *preferred* technology.

In the following sections summary remarks will be made with regard to:

1 Trends in the GPS technology – user hardware and software.
2 Institutional issues at the national and international level, such as datum definition, new applications, GPS regulation, etc.
3 Announced improvements to the GPS space and control segment.
4 Plans in the first decade of the 21st century for the development and deployment of new satellite-based positioning systems.

15.1. GPS instrumentation and techniques

It is possible to identify several user equipment/technique trends in GPS that are based on historical developments:

- Reduction in size, power consumption and cost of basic single-frequency GPS receivers, especially for navigation-type receivers intended for general use. *This is primarily the result of new receiver chip designs.*
- Improvements in receiver firmware that will lead to increased reliability, faster sampling rates, lower noise, and multipath-resistant observations.
- Increased interest in so-called 'middle tier' GPS receivers based on low-cost navigation instruments but which have carrier phase data processing capabilities added (and can be used to smooth pseudorange measurements as well).
- Increased capabilities of survey receivers, continuing the trend to ubiquitous dual-frequency measurement capability, increased internal memory/processing power, and greater automation of operation. *Real-time operation will be the norm for most surveying applications.*
- Signal processing improvements to permit high precision pseudorange measurement, both on L1 (C/A-code) as well as on L2 without knowledge of the P(Y)-code. *Currently such capabilities are only found in top-of-the-line geodetic receivers.*
- Increased use of 'development kits' to permit customization of systems for specific applications. *Hence GPS will become increasingly integrated within complex mobile positioning and control systems.*
- Better antenna design, resulting in an overall decrease in susceptibility to multipath reflections.
- Software improvements, ranging from increased 'targeting' at the application level, to increased automation and user-friendliness as a general principle. In parallel with increased integration of GPS at the hardware level, GPS data processing software will be absorbed into more complex software systems.
- Deployment of specialist GPS receivers for base station operations to support a range of navigation, surveying and geodetic applications. *This will be coupled with improvements in network communications and database technologies.*
- Increased efficiency in carrier phase-based positioning as a result of a range of improvements in ambiguity resolution algorithms. *The 'Holy Grail' is reliable single-epoch AR.*
- Integration of GPS with other navigation/positioning technologies such as gyroscopes, magnetic compasses, tiltmeters, accelerometers, vision systems, laser scanners, and microwave signals from pseudolites and non-GPS satellites, as well as integration with other sensors in general.

Many of these trends have their genesis in developments in several crucial areas. These are briefly discussed below.

15.1.1. Some GPS hardware trends

There are two basic observations to be made: (a) the GPS receivers (and ancillary equipment) will continue to improve, and (b) integrated sensors (in which GPS is but one component) will become more common. Both developments will not only contribute to traditional surveying and mapping applications, but will also be able to address many other applications that require high precision positioning information. The dramatic improvements in the *productivity* for carrier phase-based techniques have been primarily the result of several innovations: improved GPS receivers, implementation of real-time algorithms, and the development of sophisticated 'on-the-fly' ambiguity resolution algorithms (OTF-AR). The top-of-the-line dual-frequency geodetic-grade GPS receivers are capable not only of making L1 and L2 carrier phase measurements, but also precise pseudorange measurements on both frequencies. Improvements in signal processing, antenna technology and tracking electronics will result in better quality measurements which can directly contribute to shortening the 'time-to-AR' from the current several tens of seconds to just a few seconds or less.

The key to the development of *next generation* receivers is new chip designs that will permit the fabrication of a complete GPS receiver on a single chip, such as those constructed of new materials such as Silicon-Germanium (SiGe). This will result in further reduction in size, power consumption and cost of the basic receiver components, which is likely (together with other parallel developments in communications, software and infrastructure) to lead to increased integration of the GPS technology with other electronic devices. Examples include the integration of GPS into standard surveying instruments, 'Personal Digital Assistants' and mobile phones, palmtop computers, laser range finders, and so on. Although GPS receivers would become the *enabling* technology within more complex integrated systems, even if attention is focused on 'geospatial sensors' these developments will be indeed significant. For example, a surveying/mapping instrument could have a GPS receiver embedded within it. This would permit all geospatial measurements to be ultimately linked through GPS to the global datum.

No doubt one of the most significant recent developments in precise GPS has been 'Real-Time Kinematic' (RTK). While real-time pseudorange-based Differential GPS (DGPS) has been available to users for well over a decade, the shift in carrier phase data processing techniques from the post-mission mode to real-time has been a relatively recent one. The continued development of RTK-GPS techniques, together with the adoption of new mobile communication technologies, augers well for the future of centimeter-level accuracy GPS for a host of time-critical applications such as engineering stake-out, machine guidance and control, and on-line deformation monitoring systems. New communication technologies also aid DGPS. Hence, precise GPS techniques become *almost* as easy as the standalone GPS positioning capability enjoyed by casual users. This is discussed further in the context of the provision of the appropriate *GPS infrastructure* to support precise GPS applications (Section 15.3).

15.1.2. Comments on GPS software trends

While it is fair to say that software is a crucial component of new GPS receivers (with internal analogue hardware operations now replaced by software), nowhere has software made as dramatic an impact as in the case of carrier phase-based techniques. From the beginning of the use of GPS as a tool of geodetic surveying, complex software has been required to convert the raw GPS measurements to useful quantities. In particular, the OTF-AR algorithm has been most responsible for the increased productivity and flexibility of modern GPS surveying techniques. However, most GPS researchers admit that the development of data processing algorithms has nearly reached its limit in terms of ensuring centimeter-level positioning accuracy with the minimum amount of tracking time. It is likely that further improvements in carrier phase-based techniques will therefore be restricted to:

- New strategies of combining data from mobile GPS users with data from permanent continuously-operating GPS reference stations.
- The transmission of additional signals by GPS satellites.
- Augmenting GPS measurements with other data, such as from a variety of ground-based or space-based systems.
- Special applications in which additional constraints can be imposed (e.g. fixed distance between moving antennas as in the case of GPS-based attitude determination systems).

The first three of these issues are discussed further in Sections 15.3–15.5.

15.2. Institutional issues

A range of what may be referred to as 'institutional issues' impact on the adoption and use of GPS technology for geospatial applications. Several of these are briefly mentioned, under a number of headings. One of these institutional issues, the installation of increasing numbers of permanent GPS receivers around the world, deserves special mention (Section 15.3).

15.2.1. GPS system control

For many years there has been a 'tug-a-war' between civilian and military users. The former wishing to have a greater 'say' in the running of GPS while the latter have sought to maintain a tight control on what is considered a strategic asset in modern warfare. The dilemma is that GPS is a 'dual-use' technology, investing significant advantages to both classes of users. The debate on 'GPS control' has been played out mainly in the US, between industry and high profile user communities (such as aviation) on the one hand and the Department of Defense on the other. However, while the rest of the world appreciates the enormous benefits of GPS, there has also been increasing uneasy with the degree of acceptance of GPS for many crucial applications.

The following comments may be made in this regard:

- *Civilian users far out-number military users*. Yet there are few means by which civilian users can influence the Department of Defense. Plans for 'GPS modernization' (Section 15.4.1) do, however, go a significant way towards addressing the needs of civilian users for increased accuracy and signal availability.
- *GPS is a global utility*. Many users outside the US are concerned that they are relying on the use of a system controlled by the military of one country, and there are now plans to develop alternative satellite-based positioning systems (Section 15.5).
- *Control of access to the highest single-receiver performance by the US military* has meant the imposition of policies of Selective Availability (SA) on 25 March 1990, and Anti-Spoofing (AS) on 31 January 1994. Both policies have resulted in extra

cost and a reduced level of service for civilian users. (SA was abandoned on 1 May 2000, while there will be no further impact of AS on civilian users when the 'GPS modernization' program has been completed – Section 15.4.1.)

15.2.2. GPS datum issues

One of the first major impacts of GPS on many countries has been on *national geodesy*. GPS is unchallenged as the supreme geodetic tool for establishing, maintaining or renovating national control networks. Under pressure from users to produce maps on the 'GPS datum' (understood by most users to be WGS 84), many countries have redefined their national geodetic datums. Although there is some confusion in the general community between WGS 84 and the International Terrestrial Reference System (ITRS), most redefined datums are in fact some realization of the ITRS (Section 7.2). Many international maps and charts are being produced on WGS 84, which for all intents and purposes is equivalent to ITRS datums at the sub-meter level. Other impacts of GPS on datum and control networks include:

- In addition to the traditional levels of national control network (first order/class, second order/class, etc.), additional high accuracy control points established using ultra-precise 'GPS geodesy' techniques now form the 'backbones' of new datums.
- There is increasing acceptance of international standards and associated services such as provided by the International GPS Service (IGS). Many geocentric datums that have been established in the last decade or so are therefore examples of the 'densification' of the ITRS using techniques being promoted by the IGS and its committees.
- Although GPS is used to establish 3-D coordinate networks, many applications still distinguish horizontal positioning/mapping from the determination of 'heights above sea level'. *Hence GPS, geoids and leveling are inextricably linked.*
- With the increased efficiency of GPS for establishing control point coordinates, there are calls in some developed countries to reduce the number of monumented control marks. Such an issue will have to be addressed as more and more countries establish permanently operating GPS reference stations from which surveyors can easily make baseline connections.
- Where monumented control marks are established they will be increasingly located where they are most needed, and not on inaccessible hilltops as in the past. However, although intervisibility of stations is not necessary for GPS, it is still necessary to have pairs of intervisible points to define azimuth for conventional ground survey techniques.

15.2.3. GPS acceptance and regulation

There are a range of issues that must be dealt with at a national, and even state/province, level if GPS is to be accepted and even promoted for specific applications. Some of these are:

- The blurring of responsibilities for the provision of GPS infrastructure such as, for example, permanent GPS reference stations to support real-time DGPS, geodesy, integrity monitoring, etc. Greater efficiencies would be won, for example, if such infrastructure were *multi-functional*, rather than having many stations being operated by different organizations and agencies.

- How to regulate GPS services provided by real-time service providers? Should there be independent *quality assurance*? Who is liable for damages due to faults in the GPS system (space and control segment), or its local augmentation, including reference station networks?

- Issues such as GPS system testing, accreditation of GPS surveyors, procedures for GPS cadastral surveys, legal traceability of GPS results, etc. will become increasingly important.

- In future GPS positioning will not only be the prerogative of the 'positioning professions' such as navigation and surveying, and that will raise a host of legal and administrative issues concerned with education, legal liability, community safeguards, etc.

- GPS and GIS are crucial technologies for the development of the so-called 'Spatial Data Infrastructure' (SDI). While there is increasing recognition of SDI at the international level, most of the groundwork is being done at the national level. Hence the challenge is to develop national guidelines and procedures for the SDI that are also compatible with international initiatives and standards.

- The challenge for government is to promote the adoption of GPS and the development of associated services (all of which contribute to a nation's economic well being) without undermining customary practices that have served the community in the past. This is generally addressed at the application or industry-specific level.

15.3. Continuously-operating GPS networks

Continuously-Operating Reference Station (CORS) networks have been established in many countries to initially support high precision geodesy-type applications. However, there is increasing interest in using such GPS *infrastructure* to support a range of other applications, including surveying.

15.3.1. CORS networks and GPS surveying

One of the defining characteristics of the next generation of carrier phase-based GPS systems may well be the loosening (or abandonment) of the requirement for user ownership/operation of GPS base station receivers. CORS networks offer the possibility of users being able to perform differential positioning using a single GPS receiver, hence *halving equipment costs*. Although this appears to offer advantages, current implementations require that data from the *nearest* GPS base station is downloaded to the user's computer before processing can be carried out (Section 14.2.2). However, in practice, often there are not sufficient GPS base stations, nor appropriate data handling or transmission services, to replace the 'do-it-yourself' approach that has been conventionally used. Of course the GPS manufacturers promote this, as it is much easier to sell a total 'turn-key' system consisting of a pair of receiver systems.

As already mentioned, current carrier phase-based GPS techniques require the reference receiver to be within 10 or so km of the survey area to ensure reliable OTF-AR, particularly for RTK-GPS techniques. However, the establishment of a network of GPS reference receivers at a density to support all (or most) GPS surveys is hardly feasible. In order to overcome the baseline length limitation, the development of network-based methods to model and predict differential errors commenced in the mid-1990s (e.g. Wübbena *et al.* 1996; Raquet and Lachapelle 2001). In essence, the carrier phase observations from all reference stations are forwarded to a data center where the differential errors are modeled for the entire region covered by the network of reference stations. Corrections to the carrier phase observations for the reference stations and users are calculated. These corrections can be broadcast to users in an analogous manner to pseudorange-based Wide Area DGPS (Section 11.3.3). Once they are applied to the carrier phase observations, the OTF-AR procedure is the same as when using a single reference station.

Several commercial real-time network-based services have been established which, *unlike* conventional RTK techniques, do not require L1 and L2 AR. Differential corrections are transmitted to users via geostationary satellites (as with WADGPS services) and although carrier phase observations are indeed processed, only the so-called 'widelane' ambiguities (formed by differencing the L1 and L2 measurements) are resolved to their integer values. As a consequence, positioning accuracy is only at the several decimeter level. However, the advantage is that distances between reference and user receivers may be up to a 1,000 km or more, and therefore current WADGPS reference station networks can be utilized for such services. Such services therefore address the accuracy 'gap' between conventional WADGPS techniques on the one hand, and true centimeter-accuracy network-based RTK-GPS systems on the other.

15.3.2. Network-based GPS positioning

The first commercial, high precision, network-based GPS positioning system has recently been announced by the Trimble Company. Commercial network-based RTK-GPS services capable of centimeter-level accuracy will be established over the next few years in several countries. As described above, in such scenarios, as far as the user is concerned, the GPS hardware and field procedures will be no different to current single reference station RTK-GPS systems, except that the distance to the CORS network stations is of the order of several tens of kilometers as opposed to 10 km or less. However, there are a number of other advantages of *network-based processing strategies* apart from permitting an increase in inter-receiver distances:

1. Single-epoch OTF-AR algorithms can be used for GPS positioning, while assuring high accuracy, availability and reliability for critical applications.
2. Rapid-static positioning is possible even using low-cost, single-frequency GPS receivers.
3. Accuracy and reliability of continuous deformation monitoring applications (for dams, buildings, bridges, etc.) is enhanced.
4. Allows for post-mission surveying/mapping via an upload of user data to a web-based data processing service such as the ones briefly reviewed in Section 14.2.3.

How can network-based techniques improve GPS surveying performance? This question can be answered in a number of ways. If the baseline length constraint is the crucial factor impacting on static or kinematic GPS, then overcoming this constraint through the use of network-based processing techniques can result in significant performance improvements. If 'improving performance' is associated with lowering receiver costs, then a multi-reference receiver network would allow the use of single-frequency GPS receivers instead of expensive dual-frequency receivers. If 'time-to-AR' is the critical performance indicator, then network-based processing strategies would contribute to faster AR (including single-epoch AR) as well as enhancing the reliability of AR, even for baselines several tens of kilometers in length. However, the implementation of a GPS positioning technique based on data from a network of GPS base stations is much more complicated than the standard single reference receiver scenario.

Rarely would the benefits of improved positioning performance be great enough for a single user that they would be compelled to establish and operate their own CORS network. Instead a *service* could be provided by a government agency, or by private companies on a fee-for-service basis. Given that CORS networks can address many applications (including non-positioning ones), there will be increasing interest in providing enhanced user services based on the combined processing of *all* reference receiver data. Therefore it is expected that such CORS-based surveying services will be established to address the GPS survey needs of a large metropolitan area, and that such services will be offered either in real-time or via post-mission processing.

15.4. GPS satellite and modernization plans

In 1996 the former US President Bill Clinton gave an assurance that GPS would be freely available well into the first decades of the new millennium (OST 1996). Since 1999 several initiatives intended to improve the system have been announced. Before discussing these, the following comments are made to set the context in which this renewed interest in satellite-based positioning can be considered:

- In 1998 the first of the Block IIR satellites became operational. These are the GPS 'replenishment' satellites that have enhanced features including inter-satellite ranging and onboard ephemeris processing capabilities. Modifications will be made to the last 12 Block IIR satellites to accommodate the extra signals proposed under the 'GPS modernization program' (see below).
- The design of the Block IIF ('follow-on') satellites is complete, and the satellites will be launched from about 2006 onwards, although efforts are underway to 'fasttrack' this.
- The process of designing the system architecture for the next generation 'GPS III' satellites has commenced.
- There are increased opportunities of the transmission of GPS-like signals from other (mainly communications) satellites.
- Although the Russian Federation's Glonass system is in dire straits as a result of low funding, it is possible that additional satellites will be launched to keep the constellation 'alive' (Section 15.5.1).

- Development work on the FAA's Wide Area Augmentation System (WAAS) is nearing completion (Section 14.1.1.2), and testing has commenced in earnest (Zeltser 2000).
- In 1999 the EU committed itself to the development of a European navigation system known as 'Galileo', to be operational by around the year 2008 (Section 15.5.2).
- In 2000, at the World Radiocommunication Conference, frequency spectrum was assured for the current and future GPS signals, and for use by the planned Galileo system.

The issue of 'GPS modernization' has recently attracted significant attention because of the dramatic improvements to GPS performance that the removal of SA has made (OoP 2000). The modernization of GPS is a difficult and complex matter involving not only the civil and military users (and their requirements), but such further issues as costs (for the modernization), spectrum definition, security, institutional concerns about management, and the future operation of GPS as a national and international resource. By 1999 most of the elements of the modernization plans had been identified, and funding of these GPS improvements had commenced.

15.4.1. GPS improvements in the next decade

Much of the information on the 'GPS modernization' plans is drawn from the excellent articles by McDonald (1999) and Shaw *et al.* (2000). The GPS improvements identified for implementation in the next decade are:

- New signals and changed signal codes.
- Abandonment of the policies of SA and AS.
- Improvements to the control segment (Section 15.4.2).
- Increased number of satellites and the power levels of the transmitted signals.
- Increased use of augmentations of the GPS system such as provided by WAAS and LAAS.

The Presidential Decision Document on GPS of 29 March 1996 (OST 1996) implied that both the L1 and L2 frequency bands would be available for civilian use, and that a third civil frequency signal known as L5 (Figure 15.1) would be transmitted (Van Dierendonck and Hegarty 2000). The former is a change to the AS policy by allowing the civilian C/A-code (or a suitable replacement code) to be added to the L2 frequency (with power boosted to that of the current L1 signal), while the latter set off an intensive search for new frequencies that would satisfy a range of requirements, including the military and the Federal Aviation Authority. The former Vice President Al Gore announced on 25 January 1999 that the third frequency had been identified at 1176.45 MHz (just below the GPS L2 frequency), in a region of the spectrum referred to as the Aeronautical Radio Navigation Service band that satisfies aviation 'safety-of-life' requirements (Zeltser 2000). This third frequency will also have a modified code design for more precise measurements (McDonald 1999; Shaw *et al.* 2000).

It is planned that the new dual-frequency civilian tracking capability will be available on the last 12 of the Block IIR satellites (and all subsequent satellite generations),

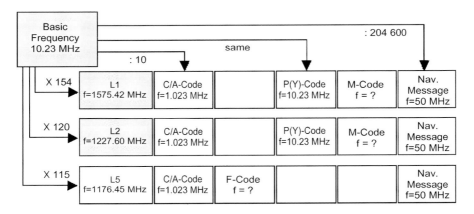

Figure 15.1 Proposed structure of the new GPS LI, L2 and L5 satellite signals (cf. Figure 8.1).

the first being slated for launch in 2003. The L5 frequency will be available for the first time on the Block IIF satellites, that have their first scheduled launch in 2006. A constellation of at least 24 satellites having the L5 signal may not be available before the end of this decade, although suitably equipped civilian users will be able to take advantage of the new signals to eliminate ionospheric delay on (some of) the signals well before then. The dual- and triple-frequency carrier phase measurements will significantly benefit 'on-the-fly' ambiguity resolution, and reliable single-epoch AR will be a reality when at least four Block IIF satellites can be tracked simultaneously.

The military dependence on the C/A-code to provide the timing information necessary to acquire the P(Y)-codes will be eliminated with the transmission of a new set of military codes, called the *M-codes* (these codes will be encrypted). Present indications are that these M-codes will be transmitted on a 'split spectrum' scheme, offset 12 MHz above and below the centers of the GPS L1 and L2 frequencies. Such a scheme will at last ensure that the civilian and military users will have separate signals as well as codes, albeit transmitted by the same constellation of GPS satellites.

These plans are welcomed, and have the potential of cementing the dominant role of GPS as the de facto standard for satellite system performance, and a crucial component of any international Global Navigation Satellite System (GNSS). Nevertheless, in the international community there are long-standing concerns about the US military's control of the system. In addition there are doubts about whether the executive and legislative branches of the US Government recognize the importance of 'GPS modernization', as well as the risks involved in delaying or discarding various aspects of the program.

15.4.2. The GPS accuracy improvement initiative

Improvements will also be made to the GPS control segment, and under the current schedule all hardware and software upgrades will be operational by about 2005 (Hay 2000). Recent improvements in the Master Control Station software used to estimate the GPS satellite orbits (and to generate the navigation message) have demonstrated that broadcast ephemeris accuracy is at the 1–2 m level (mostly). The addition of six ground stations of the National Imagery and Mapping Agency to the GPS tracking network will improve the quality of the GPS tracking measurements and the computed orbit and satellite clock error parameters further. More frequent uploads to the satellites are also planned (two per day to each satellite). Several other improvements will also be made, including an upgrade of the monitor station clocks, monitor station multipath mitigation, and improved tropospheric delay prediction. All of these complement the improvements in the current and future constellations of GPS satellites referred to in Section 15.4.1.

15.5. Other satellite-based positioning systems

There are plans by some countries to develop new satellite systems to complement (or compete) with GPS within the next decade or two. The most advanced proposals are those of the EU. However there is already an operational satellite positioning system that has been developed by the Russian Federation, the GLONASS system.

15.5.1. The Glonass system

Glonass was developed for the Russian military, and has the following characteristics (Glonass 2001):

- 24 satellites, distributed in three orbital planes.
- 64.8° inclination, 19,100 km altitude (11 h 15 min period).
- Dual-frequency (L1 in the range: 1597–1617 MHz; L2 in the range: 1240–1260 MHz).
- Each satellite transmits a different frequency on L1 ($= 1602 + K \times 0.5625$ MHz; $K \in [-7,24]$) and L2 ($= 1246 + K \times 0.4375$ MHz; $K \in [-7,24]$).
- Spread-spectrum pseudo-random noise code signal structure.
- Global coverage for navigation based on simultaneous pseudo-ranges, with an autonomous positioning accuracy of better than 20 m horizontal, 95% of the time.
- A different datum and time reference system to GPS.
- A Precise Positioning Service and a Standard Positioning Service, as in the case of GPS (Section 7.3.2).

Although some of the characteristics of Glonass are similar to GPS, there are nevertheless significant technical differences. In addition, the level of maturity of the user receiver technology and the institutional capability necessary to support the Glonass space and control segment are significantly less than in the case of GPS. Glonass was declared operational (with 24 satellites in orbit) in 1996, yet at the time of writing of this manual (mid 2001) the number of operational satellites had fallen to just seven. In fact Glonass will continue to be viewed by many user communities as a technically inferior system to GPS, a system with which there are many 'question marks' regarding its long-term viability. Yet to dismiss Glonass as a serious candidate for a 21st century satellite positioning technology because it cannot *compete* with GPS technology is too simplistic an analysis. Although Glonass has the potential to rival GPS in coverage and accuracy, this potential is unlikely to be realized in the medium term. For the foreseeable future Glonass should be considered a *complementary* system to GPS, as the

tracking of both GPS and Glonass satellite signals improves the accuracy and reliability of positioning.

The development of integrated GPS–Glonass receivers which *measure carrier phase*
offers special challenges. Another challenge is that the signals to the different Glonass
satellites are of different frequency, making the standard GPS data processing strategies inappropriate. However, the extra satellites that can be tracked to make precise
positioning a more robust procedure. During the last few years several research groups
have been researching optimal Glonass data processing techniques. In addition,
several companies have developed sophisticated GPS–Glonass receivers capable of
undertaking all types of satellite surveys, including real-time kinematic systems.
However, the uncertainty surrounding Glonass's future (the last satellite launches
were in October 2000) is stifling much needed market investment in new generation
receiver hardware.

15.5.2. The Galileo system

For many years Europe[1] has been drawing attention to the fact that the world is
increasingly embracing a crucial technology that is owned and controlled by one
country (and the military agency of that country). They have been scathing in the
criticism of this trend and have hardly acknowledged that even their own citizens (and
many of their private companies and government agencies) have enjoyed the benefits
of the GPS user technology without having to pay for it. Initiatives such as the
European Geostationary Navigation Overlay System (EGNOS), which is designed
to augment the GPS system for civil aviation applications in the European region,
were also intended to gain for Europe some degree of independence from GPS 'hege-
mony' and to 'kick start' their hi-tech navigation industries.

By 1999 the concept of a European-designed and funded satellite-based navigation
system known as *Galileo* had finally been proposed. By the year 2000 details started to
emerge in publications and at conferences. The reader is referred to Gallimore and
Maini (2000), and to articles in magazines such as *GPS World* and *Galileo's World* for
up to date information. As of mid-2001 the following was known about Galileo:

- The system will be independent of both GPS and Glonass, and will not use any of
  the frequencies already allocated to these navigation systems.
- Funding of the order of US$3 billion is required, some provided by the European
  Commission and the European Space Agency, and the rest expected to be
  provided through a business model known as 'Public–Private Partnership' (PPP).
- There will be several levels of service, some free-to-users (similar to the standard
  positioning service for GPS and Glonass), and others will attract fees.
- The aviation market is being particularly targeted as it is felt it will support
  Galileo because of the safety-critical features that such a service will provide.
- The Galileo system may be operational in 2008.

However, despite the rhetoric of creating a new satellite-based navigation system
that will complement GPS (and Glonass) there are still many disturbing issues that
have not been resolved. For example, one of the prime criticisms of GPS has been

---

[1] As separate countries, or through the EU.

that it is a 'military system'. Yet recent announcements by the European officials hints that Galileo will have a military 'dimension', and that it may have 'levers that can be pulled' if European security is at risk! Another concern is that the failure of European industry to embrace the PPP model could lead to the entire system being underwritten by public funds and the subsequent *strong encouragement* for industry to purchase Galileo equipment even if GPS already addresses their needs. There has even been some discussion concerning a 'tax' on user navigation equipment that could be levied to help fund Galileo. There are other ill-defined aspects of Galileo, including:

- The number of satellites in the Galileo constellation.
- The orbital characteristics of the constellation.
- The frequencies and ranging codes of the signals.
- The inducements that will be offered to purchase Galileo user equipment.

However, despite assurances that the Galileo system will *complement* GPS and Glonass, there is little evidence to date that in fact this will be the case. Europe claims that it can succeed in this venture, as it has in developing the Ariane launchers, the Airbus aircraft and the GSM wireless communications standard. It remains to be seen whether the additional signals from the Galileo satellites will be able to be tracked, and the measurements processed, in combination with those of GPS within the next generation (or two) of satellite positioning receivers. That is the outcome users all over the world would like to see. GPS has evolved over two decades into a tremendous positioning system and a remarkable surveying/mapping tool. Within a decade a modernized GPS system will share the stage with other satellite-based positioning systems.

References

Gallimore, J. and Maini, A., 'Galileo: the public-private partnership', *GPS World*, 11(9), 2000, 58–63.

GLONASS, *Glonass information web page*, 2001, http://www.rssi.ru/sfcsic/english.html.

Hay, C., 'The GPS accuracy improvement initiative', *GPS World*, 11(6), 2000, 56–61.

McDonald, K., 'Opportunity knocks: will GPS modernization open doors?' *GPS World*, 10(9), 1999, 36–46.

Office of Science and Technology USA, *U.S. Global Positioning System Policy, US Presidential Decision Directive*, March 29, 1996, http://www.navcen.uscg.gov/pubs/gps/white.htm.

Office of the US President's Press Secretary, *Statement by the President Regarding the United States' Decision to Stop Degrading Global Positioning System Accuracy*, May 1, 2000, http://www.ngs.noaa.gov/FGCS/info/sans_SA.

Raquet, J. and Lachapelle, G., 'RTK positioning with multiple reference stations', *GPS World*, 12(4), 2001, 48–53.

Shaw, M., Sandhoo, K. and Turner, D., 'Modernization of the Global Positioning System', *GPS World*, 11 (9), 2000, 36–44.

Van Dierendonck, A. J. and Hegarty, C., 'The new L5 civil GPS signal', *GPS World*, 11(9), 2000, 64–71.

Wübbena, G., Bagge, A., Seeber, G., Böder, V. and Hankemeier, P., 'Reducing distance dependent errors for real-time precise DGPS applications by establishing reference station networks', *Proc. 9th Int. Tech. Meeting of The Satellite Division of The U.S. Institute of Navigation*, Kansas City, Missouri, 17–20 September, 1996, 1845–1852.

Zeltser, M. J., 'The status of aviation-related improvements', *U.S. Institute of Navigation Newsletter*, 10(1), Spring, 2000, http://www.ion.org/newsletter/v10n1.html.

Part 3

Remote sensing

Chapter 16

Photogrammetric and remote sensing considerations

Don L. Light and John R. Jensen

16.1. Overview

It is possible to obtain qualitative and quantitative information about an object or area on the Earth's surface (or other celestial body) without being in direct physical contact with the surface. This task can be accomplished using specialized photographic camera systems or by the use of optical-mechanical remote sensing detectors. Normally, the term *photogrammetry* is reserved for the collection and interpretation of photographic images. The term *remote sensing* is used generally when we speak of collecting imagery using optical-mechanical sensor systems often in regions beyond the visible and near-infrared portion of the electromagnetic spectrum.

Chapters 4 and 6 reviewed the fundamental geometric equations and energy/matter interactions that are so important when obtaining aerial photography and other types of remote sensor data for earth resource applications. This chapter formally defines photogrammetry and remote sensing and introduces the breadth of the fields. Subsequent chapters review:

- Additional principles concerning the interaction of electromagnetic energy and earth surface materials;
- How we collect the remote sensor data;
- The digital image processing software and hardware required to interpret the digital remote sensor data;
- Methods of analyzing the remote sensor data;
- Methods for assessing the accuracy of thematic information extracted from remote sensor data;
- Representative applications of the use of remote sensor data, and
- A perspective on the future of remote sensing.

16.2. Photogrammetry

Photogrammetry was initiated about 1850 in Europe when the French Colonel A. Laussedat began to use photographs for obtaining measurements of the earth (Hallert 1960). Although the Europeans were using photogrammetric principles in the late 1800s, the word 'photogrammetry' did not come into general usage in the US until about 1934 when the American Society of Photogrammetry was founded. The term 'photogrammetry' is derived from the Greek words, *photos* meaning 'light', *gramma* meaning 'something drawn or written', and *metron* meaning 'to measure'. Combining these three root words originally signified measuring graphically by means of light. The original official definition of *photogrammetry* as given by the *Manual of Photogrammetry* was (American Society of Photogrammetry, 1944, 1952, 1966):

The science or art of obtaining reliable measurements by means of photographs.

In the fourth edition, the American Society for Photogrammetry and Remote Sensing (1980) expanded the definition of photogrammetry to be:

The art, science and technology of obtaining reliable information about physical objects and the environment through processes of recording, measuring, and interpreting photographic images and patterns of electromagnetic radiant energy and other phenomena.

Photogrammetry is frequently divided into specialties according to the type of photographs or images used, or the location from which they were taken. Instrumentation using photos taken from the ground surface is called *terrestrial* photogrammetry. *Aerial* or *space photogrammetry* connotes the use of traditional analog photographs (ASP 1960; Colwell 1997) or digital photographs (Light 1996, 1999; Hinz 1999) that have been taken from any airborne or spaceborne platform, and these may be either *vertical*, or *oblique* images. The geometry and characteristics of a typical aerial photograph are shown in Figure 16.1. Notice how the negative is a reversal in both tone and geometry of the real-world object space. Stereo-photogrammetry means that overlapping pairs of photography are obtained and measured, or interpreted in a stereoscopic viewing device. This provides a three-dimensional view and creates the illusion that the observer is viewing a relief model of the terrain. Stereo-photogrammetry is essential for measuring the z-coordinate and the stereoview aids in interpretation of the features residing in the imagery. Simple photogrammetric measurements are summarized in Chapter 19.

Both airborne and spaceborne photogrammetry embrace two broad categories of practice. The first is the classical category, in which quantitative measurements are made, and from which ground positions, elevations, distances, areas, and volumes can be computed, or from which planimetric and topographic maps can be drawn. This category is often called precision or *metric photogrammetry*. The most common applications of metric photogrammetry are:

- The production of topographic maps that contain isolines (contours) of equal elevation above datum, and the hydrology and transportation network;
- The extraction of Digital Elevation Models (DEM) that contain the elevation, shape, size, and coordinates of the terrain surface in a matrix of equally spaced x,y,z – coordinates;
- The extraction of planimetric building infrastructure information including the actual outline (perimeter footprint) of buildings, and
- The preparation of orthophoto image map products that have the metric qualities of a line map coupled with the image qualities of an aerial photograph.

The second category is called *photo interpretation* in which photographs are evaluated in a qualitative manner in order to evaluate timber stands, detect water pollution, classify soils, interpret geological formations, identify crops,

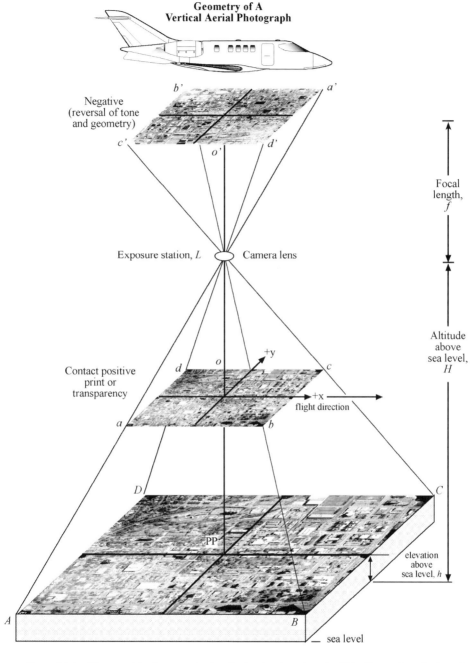

**Geometry of A
Vertical Aerial Photograph**

Negative
(reversal of tone
and geometry)

Focal
length,
f

Exposure station, *L*   Camera lens

Contact positive
print or
transparency

+y

o

+x
flight direction

Altitude
above
sea level,
H

PP

elevation
above
sea level, *h*

sea level

Figure 16.1   Most aerial photography is obtained using aircraft. This illustration depicts the geometry and coordinate characteristics of a 1:6,000-scale vertical aerial photograph of downtown Columbia, SC shown in perspective. The negative is a reversal of both the tone and geometry of the real-world while the positive print maintains the same geometry and tonal characteristics as the real-world (after Jensen, 2000).

obtain military intelligence, etc. (Colwell 1984). Photo interpretation coupled with fundamental photogrammetric analysis often results in the creation of thematic maps of:

- Land use (categories describing the human use of the land, e.g. recreation);
- Land cover (categories describing the actual ground surface cover, e.g. forest); and/or
- Biophysical information such as the identification of areas of potential stress in a color-infrared photograph of an agricultural field.

Since about 1940, photogrammetrists have transitioned from projection instruments, analogue and analytical plotters (computer assisted plotters), to soft-copy, computer driven workstations (Helava 1988). Today most photogrammetry problems are processed using Personal Computers (PCs) with software that handles photogrammetric problems using purely mathematical methods (Leberl 1994; Roberts 1996). The ever-growing power of computers since the 1950s has favored the development of so-called 'analytical photogrammetry'. In analytical photogrammetry, as distinguished from instrumental, a mathematical model is constructed to represent the geometric relations between points in the object space, the perspective center in the lens, and the image points in the imagery or photographs. These three points fall on a straight line. They are said to be co-linear, giving rise to the *colinearity conditions* for modern analytical photogrammetry (the colinearity equation is presented in Chapter 4). The computer's application of numerical analysis to this math-model results in the solution of problems for camera calibration, space intersection, and triangulation of groups, strips, or blocks of aerial, space or terrestrial photographs (Wolf and Dewitt 2000).

16.2.1. Airborne reconnaissance systems

The number of airborne reconnaissance systems in operation is substantially greater than the number of satellite systems. Airborne systems can also operate over a broad range of the electromagnetic spectrum. They can change altitude and sensors, where a satellite system is fixed once in orbit. Airborne sensors generally achieve higher spatial resolution, many at the submeter level. Airborne reconnaissance platforms can vary their altitude, fly under high clouds and with changing priorities are easily re-locatable. Characteristics of both airborne and satellite systems are characterized in Table 16.1.

Airborne military photo-reconnaissance usually means aircraft that can fly high (>30,000 ft above ground) and fast (>420 knots) to obtain reconnaissance imagery in a short time over the target, preferably undetected. Photo-reconnaissance for monitoring forest fires, floods and natural disasters can utilize ordinary aircraft when being detected is not a problem. Military photo-reconnaissance generally relies on fast aircraft and long focal length film or digital cameras. Some of the US aircraft used in aerial reconnaissance over the years are summarized in Table 16.2.

16.3. Remote sensing

The term *remote sensing* was first introduced in the early 1960s by the staff of the Geography Branch of the Office of Naval Research (ONR) (Pruitt 1979; Jensen 2000).

Table 16.1 Airborne versus satellite reconnaissance data collection characteristics

Characteristics	Airborne	Satellite
Altitude	Low to 30,000 m	175–1000 km
Swath width	Narrow (many strips)	Usually wide (large areas)
Resolution	Can be varied to meet needs	Fixed by satellite orbit (except for radar)
Illumination	Variable non-sun-synchronous	Usually sun-synchronous
Revisit	Frequent and flexible	Fixed by orbit-days, unless it has an off-nadir pointing capability
Geographic responsiveness	Re-locatable	Fixed by orbit-days
Flight path	Variable, user-defined	Fixed by orbit
Access	Intrusive	Non-intrusive

Aerial photo interpretation had become very important in World War II. The space age was just getting underway with the 1957 launch of Sputnik (USSR), the 1958 launch of Explorer 1 (US), and the collection of photography from the then secret CORONA program initiated in 1960. The Geography Branch of ONR was expanding its research using instruments other than cameras (e.g. scanners, radiometers) and into regions of the electromagnetic spectrum beyond the visible and near-infrared regions (e.g. thermal infrared, microwave). In the late 1950s it became apparent that the prefix 'photo' was being stretched too far in view of the fact that the root word, photography, literally means 'to write with (visible) light' (Colwell 1997). Pruitt (1979) wrote:

> The whole field was in flux and it was difficult for the Geography Program to know which way to move. It was finally decided in 1960 to take the problem to the Advisory Committee. Walter H. Bailey and I pondered a long time on how to present the situation and on what to call the broader field that we felt should be encompassed in a program to replace the aerial photointerpretation project. The term 'photograph' was too limited because it did not cover the regions in the electromagnetic spectrum

Table 16.2 Reconnaissance aircraft often used for photogrammetry and remote sensing

Aircraft	Service ceiling (ft.)	Speed (mph)	Range (mi.)
Cessna Skywagon	19,500	160	1,200
Beechcraft Bonanza	16,000	195	1,290
Cessna Super Skymaster	29,300	190	1,550
Beechcraft Queen Air	26,800	225	1,500
Grumman Mohawk	29,500	305	1,400
Rockwell Commander	45,000	500	1,840
Gates Learjet	45,000	510	1,500
Fairchild Hiller Porter	25,500	175	600
Lockheed NP3A Orion	30,000	380	2,400
Lockheed RC 130	35,000	368	2,500
Boeing RC 135	45,000	530	4,600
Martin RB 57	62,000	460	3,300
McDonnell Douglas RF-4	50,000	1,390	2,300
Lockheed U-2	70,000	490	2,200
Lockheed SR71	>80,000	2,070	2,000

beyond the 'visible' range, and it was in these non-visible frequencies that the future of interpretation seemed to lie. 'Aerial' was also too limited in view of the potential for seeing the Earth from space.

Based on these discussions, the term *remote sensing* was promoted in a series of symposia sponsored by ONR at the Willow Run Laboratories of the University of Michigan in conjunction with the National Research Council throughout the 1960s and early 1970s and has been in use ever since (Estes and Jensen 1998). Remote sensing was formally defined by the American Society for Photogrammetry and Remote Sensing (ASPRS) as (Colwell 1983):

> The measurement or acquisition of information of some property of an object or phenomenon, by a recording device that is not in physical or intimate contact with the object or phenomenon under study.

In 1988, ASPRS adopted a combined definition of photogrammetry and remote sensing (Colwell 1997):

> Photogrammetry and remote sensing are the art, science, and technology of obtaining reliable information about physical objects and the environment, through the process of recording, measuring and interpreting imagery and digital representations of energy patterns derived from non-contact sensor systems.

Remote sensing imagery may be acquired, not only through the use of conventional cameras, but also by recording the energy from the scene through one or more special sensors. These special sensors usually operate by electronic scanning and can record the existence of reflected or emitted radiation outside the normal visual range of film and standard cameras. Today's remote sensors can record multispectral (up to 20 spectral bands in the ultraviolet, visible, near-infrared, middle-infrared portions of the spectrum), hyperspectral (more than 20 spectral bands), active microwave energy (radar), passive microwave energy, thermal infrared energy, and active laser energy (Lidar).

16.3.1. Satellite remote sensing

Remote sensing systems may be flown onboard suborbital aircraft (e.g., Figure 16.2) and on orbital satellites (e.g., Figure 16.3). For example, NASA's Jet Propulsion Laboratory personnel routinely acquire detailed spectral reflectance imagery in 224 bands (each 10 nm in dimension) using the Airborne Visible Infrared Imaging Spectrometer (AVIRIS) flown at altitudes ranging from 3.5–20 km above ground level. This results in imagery at 3.5×3.5 m or 20×20 m spatial resolution. Conversely, an artist's rendition of the Enhanced Thematic Mapper plus (ETM$^+$) sensor flown onboard Landsat 7 launched in April, 1999, is shown in Figure 16.3. Common characteristics of orbital remote sensing systems include sun-synchronous orbits, wide swath, fixed resolutions, and fixed nadir revisit periods (refer to Table 16.1). A sun-synchronous orbit gives predictable global coverage with repetitive access over the same area at the same local sun time depending on its repeat cycle. A satellite platform is generally more stable than an aircraft platform that is buffeted by atmospheric turbulence. Some new commercial satellites are using or plan to use high latitude (polar) orbits to achieve more frequent repeat cycles.

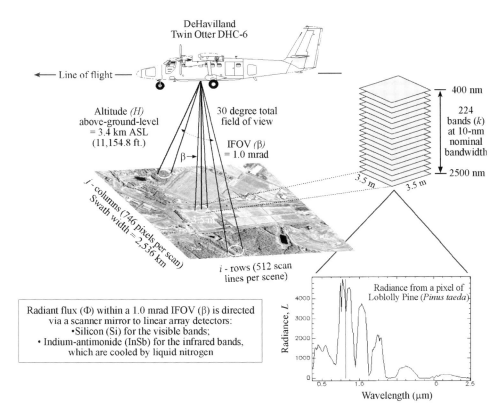

Figure 16.2 NASA Jet Propulsion Laboratory's Airborne Visible Infrared Imaging Spectrometer (AVIRIS) obtains imagery in 224 bands. It normally collects data at 20 km above ground level resulting in 20×20 m² pixels. However, sometimes it is flown at lower altitudes (e.g., 3.5 km) using a De Havilland DHC-6 Twin Otter as in this example where a portion of the Savannah River Site, SC is imaged at a spatial resolution of 3.5×3.5 m². The 224 detectors record energy in the 400–2500 nm region of the spectrum that are useful for carefully monitoring the spectral characteristics of earth surface materials (Jensen et al., 2000)

16.4. Relationship of photogrammetry and remote sensing to the other mapping sciences (surveying, GPS, cartography, GIS, geography)

Photogrammetry is largely concerned with *measuring* objects in images and converting the object being measured to a map or some intelligent data about the objects of interest that are useful to man. The body of knowledge comprising remote sensing brings together physical principles and properties of objects in our environment as expressed in the energy-matter reactions of the electromagnetic spectrum. Further, remote sensing correlates these principles to the practices, subject matter, and concepts of earth science, life, and man's cultural environment. It is therefore in the interaction between these disciplines that this new and interesting field of remote sensing is evolving.

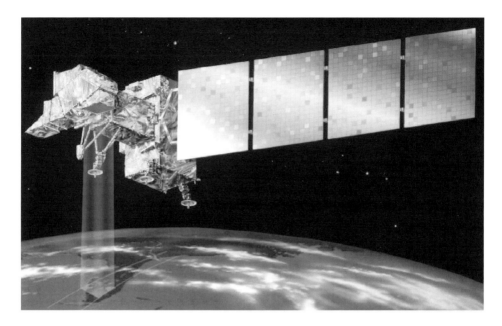

Figure 16.3 The Enhanced Thematic Mapper Plus (ETM$^+$) onboard the Landsat 7 satellite obtains spectral measurements in eight regions with spatial resolution ranging from 15×15 m^2 (panchromatic) to 30×30 m^2 (visible, near- and middle-infrared) to 60×60 m^2 (thermal infrared).

At first sight remote sensing may appear to be merely an extension of aerial or space surveying and photogrammetry. However, remote sensing generally involves optical-mechanical instrumentation that uses both the photographic and non-photographic parts of the electromagnetic spectrum. Couple this with the new techniques available for interpreting these data, remote sensing has become a special body of applied science in itself. Although remote sensing is based on principles of physics and mathematics including photogrammetry, it is also highly related to the other mapping sciences such as surveying/Global Positioning System (GPS), geography/cartography, and Geographic Information Systems (GIS). In fact, many of these other disciplines are often employed to complete a remote sensing project. Therefore, it is instructive to point out that the other mapping sciences are also based on solid principles, and each one has its own specific definition. The definitions that follow are, in some cases, slightly modified from the conventional definition to conform with today's technology. The definitions are derived from, *Launchspace: The Magazine of the Space Industry* (Space Imaging, 1999) and the American Society of Photogrammetry, *Manual of Photogrammetry* (1980, 3rd ed.).

16.4.1. Surveying

There are basically four types of surveying that relate to photogrammetry and remote sensing:

1   Plane Surveying – A survey in which the surface of the earth is considered a plane

rather than an ellipsoid. For small areas, precise results may be obtained with plane surveying methods, but the accuracy and precision of such coordinates and distances will decrease as the area surveyed increases in size, because of the planar assumption.

2 Geodetic Surveying – A survey in which the figure and size of the earth is considered in the computation of coordinates. It is applicable for large areas and long lines and is used for the precise location of basic points suitable for controlling other surveys and photogrammetric projects.

3 Cadastral Surveying – A survey relating to land boundaries and subdivisions of real estate, made to create units suitable for transfer, or to define the limitations of land titles. Such surveys are usually termed *land survey* because they define property boundaries rather than using coordinates.

4 GPS – The GPS is a satellite navigation and positioning system with which the three-dimensional position and the velocity of a user at a point on or near the earth can be determined in real time (Chapter 7).

16.4.2. Geography and cartography

Geography is defined as the '*science dealing with the areal differentiation of the earth's surface, as shown in the character, arrangement, and interrelations over the world of such elements as climate, elevation, soil, vegetation, population, land use, industries, national and political boundaries, and of the unit areas formed by the complex of these individual elements.*' Physical geography is the study of a number of earth sciences, which give us a general insight into the nature of man's environment. Cultural or human geography is concerned with an analysis of man's activities on the earth and how these activities create a cultural landscape. Remote sensing imagery is an essential element of geographic inquiry as it provides valuable information on the spatial distribution of land use and land cover, on the biophysical characteristics of ecosystems, and how landscapes change through time.

Geographers and other scientists use cartographic principles to scale the spatial characteristics of large and small land areas or all of the earth or another celestial bodies in a two- or three-dimensional map format. Robinson et al. (1978) define cartography as the '*art, science and technology of making maps, together with their study as scientific documents and works of art. In this context maps, plans, charts, sections, three-dimensional models and globes represent a scale model of the earth or any celestial body at any scale.*' Much of the information derived from photogrammetry or remote sensing ends up in a cartographic (map) format.

16.4.3. GIS

GIS get more powerful and useful in processing remote sensing and other geographic data (refer to Section 2). *Launchspace*'s remote sensing glossary defines GIS as '*an organized collection of computer hardware, software and digital geographic data designed to provide multiple layers of geographically-referenced information. GIS databases are used to efficiently capture, store, update, manipulate, analyze and display all forms of georeferenced information.*' The GIS utilizes georeferenced vector

Table 16.3 Relationship of selected applications to the mapping sciences[a]

Application	Photo-grammetry	Remote sensing	In situ surveying/GPS	Geography/cartography
Agriculture				
General (presence/absence)	H	H	M	M
Precision	M	H	H	M
Field biophysical characteristics	M	H	M	H
Forestry, rangeland, wetland				
General	M	H	M	M
Stand biophysical characteristics	H	H	H	H
Water resources				
Water spatial distribution	H	M	M	H
Water quality	M	H	H	H
Supply and demand	M	M	L	H
Geological/soils				
Soils	M	M	M	H
Surficial geology	M	H	H	H
Subsurface geology	L	M	M	M
Geomorphology	H	H	L	H
Defense maps and charts				
Aeronautical	H	M	H	H
Nautical	H	M	H	H
Trafficability (ability to traverse)	H	H	H	H
Tactical (real-time tactics)	H	H	H	H
Urban planning				
Land use/ land cover	H	H	M	H
Demographic characteristics	H	M	H	H
Socioeconomic characteristics	H	M	H	H
Engineering				
Topographic mapping and DEMs	H	M	H	M
Transportation infrastructure	H	M	M	M
Utilities	H	M	M	M
Cadastral				
Property lines for tax mapping	H	L	H	H
Building footprint	H	M	H	H
Natural disaster assessment				
Pre-disaster assessment	H	M	H	H
Post-disaster assessment	H	H	H	H

[a] H, suggests this mapping science is generally indispensable to the successful data collection, monitoring, or modeling of this phenomena; M, denotes that the application requires expertise in this mapping science; L, suggests that a mapping science plays a relatively minor role in this endeavor.

and raster (including remotely sensed imagery) to present various map themes in a digital or graphic format. Remotely sensed information often realizes its full potential when placed in a GIS where it can be interrogated and modeled in relation to other spatially distributed data (Estes and Jensen 1998).

16.4.4. Considerations

Table 16.3 identifies selected applications that are of great interest to commercial, public, and military personnel and scientists. Note how the mapping sciences have a high, medium, or low contribution to each of the applications. Thus, a broad knowledge of the mapping sciences in addition to an in-depth understanding of the systematic body of knowledge under investigation (e.g. soils, agronomy, forestry, geography, hydrology) is indispensable for successfully completing most application projects.

16.5. When is a photogrammetric approach the best solution?

The photogrammetric quantification process is significant because the interpreter's concern for *what* is present in imagery is almost always followed by a concern for *where* interpreted items are on the ground and over *what geographic areal extent*. Perhaps a more simplistic way to determine whether the project is a photogrammetry job or a remote sensing task is to look at the photogrammetric process and recognize that the most common products obtainable from a pair of stereo-photos absolutely oriented to the grid system in a photogrammetric plotter are basically positionally accurate items such as:

- Extraction of the man-made features in the stereo model, especially building and transportation system infrastructure and any demographic or socioeconomic characteristics of the population.
- Classification and delineation of the exact shoreline boundaries of water bodies such as lakes, rivers and streams.
- Classification and delineation of detailed vegetation characteristics especially for timber cruising and precision agriculture applications.
- Production of very accurate topographic line maps (including contours) or digital elevation models (x,y,z coordinates).
- Orthorectified georeferenced photo maps.
- Detailed reconnaissance information for military strategic planning and tactics.

If the project calls for any of these items, it generally belongs to the province of photogrammetry and one should use photogrammetric techniques to extract the required information. However, by far the most common use of photogrammetry is in the preparation of topographic maps. Virtually all of the US Geological Survey (USGS) and National Imagery and Mapping Agency (NIMA) quadrangle maps and most precise digital elevation models are compiled photogrammetrically.

16.6. When is a remote sensing approach the best solution?

Although photogrammetry and remote sensing are highly related, a common distinction may be that photogrammetry deals more specifically with where an object is in the x,y,z coordinate system whereas remote sensing is especially concerned with the biophysical characteristics of the phenomena at the x,y,z location. The remote sensing systems record data in a tremendous variety of regions both within and outside the optical windows traditionally used when collecting aerial photography. Remote sensing systems are specifically designed to quantitatively measure the energy/matter interactions. From the different responses recorded by the sensor, the analyst may make a biophysical assessment about the characteristics of the object or area of interest, e.g. land use/cover, temperature, surface roughness, biomass, moisture content (Thenkabail *et al.* 2000).

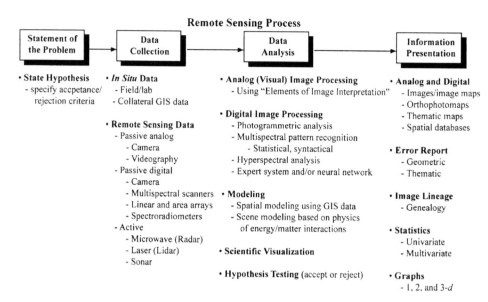

Figure 16.4   The typical sequence of events used in most remote sensing applications (adapted from Jensen 1996, 2000).

There is a typical *remote sensing process* that is followed when conducting most remote sensing projects (Jensen 1996, 2000). The major components of the process are summarized in Figure 16.4, including:

1 The hypothesis to be tested is defined including specific acceptance and rejection criteria;

2 (a) The required *in situ* data necessary to calibrate the remote sensor data and judge its geometric and thematic accuracy are collected, (b) using a variety of remote sensing instruments, hopefully, at the same time as the *in situ* data;

3 Both the *in situ* and remotely sensed data are processed using (a) analog image processing, (b) digital image processing (including photogrammetric techniques), (c) modeling, (d) scene *n*-dimensional visualization, and (e) hypothesis testing statistics, and

4 The information extracted are summarized and presented in a format (graphics, tables, graphs, etc.) that effectively communicates the desired principles.

Hopefully, the information is used to make wise natural and cultural resource management decisions.

16.7. Who are the users?

Over the past few decades there has been a significant rise in the availability of (a) digital large scale base maps (e.g. USGS Digital Line Graphs), (b) airborne (e.g. USGS National Aerial Photography Program imagery) and satellite (e.g. IKONOS imagery) remotely sensed data (although still relatively expensive), (c) a significant increase in the utility, power and ease-of-use of digital image processing and GIS, and (d) computer processing capability. This increase in technology has been accompanied by an increased awareness and understanding of the role that these spatial data play in earth resource management and planning. For example, geography departments nationwide are graduating record numbers of people with specialized training in remote sensing and GIS. Therefore, it is not surprising that increasing numbers of public, commercial, academic, and military organizations as well as private individuals are using the technology to manage and plan for the effective use of our natural and cultural resources. These conditions have contributed to making photogrammetry, remote sensing and GIS a multi-billion dollar business.

16.7.1. Public organizations

In the 50 US states there are 3,536,342 square miles of land area and 251,083 square miles of water area giving a total of 3,787,425 square miles with 12,383 miles of shoreline. Within this vast amount of territory there are about 3,141 counties, and about 600 cities with populations over 50,000. Government organizations of all kinds, from towns, cities, counties, and states to federal agencies, routinely utilize information (especially maps) derived from remotely sensed data using photogrammetric and digital image processing techniques to coordinate disaster relief, manage and map the land, and understand and solve environmental, engineering and social issues (e.g. Jensen and Cowen 1999; Space Imaging 2000).

16.7.1.1. US federal framework data

US federal agencies have adopted a *Framework* concept so that spatial information purchased at taxpayer expense can be shared at all levels of government, thereby reducing redundant expenditures of public funds. Framework data consist of the three foundation layers of geographic information and four key thematic layers (Figure 16.5). Much of the information in these framework layers are derived using photogrammetric and/or remote sensing technology. Building upon the seven framework data layers, an agency usually compiles additional information on other thematic topics (e.g. soils, cultural and demographic variables) and then models all the information in a GIS.

The USGS is the agency responsible for the primary Topographic Map Series of the US at 1:24,000 scale. There are 53,689 quadrangles each covering 7.5 × 7.5 min of latitude and longitude in the lower 49 states and 2,674 primary series, 1:63,360 scale, quadrangles in Alaska. Building and maintaining this number of maps is a sizeable task that contributes greatly to photogrammetry, remote sensing and GIS business.

16.7.1.2. US global change research program and the earth observing system

The International Geosphere – Biosphere Program (IGBP) and the US Global Change Research Program (USGCRP) call for scientific research to 'describe and understand the interactive physical, chemical, and biological processes that regulate the total Earth system' (CEES 1991; King 1999). Space-based remote sensing is an integral part of these research programs because it provides the only means of observing global ecosystems consistently and synoptically. NASA's Earth Science Enterprise (formerly Mission to Planet Earth) is the name given to the coordinated international plan to

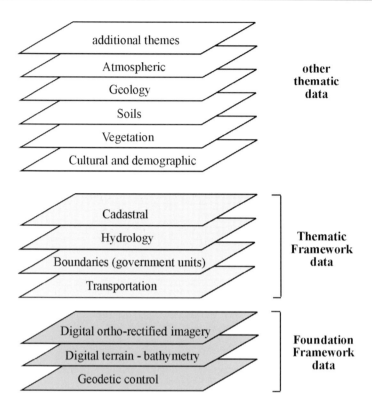

Figure 16.5   Remotely sensed data are indispensable in the creation of the thematic and foundation framework data layers. In addition, photogrammetric and remote sensing digital
image processing techniques can be used to extract other thematic data that may be
synergistically analyzed with the framework data (adapted from Federal Geographic
Data Committee).

provide the necessary satellite platforms and instruments, an Earth Observing System
Data and Information System (EOSDIS), and related scientific research for IGBP. In
particular, new satellite remote sensing instruments will include:

1   A series of near-term Earth probes to address discipline-specific measurement
     needs;
2   A series of multipurpose polar orbiting platforms, initiated in 1988, to acquire 15
     years of continuous Earth observations, called the Earth Observing System (EOS),
     and
3   A series of geostationary platforms carrying advanced multidisciplinary instru-
     ments to fly sometime after the year 2000, called the Geostationary Earth Obser-
     ving System (NASA 1998; King 1999).

Asrar and Dozier (1994) conceptualized the remote sensing science conducted as
part of the Earth Science Enterprise. They suggested that the Earth consists of two
subsystems, (1) the physical climate, and (2) biogeochemical cycles, linked by the
global hydrologic cycle. The physical climate subsystem is sensitive to fluctuations

in the Earth's radiation balance. Human activities have caused changes to the planet's radiative heating mechanism that rival or exceed natural change. For example, increases in greenhouse gases between 1765 and 1990 have caused a radiative forcing of 2.5 W/m^2. If this rate is sustained, it could result in global mean temperatures increasing about 0.2–0.5°C per decade during the next century. Volcanic eruptions and the ocean's ability to absorb heat may impact the projections. The following important questions are being addressed using remote sensing (Asrar and Dozier 1994; King 1999):

- How do clouds, water vapor, and aerosols in the Earth's radiation and heat budgets change with increased atmospheric greenhouse-gas concentrations?
- How do the oceans interact with the atmosphere in the transport and uptake of heat?
- How do land-surface properties such as snow and ice cover, evapotranspiration, urban/suburban land use, and vegetation influence circulation?

Global change research is addressing the following questions:

- What role do the oceanic and terrestrial components of the biosphere play in the changing global carbon budget?
- What are the likely effects on natural and managed ecosystems of increased carbon dioxide, acid deposition, shifting patterns of precipitation, and changes in soil erosion, river chemistry, and atmospheric ozone concentrations?
- Can we predict changes in the global hydrologic cycle using present and future observation systems and models?

The US EOS AM-1 (the *Terra* satellite launched in December, 1999) contains five remote sensing instruments designed to address many of the previously mentioned research topics. The *Terra* sensors use new remote sensing technology. For example, the Moderate Resolution Imaging Spectrometer (MODIS) has 36 hyperspectral bands from 0.405–14.385 μm that collects data at 250 and 500m and 1 km spatial resolutions. MODIS views the entire surface of the Earth every 1–2 days, making observations in 36 co-registered spectral bands, at moderate resolution (0.25–1 km), of land- and ocean-surface temperature, primary productivity, land-surface cover, clouds, aerosols, water vapor, temperature profiles, and fires (NASA 1998). Additional information about MODIS and the other Terra sensors are summarized in Chapter 18.

16.7.1.3. Military

Airborne or spaceborne *photo-reconnaissance* is a mission undertaken to obtain imagery intelligence about the activities and resources of an adversary or potential enemy or to secure data concerning meteorological, hydrographic, or geographic characteristics of a particular area. The military services are major users of remotely sensed imagery. The NIMA is responsible for mapping and charting the countries of the world except the US which is mapped by the USGS. The National Oceanic and Atmospheric Administration (NOAA) provides Nautical and Aeronautical Charts of the US. Major products provided by NIMA to the four US military services are Airforce aeronautical charts, Navy nautical charts, Army topographic maps, and

Marine Corps topographic maps and coastal charts. Another key product used by the military services is Digital Topographic Elevation Data (DTED). DTED is a grid of x,y,z – coordinates spaced at 10, 30 or 100 m to describe the topography of the Earth. The civil community calls the same type of data Digital Elevation Models (DEMs).

16.7.1.4. Academic institutions

Professors in a tremendous variety of systematic disciplines integrate remote sensing and GIS technology into the curriculum. In fact, the *Green Book* (EOM 1999) states that 47 per cent of the commercial vendors consider 'Education and Research' to be a primary market for the geo-technologies. Laboratory practice, research and thesis projects require imagery for analysis. The university departments most likely to offer mapping science related courses are Geography, Geology, Forestry, Civil Engineering, and Geodetic Science. There are approximately 100 colleges and universities that offer Mapping Science related courses in the US and the number is increasing daily.

16.7.2. The private sector

Private companies use remotely sensed imagery and GIS for facilities management, mapping, resource location, marketing, and to produce products for military establishments. For example, some farming cooperatives use imagery data of their land to improve crop yield sometimes through precision agricultural practices. Citizen's groups and societies interested in sustainable development or protecting the environment often map watersheds to study man's effects on their domain.

The American Society of Photogrammetry and Remote Sensing has 148 sustaining members. Each of these private sector companies play a role in sustaining the economics that drive the photogrammetry and remote sensing industry. The *Green Book* lists hundreds of commercial companies engaged in photogrammetry and remote sensing or in providing software and hardware to the industry. The book summarizes the per cent of vendors and what they consider to be the major markets, including:

1 Environmental monitoring
2 Cartography/map making
3 Land and Forest Management
4 Defense/Military
5 Exploration
6 Mining
7 State/local government

In the US, Title V of the 1992 Land Remote Sensing Policy Act expanded upon previous legislation authorizing and encouraging the commercialization of remote sensing satellite systems. The 1992 Act authorized the Secretary of Commerce to license the operation of private remote sensing space systems. The Department of Commerce delegated this authority to NOAA. The responsibility is to be exercised in consultation with the Department of Defense, the Department of State and other interested agencies such as the Departments of Interior and Agriculture.

Following the 1992 Act, was a 1994 Presidential Decision Directive, which paved the way to license the first high-resolution electro-optical systems to produce imagery at Ground Sample Distance (GSD) of 1 m. US policy recognizes the fact that economic and national security interests are advanced through support of a robust industry, which pays dividends in aerospace and information technology.

Since 1992, NOAA has licensed 11 high resolution electro-optical (panchromatic and multispectral) systems. By 1999, NOAA had also issued licenses for Synthetic Aperture Radar (SAR) and hyperspectral systems. At the time of the original Directive, the best commercial systems were Landsat and SPOT medium spatial resolution (10–30 m) electro-optical systems. The challenge ahead was presented by the unique attributes of radar (all weather) day/night capabilities, and hyperspectral sensors that can be designed to record 240 or more spectral bands well into the thermal infrared portion of the spectrum. Wooldridge (1999) points out that the stakes are high and the US government wants to achieve a proper balance between national security needs, and the economic promise to industry to foster and encourage this $4 billion industry.

Based on these licenses, EOSAT, Inc., launched Landsat 6 with its enhanced thematic mapper in 1993. Unfortunately, it failed to achieve orbit. EarthWatch, Inc. launched Earlybird in December, 1997. Unfortunately, all communication with the satellite was lost. Space Imaging, Inc. launched IKONOS on 27 April 1999 and it failed to achieve orbit. A second IKONOS satellite was launched on 24 September 1999. The IKONOS sensor system has a 1×1 m panchromatic band as well as four 4×4 m multispectral bands (Section 3.4). Similar sensor systems are scheduled to be launched by EarthWatch, Inc. (Quickbird) and ORBIMAGE, Inc. (Orbview 3) in 2001–2002. ORBIMAGE also plans to launch a hyperspectral satellite remote sensing system in 2001–2002 (Orbview 4). There is considerable debate over the potential ramifications of commercial satellite imagery (Sheffner and Stoney 1999; Dehqanzada and Florini 2000). Fortunately, photogrammetric engineering companies continue to collect high spatial resolution analog and digital photography (usually with spatial resolutions of <0.5 m) for extensive civil engineering applications throughout the world (Jensen and Cowen 1999; Light 1999; Maas 1998).

16.8. National and international societies devoted to photogrammetry and remote sensing

The science and practice of photogrammetry and remote sensing is prevalent through-out the world. There are 99 countries that are members of the International Society of Photogrammetry and Remote Sensing. Thirty of these countries and their national organizations are listed in Table 16.4.

The International Society for Photogrammetry and Remote Sensing meets every 4 years in a different city. The society is divided into seven commissions that carry out the activities that merit national and international attention:

Commission I	Sensors, platforms and imagery.
Commission II	Systems for data processing, analyzing and representation.
Commission III	Theory and algorithms.
Commission IV	Mapping and GIS.
Commission V	Close-range photogrammetry and machine vision.
Commission VI	Education and communications.
Commission VII	Resource and environmental monitoring.

Table 16.4 Thirty of the 99 countries that are members of the international society of photogrammetry and remote sensing

Country	Association
Argentina	Argentine Association for Photogrammetry and Related Science
Australia	Remote Sensing and Photogrammetry Association of Australia
Belgium	Societe Belge de Photogrammetric, de Teledetection et de Cartographic
Brazil	Sociedade Brasileira de Cartografia
Canada	Canadian Institute of Geomatics
China	Chinese Society of Geodesy
China-Taipei	Chinese Taipei Society of Photogrammetry and Remote Sensing
Denmark	Danish Society for Photogrammetry and Surveying
Egypt	Egyptian Committee of Surveying and Mapping
Finland	Finnish Society of Photogrammetry and Remote Sensing
France	Societe Francaise de Photogrammetrie
Germany	Deutsche Gesellschaft Fur Photogrammetrie und Fernerkundung
Greece	Hellenic Society for Photogrammetry and Remote Sensing
India	Indian Society of Remote Sensing
Indonesia	Ikatan Surveyor Indonesia
Israel	The Israeli Society of Photogrammetry and Remote Sensing
Italy	Societa Italiana di Fotogrammetria e Topografia
Japan	Japan Society of Photogrammetry and Remote Sensing
Korea	Korean Society of Geodesy, Photogrammetry and Cartography
Mexico	Sociedad Mexicana de Fotogrammetria, Fotointerpretacion y Geodesia
Netherlands	Netherlands Society for Earth Observation and Geo-Informatics
Russian Federation	National Committee of Russia
Saudi Arabia	Military Survey Department
South Africa	The South African Photogrammetry and Geo-Information Society
Spain	Spanish Society for Cartography Photogrammetry and Remote Sensing
Sweden	The Swedish Society for Photogrammetry and Remote Sensing
Switzerland	Swiss Society of Photogrammetry, Image Analysis and Remote Sensing
Thailand	The Royal Thai Survey Department
UK	United Kingdom National Committee for Photogrammetry and Remote Sensing
US	American Society for Photogrammetry and Remote Sensing

Each commission has a president and the commission is expected to convene a symposium on their subject midway between the 4 year period separating the International Congress. The year 2000 congress was held in The Netherlands. The American Society of Photogrammetry and Remote Sensing is organized along the same lines. In addition, there are numerous other organizations that conduct photogrammetric and remote sensing related research such as the Remote Sensing Society (UK), IEEE Geoscience and Remote Sensing Society, Alliance for Marine Remote Sensing Association, Remote Sensing Specialty Group of the Association of American Geographers, Federation of American Scientists (image intelligence section), Canadian Remote Sensing Society, European Association of Remote Sensing Laboratories, etc.

16.9. Summary

When a cultural or natural resource application requires that information be extracted from very high spatial resolution imagery on a timely basis, then aerial photography is usually the sensor of choice. The most accurate information is extracted from such imagery using stereoscopic photogrammetric techniques. When the application requires that biophysical information be extracted from geometrically accurate multi- and hyperspectral imagery, then remotely sensed imagery may be required. Remotely sensed data are primarily analyzed using digital image processing techniques in conjunction with GIS modeling. There is a vast network of commercial, public, academic, and military users of aerial photography and remotely sensed data. The following chapters summarize how the remotely sensed imagery are obtained, analyzed, and assessed for accuracy. In addition, several real-world applications are provided.

References

ASP, *Manual of Photographic Interpretation*, American Society of Photogrammetry, Falls Church, 1960.

ASP, *Manual of Photogrammetry*, American Society of Photogrammetry, Falls Church, 1944, 1952, 1966, 1980.

Asrar, G. and Dozier, J., *EOS: Earth Strategy for the Earth Observing System*, American Institute of Physics, Woodbury, 1994.

CEES, *Our Changing Planet: the 1992 US Global Change Research Project*, Office of Science & Technology, Washington, DC, 1991, 21 pp.

Colwell, R. N., *Manual of Remote Sensing* (2nd ed.), American Society for Photogrammetry & Remote Sensing, Bethesda, MD, 1983.

Colwell, R. N., 'From photographic interpretation to remote sensing', *Photogrammetric Engineering & Remote Sensing*, 50(9), 1984, p. 1305.

Colwell, R. N., 'History and place of photographic interpretation', *Manual of Photographic Interpretation* (2nd ed.), Philipson, W. K. (ed.), American Society for Photogrammetry & Remote Sensing, Bethesda, MD, 1997, pp. 33–48.

Dehqanzada, Y. A. and Florini, A. M., *Secrets for Sale: How Commercial Satellite Imagery Will Change the World*, Carnegie Endowment for International Peace, Washington, DC, 2000, 47 pp.

EOM, 'Green book 2000 - Who's who in the geo technologies', *EOM*, 8(10), 1999, 17–37.

Estes, J. E. and Jensen, J. R., 'Development of remote sensing digital image processing systems and raster GIS', in *The History of GIS*, Foresman, T. (ed.), Longman, New York, 1998, pp. 163–180.

Hallert, B., *Photogrammetry: Basic Principles and General Survey*, McGraw-Hill, New York, 1960, p. 1.

Helava, U. V., 'On system concepts for digital automation', *Photogrammetria*, 43(2), 1988, 57–71.

Hinz, A., 'The Z/I imaging digital aerial camera system', *Photogrammetric Week 99*, Fritsch, D. and Spiller, R. (eds.), Wichmann, Heidelberg, 1999.

Jensen, J. R., *Introductory Digital Image Processing: a Remote Sensing Perspective*, Prentice Hall, Saddle River, NJ, 1996, 318 pp.

Jensen, J. R., *Remote Sensing of the Environment: an Earth Resource Perspective*, Prentice Hall, Saddle River, NJ, 2000, 544 pp.

Jensen, J. R. and Cowen, D. C., 'Remote sensing of urban/suburban infrastructure and socio-economic attributes', *Photogrammetric Engineering & Remote Sensing*, 65(5), 1999, 611–622.

Jensen, J. R., Filippi, A. and Schill, S., *Hyperspectral Analysis of Hazardous Waste Sites on the Savannah River Site using AVIRIS Data*, EOCAP Report: NASA Stennis Space Center, MTL, Inc., OH, 2000, 55 pp.

King, M. D., *EOS Science Plan*, NASA, Washington, DC, 1999, 397 pp.

Leberl, F., 'Practical issues in softcopy photogrammetric systems', *Graz University of Technology, Graz Austria, Proceedings*, ASPRS, Washington, DC, August 26–29, 1994, pp. 223–230.

Light, D.L., 'Film cameras or digital sensors? The challenge ahead for aerial imaging', *Photogrammetric Engineering and Remote Sensing*, 62(3), 1996, 285–291.

Light, D.L., 'An airborne digital imaging system, *Proceedings, The International Symposium on Spectral Sensing Research, ISSSR*, Las Vegas, NV, October 31–November 4, 1999.

Maas, H., 'Airborne digital cameras', *GIM International*, 12, 1998, 77–79.

NASA, *NASA's Earth Observing System - EOS AM-1*, NASA Goddard Space Flight Center, Greenbelt, 1998, 34 pp.

Pruitt, E., 'The office of naval research and geography', *Annals of the Association of American Geographers*, 69(1), 1979, 106.

Robinson, A., Sale, I. and Morrison, J., *Elements of Cartography*, John Wiley & Sons, New York, 1978, pp. 1–3.

Sheffner, E. and Stoney, W., 'Are remote sensor birds finally flying?' *GeoWorld*, 12(11), 1999, 40–42.

Space Imaging, 'Remote sensing glossary', *Launchspace Magazine*, Launchspace Publications, Mclean, VA, 1999.

Space Imaging, 'State and local government - urban or rural, big or small, governments are discovering the vast power of Earth imagery', *Imaging Notes*, Space Imaging, Inc., Thornton, 15(4), 2000, p. 1.

Thenkabail, P. S., Smith, R. B. and De Paww, E., 'Hyperspectral vegetation indices and their relationships with agricultural crop characteristics', *Remote Sensing of Environment*, 71, 2000, 58–182.

Wolf, P. R. and Dewitt, B. A., *Elements of Photogrammetry with Applications in GIS*, McGraw-Hill, New York, 2000.

Wooldridge, C., 'Commercial remote sensing', *Backscatter*, August, 1999, 4–6.

Chapter 17

The remote sensing process: how do we collect the required *in situ* and remotely sensed data?

Minhe Ji

17.1. Introduction

The physics of remote sensing were summarized in Chapter 6 including characteristics of the electromagnetic spectrum, how electromagnetic energy interacts with the atmosphere and with the terrain (absorption, scattering, reflection, refraction), and radiometric definitions (e.g. radiance and percent reflectance). The fundamental characteristics of the *remote sensing process* were introduced in Chapter 16. This chapter is concerned with the remote sensing process, providing additional insight into decisions that must be made when conducting a remote sensing project.

17.2. Identify the remote sensing problem(s) to be addressed and objectives of the study

The importance of a problem statement cannot be overemphasized. A carefully-crafted problem statement represents the foundation upon which *in situ* and remote sensing data requirements are identified and financial expenditures justified. Careful problem definition saves time and money over the life of a project. Quality remote sensing projects have carefully crafted problem statements. Curran (1987) identified three types of logic that may help formalize the process: *inductive*, *deductive*, and *technologic* (Figure 17.1). *Inductive* logic is a process of learning through exploration. It involves observation, classification, generalization, and theory formulation. For example, *in situ* field spectroradiometer spectral reflectance measurement can be considered a typical application of inductive logic. After making multiple observations of the spectral reflectance characteristics for a particular ground feature (e.g. corn canopy), the data are summarized to construct a model that represents the spectral response pattern of corn. The totality of spectral models for different types of ground features (e.g. corn, soybean, wheat, cotton) is called a spectral library and may be considered to be a body of knowledge useful for spectral analysis in remote sensing.

Conversely, problem statements based on *deductive* logic start from a theory as the proposed solution to the problem at hand, usually specified in the form of a *hypothesis*. A hypothesis clearly states a position that can be either verified or falsified statistically (e.g. there is no significant relationship between the amount of near-infrared energy reflected from a corn field and the amount of biomass present). Data (observations) are then collected, empirical models constructed, and the hypothesis is tested at specific confidence intervals (Jensen 2000).

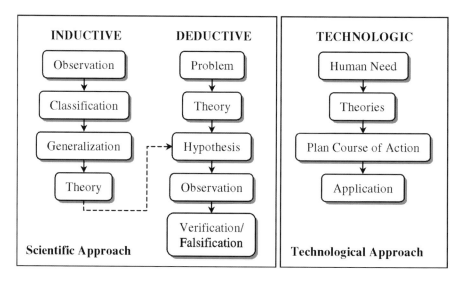

Figure 17.1   Types of logic used in remote sensing projects. Note that technologic logic is a common practice of deductive logic without explicit hypotheses.

There is a third type of logic that is often used when conducting remote sensing related projects. A problem statement may be structured using *technological* logic, which is based on human need rather than scientific inquiry. This type of logic blends "inputs from science, economics, aesthetics, law, logistics and other areas of human endeavor" (Curran 1987). For example, a land use inventory using a combination of remote sensing, GIS, and other technologies may be conducted to satisfy the need for land use planning, pollution monitoring and control, or many other environmental and socioeconomic applications.

It is important to point out that the nature of the problem is project dependent. While there may be similarities in the way that various remote sensing related projects are organized, each project requires a carefully crafted problem statement with very specific objectives. A project related to forest resource management will have a problem statement and objectives that are significantly different from an urban sprawl monitoring project. For example, Figure 17.2 identifies some of the characteristics that might be considered when attempting to conduct an inventory of the H. J. Andrews Forest in the state of Washington. Different problems may require different variables to be observed, which in turn may lead to the use of different remote sensing and *in situ* data collection approaches. The more carefully refined the problem statement, the easier it is to identify the variables that must be measured.

17.3. Identify the variables to be measured using *in situ* techniques

Remote sensing fulfills its potential best when used with other spatial information. Consequently, two types of data normally must be collected to satisfy the data requirements of a remote sensing-related application: (1) ground reference *in situ* data, and (2) remotely sensed data. Ground reference or *in situ* data may include maps, results

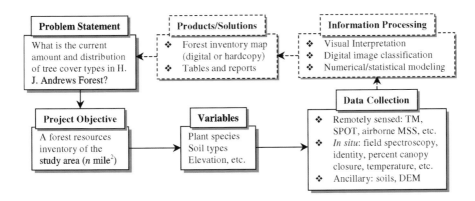

Figure 17.2 Variable identification. A simplified example of identifying variables based upon the problem statement and the project objective for a forest inventory project using remote sensing technology.

obtained from questioning inhabitants, aerial photographs, Census information, and data from scientific field measurements. Strictly speaking, *in situ* data should only refer to those data collected from field measurements. However, it is common practice to refer to all ground reference data as being *in situ, collateral,*or *ancillary* data. *In situ* data are collected (1) to provide direct information to solve the problem at hand, (2) to calibrate the remote sensor data, and/or (3) to verify that remote sensing derived information is accurate.

17.3.1. Characteristics of variables to be measured in the field

Variables selected for *in situ* measurement in most remote sensing related projects are usually based on two criteria: (1) the variable must measure some important socio-economic or biophysical variable, and (2) some aspect of the variable must be able to be related to the remote sensing measurement. For example, *in situ* measurement of the kinetic (true) temperature of urban areas using thermometers is common when studying the urban heat island effect in cities. The kinetic temperature can be related in a quantitative manner to the radiant (apparent) temperature of the same terrain recorded by a thermal infrared remote sensing system. Similarly, Leaf-Area-Index (LAI) and/or percent canopy closure of a forest canopy can be measured in the field and then related to the brightness values of pixels for modeling vegetation biomass or ecosystem dynamics (Wessman 1991).

There are three properties of data that must be considered when selecting variables for *in situ* measurement. First, *in situ* data can be either a *direct indictor* (such as temperature) of the phenomenon under investigation or an *indirect measure* (such as percent canopy closure) that serves as a good predictor for the phenomenon of interest (such as wildlife habitat suitability). Second, the *in situ* data must be able to be collected at an appropriate *scale of measurement* (i.e. nominal, ordinal, interval, or ratio). The scale of measurement dictates the kind of modeling that can be applied to the *in situ* data. For example, *in situ* data about wetland type (nominal scale) or LAI (ratio scale) can be collected. The type and LAI information together can support

Table 17.1 Three levels of *in situ* measurement in support of remote sensing-related investigations

Level	Data to be collected	Example variables
1	Target identity or condition plus geographical position (and extent for linear or areal features)	Land cover, soil moisture, vegetation type, temperature
2	Quantitative spectral data plus data from level 1	Field spectrometric measure of vegetation type and content
3	Quantitative measurement of latent variables plus data from level 2	Vegetation chlorophyll content, soil type, species diversity, water chemistry

wetland modeling, while the type information alone can only yield conventional wetlands classification (Ji and Jensen 1999).

Third, *in situ* data can be either *time-stable* or *time-critical* (Lillesand and Kiefer 2000). Time-stable measurements refer to those from a target that do not change appreciably with time. Geologic features and soil types are two phenomena that are less likely to change between the time remotely sensed data are collected and the time *in situ* data are acquired. In contrast, time-critical measurements are made from the rapidly changing ground conditions, such as crop growth, water pollution during a storm event, and diurnal temperature. The ideal solution to time-critical *in situ* data collection is to have the remote sensing and *in situ* data collected at the same time.

Depending upon the specific modeling requirements, *in situ* measurements in support of remote sensing-related investigations are obtained at one of three levels (Table 17.1). First, the geographic location, identity and/or condition of the target (e.g. land use/land cover, soil moisture content) may be collected as *in situ* data. Second, *in situ* spectral reflectance characteristics of the target are obtained in addition to information about type and condition. Such data are used for removing from the remotely sensed data the spectral effects of solar geometry and atmosphere for accurate modeling or classification. At the third level, latent variables that are to be related to the remotely sensible variables are also measured as *in situ* data. This might include water chemistry which is not directly measurable using remote sensing but can be inferred by monitoring the absorption characteristics of chlorophyll content or other variables.

Irrespective of the level that the *in situ* data are collected, there is one essential variable that cannot be neglected – the spatial location and/or geographic extent of the *in situ* data. Jensen (1996) pointed out that it is important to acquire very accurate information about the exact location of *in situ* data so that the data can be related to the corresponding location in the remotely sensed imagery. If the *in situ* data to be collected are linear (e.g. along a stream) or occupy a polygon in space (e.g. a corn field), then the linear and/or geographical extent of the targets must also be obtained. Global Positioning System (GPS) receivers are the ideal tool for gathering such positional data due to ease of use and affordability (see Part 2 on GPS). Depending on the model and data processing methods, the level of accuracy that a GPS receiver can offer ranges from sub-centimeter to hundreds of meters. The locational accuracy of *in situ* measurements should be at least compatible with, if not superior to, the spatial resolution of the remote sensing imagery selected for the project. For example, if SPOT 10 × 10 m panchromatic data are being used in the project (refer to Chapter 18), then the accuracy of any *in situ* measurement coordinates should be less than 5 m in *x* and *y*.

17.3.2. Problems associated with the in situ measurement devices

Because remotely sensed data are used by scientists and practictioners in a large number of disciplines, there are a variety of *in situ* measurements devices that are used to support remote sensing investigations. Examples include thermometers for temperature measurement, hand-held tape measures for computing a tree's Diameter-at-Breast-Height (DBH), and semi-hemispherical mirrors for canopy closure measurement. Jensen (1996) suggested that the techniques for *in situ* data collection should ideally be learned in the social, physical and natural science courses most related to the specific field of study, such as geography, chemistry, biology, forestry, soil science, hydrology, or meteorology. Even for the same variable, however, it is possible that several different methods and instruments can be used to collect the *in situ* data. For example, Martens *et al.* (1993) compared four different devices (i.e. ceptometer, line quantum sensor, plant canopy analyzer, and hemispherical photographs) for measuring LAI in a needle-leaved forest and a broad-leaved orchard and found no consistent results. Consequently, remote sensing practitioners should consult the experts in the systematic science fields for the appropriate selection and use of specific *in situ* measurement devices.

One of the most common *in situ* measurement devices used in remote sensing-related projects is the spectroradiometer. A spectroradiometer is used either in the lab and/or in the field to measure the reflectance and/or emittance characteristics of surface materials. For example, the radiation reflected or emitted from soils, vegetation, and water can be recorded by a spectrometer as a function of wavelength and used to prepare spectral reflectance curves. The ground-based spectral reflectance curves may then be compared with spectral reflectance (or emittance) curves derived using remote sensing techniques. *In situ* field measurements are generally preferred to laboratory measurements because many factors of the natural environment that influence remote sensor data (e.g. atmospheric haze, soil background effects) are difficult to duplicate in the laboratory.

17.3.3. Calibrating in situ spectral reflectance measurement devices

In situ data collection instruments must be very carefully calibrated, especially when obtaining measurements expressed as a percentage such as percent canopy closure and percent spectral reflectance. For example, as discussed in Chapter 6, percent spectral reflectance is the ratio of reflected radiance to incident radiance. Therefore, the spectroradiometer measuring *in situ* spectral reflectance must be calibrated using the incident radiance. Because many remote sensing investigations require the collection of *in situ* spectral reflectance measurements, it is instructive to provide some additional information on how these data are calibrated so that they can be used in conjunction with the remote sensing derived spectral reflectance measurements.

Jackson and Slater (1986) outlined a three-step procedure to obtain *in situ* spectral reflectance measurements using a field spectroradiometer. In their example, a radiometer mounted on a truck is first aimed at a *calibration panel* of known, stable reflectance to quantify the incoming radiation (or irradiance) at the measurement site. This calibration panel is an ideal, perfectly diffuse (i.e. Lambertian) surface. It scatters energy equally well in all directions, irrespective of the angle of the incident radiation. Next, the instrument is suspended over the target of interest and the radiation reflected by the object is measured. Finally, the spectral reflectance factor of the target is computed by ratioing the reflected energy measurement in each band of observation to the incoming radiation measured in each band. Additional information on *in situ* spectroscopic measurement and calibration procedures can be found in Thome *et al.* (1997).

17.3.4. Placing people in the field to collect in situ data

Collecting *in situ* data involves sending staff to work in the field, which can be expensive and time-consuming. To minimize the amount of fieldwork, assure data quality, and fulfill the objectives effectively and efficiently, planning and scheduling are necessary components of the task. Planning includes determining the number of field samples that need to be collected, randomly (or systematically) locating these samples throughout the study area, collecting the data, and reducing the data.

The necessary field sample size is a function of the size of the study area, the number of classes involved in the information extraction, the desired level of accuracy, and the associated confidence level in the numerical/statistical modeling. In general, sample sizes of 50–100 for each cover type are recommended, so that each category can be assessed individually (see Chapter 21 for additional information).

The identification of the location of samples in the field may be accomplished using a variety of techniques. Depending on the requirement of the project and the nature of the analysis, samples may be distributed using random sampling, stratified random sampling, or knowledge-based sampling. Simple random sampling is adequate when there is no *a priori* knowledge of the study area or when classes are equally important and occupy roughly the same amount of area. In comparison, stratified random sampling using classes as strata can accommodate the size variation among classes and ensure no statistical bias against small classes (Congalton 1988). Lillesand and Kiefer (2000) recognized that these two methods could produce sample locations that are either difficult or expensive to access or less likely to be accurately positioned without using a GPS receiver. This is especially the case when measuring non-man-made structures such as dense forest stands or wetlands. They suggested that under certain circumstances the sample should be taken only a few pixels away from field boundaries or transportation lines to overcome these problems. Knowledge-based sampling uses existing knowledge of the target and study area to help deploy the sample points adequately. For example, knowing the general distribution pattern of a thermal plume in a water body can help in laying out temperature transects for *in situ* measurement. Chapter 21 provides detailed information about how to assess the accuracy of thematic maps derived using remote sensor data.

Careful site selection minimizes the ambiguity of field samples. Since remotely sensed data often have a spatial resolution much larger than the actual size of the sample location, it is important that the site has feature homogeneity over the ground dimension of the image pixel. The size of feature homogeneity should ideally be much larger to minimize adjacency effects (e.g. scattering of light from adjacent features) when field spectroscopic data are collected for image spectra calibration (Slater *et al.* 1991). Locations with mixed ground features should be avoided to maintain the purity of the spectral data for remote sensor data calibration.

Several important factors must be considered when determining how many staff should be involved and when the staff should be sent to the field. For time-critical variables such as water pollution during a storm event, sending out the maximum number of staff to the field can minimize the time spent collecting all the field samples. On the other hand, the time-stable variables such as geological features impose less pressure on the length of time for *in situ* data to be collected. Therefore, sending fewer but well-trained staff to work in the field can ensure data quality and data consistency. Equipment calibration, staff training in the use of measurement equipment, site accessibility, and logistics for traveling are among other factors to be considered when acquiring field data.

17.3.5. Reducing in situ data

Due to the stochastic nature of many environmental variables, multiple observations may have to be made at each sample location to ensure that the essence of the variable is captured by the measurements. On the other hand, multiple measurements at the same location result in redundancy. The excessive data need to be reduced to produce for each sample location a value deemed to be representative of the true value. Among an array of methods for data reduction, two are commonly used: *averaging* and *graphing*. The averaging method takes the mean of all the observed values for a sample point and uses the mean as the representative value of the sample point. This method is used when either the range of the observed values or the number of observed values is small. When the range is great for a large number of observed values, the graphing method may be more appropriate. This method consists of three steps. First, all the values are plotted as a histogram and possible outliers are identified either visually or using some thresholding method. Second, unusual outliers are removed from the data set. Third, the mean of the remaining values is computed and used as the representative value of the sample point. In cases where the number or variance of the observed values varies from point to point, the combination of the two methods can be used to produce quality yet consistent results.

Another situation where *in situ* data need to be reduced is when the spectral resolution of the field measurement (e.g. a spectroradiometer) does not match that of a chosen remote sensor. If a spectroradiometer is used in the field, for example, it may have a spectral sampling interval much higher (e.g. 10 nm from 500 to 510 nm) than that from the remote sensing system (e.g.100 nm from 500 to 600 nm). Spectral band reduction and/or spectral matching algorithms are available in commercial digital image processing packages (e.g. ENVI, Research Systems, Inc.) and from public sources (e.g. the Low Resolution Transmittance (LOWTRAN) and Moderate Resolution Transmittance (MODTRAN) radiative transfer codes developed by the US Air Force) (Berk *et al.* 1989).

17.3.6. Using the in situ data to calibrate remote sensor data

In situ data may be used to calibrate remote sensor data. Two common uses of *in situ* data for calibration are: (1) in the class training stage of supervised land-use and land-cover classification, and (2) the statistical modeling of a ground condition through correlation and/or regression analyses. Details about supervised digital image processing of remote sensor data are provided in Chapter 18. The *in situ* data necessary for a supervised classification usually consist of the identity (nominal class) and x,y location of land-use and land-cover types. These data are then used to select corresponding pixels in the remote sensing image. In the classification training stage, the brightness values of selected image pixels are statistically summarized as the so-called *spectral signature* to represent the cover type identified by the *in situ* data. The spectral signature of the cover type can then be used to statistically infer the identity of other pixels in the image that have similar spectral characteristics (Jensen 1996).

When modeling ground biophysical conditions (e.g. percent canopy closure, soil moisture, percent impervious cover, etc.), numerical *in situ* data are commonly used as the dependent variable and related to the brightness values of the corresponding image pixels (or derived indices) to determine if a correlation exists between the two variables. Once the model is established, it can be used to statistically infer the ground condition for the entire image area with a specified confidence level. For example, Jensen *et al.* (1991) correlated wetland vegetation type, maximum canopy height, and percent canopy closure with selected vegetation index transformations derived from SPOT multispectral data. First, image brightness values of 17 sample sites were converted to radiance using absolute calibration coefficients. Selected vegetation indices were then computed using the radiance values. Statistical analyses were performed to compute the correlation between radiance, the vegetation indices, and the *in situ* data. Regression analyses were used to identify the most appropriate model to estimate and map specific vegetation variables including biomass.

17.4. Identify the variables to be measured using remote sensing techniques

In its most simple format, remotely sensed data can be used as a backdrop (background) image for landscape visualization in a GIS. In such instances, the remotely sensed image is usually overlaid with thematic information (e.g. road network, hydrography, political boundaries). However, the most common practice in most earth resource applications is to use remotely sensed data to measure, either directly or indirectly, certain variables or indices, such as land cover, vegetation canopy type and biomass, urban land cover temperature, or traffic volume in rush hours. Jensen (1996) identified two classes of variables that can be remotely sensed: (1) *biophysical* variables that are directly measurable, and (2) *hybrid* variables that may not be directly observable, but can be derived indirectly from a combination of remotely sensed data analysis and *in situ* investigation.

Biophysical variables encompass some biological or physical characteristic of the Earth's surface and in many instances can be directly measured by a remote sensor. When identifying urban heat islands, for example, the apparent temperature of man-made structures, as well as the ambient temperature of the surrounding areas, can be

Table 17.2 Selected biophysical variables and typical remote sensing systems that may be used to collect the data (after Jensen 1996, 2000)

Biophysical variable	Remote sensor that can be used to collect the data
Planimetric location (**x**, **y**)	Analog and digital aerial photography, ETM+, SPOT, RADRSAT, MODIS, ASTER
Elevation/bathymetry (**z**)	Aerial photography, SPOT, RADARSAT, ASTER, Intermap Star-3i Interferometric SAR, LIDAR
Vegetation color, chlorophyll concentration and/or content (biomass)	Aerial photography, ETM+, CASI, ATLAS, SPOT, ERS-1 Microwave, ASTER, MODIS
Foliar water content	ERS-1 Microwave, ETM+, Mid-IR, RADARSAT
Percent canopy closure	Aerial photography, ETM+, SPOT, AVIRIS, CAMS, ATLAS
Soil moisture	ALMAZ, ETM+, ERS-1 Microwave, RADARSAT, Intermap Star-3i Interferometric SAR, ASTER
Surface temperature	AVHRR, ETM+, Daedalus, ASTER, MODIS, ATLAS
Surface roughness	Aerial photography, ALMAZ, ERS-1 Microwave, RADARSAT, ASTER
Evapotranspiration	AVHRR, ETM+, SPOT, CASI, MODIS, AVIRIS, Daedalus
Water and ground color	Aerial photography, ETM+, SPOT, AVIRIS, CAMS, ATLAS

directly measured using a thermal infrared imaging system. As previously discussed, the remote sensing derived apparent temperature can be calibrated by the *in situ* true (kinetic) temperature measurement on the ground. Remote sensing derived temperature maps may then be directly incorporated into physical science models for further analyses.

Another example is the derivation of vegetation amount (or biomass). Through analysis of multispectral remote sensor data, a vegetation index can be calculated for each pixel in the study area. The value of the index indicates the amount of vegetation present (g/m^2) and thus can be included in numerical or statistical models. Selected biophysical variables and the remote sensing systems that can be used to measure them are listed in Table 17.2. The characteristics of many of these remote sensing systems are summarized in Chapter 18. It is useful for remote sensing practitioners to compile a list of their most important variables and the remote sensing systems that may be used to derive the required information.

Hybrid variables are human constructs that are derived through the comprehensive evaluation of multiple biophysical variables and, sometimes, data from sources other than remote sensing. For example, vegetation stress may be detected by analyzing a number of biophysical variables, such as a plant's chlorophyll absorption characteristics measured in the blue and red portion of the electromagnetic spectrum, its temperature measured in the thermal infrared portion of the spectrum, and/or moisture content measured using passive microwave techniques. Lake water contamination may be mapped by evaluating such variables as suspended sediment, nutrient level, chlorophyll content, and water chemicals, some of which may not be obtained from remote sensing data but from other sources. Land suitability is a typical product of a multiple-factor evaluation process. Finally, a land cover classification map derived from remotely sensed data may be based on the evaluation of several of the fundamental biophysical variables at one time (e.g. object color, location, height, and temperature). It is up to the user to identify suitable hybrid and biophysical variables that are required for a specific project and to determine whether the data must be collected using a field survey or using remote sensing technology.

17.4.1. Measurement scale of remotely sensed variables

As previously discussed in the section on *in situ* measurement, the most commonly recognized scales of measurement are *nominal*, *ordinal*, *interval*, and *ratio*. Nominal and ordinal scale data are commonly termed *categorical*, with nominal scale providing naming or labeling to individual objects in question and ordinal scale ranking objects with no specified numerical difference between the classes. Examples of these two scales are land cover types and land suitability, respectively. Classification and scoring are two common methods for deriving such categorical information as land-use types and suitability ranks.

In contrast, both interval and ratio scales are *numerical*. The fundamental difference between the two is that interval data are scaled by an arbitrary origin, whereas ratioed data are scaled to an absolute origin. If we are using remote sensing to measure the temperature of a field of cotton, for example, it is measured on an interval scale since the value zero is ambiguous between the two major measuring systems (Fahrenheit and Celsius). In such a system, it is possible to perform addition and subtraction with the values, but a multiplication or division makes no sense. This is because the ratio bears no physical meaning. Conversely, if we were measuring the spatial distribution of precipitation using remote sensing techniques, it is a ratio scaled variable that has zero as its absolute origin. Any arithmetic operation is meaningful when applied to such data. A detailed discussion of these four measurement scales in representing spatial data in both hardcopy and digital formats and their ability to be used in a quantitative manner is found in Ji (1993).

Biophysical variables derived from using remotely sensed data are inherently of interval and ratio scale. Therefore, biophysical variables are suitable for being directly incorporated into quantitative analyses and numerical modeling. Spatial reasoning for decision making with expert systems also prefers the remotely sensed data be in raw form rather than data that have been transformed to either nominal or ordinal scale. In contrast, depending on the requirements of the project, a hybrid variable can be of any scale of measurement.

17.4.2. What region(s) of the electromagnetic spectrum should be used to collect the required remotely sensed data?

Most of the information our eyes receive about the external world is in the form of blue, green, and red reflected light. However, this is only a small portion of the available electromagnetic energy moving about in the environment. Remote sensing as an extension to human vision measures electromagnetic energy in the visible and many other spectral regions to capture information about the Earth's surface. By collecting remote sensor data about the biophysical and/or hybrid variables for a project, we implicitly assume an important relationship between electromagnetic energy and these variables. This relationship can be explored in the form of two important questions:

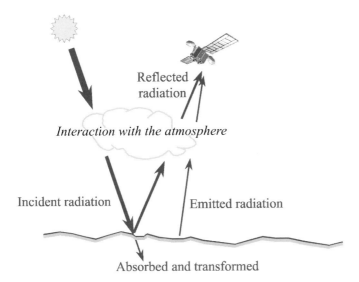

Figure 17.3   The process of remote sensing. The process of remote sensing may be envisioned as the journey of electromagnetic energy starting from its source, traveling through the atmosphere, interacting with ground features, returning to the atmosphere as either reflected or re-emitted energy, and finally arriving at the remote sensor. Of course, the remotely sensed data must then be interpreted and related to ground observations (*in situ* measurement) to change it into valuable information.

1   'how is reflected or emitted electromagnetic energy from the Earth's surface transmitted through the atmosphere and eventually recorded by a remote sensing system?'
2   'which part of the electromagnetic spectrum should be sensed to obtain remote sensing information about the variable in question?'

   These are the fundamental issues the earth resource analysts must understand before making a decision about remote sensing data collection.

   The process of remote sensing, whether it is spaceborne or airborne, can be envisaged as a journey of electromagnetic energy from its source (usually the Sun) to the Earth's surface and then to the sensor (Figure 17.3). Throughout the journey, the level and type of energy are altered while interacting with the atmosphere and the matter on the Earth's surface as discussed in Chapter 6. As such, important information about different types and conditions of surface materials are encapsulated in the characteristics of the energy that is either reflected or emitted from the Earth's surface. In the end, the information-latent energy is intercepted and recorded by the remote sensor.

17.4.3. Sources of electromagnetic energy

Electromagnetic energy recorded by remote sensing instruments may come from entirely natural sources or from man-made artificial sources. For example, Table 17.3 identifies both natural and man-made sources of illumination often used in remote sensing projects. In comparison with the remote sensing systems using the

Table 17.3 Natural and artificial energy sources, their characteristics for remote sensing purposes, and selected remote sensing systems that utilize the energy sources

	Energy sources for remote sensing	
	Natural	*Artificial*
Spectral regions	Solar energy (microwave, radiant heat, infrared, visible, ultraviolet)	Longwave microwave, laser (visible region) profiling and imaging
Type of remote sensing system	Passive, reflection-based in shortwave regions (daytime only) and emission-based in longwave regions (both daytime and nighttime)	Active, reflection-based (both daytime and nighttime)
Sample remote sensor	Landsat (MSS and ETM+), SPOT, AVHRR, ERS, SEASAT, AVIRIS, MODIS, RADARSAT, aerial photographic or digital cameras	AirSAR, RADARSAT, LIDAR

solar energy, artificial energy based remote sensors are limited by fewer choices of wavelengths, but are more flexible in choosing time, day, and illumination angles (Sabins 1997).

17.4.4. What happens when EMR interacts with earth surface features?

Chapter 6 summarized the energy/matter interactions that take place when electromagnetic energy interacts with (a) the atmosphere, and (b) the target of interest. The radiation budget and fundamental radiometric concepts, such as hemispherical reflectance and radiance were introduced. It is instructive here to provide some additional information about the energy/matter interactions at the target and how we must understand these relationships in order to select the optimum remote sensing system for the task at hand.

The spectral reflectance and emission characteristics of a surface feature, known as its *spectral response pattern* or *spectral signature*, is one of the most useful keys to discriminating among different types of targets found in remotely sensed imagery. Since most remote sensing systems record reflected EMR, measuring the radiation reflected from targets is generally of great interest. Therefore, r_λ or Pr_λ (percent reflectance, equal to $r_\lambda \times 100$) is commonly used to represent the spectral response pattern.

17.4.5. Reflective characteristics of surfaces and materials (soil, rock, water, and vegetation)

The characteristics of reflected energy can be examined from two perspectives: the physical reflection with respect to the surface roughness of the target, and the material reflection in relation to the chemical components and internal structure of the target materials. The way surfaces and materials reflect radiant energy provides a basis for both human interpretation and computer-assisted identification of different terrain features from remotely sensed data.

17.4.5.1. Physical reflection

The physical reflection of radiant energy may be best understood from two extreme ways in which energy is reflected from a target: *specular* reflection and *diffuse* reflection as discussed in Chapter 6. A perfect specular reflector can be conceived of as an extremely smooth surface that induces a mirror-like reflection, where almost all of the incident energy is directed away from the surface in a single direction. When specular reflection occurs, there may be either no or total reflected energy transmitted to the remote sensor. For example, a tranquil water surface may produce a perfect specular reflection when the remote sensing device views the water from certain angles. In contrast, a perfect diffuse reflector is such a rough surface that the incident energy is reflected almost uniformly in all directions. In reality, most surface materials are neither perfect specular reflectors nor perfect diffuse reflectors and thus have their physical reflection properties somewhere in between.

While the *roughness* of the surface serves as the determinant of the characteristics of physical reflection, it is in essence relative to the wavelength of the incident radiation. When the wavelengths are much smaller than the surface variations or the particle sizes that make up the surface, diffuse reflection will dominate. For example, fine-grained sand would appear fairly smooth to long wavelength microwaves but would appear quite rough to the visible wavelengths. Roughness also may be related to the angle of observation (Curran 1985). Objects that appear to be equally rough from all angles of observation are called *Lambertian* surfaces. Since the majority of terrain features are non-Lambertian, remote sensors must record their angle of observation and provide such information to the end user.

When describing the amount of energy physically reflected from a surface it is possible to distinguish between *hemispherical* and *bi-directional* reflection. Hemispherical refers to an angle of incidence or collection of radiant flux over a hemisphere, whereas bidirectional refers to the incidence or collection of radiant flux from one

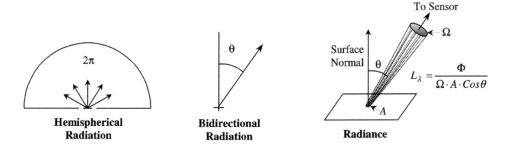

Figure 17.4 The concepts of hemispherical radiation, bidirectional radiation, and radiance. Related to hemispherical radiation are the concepts of *radiant energy*, *radiant density*, *radiant flux*, and *radiant exitance*. Radiant energy is the total energy radiated in all directions. It is called radiant density when normalized by area, radiant flux when normalized by time, and radiant exitance (i.e. radiant flux density) when normalized by both area and time. Related to bidirectional radiation is *radiance* (L_λ), the total radiant flux (Φ) radiated by a unit area (A) per solid angle (Ω) of measurement, which describes what is actually measured by a remote sensor.

direction only (Figure 17.4). The spectral measures of a_λ, t_λ, r_λ and Pr_λ (Chapter 6) in remote sensing literature are hemispherical, as they were derived from laboratory studies. In comparison, spectral reflectance recorded as brightness values in a remotely sensed image are *bi-directional*, as remote sensors can only be located at a single point in space and only look at a relatively small portion of the Earth's surface at a single instant in time. It is therefore important to develop the concepts of radiant flux density, E_λ, and radiance, L_λ, to further our understanding of the physical reflection of incident radiation (Figure 17.4). Discussions of these concepts are documented in Chapter 6 and Curran (1985) and Jensen (2000).

17.4.5.2. Material reflection and the general spectral signature

The material reflection varies among different materials and their conditions. This variation is usually measured in terms of hemispherical percent spectral reflectance, Pr_λ, which is mapped out across useful portions of the electromagnetic spectrum as *spectral reflectance curves*. In Figure 17.5, the spectral reflectance curves of vegetation, soil, rock, and water demonstrate the spectral variation that exists among numerous types of materials on the Earth's surface. The troughs of a curve are known as *absorption features* because they represent spectral regions with more complete absorption of incident energy. The ridges of a curve are known as *reflectance peaks* and are valuable clues for discriminating between different surface materials. Over the past decades reflectance spectra of numerous materials have been measured, documented, and published in the field of remote sensing. The spectral behavior of various materials have been researched and their spectral reflectance curves archived in spectral databases (e.g. Clark *et al.* 1990; Price 1995). Referencing the spectra database, visual interpreters and digital image analysts may identify the spectral regions that best separate ground features of interest and select appropriate spectral bands for remote sensing and *in situ* data collection.

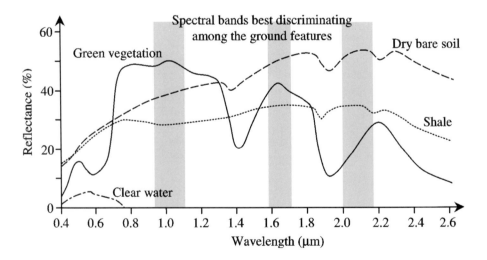

Figure 17.5 Generalized spectral reflectance curves of vegetation, soil, rock, and water (modified from Swain and Davis 1978).

Vegetation: a vegetation canopy mainly consists of plant leaves, each of which in turn consists of layers of structural fibrous organic matter, within which are pigmented, water-filled cells and air spaces. These three features – pigmentation, physiological structure, and water content – have major effects on the reflectance, absorbance, and transmittance properties of plant leaves. The major differences in leaf reflectance between species are dependent on leaf thickness, which affects both pigment content and physiological structure. Responsible for photosynthesis, chlorophyll *a* and *b* strongly absorb radiation between 0.43 and 0.45 μm in the blue wavelength region and between 0.65 and 0.66 μm in the red wavelength region. The chlorophyll pigments absorb considerably less incident green wavelength energy (0.5–0.6 μm). Therefore, summer leaves appear 'greenest' when chlorophyll content is at its peak, whereas autumn leaves appear red or yellow when less chlorophyll in the leaves results in less absorption and proportionately more reflection of red wavelength energy. A healthy leaf's internal physiological structure has discontinuous refractive indices that act as excellent diffuse reflectors for near-infrared wavelengths. For this reason, the near-infrared reflectance is often used in remote sensing to determine how healthy (or unhealthy) vegetation may be. The water content in leaves creates three major water absorption bands (near 1.4, 1.9, and 2.7 μm) that typically appear in the spectral reflectance curve of healthy green vegetation. This feature is commonly used to access plant phenological (growth) stages or stress conditions. With less moisture in plant leaves, the troughs caused by water absorption tend to flatten out.

Soil: the transmittance of a soil surface is significantly low compared to reflectance and absorptance of the incident radiant flux. Different soils appear similar in their spectral reflectance curves, with identifiable distinctions related to moisture content, organic content, texture, structure, and iron oxide content (Stoner and Baumgardner 1981). Since these factors are highly correlated, soil moisture, texture, and structure are widely used as the determinants of spectral properties of different soil types (taxonomy). Surface roughness of a soil is related to the soil structure and texture, which influences the spectral behavior of the soil. A clay soil, for example, tends to have a strong structure with a rough surface on plowing, which serves as a near perfect diffuse reflector to most of the wavelengths used by remote sensing. In contrast, the weak structure of a sandy soil may lead to a smooth surface on plowing, which tends to behave like a specular reflector. Soil moisture tends to have the similar absorption effect as healthy vegetation and generate troughs in the water absorption bands, as well as the overall spectrum, thus resulting in overall low reflectance across the spectrum. Sandy soils tend to have a low moisture content; as a result they usually appear bright in the visible, near-infrared and mid-infrared regions. In comparison, clay soils, which contain relatively high moisture content, may absorb more incident radiant flux and return relatively low reflective radiation back to the remote sensor. As a result of water and hydroxyl absorption, the decrease of reflectance is more rapid in the near-infrared region and especially the mid-infrared region with the increase of soil moisture than in the visible region (Curran 1985).

Rock: the spectral behavior of rocks is similar to that of soils, i.e. with close to zero transmittance. Little spectral research is published on rocks but a great deal on minerals, and the majority of rocks are made primarily of silicates with any combination of carbonates, sulfates, and oxides. The major determinants of the spectral reflectance characteristics of rocks include mineral types and molecular water content contained by the rock. Most important rock or mineral spectra are located in the spectral regions of 0.6–2.5 μm and 8–14 μm. The near infrared region of 1.0–2.5 μm may be used to identify distinguishable features associated with hydroxyl ions or water molecules either bound in the structure or present in fluid inclusions. The middle- and thermal-infrared region of 8–14 μm provides information directly related to the chemistry and structure of the material. Since rocks are assemblages of minerals, their spectral response characteristics are composites of those for each of their constituent minerals (Sabins 1997).

Water: the overall energy absorption and transmission are very high when the incident radiant energy interacts with water, resulting in the lowest reflectance over the spectral curve. Clear water absorbs and transmits longer wavelength visible and near-infrared radiation more than shorter visible wavelengths, which gives rise to the bluish appearance of lakes and oceans. Water absorption of the near- and middle-infrared wavelengths is so complete that on an infrared color composite image clear deep water appears dark blue or even black. This unique spectral characteristic allows remote sensing practitioners to monitor the condition of water bodies. For example, suspended sediment present in the upper layers of the water body tends to increase the reflectance in visible and near-infrared wavelengths; thus it makes an excellent indicator for quantifying such conditions in a lake after storm run-off. Remote sensing practitioners also examine the algae condition of a water body by monitoring the change of reflectance in the green wavelengths, since chlorophyll in algae absorbs more of the blue wavelengths and reflects the green, making the water appear greenish in color when algae is present. Other factors that contribute to the spectral variability of water include water depth and its physical conditions, such as turbidity and surface roughness.

As demonstrated by the previous examples, different materials in general tend to exhibit different spectral reflectance curves. The spectral characteristics of each material type is termed its general *spectral signature*, indicating that similar spectral response patterns may be derived from the same material in repeated laboratory experiments. Unfortunately, remotely sensed spectral signatures can vary over time and space for any given material due to variable and complex physical conditions present in the environment. *The variability of a spectral signature renders the process of digital image processing stochastic and interactive rather than deterministic and automatic.* Therefore, knowing where to 'look' spectrally and understanding the factors that influence the spectral response of the features of interest are critical to correctly interpreting the interaction of electromagnetic radiation with the surface and selecting appropriate spectral regions for remote sensing data collection.

17.4.6. Are we dealing with a reflectance or emission (exitance) data collection problem?

Remote sensing systems distinguish between two types of radiant energy as a result of energy–matter interaction: (1) reflected, and (2) emitted energy. While the simple radiation budget equation (Chapter 6) provides a theoretical basis for reflected energy and spectral reflectance patterns of various ground features, it does not address the fact that the absorbed solar energy is often transformed via the blackbody effect into thermal infrared energy (e.g. 8–14 μm) and re-emitted into the atmosphere. It is this energy that makes thermal imaging and short-wavelength microwave sensing possible. Emissive thermal infrared wavelengths useful for remote sensing only range from 3 to 14 μm, with two atmospheric windows ranging from 3 to 5 μm and 8 to 14 μm. For the passive remote sensing systems that measure the naturally available energy, the reflective wavelengths can only be sensed during the daytime while the solar energy is illuminating the Earth. In comparison, the emissive wavelengths can be detected day and night, as long as the amount of energy is sufficiently large to be recorded. This advantage allows emission-based remote sensing to be of value for a variety of applications. In agricultural applications, for example, long wavelength radiation emitted by plant canopies has been used extensively in the remote sensing of surface energy budgets and soil moisture conditions (Sud and Smith 1984), the estimation of evapotranspiration (Reginato et al. 1985) and assessment of plant stress (Jackson 1986). Other applications include forest fire detection and mapping, wildlife inventory, night vision and target tracking, thermal insulation assessment on energy efficiency, and urban heat island analysis.

Remote sensing of thermal radiation measures the radiant temperature T_{rad} of targets, which can be used to either discriminate between different targets or determine a target's certain characteristics. Lillesand and Kiefer (2000) suggest that T_{rad} is a function of four factors, i.e. *emissivity*, ε, *kinetic temperature*, T_{kin}, and *thermal properties* (including *thermal capacity*, c, *thermal conductivity*, K, and *thermal inertia*, P).

A target's emissivity (ε) is governed by *Kirchhoff's Radiation Law*, which states that for excellent emitters (e.g. blackbodies), the hemispherical absorptance equals the hemispherical emissivity. Therefore, assuming no transmission occurring during the interaction, emissivity becomes the complement of reflectance according to the simple radiation budget equation found in Chapter 6. Unfortunately, real-world objects are seldom blackbodies. Rather, they function as 'gray bodies' that only re-emit a proportion of the energy they receive, resulting in an emissivity smaller than their absorptance. Several gray bodies and their emissivities are summarized in Table 17.4.

Table 17.4 Emissivity of selected ground materials (after Hatfield 1979; Sabins 1997)

	Emissivity ε (where 1.00 is the emissivity of a blackbody)
Rural scene	
Vegetation with a closed canopy	0.99
Water	0.98
Vegetation with an open canopy	0.96
Wet loamy soil	0.95
Dry loamy soil	0.92
Sandy soil	0.90
Organic soil	0.89
Urban scene	
Tar/stone	0.97
Plastic and paint	0.96
Building bricks	0.93
Wood	0.90
Stainless steel	0.16

The significance of an object's emissivity in thermal infrared remote sensing lies in the estimation of a target's actual temperature, T_{kin} from the remotely sensed T_{rad}. Kinetic temperature, T_{kin}, is a target's internal temperature as recorded by a direct thermal sensing device such as a thermometer, which is positively related to the remotely sensed radiant temperature, T_{rad}. The relationship is expressed as:

$$T_{rad} = \varepsilon^{1/4} T_{kin} \tag{17.1}$$

Precise estimation of kinetic temperature using equation (17.1) has proven difficult, as emissivity to an accuracy of 0.02 is required before T_{kin} can be estimated to an accuracy of 1°C. Curran (1985) attributed the difficulty to the great spatial variability and wavelength variability of a material's emissivity. Thermal infrared remote sensing tends to perform best when estimating the magnitude of temperature differences within the environment rather than measuring actual temperatures, for which *in situ* data may be required.

A target's thermal properties determine the target's internal heat distribution and the variation of a target's temperature with time and depth. Thermal capacity, c, is a measure of a target's ability to store heat. Targets with a high c include water bodies, which hold more heat than vegetation or soil. Thermal conductivity, K, refers to the rate at which heat can pass through a target. This property is responsible for urban heat island effects, because compared to rural areas, urban areas are good conductors of heat, which store heat at depth during the daytime and have a long lasting reemission during the night. Thermal inertia, P, measures the thermal response of a target to temperature changes. It is a function of thermal capacity, c, thermal conductivity, K, and density D, and is computed as:

$$P = \sqrt{c \times D \times K} \tag{17.2}$$

Targets with a low P reach the peak heat during the day and cool down to low temperature at night, whereas targets with a high P will have a small diurnal temperature range.

17.4.7. What should be the resolution of the remote sensor data?

Remote sensing systems are used to obtain data about the Earth's resources for spatial and temporal modeling using visual or computer-based numerical and statistical methods. This statement implies that remotely sensed data must be spectrally separable and spatially discernable among important ground features, frequent enough in their collection for change detection over a required time period, and of a sufficiently large range of magnitude for numerical and statistical processing. A remote sensing project usually requires a full set or subset of these required components to meet the objectives. In order to collect the required data, one must consider each of the following four types of remote sensing resolution: *radiometric*, *spatial*, *spectral*, and *temporal*.

Radiometric resolution refers to the ability of the remote sensor to quantify incoming radiance reflected or emitted from the target. While the incoming radiance is sensed as levels of energy, the analog form must be converted into digital readings and recorded as brightness values (or digital numbers) for individual pixels. In practice, the radiometric resolution of a sensor is usually known as the dynamic range in bits, i.e. the range of the brightness values the remote sensor can support (Jensen

1996). As an inherent design parameter, for example, the dynamic range of the US Landsat Multispectral Scanner (MSS) was originally 6 bits (i.e. a value range of 0–63) for its four visible and near-infrared spectral bands. In comparison, the Landsat Thematic Mapper (TM) has an 8-bit dynamic range (i.e. a value range of 0–255 for each spectral band) for its six reflective spectral bands. Several new remote sensing systems, such as Space Imaging's IKONOS-2, record data in 11-bits (values from 0 to 2047), providing increased radiometric resolution.

Spatial resolution is usually expressed as the minimum distance between two objects that will allow them to be differentiated from one another in an image (Jensen 1996; Sabins 1997). It indicates the power of the sensor to resolve the finest detail of ground features and is a function of detector size, focal length, sensor altitude, and system configuration. For aerial photography the spatial resolution is determined by the size of the photosensitive silver grain on the film and is usually measured as the number of line-pairs-per-millimeter (lppm) that can be resolved on the photo. For digital sensors it is given as the dimensions, in meters, of the ground area that falls within the Instantaneous Field of View (IFOV) of a single detector within an array – or pixel size (Logicon 1997). Spatial resolution is an important factor to consider in determining the suitability of a particular remote sensor for a project with a given scale. Lillesand and Kiefer (2000) indicated that the spatial sampling interval should be at least one-half of the size of the target to be remotely sensed. Jensen (2000) summarized the major spatial and temporal resolution requirements for many urban/suburban applications. Figure 17.6 shows the differences in spatial resolution among some well-known sensors and the scales of environmental projects they are suitable for.

Spectral resolution refers to the number and size of the spectral regions, or bands, of

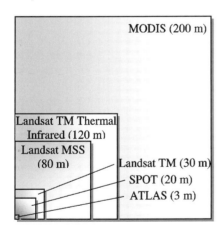
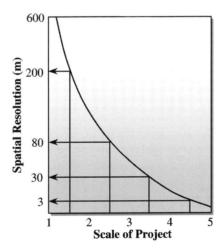

Figure 17.6 Spatial resolution of selected remote sensors and five project scales: 1, global (AVHRR: 1.1 km); 2, continental (MODIS, Landsat MSS: 200–80 m); 3, biome (Landsat MSS, TM, SAR: 80–30 m); 4, region (TM, SPOT, high altitude airborne sensors and aerial photography: 30–3 m); and 5, plot (commercial small satellites, high and low altitude aircraft: 3–1 m) (modified from Jensen 1996).

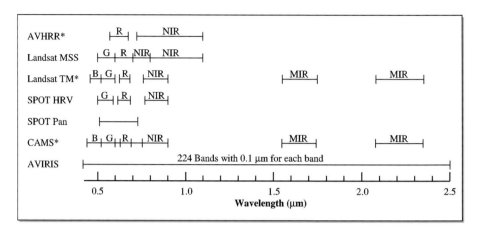

Figure 17.7 Spectral resolution of selected spaceborne and airborne remote sensors. Note that only reflective bands are documented. AVHRR has two thermal bands. Landsat TM has one thermal band. NASA CAMS has one thermal band.

the electromagnetic spectrum a sensor can utilize to observe the Earth's surface. High spectral resolution may be defined as the ability to sense many relatively small bands within a given portion of the electromagnetic spectrum. Remotely sensed data with a finer spectral resolution may form a spectral response curve that more closely approximates the real spectral curve of an object or area of interest. In comparison, the data with a coarser spectral resolution might omit some of the important details that may be useful for distinguishing between spectrally ambiguous targets. The Landsat Thematic Mapper (TM) has seven spectral bands covering a spectral range from 0.4 to 12.5 μm. It thus provides more spectral bands with which to characterize the Earth's surface features than the French SPOT High Resolution Visible (HRV) sensor which has only three spectral bands covering the range between 0.5 and 0.89 μm. NASA's Airborne Visible and Infrared Imaging Spectrometer (AVIRIS) covers the range between 0.4 and 2.5 μm with 224 contiguous spectral bands, each of which has a bandwidth of 0.01 μm (10 nm), making it an ideal hyperspectral remote sensing device to spectrally study ground surface features. The spectral resolution of many well-known remote sensors is illustrated in Figure 17.7. Sensors that record energy in a few bands of the spectrum are called *multispectral* remote sensing systems. Sensors that record energy in hundreds of bands are called *hyperspectral* remote sensing systems.

Temporal resolution refers to the time interval at which a remote sensor repeats data collection at the same location. This sensor property becomes important when the spatial phenomenon must be investigated over an extended or very specific period of time, such as long-term urban sprawl at the urban–rural fringe, seasonal crop growth, multi-year forest management, or recent natural disasters. Many satellite remote sensing systems pass over the same spot on the Earth at systematic time intervals that range from days to weeks depending on their orbital height and swath width (Jensen 1996). Airborne remote sensors may obtain repeated coverage at the required time interval, if there are sufficient resources available for data collection. Airborne remote sensing systems, however, provide the greatest flexibility as they are not constrained to specific days of collection, but can be ordered on demand. Often orbital remote sensors may not generate useful images at the optimum time interval because the success of the data collection is largely weather-dependent. Cloudy and foggy days may hamper the remote sensing of most reflective and emissive wavelengths, making it an art to match the dates for which quality images are available with the requirements of the project.

The information content of remote sensor data is directly proportional to the spatial, spectral, temporal, and radiometric resolution characteristics of the data. The implication is that the power of discriminating between different ground features or extracting valuable biophysical information about them can be increased by carefully taking into consideration each of these resolutions. Some tradeoffs, however, must be taken into account when carrying out practical remote sensing projects. Jensen (1996) suggested that the need for increased storage space, more powerful data-processing tools (hardware and software), and more highly trained personnel to perform or guide analysis could be the downside of increases in resolution. The costs of data may also increase dramatically, as the amount of data increases. A doubling of spatial resolution (e.g. from 10 to 5 m) for instance will result in a four-fold amount of data, which leads to a great increase in cost per unit area (km^2 or $mile^2$). Even with today's rapid improvement of digital storage media and computing technology, it is still important to determine the minimum resolution requirements for a given task from the outset. A thorough cost/benefit analysis of data resolution for a specific project may reduce unnecessary data collection and avoid time wasted processing excessive data. It will also help to avoid the problem of too little data to allow completion of the task.

17.5. What are the bureaucratic problems associated with the project?

Several bureaucratic problems routinely impact a project that involves *in situ* and remote sensing data collection. First, the purchasing department may not understand that the *in situ* and remote sensing data must be acquired during a very specific time interval (window of opportunity). Therefore, it is imperative that purchase orders be in place for both types of data collection well in advance so that they do not preclude data collection during the optimum window. Second, most purchasing departments must be educated so that they understand that certain types of remote sensor data are only available from certain photogrammetric or remote sensing vendors. It is not like ordering a personal computer. In many instances, there are just a few potential sources of remote sensor data (public and/or private) that meet the spatial, spectral, temporal, and radiometric requirements of the project. Third, do not assume that once the remote sensor data are collected that they will be provided on a timely basis. Document how soon after data are collected that they will be available for analysis. Fourth, to keep costs down administrators often request that studies be carried out using data that were purchased for other purposes. This is precisely why some remote sensing related projects are unsuccessful. It is imperative that both the *in situ* and the optimum remotely sensed data be acquired for the specific task at hand. Many a remote sensing project has failed because it was based on the flawed assumption that just any remotely sensed data would do and that new remote sensor data in conjunction with *in situ* measurement simply were not necessary. Quite often the administrative monetary constraints are tied to year-end funding surpluses or expenditure deadlines that cause remote sensing investigations to be carried out in a haphazard manner, as opposed to a well-thought out project.

17.6. What remote sensing systems are available to collect the required information?

Numerous remote sensing systems exist today. The sensor systems may be generally grouped into three types: spaceborne digital sensors, airborne digital sensors, and aerial photography. Each of these three system types offers a different set of features in terms of spatial, spectral, temporal, and radiometric resolutions. Earth resource analysts should be familiar with many of the remote sensing systems and their characteristics to know the options available for various projects. Detailed information about commonly used remote sensing systems is provided in Chapter 18.

17.7. How do I process the remote sensor data to turn it into useful information?

Various analog and digital methods of image processing are available to turn remote sensor data into useful information. These methods and approaches are introduced in Chapter 19.

References

Berk, A., Bernstein, L. S. and Robertson, D. C., *MODTRAN: A Moderate Resolution Model for LOWTRAN 7, Final Report, GL-TR-0122*, AFGL, Hanscomb ADB, MA, 1989, 42 pp.

Clark, R. N., et al., 'High spectral resolution reflectance spectroscopy of minerals', *Journal of Geophysical Research*, 95, 1990, 12653–12680.

Congalton, R. G., 'Using spatial autocorrelation analysis to explore the errors in maps generated from remotely sensed data', *Photogrammetic Engineering & Remote Sensing*, 54(5), 1988, 587–592.

Curran, P. J., *Principles of Remote Sensing*, Longman, New York, 1985, 282 pp.

Curran, P. J., 'Remote Sensing methodologies and geography', *International Journal of Remote Sensing*, 8, 1987, 1255–1275.

Hatfield, J. L., 'Canopy temperatures: the usefulness and reliability of remote measurements', *Agronomy Journal*, 71(4), 1979, 889–892.

Jackson, R. D., 'Remote sensing of biotic and abiotic plant stress', *Annual Review of Phytopathology*, 24, 1986, 265–287.

Jackson, R. D. and Slater, P. N., 'Absolute calibration of field reflectance radiometers', *Photogrammetric Engineering & Remote Sensing*, 52(2), 1986, 189–196.

Jensen, J. R., *Introductory Digital Image Processing*, Prentice Hall, Saddle River, NJ, 1996, 318 pp.

Jensen, J. R., *Remote Sensing of the Environment: an Earth Resource Perspective*, Prentice Hall, Saddle River, NJ, 2000, 550 pp.

Jensen, J. R., Lin, H., Yang, X., Ramsey, E., Davis, B. and Thoemke, C. W., 'The measurement of mangrove characteristics in southwest Florida using SPOT multispectral data', *Geocarto International*, 6(2), 1991, 13–21.

Ji, M., 'Interval and ratio scaled geographical data: confusions, distinctions, and cartographic and GIS representations', *Cartography & GIS*, 7(3), 1993, 49–54.

Ji, M. and Jensen, J. R., 'Effectiveness of subpixel analysis in detecting and quantifying urban imperviousness from Landsat Thematic Mapper imagery', *Geocarto International*, 14(4), 1999, 1–9.

Lillesand, T. M. and Kiefer, R. W., *Remote Sensing and Image Interpretation* (4th ed.), John Wiley & Sons, New York, 2000.

Logicon Geodynamics, Inc., *Multispectral Imagery Reference Guide*, Logicon Geodynamics, Inc., VA, 1997.

Martens, S. N., Ustin, S. L. and Rousseau, R. A., 'Estimation of tree canopy leaf area index by gap fraction analysis', *Forest Ecology and Management*, 61, 1993, 91–108.

Price, J. C., 'Examples of high resolution visible to near-infrared reflectance spectra and a standardized collection for remote sensing studies', *International Journal of Remote Sensing*, 16, 1995, 993–1000.

Reginato, R. J., Jackson, R. D. and Pinter, P. J., Jr., 'Evapotranspiration calculated from remote multispectral and ground station meteorological data', *Remote Sensing of Environment*, 18, 1985, 75–89.

Sabins, F. F., *Remote Sensing: Principles and Interpretation* (3rd ed.), W. H. Freeman, New York, 1997, 494 pp.

Slater, P. N., Bigger, S. F. and Palmer, J. M., 'Ground-reference site and on-board methods for sensor absolute calibration in the 0.4 to 2.5 micron range', *Proceedings of IGARSS 1991 - International Geoscience and Remote Sensing Seminar*, ESA Publications, Noordwijk, The Netherlands, 3, 1991, pp. 1349–1351.

Stoner, E. R. and Baumgardner, M. E., 'Characteristic variations in the reflectance of surface soils', *Soil Science Society of America Journal*, 45, 1981, 1161–1165.

Sud, Y. C. and Smith, W. E., 'Ensemble formulation of surface flux and improvement in evapotranspiration and cloud parameterizations in a GCM', *Boundary-Layer Meteorology*, 39, 1984, 185–210.

Swain, P. H. and Davis, S. M. (eds.), *Remote Sensing: the Quantitative Approach*, McGraw-Hill, New York, 1978, 396 pp.

Thome, K. J., Crowther, B. G. and Biggar, S. F., 'Reflectance- and irradiance-based calibration of Landsat-5 Thematic Mapper', *Canadian Journal of Remote Sensing*, 23, 1997, 309–317.

Wessman, C. A., 'Remote sensing of soil processes', *Agriculture, Ecosystems and Environment*, 34, 1991, 479–493.

Chapter 18

What remote sensing system should be used to collect the data?

John D. Althausen

18.1. Introduction

Several options must be carefully considered when selecting a remote sensing system to study Earth features. This chapter summarizes these options and presents opportunities that currently exist for the remote sensing analyst. When selecting a candidate remote sensing system to carry out a study, the following aspects should be considered:

- Photo/imagery type (analog and digital)
- Platform type
- Ordering specifications
- Data structure and format
- Sensor type
- Pricing

18.2. Photo/imagery type

One of the first concerns that must be addressed is the selection of imagery type. Imagery type, as referred to in this context, differentiates between hardcopy and digital (softcopy) formats. The project that is to be carried out by the remote sensing analyst normally dictates the imagery type needed. If image maps or photographs are needed for manual, visual interpretation then the selection is generally hardcopy imagery. If digital image processing of the remote sensor data is necessary then digital imagery is preferred.

18.2.1. Analog photographic prints

Analog hardcopy aerial photographs are still the most popular type of data used to visualize the Earth's landscapes, other than traditional topographic maps. Aerial cameras may expose black and white film through a Kodak Wratten 12 minus-blue (yellow) filter that records only green and red wavelength energy (Lillesand and Kiefer 2000). This is generally referred to as black and white panchromatic film. Near-infrared energy may be captured on black and white infrared film. Cameras may also expose natural color film that records blue, green, and red wavelength energy or false-color infrared film that records green, red, and near-infrared energy through a Wratten minus-blue (yellow) filter. The image analyst may request that the exposed film be processed to yield positive prints or diapositive film (translucent slide type material). Diapositive film is useful when the photography is to be backlit for visual photointerpretation or if it is to be scanned to eventually produce digital remote sensor data. For photographic processing, direct contact printing is preferred, with the negative size generally being 23 × 23 cm.

One of the most common remote sensing applications is the preparation of hardcopy image maps (sometimes referred to as photomaps). Image maps take advantage of (a) the bird's eye view of the terrain from a vertical perspective, and (b) the broad geographic coverage that remote sensing can provide. Image maps are very useful when topographic maps are unavailable for an area or not current. Image maps can be customized for the analyst by size, scale, color (band) selection and view perspective. Under most conditions, the ancillary information on an image map mimics the traditional topographic map. Additional information that may be present includes platform/sensor name, date of imagery collection, and a color-chip legend (indicating typical land covers and land uses).

Current mapping software, like ERDAS Imagine™, provides the templates to customize image maps to look and feel like standard US Geological Survey (USGS) and National Imagery and Mapping Agency (NIMA) products, with reference grids and elevation contours draped over the imagery (ERDAS 1999). Features that need to be enhanced, and sensor spectral sensitivities, will dictate the type of information conveyed on the image map and whether it should be a true color or false-color composite.

Typical interpretation tasks for both hardcopy photographs and image maps include feature detection, identification, description, and technical analysis. Imagery interpretability rating systems are commonly developed by agencies to assess the usefulness of photography and image maps. Rating systems are subjective and depend on the ability of the remote sensing analyst and the smallest element that can be resolved (Jensen and Cowen 1999).

18.2.2. Digital imagery

Aerial photography and other types of hard-copy imagery such as thermal infrared or radar images can be converted into a digital format through the process of image digitization. When such data are supplied and analyzed in the raster (matrix) environment it is generally labeled *digital imagery*. Typical image processing techniques that are carried out on digital imagery include radiometric and geometric correction, land cover/land use classification, and biophysical information extraction. Visual and digital image processing techniques are discussed in Chapter 19.

18.3. Data structure and format

Spatial modeling usually incorporates two data structures: raster and vector. Raster-based structures are defined as pixel elements in a grid-based system that are effective for mapping large, homogeneous regions, while vector-based structures, in remote sensing, are considered point-line systems that are utilized for discrete, localized mapping. Most remote sensor data are collected in raster format (e.g. Landsat ETM+ or SPOT XS) although remote sensing systems like LIDAR and CASI can collect discrete point data when in a profiling mode. There are benefits and costs associated with using the two different approaches to structuring data files. In general, raster data take up considerably more disk space and require more processing time than vector data.

18.3.1. Image format

When handling raster files, the remote sensing analyst must be aware of the particular file format the imagery is stored in. The most common formats are Band-Sequential (BSQ), Band-Interleaved-by-Pixel (BIP), and Band-Interleaved-by-Line (BIL). BSQ stores each band as a separate file that allows for each band to be analyzed as a separate and individual entity. Remotely sensed data are normally stored in this format so that each spectral band can be accessed separately. BIP stores all data layers of one pixel as adjacent members of the data set. It is a good format if the remote sensing analyst is looking at information at specific locations within numerous data bands or if there are hundreds of bands present as with hyperspectral remote sensor data. The BIL format falls in between BSQ and BIP. Rows from each band are stored adjacent to each other and thus different features of a row that correspond to each other are located near each other in the data set. BSQ, BIP, and BIL formats do not compress the raw data in their native formats.

Another type of file format is the Tag Image File Format (TIFF). TIFF is the most common format for storing digital photographs and images. This type of file format is a standard in the field of graphics and was developed in 1986 by Aldus Corporation. The purpose of TIFF is to describe and store raster images. The TIFF file format can handle both grayscale and full three-band color images and thus is useful for certain applications of remotely sensed data. A recent addition to the raw TIFF format is the introduction of georeferencing information. This newer TIFF format is called GeoTIFF. The georeferencing metadata makes GeoTIFF a popular file format for transferring rectified imagery and has become the standard within the USGS for storing and transferring their Digital Orthophoto Quadrangles (DOQs).

Companies that develop digital image processing software also have their own file formats. The file formats maintain the raw data and include header files that contain information about the raw data (number of bands, rows, and columns), image statistics, georeferencing information, and sensor parameters. Most company-specific formats can be converted and exchanged through import and export modules that are supplied with digital image processing software. Many of the most important digital image processing systems are described in Chapter 20.

18.4. Platform type

Platform type in this section is defined as either an aircraft-based or satellite-based data collection system. Described below are the advantages and disadvantages the remote sensing analyst must take into account when selecting either platform.

18.4.1. Advantages of an aircraft platform

There are many airborne remote sensing systems available to the image analyst.

Aircraft platforms offer the possibility of low flying altitudes, high resolution imagery (down to 10×10 cm pixels), frequent revisits, and variable flight paths. Overall, collection parameters for aircraft are more flexible than satellite platforms, with the aircraft platform being more agreeable to changing mission plans and relocation. With many of the aircraft systems, there is a potential for simultaneous collection of Differential GPS (DGPS) and inertial navigation information, along with the imagery, thus providing the possibility for accurate georeferencing.

18.4.2. Disadvantages of an aircraft platform

The most obvious disadvantage of selecting an aircraft platform is that the lower flying altitude will limit the area of coverage. Unlike a spaceborne sensor system, that can have areal coverage on the order of 100s to 1,000s of square kilometers, an aircraft system must have multiple flight lines to cover a large area similar to a satellite's. Mosaicing of multiple photographs or flight lines thus becomes an important factor when selecting an aircraft platform. Though imaging systems onboard aircraft are improving, the stability of these systems in terms of geometric distortions is always a concern. The analyst must always take into consideration the effects of aircraft roll, pitch, and yaw on the resulting remote sensing data collection.

18.4.3. Advantages of a satellite platform

Satellites orbit in repetitive geo-synchronous or sun-synchronous orbits. The orbits are relatively stable and result in remote sensor data that does not have as much geometric distortion in it. A satellite's altitude, orbit, and path are generally fixed with each mission, and thus provide data sets that have similar spatial resolutions, swaths (or fields-of-view for geo-synchronous satellites), and illumination. Many new satellite borne sensor systems have the capability to acquire images of the Earth not just below the satellite at nadir, but at locations off-nadir. This off-nadir pointing capability is especially useful when performing disaster assessment.

18.4.4. Disadvantages of a satellite platform

In general, there are relatively few limitations to consider when selecting a satellite platform. In the past, a lack of available high-resolution data (because of high, fixed orbits) was a concern, but today's new generation of satellite-based sensors offer high spatial resolution data collection (ground resolutions $<1 \times 1$ m) at relatively high, fixed orbits. Timeliness may also be a perceived as a limitation, but currently there are several satellite-based systems that provide near real-time data. Unfortunately, when a satellite remote sensing system has a problem it is not possible to retrieve and fix it like aircraft-based sensor systems.

18.5. Sensor type

This section identifies many of the available remote sensing systems that are operating, or are expected to be operating in the near future. It also identifies some of the historical data sets (e.g., National High Altitude Photography Program and Landsat

Multispectral Scanner) that are available in archives that may be of use to the remote sensing analyst.

18.5.1. Aircraft remote sensing systems

The discussion begins with descriptions of aircraft mapping programs and collection systems and concludes with an overview of many of the satellite-based sensor systems.

18.5.1.1. National High Altitude Photography Program (NHAP)

The Soil Conservation Service of the US Department of Agriculture carried out the National High Altitude Photography (NHAP) program between 1980 and 1987. This was a cooperative effort by federal agencies to acquire cartographic quality 23×23 cm false-color infrared and panchromatic aerial photographs. The flight lines ran north-south through the center of USGS 7.5 min topographic quadrangles and each false-color infrared photo at 1:58,000 covered the geographic area of one quad. One full frame of false-color infrared imagery covered approximately 69 square miles (176 km^2). The panchromatic photography were acquired at 1:80,000. Archived NHAP data are available and may be utilized by the remote sensing analyst for an historical perspective of a region. The National Cartographic Information Centers (NCIC) in each state can be contacted to determine coverage or a search may be conducted via the internet.

18.5.1.2. National Aerial Photography Program (NAPP)

The requirements for higher spatial resolution aerial photography by many federal agencies led to the replacement of the NHAP by the National Aerial Photography Program (NAPP) in 1987. This false-color infrared coverage is flown at an altitude of 20,000 feet above ground level and the frames are centered on quarter sections of USGS quadrangles. The photography has a nominal scale of 1:40,000 and one full frame covers approximately 32 square miles (83 km^2). The intent is to re-photograph areas every 5 years for the entire conterminous US. Many of the recent NAPP scenes, especially from the 1993 to 1995 aerial campaigns, are available as Digital Ortho-photo Quads through the USGS and numerous state agencies. An example of NAPP color-infrared photography is shown in Figure 18.1 (see plate 8).

Both the NHAP and NAPP missions obtained both leaf-off and leaf-on aerial photography depending upon mission requirements. If the use of the photography was primarily for contour mapping, then the requirements of the photography were better met if the ground surface can be seen. This necessitates data collection in the early spring when the leaves are off the trees. If the need is to map the different types of vegetation based on their spectral characteristics, then of course aerial photographs that capture trees 'with leaf' are more useful.

18.5.1.3. Aircraft videography

A relatively recent development in the field of remote sensing is real-time true color and false-color infrared video. This technology allows an interpreter to view the color

Table 18.1 Airborne Data Acquisition and Registration System 5500 characteristics

Sensor parameter	ADAR 5500 specification
Imaging frame	1500 × 1000 pixels
Instantaneous field of view	0.44 mrad
Spatial resolution	Dependent upon flight mission
Spectral filters	Four programmable bands between 0.400 and 1000 μm (blue to near-infrared)

or false-color infrared imagery as it is being recorded, or in 'real time'. The impact of this development is of major significance to many fields that study the Earth's surface condition. The main difficulties with videography are rectifying and registering information gathered from this utility in a GIS format and carrying out radiometric corrections across the entire series of frames. Also, the individual frames of videography contain relatively few picture elements. This means that to acquire high spatial resolution imagery it is necessary to fly many flightlines at low altitudes. Videography imagery is notoriously difficult to edgemap and place in a controlled mosaic format. However, it is very inexpensive with the major mission expense being the mobilization of the aircraft.

18.5.1.4. Digital photography – Airborne Data Acquisition and Registration System

Positive Systems, Inc., supports two Airborne Data Acquisition and Registration Systems (ADAR 3000 and ADAR 5500) that may be of interest to the remote sensing analyst. Both systems are capable of providing high-resolution digital imagery. The ADAR 3000 is a single camera system that is capable of acquiring digital panchromatic, true color, or color-infrared photographs. The ADAR 5500 system (Table 18.1) is a four-camera system that provides a four-band multispectral image. Data collected by the ADAR 3000 system is ideally suited for GIS and Computer-Aided Design (CAD) applications, while data collected by the ADAR 5500 system can be digitally processed and utilized for environmental and wetlands monitoring, forestry applications, and precision agriculture. Data from both systems are spatially and spectrally of high fidelity and can provide the remote sensing analyst with a seamless mosaic of an area at a spatial resolution of 0.5 × 0.5 m with spectral sensitivity in the visible and near-infrared regions. An example of ADAR 5500 imagery is shown in Figure 18.2 (see plate 9).

18.5.1.5. Digital photography – Airborne Multispectral Digital Camera

Daedalus Enterprises, Inc. is internationally recognized as a provider of airborne multispectral imagers for environmental remote sensing. One of its latest systems (developed in conjunction with the Environmental Research Institute of Michigan-International, Inc.) is the Airborne Multispectral Digital Camera (AMDC). The AMDC offers digital photography in numerous imaging formats that includes panchromatic, true color, and false-color near-infrared. Applications for this system include hydrologic mapping, precision agriculture, forest inventory, wetlands analysis, and environmental impact. Table 18.2 lists the general specifications of this digital camera system.

Table 18.2 Daedalus Inc., Airborne Multispectral Digital Camera characteristics.

Sensor parameter	AMDC specification
Imaging frame	2000 × 2000 pixels
Instantaneous field of view	0.32 mrad
Spatial resolution	Dependent upon flight mission
Spectral filters	0.410–0.486 μm (blue)
	0.508–0.559 μm (green)
	0.605–0.700 μm (red)
	0.790–1.000 μm (near-infrared)
	0.410–0.650 μm (panchromatic)

18.5.1.6. Aircraft multispectral imagery – Daedalus Airborne Multispectral Scanner

The Daedalus Airborne Multispectral Scanner (AMS) is a popular multispectral imaging system that has been used by numerous international government and environmental agencies. Very similar in specifications to two earlier Daedalus scanners (DS-1260 and DS-1268), the AMS offers an eight-channel VIS/NIR spectrometer and a dual element thermal infrared detector (Table 18.3). The AMS can thus acquire true color, false-color near-infrared, and thermal imagery. Applications include wetlands delineation. An example of AMS imagery collected over the Dry Tortugas in Florida is shown in Figure 18.3 (see plate 10).

Besides the AMS and the previously mentioned AMDC, Daedalus systems also include the Airborne Thematic Mapper (ATM), Airborne Hyperspectral Scanner (AHS) and Multispectral Infrared and Visible Imaging Spectrometer (MIVIS). The ATM has bandwidths similar to the Landsat Thematic Mapper and SPOT multispectral systems and has been utilized to simulate these two systems to fill in gaps of data. AHS offers 48 spectral bands of data, spatially co-registered and sensitive from the mid-blue to thermal regions. The MIVIS is a hyperspectral spectrometer with up to 128 bands. The fine bandwidths of this system allow it to be used in sensitive environmental studies. All Daedalus systems can be flown on a wide variety of aircraft (Learjets, ER-2s, etc.) and at varying altitudes, such that mapping scales can vary from 1:1,000 to 1:100,000. Any of these systems is a prospect for the remote sensing analyst if he or she can find an aerial survey firm to fly the specified Daedalus system. In the US, numerous government agencies and aerial mapping firms have flown these systems on their aircraft.

Table 18.3 Daedalus, Inc., Airborne Multispectral Scanner characteristics

Sensor parameter	AMS specification
Instrumentation	Cross-track scanner
Imaging swath	714 pixels
Instantaneous field of view	2.5 mrad
Spatial resolution	Dependent upon flight mission
Spectral bands	0.42–0.45 μm/0.45–0.52 μm (blue)
	0.52–0.60 μm (green)
	0.60–0.63 μm/0.63–0.69 μm (red)
	0.69–0.75 μm/0.76–0.90 μm/0.91–1.05 μm (near-infrared)
	3.0–5.5 μm/8.5–12.5 μm (thermal)

Table 18.4 Airborne Imaging Spectroradiometer for Applications characteristics

Sensor parameter	AISA specification
Instrumentation	Pushbroom linear array
Imaging swath	364 pixels
Instantaneous field of view	1.0 mrad
Spatial resolution	Dependent upon flight mission
Spectral bands	10–70 programmable bands between 0.430 and 0.900 μm (blue to near-infrared) and now middle-infrared

18.5.1.7. Aircraft hyperspectral imagery – Airborne Imaging Spectroradiometer for Applications

Spectral Imaging Ltd. and 3Di, LLC (Easton, MD) have developed and marketed the Airborne Imaging Spectroradiometer for Applications (AISA). It offers the remote sensing analyst a hyperspectral data set extending from the visible into the near and middle-infrared. AISA (Table 18.4) has been used by 3Di and its customers in a number of applications including precision agriculture, wetlands delineation, and coral reef mapping.

18.5.1.8. Aircraft hyperspectral imagery – Compact Airborne Spectrographic Imager-2

The Compact Airborne Spectrographic Imager-2 (CASI-2) is a hyperspectral imager that utilizes a push-broom spectrograph. It is a programmable system that offers the user a choice of 288 different visible and near-infrared wavelengths. Flown in an aircraft, CASI can supply spatial resolutions from 0.5 to 3 m. This type of remote sensing system has been utilized in a number of different environmental applications including forestry, agriculture, and wetland mapping. With the ability to select specific wavelengths, CASI-2 offers the remote sensing analyst a potential to study environmental features, like vegetation health and stress. Table 18.5 summarizes the characteristics of CASI-2.

18.5.1.9. Aircraft hyperspectral imagery – Airborne Visible-Infrared Imaging Spectrometer

The NASA Jet Propulsion Lab developed and operates the Airborne Visible-Infrared Imaging Spectrometer (AVIRIS), a hyperspectral whiskbroom imaging system. The sensor is typically flown out of the NASA Ames Research Center onboard an ER-2 aircraft. Data collection is generally obtained at 20 × 20 m resolution when the

Table 18.5 Compact Airborne Spectrographic Imager-2 characteristics

Sensor parameter	CASI-2 specification
Instrumentation	Pushbroom linear array
Imaging swath	512 pixels
Instantaneous field of view	1.34 mrad
Spatial resolution	Dependent upon flight mission
Spectral bands	19–288 programmable bands between 0.400 and 1.000 μm (blue to near-infrared)

Table 18.6 Airborne Visible-Infrared Imaging Spectrometer characteristics

Sensor parameter	AVIRIS specification
Instrumentation	Whiskbroom linear array
Imaging swath	614 pixels
Instantaneous field of view	1.0 mrad
Spatial resolution	Dependent upon flight mission
Spectral bands	224 bands between 0.400 and 2.500 μm (blue to mid-infrared)

aircraft is flown at 20 km altitude. AVIRIS (Table 18.6) collects 224 bands of imagery, at 10 nm intervals, between 0.400 and 2.500 μm. Applications involving AVIRIS data are numerous and include ecological assessment, coastal/marine mapping, geology/ mineralogy, and atmospheric profiling. Appendix A provides websites for data from AVIRIS and other sensors.

18.5.1.10. Aircraft laser – Airborne Oceanographic LIDAR-3

The first Airborne Oceanographic LIDAR (LIght Detection And Ranging) system (AOL) was developed by NASA in the late 1970s. The current system, AOL-3, is flown for validating data collected by the Moderate-Resolution Imaging Spectro-radiometer (MODIS). Another application to which the AOL-3 is well suited is imaging of fluorescence on the ocean surface of dissolved organic material and chlor-ophyll. Imaging spectrometers are also flown by NASA to collect ocean color infor-mation simultaneously with the AOL-3. Table 18.7 summarizes the basic characteristics of the AOL-3 system.

18.5.1.11. Aircraft laser – Airborne Topographic Mapper

Developed and marketed by NASA, the Airborne Topographic Mapper (ATM) is a LIDAR instrument designed to conduct high-resolution surface topographic mapping. As part of a joint effort with NOAA, NASA has used the ATM to map beach profiles from Delaware to South Carolina. When collected with DGPS, ATM instruments (Table 18.8) have collected elevation information at accuracies approaching ± 10 cm. Numerous commercial firms now offer digital elevation model information derived from LIDAR systems.

Table 18.7 Airborne Oceanographic LIDAR-3 characteristics

Sensor parameter	AOL-3 specification
Instrumentation	Dual wavelength laser fluorospectrometer
Pulse width	12 ns
Instantaneous field of view	1.0 mrad
Spectral bands	0.355 μm (ultra-violet) 0.532 μm (green)

Table 18.8 Airborne Topographic Mapper characteristics

Sensor parameter	ATM specification
Instrumentation	Laser transmitter system
Pulse width	7 ns
Instantaneous field of view	2.1 mrad
Spectral bands	0.523 μm (green)

18.5.2. Satellite high-resolution multispectral imagery

18.5.2.1. Space Imaging, Inc., IKONOS-2

The recently launched IKONOS-2, by Space Imaging, Inc., is the civilian world's first high spatial resolution satellite remote sensing system. After IKONOS-1 failed to achieve orbit on April 27, 1999, the successful launch and deployment of IKONOS-2 (September 24, 1999) jump-started the high-resolution, commercial satellite imaging industry (Fritz 1996). The Kodak camera system onboard IKONOS-2 has one panchromatic band at 1 × 1 m spatial resolution and four multispectral bands at 4 × 4 m resolution. IKONOS-2 has a small Earth footprint of 11 km² (the size of a US township). Revisit times are 3 days for the panchromatic sensor and 1.5 days for the multispectral sensor. The system can view both along-track and cross-track, and has fore and aft imaging capabilities for stereo viewing. Data from IKONOS-2 is useful for applications in urban planning, natural resource investigations, and disaster management. Table 18.9 summarizes the basic characteristics of the IKONOS-2 system. Several other high-resolution satellite sensor systems are scheduled for launch including, Orbimage, Inc.'s OrbView-3 and OrbView-4 and EarthWatch, Inc.'s Quickbird.

18.5.3. Satellite moderate-resolution multispectral imagery

18.5.3.1. Landsat Multispectral Scanner

The Multispectral Scanner (MSS) was an optical-mechanical system that had a ground resolution of 79 × 56 m and recorded energy in four broad spectral bands (green, red,

Table 18.9 IKONOS-2 characteristics

Sensor parameter	HRV/HRVIR specification
Instrumentation	Kodak linear array digital camera
Imaging swath	11 km
Repeat coverage	1.5–3 days
Spatial resolution	4 × 4 m (multispectral); 1 × 1 m (panchromatic)
Spectral bands	0.45–0.52 μm (blue)
	0.52–0.60 μm (green)
	0.63–0.69 μm (red)
	0.76–0.90 μm (near-infrared)
	0.45–0.90 μm (panchromatic)

Table 18.10 Landsat Multispectral Scanner characteristics

Sensor parameter	MSS specification
Instrumentation	Cross-track scanner
Imaging swath	185 km
Repeat coverage	18 days (Landsat-1, 2, 3); 16 days (Landsat-4 and 5)
Spatial resolution	79 × 56 m
Spectral bands	0.5–0.6 μm (green)
	0.6–0.7 μm (red)
	0.7–0.8 μm (near-infrared)
	0.8–1.1 μm (near-infrared)

and two near-infrared). The MSS was flown onboard Landsat 1 through Landsat 5 (Table 18.10). Geographic coverage was approximately 185 × 170 km. Since this sensor collected data between 1972 and 1992, the remote sensing analyst may utilize this data set for historical analysis to supplement current data collection.

18.5.3.2. Landsat Thematic Mapper

Landsat 4 and 5 carried the Thematic Mapper (TM) sensor payload in addition to the MSS. The TM had three visible bands (30 m spatial resolution), three short-wave infrared bands (30 m spatial resolution), and one thermal band (120 m spatial resolution) (Engel and Weinstein 1983). The TM short-wavelength infrared bands allowed for vegetation discrimination, geologic interpretation, and soil moisture differentiation. The utilization of TM data for various interpretive purposes is facilitated by the system's unique combination of bandwidths, spatial resolution, and geometric fidelity. Table 18.11 summarizes the basic characteristics of the TM system.

18.5.3.3. Landsat Enhanced Thematic Mapper Plus

Landsat-7 Enhanced Thematic Mapper Plus (ETM+) is an environmental satellite launched on April 15, 1999. Landsat-7 is a continuation of the Landsat satellite series which had a major setback when Landsat-6 was lost on October 5, 1993, when it did not achieve orbit (Table 18.12). The ETM+ sensor collects eight bands of data from

Table 18.11 Landsat Thematic Mapper characteristics

Sensor parameter	TM specification
Instrumentation	Cross-track scanner
Imaging swath	185 km
Repeat coverage	16 days
Spatial resolution	30 × 30 m (bands 1-5, 7); 120 × 120 m (band 6)
Spectral bands	1: 0.45–0.52 μm (blue)
	2: 0.52–0.60 μm (green)
	3: 0.63–0.69 μm (red)
	4: 0.76–0.90 μm (near-infrared)
	5: 1.55–1.75 μm (mid-infrared)
	6: 10.4–12.5 μm (thermal-infrared)
	7: 2.08–2.35 μm (mid-infrared)

Table 18.12 Timetable of Landsat launches dating back to July 23, 1972

System	Launch (date)	End of service (date)	Sensor	Resolution (m)	Original communications	Altitude (km)	Revisit time (days)
Landsat 1	7/23/72	1/6/78	RBV	80	Direct downlink with recorders	918	18
			MSS	80			
Landsat 2	1/22/75	2/25/82	RBV	80	Direct downlink with recorders	918	18
			MSS	80			
Landsat 3	3/5/78	3/31/83	RBV	30	Direct downlink with recorders	918	18
			MSS	80			
Landsat 4	7/16/82	MSS operational	MSS	80	Direct downlink TDRSS	705	16
		TM – 8/93	TM	30			
Landsat 5	3/1/84	Still Active	MSS	80	Direct downlink TDRSS	705	16
			TM	30			
Landsat 6	10/5/93	10/5/93	MSS	80	Direct downlink with recorders	705	16
			ETM	30/15			
Landsat 7	4/15/99	Still Active	ETM+	30/15	Direct downlink with recorders	705	16

the visible to thermal region (Table 18.13). The data are quantized to the best 8 of 9-bits with a range of values from 0 to 255. Like its predecessor, the TM, the spatial resolution for the visible and short-wave infrared bands (1–5 and 7) for ETM+ is 30 × 30 m. For historical continuation of the Landsat band numbering sequence, ETM+ Band 6 is designated as the thermal band. The spatial resolution of the thermal band was improved, over the previous TM thermal band, to 60 × 60 m. Band 8 is the new panchromatic band. It has spatial resolution of 15 × 15 m and offers improved mapping in urban areas for Landsat data users. Figure 18.4 (see plate 11) illustrates the high-quality imagery available from the Landsat-7 ETM+.

Table 18.13 Landsat Enhanced Thematic Mapper Plus characteristics

Sensor parameter	ETM+ specification
Instrumentation	Cross-track scanner
Imaging swath	185 km
Repeat coverage	16 days
Spatial resolution	30 × 30 m (bands 1-5, 7); 60 × 60 m (band 6); 15 × 15 m (band 8)
Spectral bands	1: 0.45–0.515 μm (blue)
	2: 0.525–0.605 μm (green)
	3: 0.63–0.69 μm (red)
	4: 0.75–0.90 μm (near-infrared)
	5: 1.55–1.75 μm (mid-infrared)
	6: 10.4–12.5 μm (thermal-infrared)
	7: 2.09–2.35 μm (mid-infrared)
	8: 0.53–0.90 μm (panchromatic)

Table 18.14 Timetable of SPOT launches since February 22, 1986

System	Launch (date)	End of service (date)	Sensor	Resolution (m)	Original communications	Altitude (km)	Revisit time (days)
SPOT 1	2/22/86	Still active	PAN	10	Direct downlink with recorders	822	26: nadir
			XS	20			1-5: off-nadir
SPOT 2	1/22/90	Still active	PAN	10	Direct downlink with recorders	822	26: nadir
			XS	20			1-5: off-nadir
SPOT 3	9/26/93	11/4/96	PAN	10	Direct downlink with recorders	822	26: nadir
			XS	20			1-5: off-nadir
SPOT 4	3/24/98	Still active	PAN	10	Direct downlink with recorders	822	26: nadir
			XS	20			1-5: off-nadir
			VEGETATION	1000			

18.5.3.4. SPOT – Systeme Pour l'Observation de la Terre

Systeme Pour l'Observation de la Terre (SPOT) was conceived and designed by the French Centre National d'Etudes Spatiales (CNES). Four satellites have been launched and three are currently operational (Table 18.14). SPOT was the first civilian satellite to include a linear array sensor, employ a push-broom scanning technique, and to have pointable optics (stereoscopic imaging). The payload, for SPOTs 1, 2, and 3 consisted of two identical High-Resolution-Visible (HRV) imaging systems, one that has two visible bands (20×20 m spatial resolution) and one near-infrared band (20×20 m spatial resolution), and another that has one panchromatic band (10×10 m spatial resolution). SPOT-4 added a mid-infrared band, with 20×20 m spatial resolution to its multispectral system (called the HRVIR – High-Resolution Visible-Infrared). Also on SPOT-4, a single red band replaced the broad panchromatic band. Table 18.15 summarizes the basic characteristics of the SPOT system.

Due to its pointability and several satellite constellations, the SPOT sensor can obtain repetitive coverage of the globe every 1–5 days. The SPOT sensor can provide the remote sensing analyst with increased frequency of coverage of areas with cloud cover problems, stereoscopic viewing, and DEM generation. Typical applications for SPOT multispectral imagery include precision agriculture, natural resource management, and wetlands inventory.

18.5.4. Satellite moderate-resolution hyperspectral imagery

18.5.4.1. Moderate-Resolution Imaging Spectroradiometer

The Moderate-Resolution Imaging Spectroradiometer (MODIS) was flown on the recently launched (December 18, 1999) TERRA satellite by NASA. This system is the first civilian-available, satellite-based hyperspectral sensor collecting data over numerous spatial resolutions and spectral wavelengths (Table 18.16). MODIS offers the remote sensing analyst a variety of uses from hyperspectral imaging of the oceans,

Table 18.15 Systeme Pour l'Observation de la Terre characteristics

Sensor parameter	HRV/HRVIR specification
Instrumentation	Pushbroom linear array
Imaging swath	60 km
Repeat coverage	1–5 days
Spatial resolution	20 × 20 m (multispectral); 10 × 10 m (panchromatic)
Spectral bands	B1: 0.50–0.59 μm (green)
	B2: 0.61–0.68 μm (red)
	B3: 0.79–0.89 μm (near-infrared)
	B4: 1.58–1.75 μm (mid-infrared) – on SPOT-4 only
	P: 0.51–0.73 μm (panchromatic) – on SPOT-1, 2, and 3;
	P: 0.61–0.68 μm (panchromatic) – on SPOT-4

terrestrial surface, and the atmosphere. One of the unique features of MODIS is its calibration system that allows for onboard conversion of raw digital numbers to reflectance or radiance values (Salomonson *et al.* 1995; Barnes *et al.* 1998). Figure 18.5 (see plate 12) is an example of coastal images obtained by the MODIS sensor.

18.5.5. Satellite moderate-resolution microwave imagery

18.5.5.1. RADARSAT-1 and -2

RADARSAT is an advanced Earth observation satellite system developed by Canada to monitor environmental change and to support resource sustainability. With the launch of RADARSAT-1 on November 4, 1995, RADARSAT provided access to the first fully operational civilian radar satellite system capable of large-scale production and timely delivery of data. Applications for its usage include ice reconnaissance, coastal surveillance (Figure 18.6, see plate 13), and soil/vegetation moisture studies

RADARSAT (Table 18.17) is developed and marketed by the Canada Space Agency and Canada Centre for Remote Sensing. RADARSAT-2 was recently launched and offers the remote sensing analyst high-resolution microwave data with multiple polarizations.

Table 18.16 Moderate-Resolution Imaging Spectroradiometer characteristics

Sensor parameter	MODIS specification
Instrumentation	Whiskbroom linear array
Imaging swath	2700 km
Repeat coverage	1–2 days
Spatial resolution	250 × 250 m (bands 1–2)
	500 × 500 m (bands 3–7)
	1 × 1 km (bands 8–36)
Spectral bands	20 bands: 0.4–3.0 μm (blue to mid-infrared)
	16 bands: 3.0–to 15 μm (thermal-infrared)

Table 18.17 RADARSAT-1 characteristics

Sensor parameter	RADARSAT-1 specification
Instrumentation	Active microwave C band
Imaging swath	50–500 km
Repeat coverage	24 days
Spatial resolution	$8 \times 8–100 \times 100$ m (multi-beam)
Polarization	HH
Depression angles	40–70°

18.5.6. Satellite coarse-resolution multispectral imagery

18.5.6.1. NOAA Advanced Very High Resolution Radiometer

NOAA's Advanced Very High Resolution Radiometer (AVHRR) is one of the oldest and most widely utilized satellites for global studies (Table 18.18). It was developed for global and repetitive (daily) monitoring of sea surface temperature, oceanic currents, terrestrial vegetation, and polar ice (Cracknell 1997). The AVHRR has one visible band, one near-infrared band, and three thermal bands (Table 18.19). The ground resolution for the sensor is 1.1×1.1 km. Since data can be globally mapped at 1 and 4 km grids, it can be utilized for input into global models and databases. NOAA AVHRR data is relatively inexpensive.

Table 18.18 Timetable of AVHRR launches since October 13, 1978

System	Launch (date)	End of service (date)	Sensor	Resolution (km)	Original communications	Altitude (km)
TIROS-N	10/13/78	1/30/80	AVHRR	1.1	Direct downlink/ recorders	833
NOAA-6	6/27/79	11/16/86	AVHRR	1.1	Direct downlink/ recorders	833
NOAA-B	5/29/80	5/29/80	AVHRR	1.1	Direct downlink/ recorders	833
NOAA-7	6/23/81	6/7/86	AVHRR	1.1	Direct downlink/ recorders	833
NOAA-8	3/28/83	10/13/85	AVHRR	1.1	Direct downlink/ recorders	833
NOAA-9	12/12/84	5/11/94	AVHRR	1.1	Direct downlink/ recorders	833
NOAA-10	9/17/86	Stand-By	AVHRR	1.1	Direct downlink/ recorders	833
NOAA-11	9/24/88	9/13/94	AVHRR	1.1	Direct downlink/ recorders	833
NOAA-12	5/14/91	12/15/94	AVHRR	1.1	Direct downlink/ recorders	833
NOAA-13	8/9/93	8/9/93	AVHRR	1.1	Direct downlink/ recorders	833
NOAA-14	12/30/94	Still Active	AVHRR	1.1	Direct downlink/ recorders	833
NOAA-15	5/13/98	Still Active	AVHRR	1.1	Direct downlink/ recorders	833

Table 18.19 Advanced Very High Resolution Radiometer characteristics

Sensor parameter	AVHRR specification
Instrumentation	Cross-track scanner
Imaging swath	2700 km
Repeat coverage	Daily
Spatial resolution	1.1 × 1.1 km
Spectral bands	1: 0.58–0.68 μm (red)
	2: 0.725–1.10 μm (near-infrared)
	3: 3.55–3.93 μm (thermal-infrared)
	4: 10.30–11.30 μm (thermal-infrared)
	5: 11.50–12.50 μm (thermal-infrared) – not on NOAA-6, 8, 10

18.5.6.2. SPOT-4 VEGETATION

VEGETATION is a very wide-angle (2,000 km-wide swath) earth observation instrument offering a spatial resolution of about 1 × 1 km and high radiometric characteristics. It uses the same spectral bands as the SPOT-4 HRVIR instruments (B2, B3, and B4) plus an additional band known as B0 (0.43–0.47 μm) for oceanographic applications and for atmospheric corrections. The VEGETATION instrument (Table 18.20) flying on SPOT-4 provides global coverage on an almost daily basis at a resolution of 1 km, thus making it an ideal tool for observing long-term environmental changes on a regional and worldwide scale.

18.5.6.3. Sea-Viewing Wide Field-of-View Sensor

SeaWiFS (Sea-viewing Wide Field-of-View Sensor) was launched aboard the SeaStar satellite (OrbView-2) on August 1, 1997. It was developed, and is operated, by Orbimage, Inc. and NASA. SeaWiFS was the long-awaited follow up to NASA's Coastal Zone Color Scanner. SeaWiFS is optimized for use over water but terrestrial sensing was given some consideration in its design (Table 18.21). Applications for SeaWiFS include phytoplankton mapping, modeling of nitrogen/carbon cycling, and delineation of major ocean features.

SeaWiFS data are available in three formats. High Resolution Picture Transmission (HRPT) data are full resolution image data transmitted to a ground station, as they are collected (1.1 × 1.1 km resolution). Local Area Coverage (LAC) scenes are also full resolution data, but recorded with an on-board tape recorder for subsequent transmis-

Table 18.20 SPOT-4 VEGETATION characteristics

Sensor parameter	VEGETATION specification
Instrumentation	Pushbroom linear array
Imaging swath	2000 km
Repeat coverage	Daily
Spatial resolution	1 × 1 km
Spectral bands	B0: 0.43–0.47 μm (blue)
	B2: 0.61–0.68 μm (red)
	B3: 0.79–0.89 μm (near-infrared)
	B4: 1.58–1.75 μm (mid-infrared)

Table 18.21 SeaWiFS characteristics

Sensor parameter	SeaWiFS specification
Instrumentation	Cross-track scanner
Imaging swath	2800 km
Repeat coverage	Daily
Spatial resolution	1.1 × 1.1 km
Spectral bands	0.402–0.422 μm/0.433–0.453 μm/0.480–0.500 μm (blue)
	0.500–0.520 μm/0.545–0.565 μm (green)
	0.660–0.680 μm (red)
	0.745–0.785 μm/0.845–0.885 μm (near-infrared)

sion during a station overpass (1.1 × 1.1 km resolution). Finally, Global Area Coverage (GAC) data provide daily sub-sampled global coverage recorded on the tape recorders and then transmitted to a ground station (4 × 4 km resolution).

18.6. Ordering specifications

Now that many of the most widely used aircraft and satellite-based remote sensing systems have been identified, the next step is to determine the requirements of the specific imagery needs of the project.

18.6.1. Aircraft film/digital photography

When ordering photography, whether hardcopy or digital, it is important to understand how the data was or will be collected by the aircraft system. For example, before any photography is processed or interpreted the user should have predetermined specifications on collection date and time, flight parameters, and camera and film specifications.

18.6.1.1. Collection date and time

For aerial imagery to be properly interpreted and analyzed, the data should be collected at a very specific time of day and year. Under normal conditions, remotely sensed data is collected during mid-morning hours. This minimizes the effects of cloud cover and other meteorological effects, allows the interpreter to take advantage of a sun angle that will supply sufficient scene brightness, and allows for the casting of sufficient shadows that can be utilized as a recognition element. The time of year will vary depending on geographical region and the information that is needed. For example, when studying wetlands productivity it is important to collect data during peak biomass. If aerial photography of the Florida wetlands was needed it would be best to collect the data in April or May which also coincides with the end of the dry season and expected lowest water levels.

18.6.1.2. Flight parameters

The scale of the photography is determined by the flying altitude of the aircraft and optics (focal length) of the camera system (refer to Section 19.2.2 for the scale equation). By knowing beforehand the focal length that will be used by the aerial mapping firm, the user can determine what flying altitude above ground level is necessary for collecting a scale of photography that is suitable for the project. Once this has been determined, flight line maps can be generated that indicate where exposures should be made, along with the amount of overlap and sidelap between individual frames of aerial photography.

After collection, the aerial mapping firm should provide the user with a detailed map showing the actual flight path of the aircraft. Maps should include topographic ground information and supply geo-referenced principal points for each photograph as well as the area covered by each photograph. Approximate scale of the photography, flying altitude above sea level, percent overlap between successive photographs and percent sidelap between adjacent flight lines should also be reported. If available, aircraft attitude information should be provided for each exposure-frame.

It is recommended that aerial mapping firms report standard flight mission parameters as specified by the American Society for Photogrammetry and Remote Sensing, including a map illustrating plotted locations of each frame of coverage from the film roll, including frame center coordinates from the on-board GPS system.

18.6.1.3. Camera and film specifications

To understand and properly interpret aerial photography, several types of information need to be provided to the user. The variables shown in Table 18.22 represent the minimal amount of information that should be provided on camera and film specifications. If the photography is going to be ortho-corrected then having this information becomes even more critical.

18.6.2. Satellite sensor imagery

Satellites and the sensors they carry have certain design features that influence how they operate, and why you might select one over another (Jensen 1996; Kramer 1996; Rees 1999). Below are some parameters that might influence your decision to select one satellite-based sensor over another. Again, as with aircraft-based systems, your decision to choose one sensor over another should be based on project requirements.

Table 18.22 Film and camera specifications that need to be identified to carry out analysis and product generation

Film parameter	Camera parameter
Film type (panchromatic, true, or false color)	Lens type
Film frame (e.g. 23 × 23 cm)	Lens angular field of view
Film product (e.g. print or diapositive)	Lens focal length
	Lens F-stop/aperture
	Camera type
	Camera shutter speed setting
	Filter types
	Filter 'T' (transmittance) value

18.6.2.1. Spatial resolution and spectral sensitivity

Satellite remote sensing systems are often summarized in terms of their spatial and spectral resolution characteristics (defined in Chapter 17). The spatial resolution (e.g. 10×10 m) of satellite-based remote sensing systems that look at the terrain from a vertical vantage point (i.e. nadir) are based on (1) the altitude of the spacecraft in orbit, and (2) the detector optical characteristics. Generally, the number of spectral bands and their bandwidth sensitivities are all decided upon pre-launch (Jensen 1996; Schowengerdt 1997). Sensor system characteristic tables like those introduced earlier in the chapter are available for all environmental satellite systems and should be evaluated closely before imagery is purchased.

18.6.2.2. Orbit parameters

All civilian remote sensing satellites map the Earth in near-polar sun-synchronous or geo-synchronous orbits. Once in orbit, satellites generally remain at a constant speed and fixed in their paths. Therefore, most satellites cannot be moved or tasked to other areas away from their orbital paths. Of course this is not the case with military surveillance satellites. There are times when earth-resource satellites are tasked to collect nighttime imagery (e.g. thermal sensors). It is important to remember that microwave satellite systems can collect data both during the day and night (Way and Smith 1991; Jensen 2000; Lillesand and Kiefer 2000).

18.6.2.3. Viewing geometry

Sensor systems can either be fixed at a certain viewing angle or have pointable optics. Most of the time, when a sensor is fixed in its viewing angle the sensor is positioned at nadir. When fixed at nadir, a sensor can only image information that is directly below the path of the satellite it is carried on. Pointable sensors can move side-to-side or aft-to-fore acquiring imagery that is not directly beneath the satellite's track. Adjustable viewing is extremely important because it can help increase repeatable coverage times and introduce the possibility of stereo imaging. However, the remote sensing analyst should keep in mind that pointing away from nadir could introduce some geometric/ dimensional problems to the pixel sizes in the imagery (Westin 1990; Seto 1991). Off-nadir viewing also introduces Bi-Directional Reflectance Distribution Function (BRDF) characteristics in the imagery which are beyond the scope of this discussion but can be very important when monitoring vegetation obtained at different viewing angles (Jensen and Schill 2000).

18.6.2.4. Revisit times

The number of days that elapse between a satellite passing directly overhead is generally referred to as revisit time. Revisit time is a function of a satellite's orbit, altitude, sensor optics, and sensor pointability. For nadir-only viewing sensors, revisit time is primarily a function of the size of the imaging swath of the system and orbital path. For a pointable optics sensor, revisit time is a function of how far off-nadir the sensor can point and the latitude where the data are being collected, as well as imaging swath and orbital path.

Table 18.23 Price comparison of recently collected imagery using several remote sensing systems[a]

System	Spectral range	Spatial resolution	Cost per km² ($)
Aerial photography	PAN or VIS/NIR	1:12,000	200[a]
ADAR 5500	VIS/NIR	1 m	100[a]
IKONOS-2	PAN or VIS/NIR	1 m or 4 m	29[b]
Landsat-7 ETM+	VIS/NIR/MIR/TIR/PAN	30 m/60 m/15 m	0.02[b]
SPOT-4	PAN or VIS/NIR/MIR	10 m or 20 m	0.70[b]
RADARSAT	Fine-Beam	10 m	1.60[b]
RADARSAT	Standard-Beam	30 m	0.30[b]

[a] Cost based on a study area of 250 km².
[b] Cost based on one full scene of imagery with no advanced post-processing.

18.6.2.5. Cost

The price of aircraft and satellite photography and imagery varies with spatial and spectral resolution, area of coverage, processing level, and age. One way to look at the relative cost of remote sensor data is to determine the cost of collecting imagery for a specific size of geographic area (e.g. per km²). Table 18.23 provides pricing for imagery collected by several of the current cameras or sensors. It is of course less expensive to purchase archived data. Unfortunately, because prices fluctuate with age it is difficult to set prices on older photography and imagery products.

Appendix A. Sensors described in Chapter 18

WWW addresses (or postal addresses) are provided to give the reader an additional resource for information and specifications on each remote sensing system.

Aircraft film photography	
NHAP	National High Altitude Photography Program.
	http://edcwww.cr.usgs.gov/glis/hyper/guide/napp
NAPP	National Aerial Photography Program.
	http://edcwww.cr.usgs.gov/glis/hyper/guide/napp
Aircraft videography	
V-STARS	Videography System by Geodetic Services, Inc.
	http://www.geodetic.com
Aircraft digital photography	
ADAR	Airborne Data Acquisition and Registration System.
	http://www.possys.com/
AMDC	Airborne Multispectral Digital Camera. Daedalus Enterprises, Inc.,
	P.O. Box 1869, Ann Arbor, MI 48106, USA
Aircraft multispectral imagery	
AMS	Daedalus Airborne Multispectral Scanner. Daedalus Enterprises, Inc.,
	P.O. Box 1869, Ann Arbor, MI 48106, USA

Aircraft hyperspectral imagery
AISA   Airborne Imaging Spectroradiometer for Applications.
                        http://www.3dillc.com/rem-hyper-aisa.html
CASI-2   Compact Airborne Spectrographic Imager-2. http://www.itres.com/
AVIRIS   Airborne Visible-Infrared Imaging Spectrometer.
                        http://makalu.jpl.nasa.gov/aviris.html

Aircraft laser
AOL-3   Airborne Oceanographic LIDAR-3.
                        http://aol.wff.nasa.gov/aolfl_pub.html
ATM   Airborne Topographic Mapper. http://aol.wff.nasa.gov/aoltm.html.

Satellite high-resolution multispectral imagery
IKONOS-2   http://www.spaceimage.com/aboutus/satellites/IKONOS/ikonos.html

Satellite moderate-resolution multispectral imagery
MSS   Landsat Multispectral Scanner.
                        http://geo.arc.nasa.gov/sge/landsat/landsat.html
TM   Landsat Thematic Mapper.
                        http://geo.arc.nasa.gov/sge/landsat/landsat.html
ETM+   Landsat-7 Enhanced Thematic Mapper Plus.
                        http://landsat.gsfc.nasa.gov/.
SPOT   Systeme Pour l'Observation de la Terre.
                        http://www.spotimage.fr/home/system/welcome.htm.
EARTH EXPLORER   The USGS Earth Explorer site. http://earthexplorer.usgs.gov

Satellite moderate-resolution hyperspectral imagery
MODIS   Moderate-Resolution Imaging Spectroradiometer. http://
                        ltpwww.gsfc.nasa.gov/MODIS/MODIS.html

Satellite moderate-resolution microwave imagery
RADARSAT   http://radarsat.space.gc.ca/

Satellite coarse-resolution multispectral imagery
AVHRR   Advanced Very High Resolution Radiometer. http://edcdaac.usgs.gov/
                        1KM/avhrr.sensor.html
VEGETATION   http://www.spotimage.fr/home/system/introsat/payload/vegetati/
                        welcome.htm
SeaWiFS   Sea-viewing Wide Field of View Sensor. http://seawifs.gsfc.nasa.gov/
                        SEAWIFS.html

References

Barnes, W. L., Pagano, T. S. and Salomonson, V. V., 'Prelaunch characteristics of the Moderate Resolution Imaging Spectroradiometer (MODIS) on EOS-AM1, *IEEE Transactions on Geoscience and Remote Sensing*, 36(4), 1998, 1088–1100.

Cracknell, A. P., *The Advanced Very High Resolution Radiometer*, Taylor & Francis, London, 1997, 534 pp.

Engel, J. L. and Weinstein, O., 'The Thematic Mapper: an overview', *IEEE Transactions on Geoscience and Remote Sensing*, 21(3), 1983, 258–265.

ERDAS, *ERDAS Imagine Field Guide*, ERDAS Inc., Atlanta, GA, 1999, 672 pp.

Fritz, L.W., 'The era of commercial earth observation satellites', *Photogrammetric Engineering and Remote Sensing*, 62(1), 1996, 39–45.

Jensen, J. R., *Introductory Digital Image Processing: a Remote Sensing Perspective* (2nd ed.), Prentice Hall, Saddle River, NJ, 1996, 318 pp.

Jensen, J. R., *Remote Sensing of the Environment: an Earth Resource Perspective*, Prentice Hall, Saddle River, NJ, 2000, 544 pp.

Jensen, J.R. and Cowen, D.C., "Remote Sensing of Urban/Suburban Infrastructure and Socio-economic Attributes", *Photogrammetric Engineering and Remote Sensing*, 65(5), 1999, 611–622.

Jensen, J.R. and Schill, S.R., 'Bidirectional Reflectance Distribution Function (BRDF) Characteristics of Smooth Cordgrass (*Spartina Alterniflora*) Obtained Using A Sandmeier Field Goniometer,' *Geocarto International*, 15(2), 2000, 21–28.

Kramer, H. J., *Observation of the Earth and its Environment: Survey of Missions and Sensors* (3rd ed.), Springer-Verlag, Berlin, 1996, 960 pp.

Lillesand, T. M. and Kiefer, R. W., *Remote Sensing and Image Interpretation* (4th ed.), John Wiley & Sons, New York, 2000, 724 pp.

Rees, G., *The Remote Sensing Data Book*, Cambridge University Press, Cambridge, 1999, 262 pp.

Salomonson, V.V., Barker, J. and Knight, E., 'Spectral characteristics of the Earth Observing System (EOS) Moderate-Resolution Imaging Spectroradiometer', *Imaging Spectrometry*, 2480, 1995, 142–152.

Schowengerdt, R. A., *Remote Sensing: Models and Methods for Image Processing* (2nd ed.), Academic Press, San Diego, CA, 1997, 522 pp.

Seto, Y., 'Geometric correction algorithms for satellite imagery using a bi-directional scanning sensor', *IEEE Transactions on Geoscience and Remote Sensing*, 29(2), 1991, 292–299.

Way, J. and Smith, E. A., 'The evolution of synthetic aperture radar systems and their progression to the EOS SAR', *Transactions on Geoscience and Remote Sensing*, 29(6), 1991, 962–985.

Westin, T., 'Precision rectification of SPOT imagery', *Photogrammetric Engineering and Remote Sensing*, 56(2), 1990, 247–253.

Chapter 19

Information extraction from remotely sensed data

Sunil Narumalani, Joseph T. Hlady and John R. Jensen

19.1. Introduction

Remote sensing data can be acquired in *analog* (hard-copy) or *digital* formats. Data in analog format include aerial photographs and images that have been acquired by airborne or satellite remote sensing systems, which are reproduced on paper print or positive transparencies (Figure 19.1a). Digital remotely-sensed data consist of a matrix of numbers, where each cell, commonly referred to as a picture element (pixel), is represented by a Brightness Value (BV) (Figure 19.1b). When an image is viewed on a computer screen, each pixel's BV is depicted in a shade of gray or color. Because of

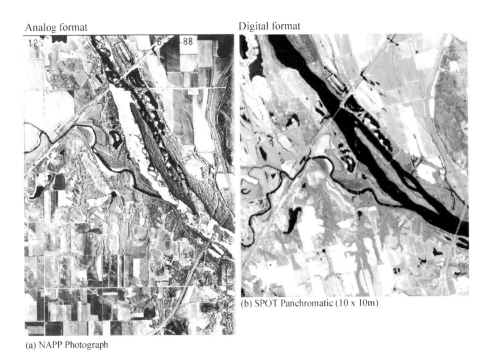

Analog format   Digital format

(b) SPOT Panchromatic (10 x 10m)

(a) NAPP Photograph

Figure 19.1 Examples of an analog and digital image of an area of the Platte River near Ashland, Nebraska. The analog image (a) is a National Aerial Photography Program (NAPP) photograph acquired in June, 1988 at a scale of 1:40,000. The SPOT panchromatic image (b) has a nominal spatial resolution of 10×10 m.

the differences in data formats, both analog and digital image processing techniques have been developed for the interpretation and analysis of remote sensing data.

19.2. Analog image processing

Analog images (i.e. aerial photographs or hard copy imagery) contain a visual record of objects, features, and patterns. To convert these data into information, an analyst must utilize photointerpretation techniques. The level of complexity involved in the photointerpretation process is dependent on the type of information that needs to be extracted for a given application. While performing photointerpretation tasks, an analyst may find that several features are easily recognizable and can be delineated or classified without difficulty. For example, the simple recognition of features and objects, such as water bodies, trees, buildings, and road networks, can be performed rather quickly and without the photointerpreter possessing a high level of expertise. However, more complex and perhaps more pertinent photointerpretation tasks involve a deeper understanding of the biophysical interactions, human influences, and fundamental elements of photointerpretation. A case in point is the soil survey maps produced by the US Department of Agriculture for every county in the US (for example, USDA-SCS 1990). These maps delineate soil types (taxonomy) derived from a combination of low altitude aerial photography and careful field investigation. However, the process transcends the simple identification and interpretation of soil types, and requires that the photointerpreter is either a soil scientist or works closely with one. An understanding of the physical and chemical properties of the soils, soil – moisture interactions, and soil taxonomy, combined with the fundamental elements of image interpretation, are used to delineate the soil types.

19.2.1. Fundamental elements of image interpretation

The visual interpretation of analog images requires that the analyst systematically utilize several characteristics of features visible on the aerial photograph. These are commonly called the *fundamental elements* used in image interpretation and include shape, size, pattern, height, shadow, tone (or color), texture, site, and association (Avery and Berlin 1992; Jensen 2000; Lillesand and Kiefer 2000).

Shape refers to the form, configuration or outline of an object. Often the shape of an object allows the photointerpreter to easily identify the feature. A classic example is the Pentagon Building near Washington, DC, whose distinct shape allows for easy interpretation.

Size is another important criterion that an image interpreter can use to identify and differentiate between objects and features. One of the simplest uses of size is in the discrimination of road networks based on road width, e.g. the identification of a divided, multiple-lane interstate, an undivided paved highway, or a narrow dirt road.

Pattern pertains to the overall spatial arrangement of objects. The pattern of the features in the landscape can offer clues to the type of cultural or natural features. For example, drainage patterns, e.g. dendritic, trellis, rectangular, and centripetal, are distinctly visible on aerial photos and can help geomorphologists or hydrologists identify the dominant drainage, soil characteristics, and structural control (faults, folds) for a given study area. Studying pattern differences in conjunction with shape and size can help an interpreter differentiate between large shopping malls *vs.* linear strip shopping malls, and office *vs.* apartment building complexes.

The *height* of or *elevation* of objects can be used in combination with shape, if stereopairs of photographs are available. Stereoscopic coverage implies that adjacent pairs of overlapping vertical aerial photographs are available, thus providing two different perspectives of the ground area. When viewed through a stereoscope, a three-dimensional effect is rendered and height differences between objects can be detected. Thus, the vertical dimension can be used to differentiate between multi-story and single-story buildings, or to classify vegetation types based on their height.

Shadows can be an aid, as well as a hindrance, to image interpretation. Shadows are useful because they can provide a profile view of objects, thus enabling their recognition, e.g. water towers, or to accentuate topographic features, if the photograph or image is acquired at the proper angle. However, shadows may be problematic, if they conceal or subdue adjacent features/objects, such that their interpretation would be either difficult or impossible. Shadows can also create a pseudoscopic inversion effect, i.e. ridgelines become valleys and vice versa, if the shadows are not oriented so that they fall toward the viewer during stereoscopic examination.

Tone or *color* refers to the brightness of an object based on its reflective characteristics within the electromagnetic spectrum. Objects and features can be identified or distinguished based on tonal characteristics in a black and white photograph or image or by their color in a true color or false-color image/photograph. For example, tonal variations can be used to differentiate soil types. Soils may be dark or light, depending upon their moisture content. Color differences can help distinguish between healthy *vs.* stressed vegetation on a color infrared image.

Texture is the frequency of changes in tone or color (also sometimes referred to as coarseness or smoothness) on the photograph or image. Because it is produced by an aggregation of features that are too small to detect, e.g. tree leaves, leaf shadows, the texture is dependent on the scale of the photograph and the homogeneity or heterogeneity of features in the photo. Analysts can use texture to discriminate between features with similar reflectance. Thus, a large area of healthy pasture recorded on 1:5,000 scale panchromatic aerial photography would have a smooth texture as compared to a mixed forest with deciduous and coniferous trees present in the same photograph.

Site relates to the geographic and topographic characteristics of the object of interest in the aerial photography or image. Consequently, knowledge about a specific location can help the analyst's decision-making process. For example, a photograph of a wetland in the southeastern US would contain certain species of aquatic macrophytes that are indigenous to the area because of the climatic and hydrologic conditions. Similarly, the morphology of a study area may help determine the soil types and thus facilitate the identification and delineation of soil taxonomy.

Association pertains to the probable occurrence of certain features in relation to others. Identification of a feature or object would tend to confirm or deny the existence of another or allow an analyst to draw conclusions based on environmental characteristics. In the Sandhills of Nebraska, an otherwise semi-arid region, a multitude of small lakes and marshes covered with lush vegetation cover exist in sharp contrast to the sparsely vegetated uplands (dunes). This is the result of a 'dynamic interaction between climatic, hydrologic, chemical, and biological processes' (Gosselin *et al.*

2000). In many interdunal valleys the water table intersects with the valley floor, thus resulting in the formation of inland fresh marshes and wet meadows. Conversely, in upland areas, the depth to groundwater is greatly increased and the sandy soils cause rapid infiltration of any precipitation. An image interpreter can therefore make specific associations between the features being interpreted and the environmental conditions that favor their existence.

The fundamental elements of analog image interpretation described above can be used to extract specific qualitative information about features and objects (e.g. building type, land cover class). However, quantitative information extraction, such as height of an object, scale of the photograph, spatial location (x, y coordinates), and area, distance, and volume measurements require knowledge of photogrammetric techniques.

19.2.2. Introductory photogrammetry

As previously defined, *photogrammetry* is the *science* and *art* of obtaining reliable and accurate measurements from photographs. The quantitative information derived can range from approximations of distances and elevations using unsophisticated instruments and basic geometric computations to extremely precise measurements using advanced, computer-driven hardware and software, and complex mathematical techniques (Chapters 4 and 16).

One of the first things a photointerpreter does is to compute the *scale* of the aerial photograph. This is usually done prior to making any quantitative photogrammetric measurements. Scale defines the relationship between a linear distance on a vertical photograph (or map product) and the corresponding actual distance on the ground (Avery and Berlin 1992). Scale may be expressed as a *representative fraction* (e.g. 1:25,000 or 1/25,000), where 1 unit of measure (e.g. inch, centimeter) on the photograph or map represents 25,000 units of the *same measure* on the ground. For example, if the linear distance between two road intersections is 0.5 inches on a 1:25,000 map, then the actual ground distance would be 12,500 inches (or 1,041.67 ft). The concept of scale can be confusing because a *larger denominator* in the representative fraction results in *smaller-scale* products (photograph or map) and *smaller denominators* results in *larger scale* products. Therefore, an aerial photograph that has a scale of 1:5,000 is a *larger-scale* image that covers a smaller ground area than a 1:40,000 scale photograph that has a *smaller scale* and covers a much larger geographic area.

The scale of an analog image can be computed in several different ways. The easiest method requires the measurement of the distance between two identifiable points on the photo, as well as the measurement of the distance between the same two points on a map (whose scale is known) using the following equation:

$$\text{PhotoScale} = \frac{\text{PhotoDist}}{\text{GroundDist}} = \frac{d}{D} \tag{19.1}$$

Scale can also be determined using information on the focal length of the camera and the flying height of the aircraft or sensor platform above the terrain:

$$\text{PhotoScale} = \frac{\text{FocalLength}}{\text{PlatformHeight}} = \frac{f}{H} \tag{19.2}$$

Equation (19.2) may be modified to include information about the local datum, e.g. Mean Sea Level (MSL). For example, a photograph may be acquired at a flying height of H m above MSL. Assuming that the terrain is relatively flat and lies h m above MSL, Equation (19.2) can be modified as follows (refer to Figure 16.1):

$$\text{PhotoScale} = \frac{\text{FocalLength}}{\text{PlatformHeight} - \text{TerrainElevation}} = \frac{f}{H - h} \qquad (19.3)$$

Notice the assumption being made about the terrain. In areas of *high relief*, or those where significant *variations in elevation* occur, a photograph would show evidence of changes in scale associated with differences in relief. To incorporate the changes in relief, an average terrain elevation value is used in place of the *TerrainElevation* parameter in equation (19.3). However, the scale would be accurate only for those areas in the photograph that are at the 'average' elevation. At all other elevations, the scale would be an approximation.

Once the scale of the aerial photograph is known, *ground distances* between features can be easily determined by relating the distance measured on the photograph to that of the scale – similar to how distances are measured on a map. However, the analyst must be cautious about the terrain, because differences in local relief can affect the scale within the photograph.

Photogrammetric techniques can also be used to compute heights of features or objects. It is instructive to define a few terms associated with an analog image, specifically an aerial photograph, prior to discussing the techniques of height measurements. Most analog images or aerial photographs are taken vertically with the sensor system looking straight down at the terrain. Unfortunately, the stability of the platform (aircraft) often leads to the acquisition of tilted images. Practically every aerial photograph contains some degree of tilt, however, the exact angle and direction of the tilt are rarely known. Small amounts of tilt often go undetected and usually tilt angles of less than 2 or 3° can be ignored (Avery and Berlin 1992). In a perfectly vertical aerial photo, i.e. no tilt displacement, the nadir point, isocenter, and principal point are identified as the same point. This is not the case in tilted photographs, where each point would be displaced from the others (Figure 19.2). The *principal point* is the geometric center of the photograph, while the *nadir* is defined as the point directly below the sensor system, and the *isocenter* lies between the nadir and principal point.

In general, aerial photographs do not show a truly top view of objects, but rather objects appear to be leaning away from the principal point. This effect is defined as *relief displacement* where, with the exception of the point directly below the camera (*nadir*) or its immediate neighborhood, all objects protruding above ground level will appear to be inclined radially away from the center. For example, at nadir the top of a tall building will appear to be pointing at the camera, but if it is near the edge of the photo, it would present a non-vertical perspective, i.e. leaning away (Figure 19.3a–c).

While relief displacement causes distortion in an image, it can also be used to measure heights of objects providing certain conditions are met. These include (a) the principal point and nadir are the same, i.e. the photo is vertical; (b) the flying height of the platform is known; (c) the top and base of the feature are clearly visible; and (d) the displacement is large enough to be measured (Avery and Berlin 1992; Jensen 2000). Given these conditions, the approximate height of a feature, using relief displacement can be measured by the following equation:

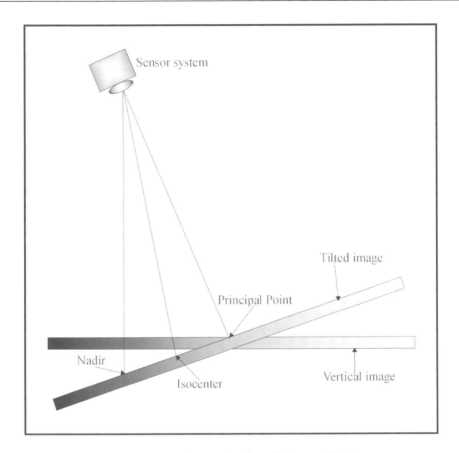

Figure 19.2 Conceptual diagram showing the effect of photographic tilt.

$$h = \frac{d}{r} \times H \tag{19.4}$$

where h is the height of object being measured, d is the length of displaced object from top to base, r is the radial distance from nadir to top of object, and H is the flying height of the platform.

The method described above enables the computation of heights using a single aerial photograph. In cases where the base of the object may not be clearly visible or the radial displacement is too small, a stereo-pair is used to determine the heights. *Stereoscopic parallax* – the apparent displacement of features caused by two different points of observation – is measured and used to compute object heights and terrain elevation using an engineering scale, stereoscope, or a parallax measuring device.

Using stereoscopic photographs and photogrammetric techniques, it is possible to derive x, y coordinates or the precise planimetric locations of features from an aerial photograph (Chapter 4). Global Positioning System (GPS) units can provide precise x, y coordinates for any point on the earth's surface. By identifying Ground Control Points (GCPs) and their respective coordinates with a GPS, an image analyst can correct an aerial photograph for any distortions and transform it into a common

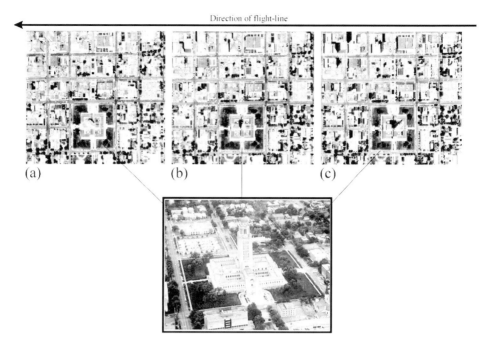

Figure 19.3   Relief displacement of the Lincoln, Nebraska capitol building along a flight-line (in the inset is an oblique view of the same building).

map projection. In addition, with Digital Elevation Models (DEMs) – the derivation of precise z values of elevation obtained at regular intervals – the image can be ortho-rectified to remove relief distortion in the photograph.

The discussion so far has focused on using analog image interpretation techniques. The methods provide an effective and often accurate means of extracting information from remotely sensed data. However, with the advent of satellites and the widespread availability of digital data, it is important that remote sensing scientists be familiar with *digital image analysis* techniques. While powerful computing technologies have facilitated the rapid processing and interpretation of digital data, it should be noted that they cannot entirely replace the human abilities of visualization and perception for performing image analysis tasks.

19.3. Digital image processing

The development of inexpensive, high-speed computers, high-resolution displays (monitors, LCDs), and large storage devices (hard disks with gigabytes of disk space, read-write CDs, optical disks), and digital image processing software has significantly improved the processing of digital remote sensor data. Fundamental digital image processing tasks include: radiometric and geometric correction of image data, image enhancement, thematic information extraction, and digital change detection techniques.

While analog techniques utilize several elements of image interpretation for the accurate interpretation of images, only a few of these elements are usually used in digital image processing. Hardin and Thomson (1992) found that the majority of digital image analyses is based on using the tone and color of a pixel in statistical pattern recognition algorithms. Efforts have also been made to incorporate texture, shape, context, and association to facilitate digital image analyses (Gong and Howarth 1992; Narumalani *et al.* 1998).

19.3.1. Image preprocessing: radiometric and geometric corrections

During the image acquisition process, various types of error may be introduced into the digital remote sensor data. These errors are caused by a number of factors including: (a) atmospheric and environmental conditions, (b) the sensor system, and (c) orbital geometry. The errors can also be categorized as *radiometric* and *geometric*. Therefore, prior to further analysis of a digital image, it is necessary to remove these errors using various image restoration methods.

19.3.1.1. Radiometric correction

Radiometric errors may be caused either by problems associated with the sensor system detectors or by environmental factors. Sensor system errors include line-start problems, line drop-out, and striping or banding. *Line-start* problems occur if a detector fails to collect data at the beginning of a scan line or if it stops collecting data somewhere along a scan line. If the problem is systematic, e.g. the BV for the first pixel in a line is offset by 40 pixels, then the image can be corrected by applying a simple horizontal adjustment. However, if random line-start problems occur throughout an image, it would require substantial interaction by the analyst to correct the data.

Line drop-out refers to missing data along a complete scan line, caused by a detector malfunction. In this case no data are collected for the entire line (data are all 0s) and appears black (0) or white (255) on the computer screen. Missing data are a serious problem and there really is no way to retrieve the data. However, the visual appearance of an image may be improved by averaging the BVs of the lines preceding and succeeding any missing lines.

Data acquired by multispectral scanning systems that sweep multiple scan lines simultaneously may contain systematic *striping* or *banding*. This problem may be caused by variations in the response of the different detectors or by offsets in the gain settings in a single band. Striping and banding is often present in imagery acquired by the Landsat Multispectral Scanner System (MSS) and the Thematic Mapper (TM). Various methods have been provided for the cosmetic removal of striping, however, it should be cautioned that if the data are to be used for biophysical information extraction, some of the procedures will affect the integrity of the data (Crippen 1989; Helder *et al.* 1992; Jensen 1996).

Radiometric errors caused by environmental factors include *atmospheric* and *topographic* effects. Whether remote sensing data are acquired by airborne or space-based systems, the ground reflectance as measured by the sensor system differs from the radiation recorded on the surface because of the intervening atmosphere. The atmosphere can selectively absorb, reflect, scatter, refract, and transmit radiation resulting in a distorting effect that varies over time (temporally) and space (spatially), as well as across the electromagnetic spectrum (spectrally). Jensen (1996) suggested that because the energy recorded by a sensor system is a true signal of the prevalent atmospheric conditions at the time of image acquisition, there is really no such thing as atmospheric error. However, many image analysts prefer to perform some type of radiometric correction to remove or minimize any atmospheric attenuation. Often, the extraction of critical biophysical information, e.g. biomass, canopy closure, is dependent on small differences in the spectral response recorded by a sensor system and even insignificant amounts of atmospheric effects may hinder or prevent these subtle differences from being discriminated.

Removal of atmospheric effects can be performed by *absolute* or *relative* radiometric correction techniques. *Absolute radiometric corrections* require that an analyst understand the precise nature of the absorption and scattering taking place, as well as their effects on the radiation transmittance. The development of radiative transfer models such as LOWTRAN, MODTRAN, and HITRAN (LOW resolution TRANsmittance, MODerate resolution TRANsmittance, and HIgh resolution TRANsmittance, respectively) can calculate atmospheric transmittance and radiance at various spectral resolutions. These algorithms consider several parameters in their computations, such as spherical refractive geometry, various types of scattering (Rayleigh, Mie, and non-selective), and profiles of atmospheric gases, aerosols, and clouds (Kneizys *et al.* 1996). Consequently, these algorithms can mathematically model almost any kind of atmosphere, which can then be used to perform radiometric corrections on the remote sensor data. The software is available on the Internet and can either be downloaded or interactively queried to perform the computations. Natsuyama *et al.* (1998) discussed other types of radiometric corrections that may be used to remove atmospheric effects from Landsat and aircraft image data (see http://www.vsbm.plh.af.mil/soft/modtran.html or http://www.vsbm.plh.af.mil/soft/hitran.html).

In cases where atmospheric conditions are unknown during the time of image acquisition (e.g. historical remote sensing data), *relative radiometric correction* techniques may be applied to *normalize* the data. This empirical approach can be used when multiple scenes need to be analyzed, e.g. change detection studies, because it attempts to standardize the atmospheric, phase angle and detector calibration conditions to a standard (baseline or reference) scene selected by the analyst. Eckhardt *et al.* (1990) provide a detailed example of this technique. The method applies regression equations to predict the pixel values of the other, i.e. 'non-standard,' multi-temporal scenes by matching BVs of selected *targets* to the reference scene. An assumption is made that the targets in the scene are constant reflectors and any variations of BVs between scenes are a function of radiometric differences. The targets must meet certain criteria before being selected, including: (a) being at approximately the same elevation as other land features in the scene, (b) containing only a minimal amount of vegetation, (c) situated in a relatively flat area, and (d) any patterns observed on the normalization targets should not change over time. Image normalization helps to minimize variations in BVs caused by radiometric factors, thus any differences in BVs between multi-temporal images hopefully would be indicative of *actual* change.

19.3.1.2. Geometric correction

Geometric errors have been categorized into systematic and non-systematic distortions. Jensen (1996) summarized six types of systematic and two types of non-systematic distortions. Systematic distortions include: (a) scan skew, (b) mirror-scan velocity, (c) panoramic distortion, (d) platform velocity, (e) earth rotation, and (f) perspective geometry including the Earth's curvature. The distortions are caused by and can be corrected through the analysis of sensor characteristics and the ephemeris data of the platform. For the most part, commercially available remote sensing data have been corrected for systematic distortion by the vendor, e.g. SPOT Image, Space Imaging, Orbital Sciences Corporation.

Non-systematic distortions are caused by changes in the attitude and altitude of the sensor platform, and can only be corrected by collecting several GCPs and implementing a geometric transformation algorithm. If the images are not corrected for non-systematic distortions, they will be planimetrically inaccurate – i.e. the x, y position of a given pixel will not be in its precise geographic location. Images can be made planimetric by implementing geometric rectification procedures that perform *spatial* (x, y) and *intensity* (BV) interpolation on the image.

To remove spatial distortion in an image, a series of x, y coordinates for well-identifiable GCPs are collected from a planimetric basemap (e.g. US Geological Survey (USGS) topographic map, Digital Orthophoto Quads (DOQs), or another rectified image) or by using a GPS unit. These coordinates are commonly referred to as *map* or *GCP coordinates*, and their units of measure, e.g. m or ft, are dependent on the projection used. For example, coordinates collected using the Universal Transverse Mercator (UTM) map projection are identified as Easting (x) and Northing (y), and measured in meters. Each map coordinate should have a counterpart *image coordinate*, i.e. each point collected from the basemap should also be identified on the image and its respective x, y recorded. The x, y of the image coordinates refers to the row/column location of the specific pixel where the GCP is identified. A mathematical relationship can be established between the map coordinates and the image coordinates using a least-squares method and thus modeling the corrections directly in the image domain (Novak 1992) or by using analytic photogrammetry.

Jensen (1995) describes the *order* of the polynomial to use in rectifying an image, which in essence refers to the highest exponent used in the equation. This is dependent on several factors such as the distortion in the imagery, number of GCPs used, and the topographic relief displacement. Areas with high topographic relief, images with significant distortion perhaps caused by the instability of the sensor platform – e.g. airborne scanner systems – or sites where few GCPs can be extracted often require higher-order transformations. Conversely, an image acquired for a relatively flat terrain with well-identifiable GCPs and from a stable platform, e.g. Landsat TM or SPOT, could utilize a first-order, six-parameter, affine transformation to rectify the image:

$$x' = a_0 + a_1x + a_2y$$

$$y' = b_0 + b_1x + b_2y \tag{19.5}$$

Novak (1992) states that this type of transformation models six types of distortion including: (a) translation in x, (b) translation in y, (c) scale change in x, (d) scale change in y, (e) skew, and (f) rotation. Once the coefficients for these equations are determined, it is possible to transfer the pixel values from the original distorted image to the output rectified image.

In a process commonly known as *re-sampling*, a new BV is interpolated for each new x, y location of the rectified output image. Consider an original, unrectified image (input image) to be x' and the geometrically correct image (output image) to be x. The analyst will need to select an intensity interpolation algorithm to derive new BVs for each x, y location in the rectified output image (Figure 19.4). There are three commonly used re-sampling schemes. The *nearest neighbor* re-sampling technique assigns the BV of the closest pixel in the input image to the output image (Bernstein 1983). Thus, pixel a in the output image x would acquire the value of pixel a' from the input image x' (Figure 19.4). This technique maintains the integrity of the data, i.e. the data are not smoothed and is computationally less intensive. However, the output image maintains a 'blocky' appearance that can be aesthetically less pleasing.

Bilinear interpolation re-sampling assigns BVs to the output image by computing the distance-weighted average of the *four* nearest pixels in the input image (Bernstein 1983). In this case, the BV for pixel a in output image x is the weighted average of input pixels labeled a' and b' (Figure 19.4). The method produces a smoother image because the BVs are averaged, but the spectral integrity of the data is compromised.

Cubic convolution re-sampling computes the new BVs using the weighted average of 16 input pixels (Jensen 1996). The output pixel a thus has a BV value based on pixels a', b' and c' (Figure 19.4). The resulting image is smoother and aesthetically appealing, however, the original data values are severely affected. In general, most remote sensing practictioners prefer to use the original, unaltered data for thematic information extraction and for other useful analyses. However, to produce good quality output for display purposes, the cubic convolution re-sampling technique is preferred.

It is now possible to place a GPS receiver and inertial system onboard aircraft during the collection of digital remote sensor data. This allows the measurement of: (a) the exact location of the aircraft in three-dimensional space, and (b) the roll, pitch, and yaw characteristics of the aircraft and the remote sensing system at the exact instant that each pixel of remote sensor data is collected. Using the appropriate algorithms, it is possible to utilize only the aircraft/sensor GPS information to geometrically rectify the digital remote sensor data to a rigorous map projection. This is very important as the procedure can minimize or eliminate GCPs that are required to rectify remote sensor data (Bossler 1997).

19.3.2. Thematic information extraction

One of the major uses of remotely sensed data is the extraction of thematic information. The thematic information may be: (a) quantitative biophysical information such as vegetation biomass, Leaf Area Index (LAI), surface temperature, and moisture content, or (b) information on various classes of land use/land cover. Biophysical information extraction is discussed in Section 19.4.3.

There are several algorithms available that can enable a user to derive *information classes* from data collected in multiple portions of the electromagnetic spectrum. At present the majority of the spectral pattern recognition algorithms can be categorized

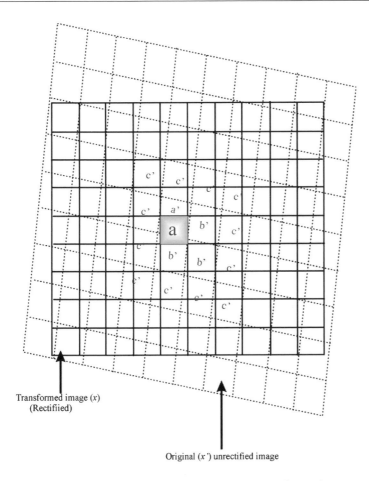

Transformed image (x)
(Rectifiied)

Original (x') unrectified image

Figure 19.4 Geometric correction and intensity interpolation. A conceptual depiction of implementing geometric correction and intensity interpolation on an image. The dashed
lines represent the original image. Using pixel 'a' as an example, new BVs are
assigned based on the re-sampling algorithm selected.

as *supervised* or *unsupervised* classification techniques. However, there are other
methods such as *hybrid* image classification, which utilize both supervised and unsupervised procedures, as well as the input of ancillary information. It is also possible
to use *fuzzy logic*, which accounts for the heterogeneous nature of the landscape and
the spectral signature recorded in a given pixel (Jensen 1996). In addition, advances in
Artificial Intelligence (AI) are allowing the development of expert systems and neural
networks that can be used for stratifying an image into classes and for object identification. Most of this discussion will be focused on the widely used supervised and
unsupervised classification techniques.

Prior to implementing any classification algorithm, it is important for an image
analyst to select an appropriate classification scheme. A classification scheme defines
the specific land use or land cover categories that are required for a given application.
An analyst may select a pre-defined classification scheme, e.g. USGS Land Use/Land

Cover Classification System (Anderson *et al.* 1976) or the US Fish and Wildlife Service Wetland Classification Scheme (Cowardin *et al.* 1979), or develop an application specific scheme. It is generally better to use a well-accepted classification scheme and adjust the various class definitions to suit the project at hand.

19.3.2.1. Supervised classification

When performing supervised classification on an image, several steps are needed including selection of training sites, refinement of training sites, selecting an appropriate image classification algorithm, and production of the final classification map. Supervised classification procedures require constant input from the analyst, especially in the refinement of training sites. Thus, the image analyst must be either very familiar with the study site or should interact with an individual who is cognizant of the land cover/land use categories visible on the image, prior to extracting training sites.

A training site is an area (or patch in landscape ecology terminology) that represents a homogeneous example of a specific land cover (or land use) to be mapped from the image. The statistical spectral characteristics of these training sites are used in the classification algorithm to classify the entire image. Therefore, the training sites selected should represent all possible land cover types and the spectral variations within each land cover type, being both *representative* and *complete*. Often, it may be necessary to obtain multiple training sites for the same land cover category. For example, multiple training sites representing the different spectral characteristics of a feature may be collected (e.g. clear water, deep water, and turbid water training sites) even though these represent a single *information* class, e.g. water.

Training sites can be extracted by outlining an *Area Of Interest* (AOI) interactively on the screen or by *seeding a pixel* using spectral and spatial search criteria to find similar pixels. For either method, it is important to acquire several training sites throughout the image for each representative class and ensure that the training data represent all possible categories that have to be extracted from the image.

After the training data have been collected, the analyst proceeds to evaluate the validity of these sites. Validation of training information is often referred to as the *training data refinement* process. Here the analyst performs a quantitative assessment of how representative the training data are by using statistical, graphical, or a combination of both techniques. This aids in: (a) reducing *data dimensionality* (i.e. identifying redundant bands and removing them from further analysis), (b) providing information on whether additional training sites need to be acquired, and (c) eliminating (or merging) training sites that are multi-modal or redundant in their spectral characteristics.

Statistical methods, such as *transformed divergence,* utilize the mean and covariance matrices of the training data to determine class separability (Mausel *et al.* 1990). Other separability measures are summarized in Jensen (1996).

Training data may also be assessed using *graphical techniques* such as coincident spectral plots, histograms, scatter plots, and co-spectral parallelepipeds (Lillesand and Kiefer 2000). These methods provide a visual display of the distribution of training data in feature space, i.e. on a band-by-band basis or from a multi-dimensional perspective, and illustrate overlapping classes, multi-modal training sites, the extent of spectral separation, and band redundancy (Figure 19.5). This particular feature space plot depicts the distribution of five training sites selected from a SPOT multi-spectral (XS) image. Note the overlap of grassland training data with those of agriculture and forest categories. Thus, even a simple, visual assessment may be useful for indicating potential problems in the training site data.

An image analyst may also utilize a combination of *statistical* and *graphical* techniques to refine the training data set. For example, quantitative information can indicate potential overlap between categories, while visual tools can be used to identify the bands where overlap is occurring. The training data refinement process may be iterated several times before yielding satisfactory results. Test classifications, e.g. on an image subset, may be used to examine the effectiveness of the training data for extract-

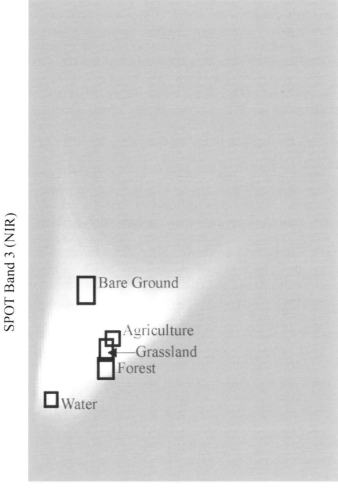

SPOT Band 2 (Red)

Figure 19.5 Co-spectral parallelepiped plot. A co-spectral parallelepiped plot for five training sites derived from the SPOT multispectral (XS) image shown in Figure 19.1b. Note the overlap of the grassland training data.

ing the thematic information. New training sites will likely need to be delineated and others eliminated after a test classification to represent the unclassified pixels or to remove confusion between spectrally similar categories.

Once the statistical characteristics of the training data have been obtained, it is possible to use a pattern recognition algorithm to classify the image into specific classes of information. Three classification algorithms for spectral pattern recognition that are the most frequently used include *minimum distance to means*, *parallelepiped*, and *maximum likelihood* classifiers.

Minimum distance to means is a fairly straightforward classification algorithm that assigns each pixel to a specific land cover class by computing the distance of an unknown pixel in multispectral feature space to the mean or average spectral value of each category (Figure 19.6a). In this example, the six clusters represent various land cover categories, including urban, forest, water, grassland, agriculture, and bare ground, with the '+'s indicating each class mean. With this algorithm, pixel '1' will be assigned to the grassland cover type because it has the *minimum* distance to the category mean. This algorithm is mathematically simple and computationally efficient. However, it does not take into account the *variance* in the spectral response of each land cover type and could lead to significant misclassifications in images where some land cover types may have a high degree of variability in their signatures.

The *parallelepiped* classifier takes into account the variance by using a one-standard deviation (or *n*-standard deviation) threshold. In effect, it forms a multi-dimensional (depending on the number of bands being used for the classification) boundary (a parallelepiped) in feature space, and applies Boolean logic to assign a pixel to a land cover category. Thus, pixel '1' would be classified as grassland using the parallelepiped decision rule (Figure 19.6b). If a pixel occurs outside of the boundaries defined by the parallelepipeds, it would remain unclassified. In addition, parallelepipeds may tend to overlap, e.g. forest and grassland, and pixels within these regions may be assigned arbitrarily. Overlap in the spectral response of different land use categories frequently occurs because the features exhibit similar spectral response patterns (e.g. healthy grassland and cropland). At the edges of the clusters, it may be difficult to separate two or more land cover types. To minimize misclassifications caused by overlapping parallelepipeds, the correlation and covariance between the different spectral classes should be considered. A *stepped parallelepiped* decision rule will eliminate much of the misclassification caused by a traditional parallelepiped algorithm.

The *maximum likelihood* classifier assigns each pixel to the land cover category for which it has the *highest probability* of being a member (Swain and Davis 1978). It assumes that the training data for each class in each band are Gaussian – i.e. normally distributed (Foody *et al.* 1992). If the histogram of a single training class exhibits multi-modal tendencies, then it may be necessary for the analyst to *train* separately on each of the modes of the histogram.

The maximum likelihood algorithm considers the variance and covariance of each category's spectral response and formulates bell-shaped surfaces called *probability density functions* for each category (Lillesand and Kiefer 2000). *Equiprobability contours* are delineated to express the sensitivity of the classifier to the covariance, with the probability values decreasing as one moves away from the core of a given cluster (Figure 19.6c). Here, pixel '1' would be classified as agriculture because it lies within the equiprobability contours generated for that land cover.

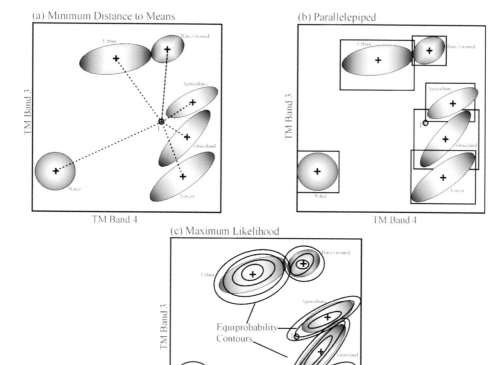

Figure 19.6 How the three classification algorithms would classify pixel 'I'. In (a) and (b), the pixel would be assigned to the grassland land cover, while the maximum likelihood (c) would classify the pixel as agriculture.

The maximum likelihood classification is usually based on the *Bayes formula*:

$$P(C_i/x) = \frac{P(x/C_i)/P(C_i)}{P(x)} \tag{19.6}$$

where the most likely class label C_i for a given feature vector x is the one with the highest posterior probability $P(C_i/x)$. Each $P(C_i/x)$ out of N possible classes is calculated and the class C_i with the highest value is selected. $P(x/C_i)$ is the class probability density, $P(x)$ is the prior probability, and $P(x)$ is the class independent feature probability density. The class probability density function for a class C_i is:

$$P(x/C_i) = (2\pi)^{-\frac{M}{2}} |V_i|^{\frac{1}{2}} e^{-\frac{1}{2}(y^T V_i^{-1} y)} \tag{19.7}$$

where M is the number of features (i.e. multispectral bands) used in the analysis; V_i is the $M \times M$ variance-covariance matrix of class C_i; $|V_i|$ is the determinant of V_i; V_i^{-1} is the inverse of V_i; y equals $x - m_i$ (where m_i is the class mean vector) and is a column vector with M components, and y^T is the transpose of y (a row vector).

19.3.2.2. Unsupervised classification

In supervised classification a priori information about the site is necessary for extracting training sites for use in subsequent analysis. Thus, an analyst provides a significant amount of input during the entire supervised classification process. In contrast, *unsupervised classification* requires a minimum amount of input from the user. The computer automatically assigns a pixel to a group or cluster based on its similarity to the spectral characteristics of the clusters that are based on user-defined criteria.

One of the most widely adopted unsupervised classification algorithms is the Iterative Self Organizing Data Analysis Technique (ISODATA) (Tou and Gonzalez 1977). Typically, the analyst identifies some initial state parameters that include: the maximum number of clusters to be identified by the algorithm, e.g. 25; the maximum number of iterations performed by the algorithm, and the maximum percentage of unchanged pixels per successive iteration.

The input parameters for ISODATA depend on the selected software. Some algorithms require additional parameters, such as: the minimum number of members in a cluster (by per cent); the maximum standard deviation within a cluster; the split-separation value to define the parametric limits of a cluster; and the minimum distance between cluster means (Jensen 1996). The basic procedures and parameter requirements described for ISODATA are generally used for other unsupervised classification algorithms. Identification of cluster means, measuring distances in feature space, and the reassignment of pixels based on their spectral characteristics, are the fundamental criteria of unsupervised classification.

After the computer has generated the specified clusters, the analyst must assign names to them, in effect converting them to *information classes*. This may be a challenging task, because it requires an understanding of the spectral characteristics of the landscape and terrain represented in the image. In addition, some clusters representing *mixed pixels* or different features with similar spectral characteristics (*mixed clusters*), cannot be assigned to a single category. The analyst may need to identify and isolate these mixed clusters and execute additional unsupervised iterations (*cluster-busting*), or extract training data for those areas containing mixed pixels or clusters and perform a limited supervised classification.

The extraction of thematic information from remotely sensed data is a *science* and an *art*. Several factors including image quality, choice of an appropriate classification algorithm, image interpretation experience, knowledge of spectral characteristics of features, and familiarity with the study area make the difference when producing a useful thematic map.

19.3.3. Hyperspectral remote sensing image analysis

Campbell (1996) defined hyperspectral remote sensing as 'an application of the practice of *spectroscopy* to examination of solar radiation reflected from the earth's surface.' *Spectroscopy* pertains to the detailed study of spectral information. Therefore, *hyperspectral remote sensing* utilizes a large number of narrow spectral bands from the electromagnetic spectrum (i.e. sensor systems with very fine spectral resolution). The number of bands in a hyperspectral sensor system can vary from tens to several hundred, with the width of each channel typically ranging from 1–20 nanometers (nm). Hyperspectral data have the advantage of using the narrow spectral bandwidth information to acquire more information about an object, than is possible with the wide bands measured by typical multispectral sensors. The concept of hyperspectral imaging can be illustrated by a hyperspectral image cube, where information for each pixel is described as a continuous (or near-continuous) spectral trace of the pixel surface (Figure 19.7).

There are several hyperspectral imaging systems in use today, as previously mentioned in Chapter 18. The Airborne Visible-Infrared Imaging Spectrometer (AVIRIS) is a typical example of an operational hyperspectral system. Developed by NASA and the Jet Propulsion Laboratory (JPL), AVIRIS was placed into service in 1989. The AVIRIS acquires images in 224 spectral bands, each 10 nm wide, and at its 'normal' flying altitude of 20 km above the terrain, each pixel has a spatial resolution of 20 m. Other examples include the Compact Airborne Spectrographic Imager (CASI) with 228 bands in the visible-near infrared region (0.40–0.90 μm) at 1.8 nm intervals, the TRW Imaging Spectrometer (TRWIS III) with 384 bands in the 0.30–2.50 μm range, and the Advanced Airborne Hyperspectral Imaging Spectrometer (AAHIS) with 288 bands in the 0.40–0.90 μm range. These sensor systems have been used on airborne platforms. The first satellite-based hyperspectral system Hyperion on the EO-1 satellite was launched in 2001. Hyperion is a 242-band instrument and sets the standard for future earth-orbiting hyperspectral sensor systems.

Because hyperspectral imaging systems provide large quantities of data in numerous narrow contiguous bands of the spectrum, traditional image processing techniques may not be directly applicable to the interpretation of these data. For example, atmospheric effects may be more evident in hyperspectral images due to the attenuation and scattering of the radiation, thus decreasing or increasing the BVs recorded. Consequently, spectral comparisons between images or sometimes even within a single

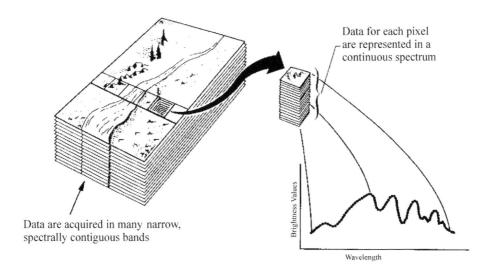

Figure 19.7 Image cube representing a hyperspectral data set and the resulting spectral curve (adapted from Vane and Goetz 1988).

image may not be feasible, and it is necessary to implement atmospheric correction procedures prior to interpreting the data (Green 1990). Conversely, the high spectral resolution may be useful for analyzing atmospheric characteristics, such as water vapor content, at the time of image acquisition.

Analyses of hyperspectral data can be performed using a variety of techniques, some of which include spectral matching, spectral unmixing, and Spectral Angle Mapping (SAM). *Spectral matching* refers to the analytical procedures used to identify the spectral response of a given pixel to that of spectra from a spectral library of known materials (Kruse *et al.* 1990). *Spectral unmixing* is based on the assumption that the spectral response pattern of a pixel is a linear mixture of spectral signatures of the various features present within the pixel (Lillesand and Kiefer 2000). Kruse *et al.* (1993) describe the SAM procedure that involves the analyses of vectors (observed reflectance) in multidimensional space (based on the number of bands).

Hyperspectral imaging has the potential to revolutionize the techniques of digital image analysis for information extraction. The lack of commercial space-borne hyperspectral sensor systems has dissuaded the common use of these data. However, this problem may be remedied in the future.

19.4. Advanced digital image understanding techniques

As stated earlier, most digital image processing algorithms utilize tone and color in statistical pattern recognition algorithms to analyze an image. Consequently, human visual interpretation is often superior to machine-based analysis because the human brain can make inferences and deductions, formulate associations, as well as utilize real-world knowledge and visual processing experience to extract meaningful information from an image. In an effort to replicate some of these abilities of the human brain, computer scientists have developed systems that incorporate AI through the implementation of expert systems and neural networks.

19.4.1. Expert system analysis of remote sensor data

Expert systems are decision support computer systems that provide the user with guidance in dealing with a problem. Grabowski and William (1993) describe an expert system as 'a system that uses computer technology to store and interpret the knowledge and experience of one or more human experts in a specific AOI.' A portion of this knowledge is 'textbook' knowledge, as well as basic data that are stored in the systems. However, much of an expert system's knowledge is based on the knowledge and experience of human experts, which is programmed into the expert system. By using this human knowledge and experience, expert systems manage and make inferences about the stored data to assist the user. One of the key features of expert systems is their flexibility and ability to store a tremendous amount of information and knowledge that can be updated on a regular basis. In recent years, with the advent of more and more remote sensing platforms collecting an ever-increasing variety and volume of data, interest in the application of expert systems in remote sensing is growing.

Information management and *image processing* are the two main areas in remote sensing where expert systems could potentially be used. Information management includes the use of expert systems to aid an analyst in selecting the most appropriate data available to meet the objectives of a given study; preparing those data for use; and the proper storage of new information and knowledge created during the processing. *Image processing* pertains to carrying out the basic functions associated with image processing as well as extracting salient information from the imagery based on generalized models and credible inferences.

Information management, in general, implies the basic handling, cataloging, and proper storage of data/information. Expert systems take information to the next level. Expert systems are being developed for the sole purpose of aiding users in determining what information is best suited for the work they wish to do and then retrieving the information from its various sources (since the information does not have to be stored within the expert system), compiling, and preparing the information for the user.

Because of the increasingly widespread use of remote sensing and the diversity of data available, e.g. from different sensor systems at various spatial resolutions, a user may be unfamiliar with the advantages and disadvantages of different types of data and the appropriate processing techniques. Furthermore, with the advent of the US based Earth Observing System (EOS), the volume of remote sensing data available to earth scientists will be staggering. It would be challenging for even the best informed scientists to maintain the pace in using not only the best and most current data, but also in an effective and appropriate manner (Star *et al.* 1987; Vetter *et al.* 1995). The same holds true for data created by the user. Each time the data are used, the user will have information to input back into storage and possibly enhance the knowledge of the expert system. When the expert system is used as a data manager, the user's efficiency and effectiveness in acquiring and using data increases.

The key to using expert systems for information management is the knowledge of what data are available and their attributes, as well as evaluating the best usage of those data and the techniques to maximize their potential applications. Thus, expert systems used for information management are not just gigantic databases or query tools, but also have the ability to infer, based on the user's goal, what data are best suited for a specific purpose.

The second major area where expert systems are being designed and used is for *image processing*. In applications where large volumes of remotely sensed data must be processed, e.g. forest inventory, expert systems can be effective in performing many of the basic tasks. For example, in a forest inventory and monitoring application an expert system can automatically request suppliers for imagery of the desired study area, perform data conversions, and preprocess (e.g. radiometric and geometric correction) the imagery for subsequent information extraction. These fundamental tasks can otherwise occupy a substantial portion of the image analyst's time.

In such cases it is important to recognize that the expert system has knowledge of the image characteristics (resolution, projection, datum), as well as the pertinent information about fundamental image processing techniques. The system then makes inferences on how to effectively utilize this knowledge base. The appropriate image operators and optimal processing parameters can then be determined through the image processing knowledge base of the expert system. This knowledge base can be interrogated and instructions given to implement or write digital image processing algorithms to complete the fundamental image processing tasks (Harrison and Harrison 1993).

It is important to note that there is a difference between expert systems for *image processing* and *Image Understanding Systems* (IUS). While an expert system for image processing may be used to sort through images and find appropriate ones for a given application, perform preliminary image processing, and possibly write algorithms for specific tasks, the expert system would not actually have knowledge *about* the image (i.e. be able to describe its contents quantitatively) or the image domain. IUS are used to describe the features and objects in an image, such as roads and buildings, and to identify changes (Uberbacher *et al.* 1996).

19.4.2. Neural network analysis of remotely sensed data

Similar to expert systems, neural networks emerged from the realm of AI and started to be used in remote sensing applications in the 1980s. The concept of neural networks or Neural Nets (NNs) date back to WW II, but like expert systems, did not progress much until computers became more powerful and were available for more elaborate research. A more precise name for NNs is Artificial Neural Networks (ANNs), which were originally processing systems based on how the brain processes data. ANNs today have become for many, simply an applied 'mathematical technique with co-incident biological terminology' (Hewiston and Crane 1994). Modern ANNs still resemble the brain in that ANNs can both generalize from only a limited amount of information and that they can 'learn.'

Untrained ANNs start off with a set of elementary inputs and expected outputs, and through *learning algorithms* develop relationships between inputs and outputs. As an ANN works through more and more problems and iterations of similar or related problems, it creates more relationships while refining others, thus 'learning' and becoming more efficient and accurate. Unlike expert systems, ANNs have to be *trained* (or calibrated) and must be given continuous information and problems to deal with to advance, whereas expert systems work from a set of knowledge passed on in their programming, generally from a live expert. This makes expert systems ready to work effectively from the start, while ANNs have to develop over time before they can be considered highly efficient.

With the availability of powerful computers at affordable prices, ANNs are increasingly being applied to remote sensing and image processing tasks. It is expected that their development and use will grow in the near future. In addition to their learning potential, ANNs have several other characteristics that make them a good candidate for image processing. Because an ANN can process data with limited experience based on learned relationships, if new data are introduced, the ANN will process the data, provided the information resembles the original training set. This is not true with expert systems, which need to be re-programmed, if a large change in their knowledge base or capabilities occurs. ANNs are also very robust in that they can still function despite missing input parameters, since they have already been trained to some degree by the original input-output relationships. ANNs can also work through noisy or clustered data and adjust well from one working environment to another (i.e. *operating flexibility*).

Furthermore, ANN processing involves a considerable amount of statistical analysis, much of it cluster analysis (Dai and Khorram 1999). Because a significant portion of digital image analysis is statistical, there is a natural benefit in using ANNs for

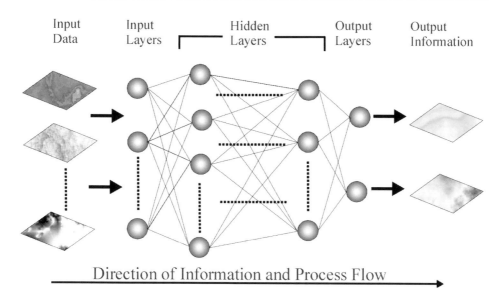

Input Data   Input Layers   Hidden Layers   Output Layers   Output Information

Direction of Information and Process Flow

Figure 19.8   Information and processing flow in a typical neural network.

processing images. This is complemented by the fact that with subsequent iterations of the same processing and analysis tasks, ANNs refine themselves to become more accurate and efficient.

An ANN can be conceptualized as consisting of three or more layers, with each layer being made up of multiple nodes (Figure 19.8). The *input* and *output* layers form the initial and terminal nodes, with one or more hidden layers comprising the processing nodes. Information flow occurs along a network of linkages between nodes. Each node can be potentially linked to other nodes in the preceding and/or succeeding layers.

Atkinson and Tatnall (1997) cite several reasons for the increase in recent years to utilize ANNs for processing remotely sensed data. They include: being more accurate than statistical classifiers, especially when the feature space is complex and the source data have different statistical distributions; can incorporate different types of data, e.g. different sensor systems; can incorporate a priori knowledge and realistic physical constraints into the analysis; and can perform more rapidly than other techniques.

ANNs have been successfully used to extract land cover information (Hepner *et al.* 1990) and vegetation variables (Kimes *et al.* 1998). Because they can be used on single and multi-source data for pattern recognition, their application in change detection studies can be very valuable. This is evident in a study by Dai and Khorram (1999), who developed an experimental change-detection system based on ANNs to produce a reliable automated change-detection system for identifying categorical changes. Furthermore, by training ANNs to identify particular features in an image domain or spectral characteristics from hyperspectral data, it is likely that information extraction would be greatly simplified.

In summary, by being able to make decisions on their own while still maintaining a high level of accuracy, expert systems and ANNs can perform various types of image processing tasks, thus improving digital image analysis. Given the capability of

AI-based systems to make inferences about a feature not only on 'text-book knowledge,' but also on the *experience* of an expert makes them effective decision support tools.

19.4.3. Biophysical information extraction

Remote sensing analysis allows an analyst to extract certain types of biological and physical (*biophysical*) information from an image. Jensen (1996, 2000) lists several biophysical variables that can be derived from remotely sensed data (Table 17.2). Biophysical variables can be divided into two categories, including those that can be directly sensed and those that are derived by analyzing more than one biophysical variable. For example, surface temperature is a direct biophysical variable that can be extracted from thermal infrared data. On the other hand, hybrid biophysical information, such as land use or land cover, may necessitate the analysis of surface color, temperature, and perhaps height or elevation.

In addition to extracting biophysical variables from remotely sensed images, the data can also be used for biophysical modeling. This involves the development of quantitative relationships between the remote sensing data and actual biophysical processes occurring on the surface. Data acquired by field, aircraft, and satellite sensor systems may be used with radiometric correction and calibration models to infer important biophysical properties and develop biophysical parameters for use with ecosystem models. Examples of biophysical modeling include LAI, plant productivity, pollution content in water bodies, physiological crop modeling, and hail damage assessments.

Lillesand and Kiefer (2000) describe three basic approaches for relating remote sensing data to biophysical processes, including *physical modeling, empirical modeling*, and a *combination of physical and empirical methods*. In *physical models* the emphasis is on the radiometric characteristics of the data, e.g. atmospheric parameters, reflectance and emission factors. For example, Myneni *et al.* (1988) modeled the interaction of solar radiant energy with vegetation canopies for an understanding of LAI, leaf angle distribution, and canopy photosynthetic rates. Similarly, theoretical reflectance models have also been used for estimating agronomic variables (Goel and Thompson 1984). An understanding of energy-atmosphere interactions with land and aquatic features can be used to model a variety of biophysical parameters for an improved perspective of terrestrial processes.

Empirical models focus on establishing mathematical relationships between data acquired by a remote sensor system and *in situ* field observations. Simultaneous observations – i.e. multi-platform remote sensing systems and on-ground field crews – provide an understanding of the links between a terrestrial process and its observation by a distant sensor system. Processes and parameters that may appear homogeneous and relevant at one level of observation become heterogeneous and insignificant at another (Cao and Lam 1997). For example, carbon fluxes occurring at a micro-scale (field-level) may be misinterpreted or misrepresented at a macro-scale (satellite image) by merely extrapolating the observations from the remote sensor data. By understanding and quantifying these relationships, remote sensing data can be used to facilitate large-area studies using biophysical models (e.g. global change).

In the *combined method* model, the analyst may utilize a theoretical reflectance model to derive corrected data values for an image and subsequently correlate this information with *in situ* measurements to derive a biophysical model. For example, Jensen *et al.* (1998) used the combined physical and empirical modeling technique to quantify the biomass and LAI of smooth cordgrass (*Spartina alterniflora*) along Murrells Inlet in South Carolina. The data were processed using a relative radiometric correction method. Statistical analyses were used to quantify the relationship between *in situ* observations and high resolution Calibrated Airborne Multispectral Scanner (CAMS) data. Numerous studies have documented how remote sensing derived vegetation indices can be correlated with *in situ* measurements to predict the spatial distribution of biomass, LAI, primary productivity and other variables (Gorte 2000).

19.4.4. Digital soft-copy photogrammetry

Digital photogrammetry utilizes digital photographs and/or images and digital image processing techniques to perform a variety of photogrammetric tasks. Data may be obtained from digital cameras, digitized photographs, and satellite remote sensing systems. Digital photogrammetry is normally used to derive accurate digital terrain data and for the production of orthophotos, i.e. terrain corrected and planimetrically accurate photographs (Kraus 1993). In addition, it can be used to extract the exact location and elevation of individual buildings, trees, and roads and for identifying specific types of land use while viewing a digital three-dimensional view of the terrain (Jensen 1995). Because the remote sensor data are in digital format, soft-copy photogrammetry can be linked directly to image processing and Geographic Information Systems (GIS), thus facilitating image-processing functions such as image overlay, rectification, topographic and thematic analyses. Furthermore, extraction of planimetric information directly from digital sources assures a higher degree of accuracy, as compared to an analog product that may be subject to distortion. Consequently, soft-copy photogrammetry techniques have the potential of carrying out an entire array of image processing and analyses tasks at a single workstation. An example of the use of soft-copy photogrammetry applied to updating tax maps is found in Chapter 22, Section 4.

19.5. Summary

This chapter introduced the fundamental concepts of extracting information from remotely sensed data and briefly discussed some of the contemporary advanced image processing techniques. The choice of an appropriate method for information extraction is dependent on several factors, including data format, data sources, image quality, experience of the analyst, type of information required, and potential use of the information. Because remote sensing data offer a unique perspective of the earth from a spatial and spectral perspective, the information remote sensing data provide can be a critical factor in decision support systems. New and more powerful applications are being sought for the use of remote sensing-derived information as inputs to biophysical modeling and quantitative analysis of natural processes. These factors, combined with the launch of new satellites, such as IKONOS and the *Terra* EOS, and the increasing use of hyperspectral data, provide a synergistic environment for furthering the development of innovative image processing techniques for the near future.

References

Anderson, J. R., Hardy, E., Roach, J. and Witmer, R., *A Land Use and Land Cover Classification System for Use with Remote Sensor Data*, U. S. Geological Survey Professional Paper 964, U.S. Geological Survey, Washington, DC, 1976, 28 pp.

Atkinson, P. M. and Tatnall, A. R. L., 'Neural networks in remote sensing', *International Journal of Remote Sensing*, 18(4), 1997, 699–709.

Avery, T. E. and Berlin, G. L., *Fundamentals of Remote Sensing and Airphoto Interpretation*, Macmillan Publishing Company, New York, 1992, 472 pp.

Bernstein, R., 'Image geometry and rectification', *Manual of Remote Sensing*, Ch. 21, Vol. 1, Colwell, R. N. (ed.), American Society of Photogrammetry, Bethesda, MD, 1983, 875–881.

Bossler, J., 'Airborne system promises large scale mapping advancements', *GIS World*, June, 1997.

Campbell, J. B., *Introduction to Remote Sensing*, The Guilford Press, New York, 1996, 622 pp.

Cao, C. and Lam, N. S., 'Understanding the scale and resolution effects in remote sensing and GIS', *Scale in Remote Sensing and GIS*, Quattrochi, D. A. and Goodchild, M. F. (eds.), Lewis Publishers, New York, 1997, 57–72.

Cowardin, L. M., Carter, V., Golet, F. C. and LaRoe, E. T., *Classification of Wetlands and Deepwater Habitats of the United States*, US Fish and Wildlife Service, Washington, DC, FWS/OBS-79/31, 1979, 103 pp.

Crippen, R. E., 'A simple spatial filtering routine for the cosmetic removal of scan-line noise from Landsat TM P-tape imagery', *Photogrammetric Engineering and Remote Sensing*, 55(3), 1989, 327–331.

Dai, X. L. and Khorram, S., 'Remotely sensed change detection based on artificial neural networks', *Photogrammetric Engineering & Remote Sensing*, 65(10), 1999, 1187–1194.

Eckhardt, D. W., Verdin, J. P. and Lyford, G. R., 'Automated update of an irrigated lands GIS using SPOT HRV imagery', *Photogrammetric Engineering and Remote Sensing*, 56(11), 1990, 1515–1522.

Foody, G. M., Campbell, N. A., Trood, N. M. and Wood, T. F., 'Derivation and applications of probabilistic measures of class membership from the maximum-likelihood classification', *Photogrammetric Engineering and Remote Sensing*, 58(9), 1992, 1335–1341.

Goel, N. S. and Thompson, R. L., 'Inversion of vegetation canopy reflectance models for estimating agronomic variables. IV. Total inversion of the SAIL model', *Remote Sensing of Environment*, 15, 1984, 237–253.

Gosselin, D. C., Rundquist, D. C. and McFeeters, S. K., Remote Monitoring of Selected Groundwater-Dominated Lakes in the Nebraska Sandhills', *Journal of the American Water Resources Association*, 2000, in press.

Gong, P. and Howarth, P., 'Frequency based contextual classification and gray level vector reduction for land use identification', *Photogrammetric Engineering and Remote Sensing*, 58(4), 1992, 423-437.

Gorte, B. G. H., 'Land-use and catchment characteristics', *Remote Sensing in Hydrology and Water Management*, Schultz, G. A. and Engman E. T. (eds.), Springer-Verlag, New York, 2000, pp. 133–156.

Grabowski, M. and William, W., *Advances in Expert Systems for Management*, Jai Press, Greenwich, 1993, 255 pp.

Green, R., 'Retrieval of reflectance from calibrated radiance imagery measured by the airborne visible/infrared imaging spectrometer (AVIRIS) for lithological mapping of the Clark Mountains, California', *Proceedings of the Second Airborne Visible/Infrared Imaging Spectrometer (AVIRIS) Workshop*, Jet Propulsion Laboratory, 1990, pp. 167–175.

Hardin, P. J. and Thomson, C. N., 'Fast nearest neighbor classification methods for multi-spectral imagery', *Professional Geographer*, 44(2), 1992, 191–201.

Harrison, A. P. and Harrison, P., 'An expert system shell for inferring vegetation characteristic prototype help system', *JJM Systems*, Task I: NASA - CR- 193411, 1993, 23 pp.

Helder, D. L., Quirk, B. K. and Hood, J. J., 'A technique for the reduction of banding in Landsat Thematic Mapper images', *Photogrammetric Engineering and Remote Sensing*, 58(10), 1992, 1425–1431.

Hepner, G. F., Logan, T., Ritter, N. and Bryant, N., 'Artificial neural network classification using a minimal training set: comparison to conventional supervised classification', *Photogrammetric Engineering and Remote Sensing*, 56(4), 1990, 469–473.

Hewiston, B. C. and Crane, R. G., *Neural Nets: Applications in Geography*, Kluwer Academic Publishers, Boston, MA, 1994, 194 pp.

Jensen, J. R., Coombs, C., Porter, D., Jones, B., Schill, S. and White, D., 'Extraction of Smooth Cordgrass (*Spartina alterniflora*) biomass and leaf area index parameters from high resolution imagery', *Geocarto International*, 13(4), 1998, 25–34.

Jensen, J. R., 'Issues involving the creation of digital elevation models and terrain corrected orthoimagery using soft-copy photogrammetry', *Geocarto International*, 10(1), 1995, 1–17.

Jensen, J. R., *Introductory Digital Image Processing: a Remote Sensing Perspective*, Prentice Hall, Saddle River, NJ, 1996, 318 pp.

Jensen, J. R., *Remote Sensing of the Environment: an Earth Resource Perspective*, Prentice Hall, Saddle River, NJ, 2000, 544 pp.

Kimes, D. S., Nelson, R. F., Manry, M. T. and Fung, A. K., 'Attributes of neural networks for extracting continuous vegetation variables from optical and radar measurements, *International Journal of Remote Sensing*, 19(14), 1998, 2639–2663.

Kneizys, F. X., Abrequ, L. W., Anderson, G. P., Chetwynd, J. H., Shettle, E. P., Berk, A., Bernstein, L. S., Robertson, D. C., Acharya, P., Rothman, L. S., Selby, J. E. A., Gallery, W. O. and Clough, S. A., *The MOTRAN 2/3 Report and LOWTRAN 7 Model*, Abreau, L. W. and Anderson, G. P., eds., prepared for Phillips Laboratory, Hanscom AFB, 1996, 261 pp.

Kraus, K., 1993, *Photogrammetry, Fundamentals and Standard Processes*, Vol. 1, Dummler, Bonn.

Kruse, F., Kieren-Young, K. and Boardman, J., 'Mineral mapping at Cuprite, Nevada with at 63-channel imaging spectrometer', *Photogrammetric Engineering and Remote Sensing*, 56(1), 1990, 83–92.

Kruse, F., Lefkoff, A. B. and Dietz, J. B., 'The spectral image processing system (SIPS) – inter-active visualization and analysis of imaging spectrometer data', *Remote Sensing of Environment*, 44, 1993, 145–163.

Lillesand, T. M. and Kiefer, R. W., *Remote Sensing and Image Interpretation*, John Wiley and Sons, New York, 2000, 750 pp.

Mausel, P. W., Kamber, W. J. and Lee, J. K., 'Optimum band selection for supervised classification of multispectral data', *Photogrammetric Engineering and Remote Sensing*, 56(1), 1990, 55–60.

Myneni, R. B., Burnett, R. B., Asrar, G. and Kanemasu, E. T., 'Single scattering of parallel direct and axially symmetric diffuse solar radiation in vegetative canopies', *Remote Sensing of Environment*, 20, 1988, 165–182.

Narumalani, S., Zhou, Y. and Jelinski, D. E., 'Utilizing geometric attributes of spatial information to improve digital image classification', *Remote Sensing Reviews*, 16, 1998, 233–253.

Natsuyama, H. H., Ueno, S. and Wang, A. P., *Terrestrial Radiative Transfer: Modeling, Computation, and Data Analysis*, Springer-Verlag, New York, 1998, 279 pp.

Novak, K., 'Rectification of digital imagery', *Photogrammetric Engineering and Remote Sensing*, 58(3), 1992, 339–344.

Star, L. J., Stoms, D. M., Friedl, M. A. and Estes, J. E., 'Electronic browsing for suitable GIS data', *Proceedings International GIS Symposium: the Research Agenda*, 1987, pp. 1–8.

Swain, P. H. and Davis, S. M., *Remote Sensing: the Quantitative Approach*, McGraw-Hill, New York, 1978, pp. 166–174.

Tou, J. T. and Gonzalez, R. C., *Pattern Recognition Principles*, Addison-Wesley, Reading, MA, 1977, 377 pp..

Uberbacher, E. C., Xu, Y., Lee, R. W., Glover, C. W., Beckerman, M. and Mann, R. C., 'Image exploitation using multi-sensor/neural networks', *Proceedings SPIE. 24th AIPR Workshop on Tools and Techniques for Modeling and Simulation*, Washington, DC, 1996, pp. 134–145.

USDA-SCS, *Soil survey of Savannah River Plant area, parts of Aiken, Barnwell, and Allendale Counties, South Carolina*, USDA - Soil Conservation Service, 1990, 127 pp.

Vane, G. and Goetz, A. F. H., 'Terrestrial imaging spectroscopy', *Remote Sensing of Environment*, 24, 1988, 1–29.

Vetter, R., Ali, M., Daily, M., Gabrynowicz, J., Narumalani, S., Nygard, K., Perrizo, W., Ram, P., Reichenbach, S., Seielstad, G. A. and White, W., 'Accessing Earth system science data and applications through high-bandwidth networks', *IEEE Journal on Selected Areas in Communications*, 13(5), 1995, 793–805.

Chapter 20

Remote sensing digital image processing system hardware and software considerations

John R. Jensen and Ryan R. Jensen

20.1. Introduction

Analog and digital remotely sensed data are used operationally in many planning and Earth and social science applications (Rencz 1999; Jensen 2000b). Analog (hardcopy) remotely sensed data such as aerial photographs are routinely analyzed visually using optical instruments such as stereoscopes. The *Manual of Photographic Interpretation* published by the American Society for Photogrammetry and Remote Sensing provides an excellent overview of optical image interpretation instruments (ASPRS 1997). Most digital remote sensor data are analyzed using a digital image processing system that consists of both hardware and software (Jensen 1996). This chapter introduces:

- Fundamental digital image processing system characteristics;
- Digital image processing (and some Geographical Information Systems (GIS)) software requirements, and
- Commercial and public providers of digital image processing systems.

Chapter 5 contains basic computer background that is complementary to the material in this chapter.

20.2. Image processing system characteristics

To successfully process digital remote sensor data, it is first necessary to understand the various algorithms and proper application of the remote sensing and GIS science technology. One may either learn the technology or hire technicians, scientists or engineers trained in the technology. Once qualified people are in place, they must select an appropriate digital image processing system that meets their needs.

Many projects require the analysis of large geographically extensive remote sensing datasets. For example, a single Landsat 7 Enhanced Thematic Mapper Plus (ETM^+) image that records spectral information for a 185×185 km area, consists of eight co-registered bands (bands 1–5 and seven at 30×30 m spatial resolution; band six at 60×60 m; band eight at 15×15 m). The entire dataset consists of 389.7 MB of data ($6{,}166$ rows $\times 6{,}166$ columns $\times 6$ bands$/1{,}000{,}000 = 228.1$ MB; $3{,}083 \times 3{,}083/1{,}000{,}000 = 9.5$ MB; $12{,}333 \times 12{,}333/1{,}000{,}000 = 152.1$ MB; $228.1 + 152.1 + 9.5 = 389.7$ MB).

Scientists are also taking advantage of higher spatial and spectral resolution remote sensor data. For example, a single Space Imaging IKONOS 11×11 km, 1×1 m panchromatic scene contains $11{,}000 \times 11{,}000$ picture elements and is 121 MB of data

Table 20.1 Historical development of the Intel family of Central Processing Units (CPUs) used in numerous IBM compatible PCs (Freedman 1995; Spooner 1999)

CPU (word size)	Clock speed (MHz)	MIPS
8088 (16)	5	0.33
8086 (16)	5–10	0.33–0.66
286 (16)	6–12	1.2–2.4
386DX (32)	16–40	6–15
486DX (32)	25–100	20–80
Pentium I	60–200	100–250
Pentium II	300–400	300–400
Pentium III	500–1,300	>500
Pentium 4	1,400–2,000	⁓ 1,500

(11000 × 11000/1,000,000 assuming one byte of data per pixel). A single 512 × 512 pixel AVIRIS subscene contains 224 bands of 12-bit data (or 1.5 bytes of data) and is 88.1 MB in size (512 × 512 × 1.5 × 224/1,000,000). Processing such large remote sensor datasets requires a significant number of computations and a significant amount of data throughput. Key components of the computer selected will dictate how fast the computations or data throughput operations can be performed and the precision with which they are made.

Scientists often populate digital image processing laboratories with Personal Computer (PC)-based digital image processing systems because the hardware and software are relatively inexpensive per unit and hardware maintenance is low. Quality PCs appropriately configured for digital image processing can cost $3,000 or less. Table 20.1 summarizes the historical development of the Intel family of central processing units (CPUs). Image processing PCs come equipped with a fast graphics card and a high-resolution color monitor. Over the past 5 years or so, the distinction between PCs and computer workstations has disappeared.

Figure 20.1 summarizes the components found in a typical digital image processing laboratory environment. Note that multiple PCs can function independently or be networked to a server as shown. Both PCs and workstations can have multiple Central Processing Units (CPUs) that allow data to be processed in parallel and at greater speed.

Ever more powerful PCs and workstations are ideal for intensive CPU-dependent tasks such as image registration/rectification, mosaicking multiple scenes, spatial frequency filtering, terrain rendering, classification, hyperspectral image analysis, or complex spatial GIS modeling. Output from intensive processing may be shared over a network by workstations and PCs.

The most efficient digital image processing environment exists when a single user works on a single PC or workstation, but may share data and processing across a fast network that links the environment. Unfortunately, this is not always possible due to cost constraints. The sophisticated PC and/or workstation lab shown in Figure 20.1 might be ideal for perhaps seven people conducting operational image processing or conducting research, but ineffective for education or short course instruction where many analysts (e.g. >20) must be served. Fortunately, the price of PCs and workstations continues to decline making it practical to configure laboratory and teaching environments so that each participant has access to their own computing environment.

Computer Systems and Peripheral Devices in A
Typical Digital Image Processing Laboratory

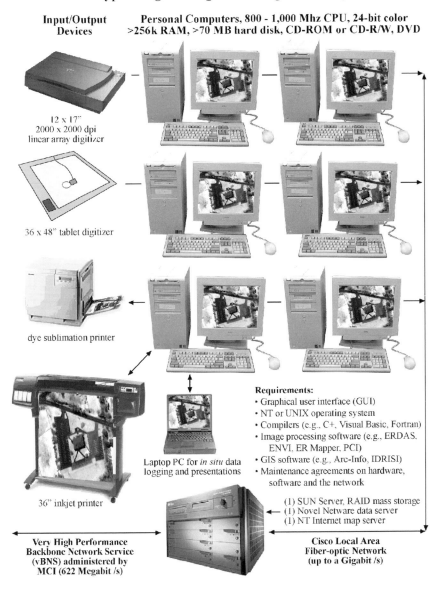

Input/Output Devices

Personal Computers, 800 - 1,000 Mhz CPU, 24-bit color >256k RAM, >70 MB hard disk, CD-ROM or CD-R/W, DVD

12 x 17"
2000 x 2000 dpi
linear array digitizer

36 x 48" tablet digitizer

dye sublimation printer

36" inkjet printer

Laptop PC for *in situ* data
logging and presentations

Requirements:
- Graphical user interface (GUI)
- NT or UNIX operating system
- Compilers (e.g., C+, Visual Basic, Fortran)
- Image processing software (e.g., ERDAS, ENVI, ER Mapper, PCI)
- GIS software (e.g., Arc-Info, IDRISI)
- Maintenance agreements on hardware, software and the network

(1) SUN Server, RAID mass storage
(1) Novel Netware data server
(1) NT Internet map server

Very High Performance Backbone Network Service (vBNS) administered by MCI (622 Megabit /s)

Cisco Local Area Fiber-optic Network (up to a Gigabit /s)

Figure 20.1 A digital image processing lab may consist of complex instruction-set-computers (CISC) and/or reduced-instruction-set-computers (RISC) and peripheral devices. In this example there are six 24-bit color desktop PCs and one laptop. The PCs communicate locally via a local area network (LAN) and with the world via the Internet. Each PC has <256 MB random access memory (RAM), <70 GB hard disk space, a CD-ROM, and often a removable storage device such as a CD-R/W, digital video disk (DVD), and/or ZIP drive. Microsoft NT or UNIX are the operating systems of choice. Image processing software and remote sensor data may reside on each PC (increasing the speed of execution), or be served by the server, minimizing the amount of mass storage required on each PC.

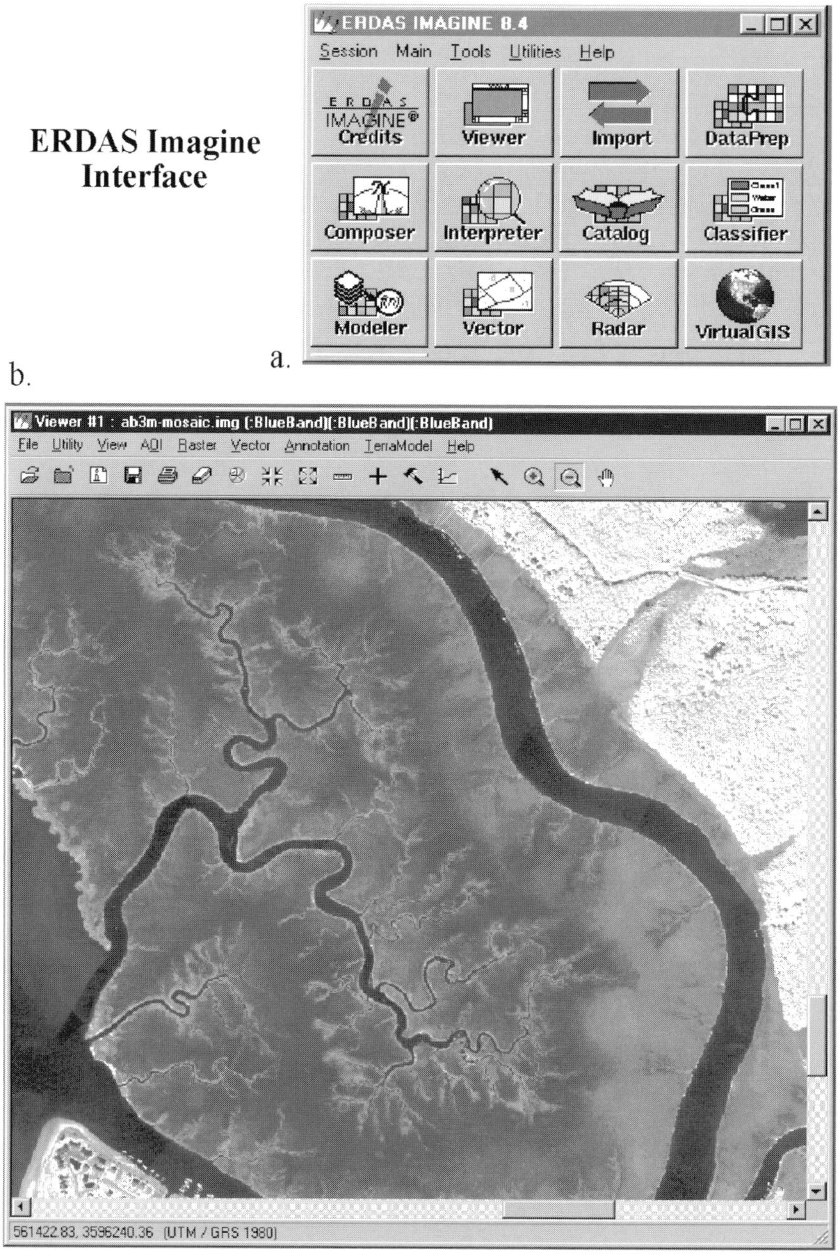

ERDAS Imagine Interface

a.

b.

Figure 20.2 The ERDAS Imagine graphical user interface (GUI) consists of: (a) point-and-click icons that are used to select various types of image processing and GIS analysis, in the (b) Imagine viewer interface.

Plate 1 The electromagnetic spectrum and its conventional partitioning into separate spectral regions for remote sensing purposes

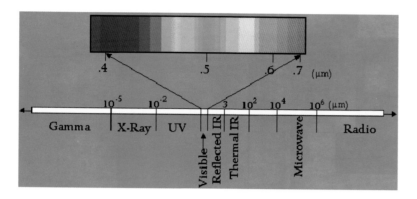

Plate 2

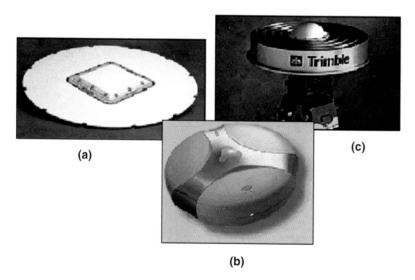

(a)

(b)

(c)

(a) Micro-centred L1/L2 antenna with ground plane
(courtesy Trimble Navigation, Ltd)

(b) L1/L2 choke ring antenna
(courtesy Trimble Navigation, Ltd)

(c) *RegAnt* zero-centred antenna mounted on choke ring in a fully weatherproof enclosure
(courtesy Javad Positioning Systems)

Plate 3

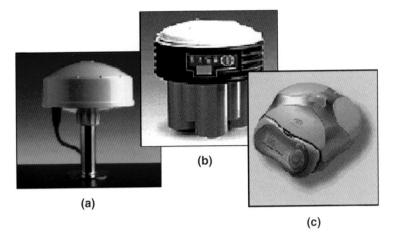

(a)   *Acutime II*, a six-channel GPS receiver-cum-antenna
(courtesy Trimble Navigation, Ltd)

(b)   An integrated *Locus SurveySystem*
(courtesy Magellan Corporation/Ashtec Precision Products)

(c)   *Odyssey*, an integrated survey-grade GPS receiver
(courtesy Javad Positioning Systems)

Plate 4 Leica System 500 GPS
survey equipment on a tripod
(courtesy Leica Geosystems Pty Ltd,
Australia)

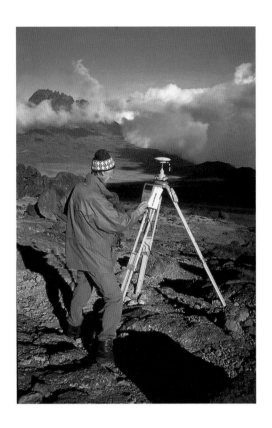

Plate 5 Trimble 4700 GPS survey
equipment on a range pole and backpack
(courtesy Trimble Navigation, Ltd)

Plate 6 Trimble 4800 GPS survey equipment on a range pole
(courtesy Trimble Navigation, Ltd)

Plate 7 Leica System 500 backpack
(courtesy Leica Geosystems Pty Ltd, Australia)

Plate 8 Digital orthophoto from color-infrared aerial photography

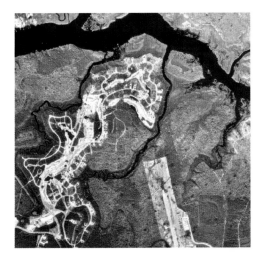

The southeast portion of the Cainhoy, SC Digital Orthophoto Quarter Quadrangle (DOQQ) created from NAPP color-infrared aerial photography acquired on February 9, 1995 (1 × 1 m spatial resolution). A golf course community and small airport lie south of the Wando River.

(courtesy U.S. Geological Survey)

Plate 9 ADAR 5500 digital photograph

A natural color ADAR 5500 digital photograph (RGB = bands 3, 2, 1) of a suburban community east of Orlando, FL acquired on November 5, 1995 at 9:00 am (1 × 1 m spatial resolution).

(courtesy Positive Systems, Inc.)

Plate 10 Daedalus airborne multispectral scanner image

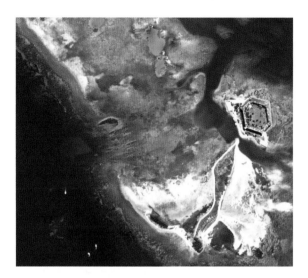

Fort Jefferson National Monument in the Dry Tortugas of Florida recorded by a Daedalus Airborne Multispectral Scanner (AMS) system (RGB = bands 5, 3, 1). The image was acquired on August 30, 1995 and reveals the water penetration capabilities of the visible bands of this sensor.

Plate 11 Landsat Enhanced Thematic Mapper Plus (ETM+) image

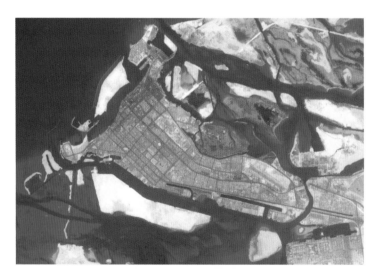

Landsat Enhanced Thematic Mapper Plus (ETM+) image of Abu Dhabi, United Arab Emirates obtained on July 11, 1999 (RGB = bands 7, 4, 3) at 30 × 30 m spatial resolution
(courtesy National Aeronautics and Space Administration)

Plate 12 MODIS image

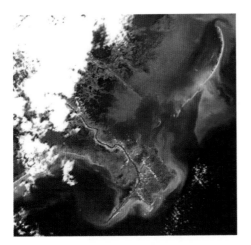

An image of the Mississippi Delta acquired by the MODIS system onboard NASA's *Terra* satellite. This scene, acquired on February 24, 2000, depicts the coastal features that surround the delta, including sediment plumes, shallow water bays, and barrier islands.

(courtesy National Aeronautics and Space Administration)

Plate 13 RADARSAT-1 image

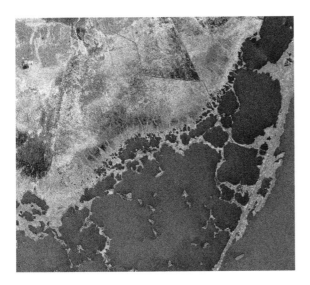

RADARSAT-1 image of the southern end of the Florida Everglades and the northeastern corner of Florida Bay. The scene was recorded on September 27, 1997 in a Standard 1 Beam mode (12.5 × 12.5 m spatial resolution).

(courtesy RADARSAT International)

Plate 14 Multi-band digital photography. Digital camera imagery of Hartford, CT, at a spatial resolution of 0.5 × 0.5 m.

(a) Color-infrared color composite (RGB = NIR, red, green)

(b) Green band

(c) Red band

(d) Near-infrared band

(a) False-color composite of multi-band digital photography of Hartford, CT. The data were obtained at a spatial resolution of 0.5 × 0.5 m.

(b–d) The individual bands of digital photography.

(courtesy of Positive Systems, Inc.)

Plate 15 Landsat Thematic Mapper false-color composite (RGB = bands 4, 3, 2) for Dawson County, Nebraska, acquired on August 8, 1997

Plate 16 Crop classification from 1997 multi-date Landsat TM data

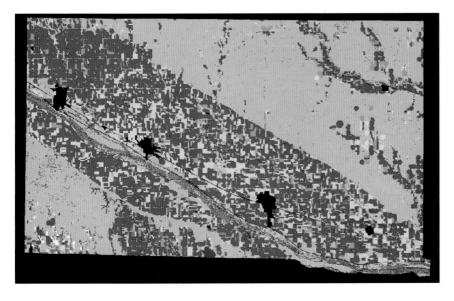

Results of the supervised agricultural crop classification of 1997 multi-date Landsat TM data for Dawson County, Nebraska. Legend: dark green = corn; light green = soybeans; orange = sorghum; yellow = alfalfa; pink = small grains; tan = range, pasture, grass; black = urban land and roads; blue = open water; red = riparian forest and woodlands; cyan = wetlands and sub-irrigated grasslands; gray = other agricultural land; white = summer fallow.

Plate 17 Landsat TM false-color composite and spectral reflectance profiles

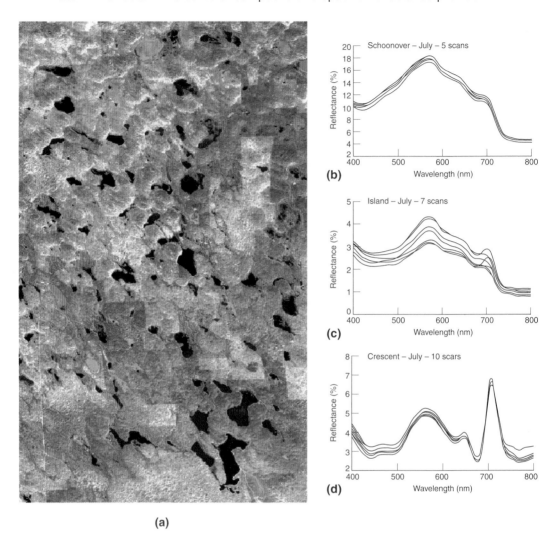

(a)

(a) Landsat TM false-color composite acquired over Northern Garden County, Nebraska, acquired on 9 October 1997.

(b–d) Spectral reflectance profiles obtained using a field spectroradiometer operating at close range over Schoonover Lake, Island Lake, and Crescent Lake during the 1994 field season.

Spectral profiles adapted from Fraser (1996)

Plate 18 Entity types for the example landfill location problem include land use, public water supplies, private wells, water table, and floodplain

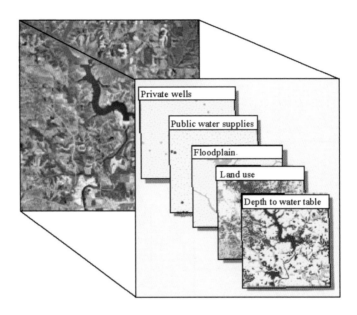

Private wells

Public water supplies

Floodplain.

Land use

Depth to water table

Plate 19 An example of USGS topographic quadrangle maps

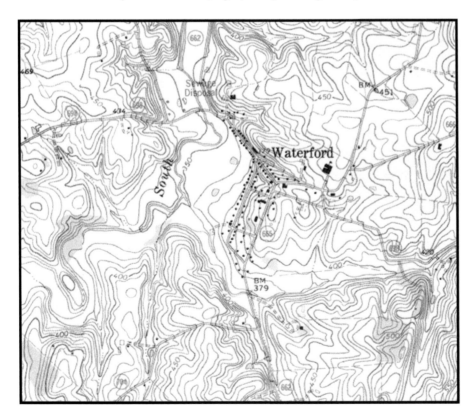

Plate 20 Overlay combines data from different data sets

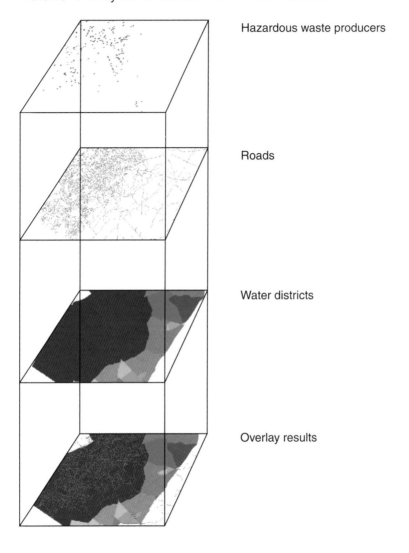

Hazardous waste producers

Roads

Water districts

Overlay results

Plate 21 Areas within 3.5 miles of an airport

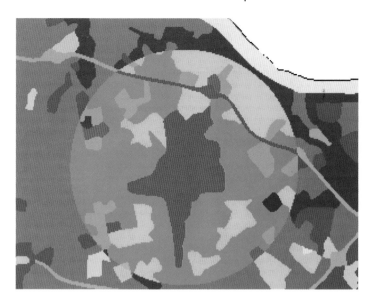

Plate 22 Existing recreational sites and half-mile buffers around streams, abandoned railroads, bike trails and utilities

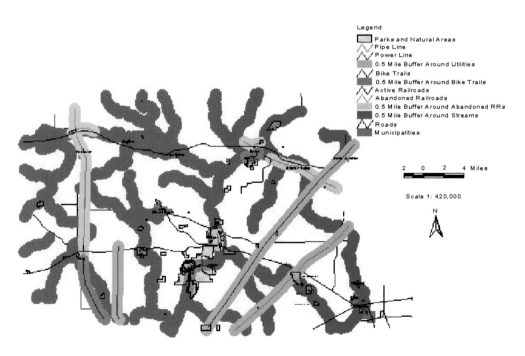

Legend

Parks and Natural Areas
Pipe Line
Power Line
0.5 Mile Buffer Around Utilities
Bike Trails
0.5 Mile Buffer Around Bike Trails
Active Railroads
Abandoned Railroads
0.5 Mile Buffer Around Abandoned RRs
0.5 Mile Buffer Around Streams
Roads
Municipalities

2   0   2   4 Miles

Scale 1: 420,000

N

Plate 23 Insurance agent on ATV mapping perimeter of a vineyard in California

Plate 24 Downloading data to GIS

Plate 25 Image of a clear-cut rendered from visualization software

Plate 26 Rendered image of a partial clearing

It was documented early on that a better scientific visualization environment for the analysis of remote sensor data takes place when the analyst communicates with the digital image processing system *interactively* using a point-and-click Graphical-User-Interface (GUI) (Mazlish 1993; Miller *et al.* 1994). Several effective digital image processing GUIs include ERDAS Imagine's intuitive point-and-click icons (Figure 20.2a,b), Research System's Environment for Visualizing Images (ENVI) hyperspectral data analysis interface (Figure 20.3), ER Mapper, IDRISI and Adobe Photoshop's interface which is very useful for processing photographs and images with <3 bands of data. Non-interactive *batch* processing is of value for time-consuming processes (e.g. re-sampling during image rectification, intensive filtering) and helps to free-up lab PCs or workstations during peak demand because batch jobs can be stored and executed when the computer is idle.

As previously mentioned, it is possible to purchase PCs, workstations, and mainframe computers that have multiple CPUs that can operate concurrently (Figure 20.4). Specially written parallel processing software can parse (distribute) the remote sensor data to specific CPUs to perform efficient digital image processing (Faust *et al.* 1991). For example, consider a 1024 node (CPU) parallel computer system. If a remote sensing dataset consisted of 1024 pixels (columns) in a line and 1024 lines in the entire dataset, each of the 1024 CPUs could be programmed to process an entire line of remote sensor data, speeding up the processing of the entire dataset by 1024 times. Similarly, if 224 bands of AVIRIS hyperspectral data were available, 224 of the 1024 processors could be allocated to evaluate the 224 brightness values associated with each individual pixel with 800 additional CPUs available for other tasks.

Even with single-CPU computers, it is possible to perform parallel processing by connecting to other computers in a network. This type of parallel processing requires sophisticated distributed processing software. In practice, it is difficult to parse a program so that multiple CPUs can execute different portions of the program without interfering with each other. However, many vendors are developing digital image processing code that takes advantage of multiple CPUs and parallel architecture.

Most CPUs now contain an *arithmetic coprocessor*, a special mathematical circuit that performs high-speed floating point operations. Sophisticated image processing software does not function well without a math coprocessor. If substantial financial resources are available, then a considerably more expensive array processor is even better. It consists of a bank of memory and special circuitry dedicated to performing simultaneous computations on elements of an array (matrix) of data in n dimensions (Freedman 1995). Remotely sensed data are collected and stored as arrays of numbers, so array processors are especially well suited to image enhancement and analysis operations. However, the software must be written to take advantage of the coprocessor. If individual programs contain no coprocessor instructions, the coprocessor is not used.

Random Access Memory (RAM) is the computer's primary temporary workspace. It consists of special bank(s) of memory chips. RAM chips require power to maintain their content. Therefore, all information stored in RAM must be saved to a hard disk (or other media) before turning off the computer. The computer should contain sufficient RAM for the operating system, image processing applications software, and any remote sensor data that must be held in memory while calculations are performed. Figure 20.1 depicts individual workstations with 256 MB of RAM. One can never

ENVI Interface

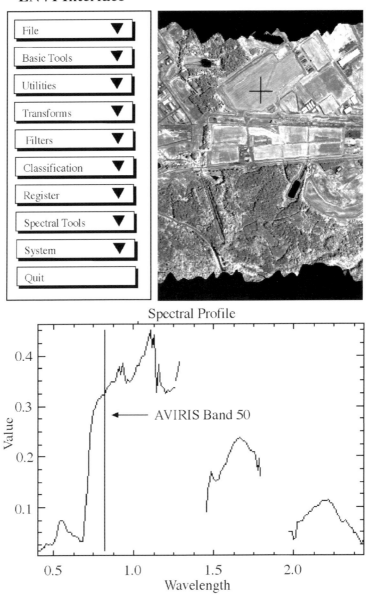

File	▼
Basic Tools	▼
Utilities	▼
Transforms	▼
Filters	▼
Classification	▼
Register	▼
Spectral Tools	▼
System	▼
Quit	

Spectral Profile

←—— AVIRIS Band 50

Value

0.4

0.3

0.2

0.1

0.5   1.0   1.5   2.0

Wavelength

Figure 20.3 Research System's Environment for Visualizing Images (ENVI) graphical user interface allows complex operations to be applied to hyperspectral data. Here, the cursor is located on a clay capped hazardous waste unit on the Savannah River site. The three bands used to make the color composite (shown here in black and white) are depicted as bars in the spectral profile plot. The spectral reflectance curve at the cursor location is displayed in the spectral profile plot. This dataset consisted of 224 bands of 20×20 m² AVIRIS hyperspectral data.

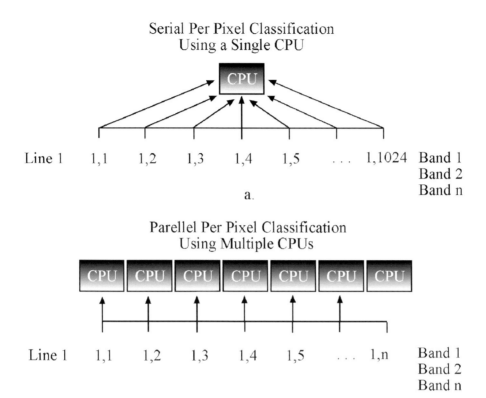

Figure 20.4 Example of serial vs. parallel per pixel image classification logic.

have too much RAM for image processing applications. Fortunately, RAM prices continue to decline while the speed and capacity of the RAM continues to increase. The material in Sections 20.2.1–20.2.5 is covered in a general way in Chapter 5 but provided here in the context of remote sensing.

20.2.1. Operating system and software compilers

The operating system is a master control program that controls all of the computer's higher order functions. It is the first program loaded into memory (RAM) when the computer is turned on, and its main part, called the kernel, resides in memory at all times (Freedman 1995). The operating system sets the standards for the image processing application programs that are executed by it. All programs must 'talk to' the operating system. The difference between a single user operating system and a network operating system is its multi-user capability. DOS, Windows 98, Windows 2000 and the Macintosh OS are single-user operating systems designed for one person at a desktop computer working independently. Windows NT, UNIX, and Linux are network operating systems that are designed to manage multiple user requests at the same time and handle the more complex network security. The operating system provides a user interface and controls multi-tasking. It handles the input and output to the disk and all peripheral devices such as printers, plotters, and color displays.

Currently, most image processing software is available on UNIX, Windows 98 (Millennium edition), and Windows NT (2000) operating systems. Linux, a free implementation of UNIX that can run on both Intel and Motorola microprocessors, has become very popular over the last few years and more image processing applications will be developed for it.

A *compiler* is the software that translates a high-level programming language such as Visual Basic, C or FORTRAN into machine language (refer to Chapter 5). A compiler usually generates assembly language first and then translates the assembly language into machine language. The high-level programming languages most often used in the development of digital image processing software are C, C++, Visual C++, JAVA, Visual Basic and FORTRAN. Many digital image processing systems provide a toolkit that programmers can use to implement their own digital image processing algorithms (e.g. ERDAS). The toolkit consists of fundamental subroutines that perform very specific tasks such as reading a line of image data into RAM or changing the color of a pixel (red, green, and blue) on the screen.

Ideally remote sensing analysts should have some background knowledge of one of these programming languages. This knowledge allows analysts to modify existing software and create stand-alone image processing applications. Visual Basic, a programming language and environment created by Microsoft, has become extremely popular for image processing applications. The intention is not to make programmers out of remote sensing scientists, but to insure that remote sensing analysis is never limited to existing software capability.

20.2.2. Mass storage

Digital image processing of remote sensing and related GIS data requires substantial mass storage resources. For example, during the early part of a remote sensing project a 389.7 MB 8-band Landsat ETM+ image may be loaded onto the hard disk to obtain an appreciation of the geographic extent of the image and its apparent quality. Therefore, the mass storage media should have rapid access times, have longevity (i.e. last for a long time), and be inexpensive (Rothenberg 1995). Digital remote sensor data (and other ancillary raster GIS data) are normally stored in a matrix Band Sequential (BSQ) format in which each spectral band of imagery (or GIS data) is stored as an individual file. Each picture element of each band is represented in the computer by a single 8-bit byte (with values from 0 to 255). The best way to make the brightness values available to the computer rapidly is to place the data on a hard disk, CD-ROM or DVD where each pixel of the data matrix may be accessed at random and at great speed (within microseconds). The cost of hard disk, CD-ROM and DVD storage per gigabyte continues to decline rapidly. It is common for digital image processing laboratories to have gigabytes of hard disk mass storage associated with each workstation as suggested in Figure 20.1. In fact, many image processing laboratories now use Redundant Arrays of Inexpensive hard Disks (RAID) technology in which two or more drives working together provide increased performance and various levels of error recovery and fault tolerance (Freedman 1995). Other storage media, such as magnetic tapes, are usually too cumbersome and/or slow for real time image retrieval, manipulation, and storage because they do not allow random access of data. However, given their large storage capacity (up to 40 GB), they remain a cost-effective way of storing data.

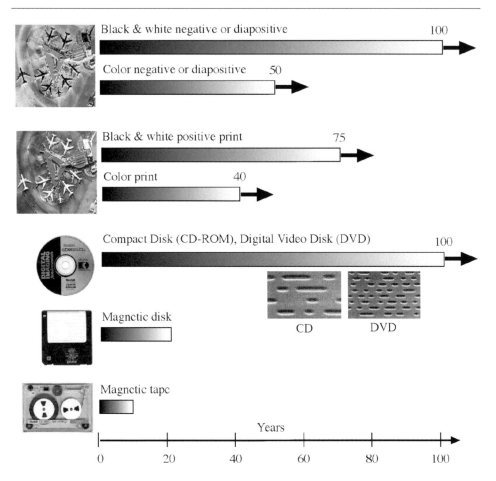

Figure 20.5 Analog photography is an excellent storage media with black and white negatives or diapositives having the longest usable life. Optical disk media such as CD-ROMs and DVDs are superior to magnetic media for storing digital remote sensor data (Jensen et al. 1996).

Figure 20.5 depicts several types of analog and digital remote sensor data mass storage devices and the average time to physical obsolescence, i.e. when the media begin to deteriorate and information is lost. Properly exposed, washed, and fixed black and white aerial negatives have considerable longevity, often more than 100 years. Color negatives with their respective dye layers have longevity, but not as long as the black and white negatives. Similarly, black and white paper prints have greater longevity than color prints (EDC 1995; Kodak 1995). Hard and floppy magnetic disks have relatively short longevity, often less than 20 years. Magnetic tape media (e.g. 3/4″ tape, 8-mm tape, and 1/2″ tape shown in Figure 20.5) can become unreadable within 10–15 years if not rewound and properly stored in a cool, dry environment (EDC 1995).

Optical disks can now be written to, read, and written over again at very high speeds (Normile 1996), and can store much more data than other portable media such as floppy disks. The technology used in re-writeable optical systems is Magneto-Optics (MO), where data is recorded magnetically like disks and tapes, but the bits are much

Table 20.2 Common CD-ROM drive speeds, seek times, and data transfer rates

Drive speed	Seek time (milliseconds)	Data transfer rate (maximum) (kB/s)
Single speed	600	150
2 ×	320	300
3 ×	250	450
4 ×	135–180	600
6 ×	135	900
8 ×	135	1.2
10 ×	135	1.6
12 ×	100–150	1.8
16 ×	100–150	2.4
24 ×	100–150	3.6
32 ×	100–150	4.8

smaller because a laser is used to etch the bit. The laser heats the bit to 150°C at which temperature the bit is realigned when subjected to a magnetic field. In order to record new data, existing bits must first be set to zero. Only the optical disk provides relatively long-term storage potential (>100 years). In addition, optical disks store large volumes of data on relatively small media. Advances in optical Compact Disc (CD) technology promise to increase the storage capacity from the current 680 MB per disc to more than 17 GB using new re-writeable DVD technology (Normile 1996). In many remote sensing laboratories, re-writeable CD-RWs have supplanted tapes as the backup system of choice. One of the few negative things about optical disks is that they are not as fast as conventional hard drives (Table 20.2). It is worth noting that DVD drives are backwards compatible and can read data from CDs.

It is important to remember when archiving remote sensor data that sometimes it may be the loss of (a) the read-write software, and/or (b) the read-write hardware (the drive mechanism and heads) that are the problem and not the digital media itself (Rothenberg 1995; Jensen *et al.* 1996). Therefore, as new computers are purchased it is a good idea to set aside a single computer system that is representative of a certain computer era so that one can always read any data stored on old mass storage media.

20.2.3. Screen display resolution

The number of rows and columns of pixels that can be displayed on a cathode-ray-tube (CRT) screen at one time is called the display resolution. Most users prefer a regional perspective when performing terrain analysis using remote sensor data. Therefore, the image processing system should be able to display at least 1024 rows by 1024 columns of pixels and preferably more on the CRT at one time. This allows larger geographic areas to be examined at one time and places the terrain of interest in its regional context.

20.2.4. Screen color resolution

The number of gray-scale tones or colors, e.g. 256, that a pixel may be displayed in on

Table 20.3 Video RAM required for different display resolutions (Webopedia.com)

Resolution	256 colors (8-bit)	65,000 colors (16-bit) (MB)	16.7 million colors (24-bit, true color) (MB)
640 × 480	512 kB	1	1
800 × 600	512 kB	1	2
1,024 × 768	1 MB	2	4
1,152 × 1,024	2 MB	2	4
1,280 × 1,024	2 MB	4	4
1,600 × 1,200	2 MB	4	6

a screen out of a palette of available colors is referred to as the CRT screen color resolution. Most digital image processing systems provide access to a tremendous number of displayable colors from a large color palette, e.g. 16.7 million. The primary reason for these significant color requirements is that image analysts must often display a color composite of several individual images at one time on the screen. For example, to display the false-color composite of the digital camera image of Dunkirk, New York shown in Figure 20.6a (see plate 14) it was necessary to place three separate 8-bit images in three distinct planes of *image processor memory*. Band 3 (0.7–0.8 μm) near-infrared data were placed in the red image processor memory plane (Figure 20.6, see plate 14). Band 2 (0.6–0.7 μm) red data were placed in the green memory plane. Green band 1 (0.5–0.6 μm) data were placed in the blue image plane. Thus, each pixel could take upon itself any of 2^{24} possible color combinations (16,777,216). Such true color systems are expensive because every pixel location is *bit mapped*, i.e. a specific location in video memory keeps track of the blue, green, and red brightness values. The network configured in Figure 20.1 has six 24-bit color workstations.

Several current remote sensing systems now collect data with 10, 11, and even 12-bit radiometric resolution with brightness values ranging from 0 to 511, 0 to 1023, or 0 to 2047, respectively. Unfortunately, despite advances in video technology, at the present time it is necessary to generalize the radiometric precision of the remote sensor data to 8-bits per pixel simply because current video display technology cannot handle the demands of the increased precision.

Video cards or adapters are boards that plug into the CPU's mother board to enable display capabilities. They contain memory so the CPU's RAM is not used to process the display information. Most adapters have graphics coprocessors (accelerators) for rapidly displaying the results of graphics calculations. The more video memory the better for image processing applications (Table 20.3).

20.2.5. Image scanning (digitization) considerations

Many projects require the analysis of analog (hard-copy) aerial photography such as color-infrared US National Aerial Photography Program (NAPP) imagery. Also, many studies make use of multiple dates of historical panchromatic black and white aerial photography, e.g. 1:20,000 US Agricultural Stabilization and Conservation Service photography obtained in the 1930s and 1940s. Such data are often digitized (translated into digital form), rectified, and then analyzed in a digital image processing system. To be of value, careful decisions must be made about the spatial and radiometric resolution of the digitized image data. Tables that summarize the relationship between input image scale, digitizer detector Instantaneous-Field-Of-View (IFOV) measured in dots-per-inch (dpi) and micrometers (μm), and output image spatial resolution in meters are found in Jensen (1995b, 1996).

Images to be used for simple illustrative purposes may be scanned at low resolutions such as 100 dpi (254 μm). For vertical aerial photography obtained at 1:20,000 scale, this yields images with a spatial resolution of 5.08×5.08 m that retain sufficient detail for orientation purposes. For a $9'' \times 9''$ photograph, this produces a 0.81 MB file ($9 \times 9 \times 100 \times 100$ dpi $= 810,000$ bytes) for black and white photographs and a 2.43 MB file for color aerial photography. This assumes that the black and white images are scanned at 8-bit radiometric resolution and that color images are scanned at 24-bit resolution. Images that are to be used for scientific purposes are usually scanned at much higher spatial resolution such as 500 dpi (12.7 μm) to retain the subtle reflectance and/or emittance information found in the original imagery. For 1:20,000 scale photography, this yields a spatial resolution of 1.02×1.02 m per pixel and results in a 20.25 MB file for a single black and white photograph and a 60.75 MB file for a single color photograph (Jensen 1995b, 1996).

Most aerial photography is collected using a $9'' \times 9''$ metric camera. The ideal scanning system digitizes the entire image and any ancillary 'titling' information on the periphery of the image at one time. Therefore, the digitizer of choice should have a field of view of at least $9'' \times 9''$. While it is possible to use inexpensive $8.5'' \times 11''$ desktop scanners, this requires that the imagery be broken up into two parts that must then be mosaicked together. This introduces radiometric and geometric error and should be avoided. Advances in the 'desktop publishing' industry have spurred the development of flatbed, $11'' \times 14''$ desktop linear array digitizers that can be used to digitize negatives, diapositives, and paper prints at 100–3,000 dpi (Foley et al. 1994). Some scanners digitize color photographs with one color filter then repeat the process with the other color filters. This can result in color misregistration and loss of image quality. Ideally, color aerial photography is scanned in a single pass and converted into three registered RGB files.

Area array Charge-Coupled-Device (CCD) digital camera technology has been adapted for hard-copy image digitization (Figure 20.1). Typical area array CCD systems digitize from 160 to 3,000 dpi (approximately 160–8.5 μm) over a $10'' \times 20''$ image area (254×508 μm). They scan the original negative or positive transparency as a series of rectangular image tiles. The scanner then illuminates and digitizes a reseau grid which is an array of precisely located cross-hatches etched into the glass of the film carrier. The reseau grid coordinate data are used to locate the exact orientation of the CCD camera during scanning and to geometrically correct each digitized 'tile' of the image relative to all others. Radiometric calibration algorithms are then used to compensate for uneven illumination encountered in any of the tile regions. Area array digitizing technology has obtained geometric accuracy of < 5 μm over 23×23 cm images when scanned at 25 μm per pixel and repeatability of < 3 μm (Jensen 1996).

20.3. Image processing and GIS software requirements

A variety of digital image processing and GIS functions are required to analyze re-motely sensed data for earth resource management and planning applications. Some of the most important functions are summarized in Table 20.4. The reader is encouraged to review textbooks on digital image processing and GIS to obtain specific information about the algorithms (Jensen 1996, 2000, 2000b; Lillesand and Kiefer 2000). It is useful to briefly identify characteristics of the most important functions.

20.3.1. Preprocessing

Remote sensor data contain radiometric and geometric error. The data must be care-fully preprocessed before reliable and useful information can be extracted from it. The person processing the data must radiometrically correct it to remove (a) system intro-duced error (e.g. systematic and random electronic noise, stripping) and/or (b) envir-onmentally introduced image degradation (e.g. scattering due to atmospheric haze, absorption due to water vapor, sunglint from water). Many remote sensing projects do not require detailed radiometric correction. However, projects dealing with water quality, differentially illuminated mountainous terrain, and subtle differences in vege-tation health, biomass, and leaf-area-index do require careful radiometric correction (Jensen 1995a,b, 2000a,b; Bishop 1998). Therefore, the application software should provide a robust suite of radiometric correction alternatives.

Analysts normally require that the information extracted from remotely sensed data be placed in a GIS. This involves rectification of the remote sensor data to a map projection using nearest-neighbor, bilinear interpolation, or cubic convolution re-sampling logic (refer to Chapter 19). The image processing software should allow Ground Control Points (GCPs) to be easily and interactively identified on the base map and in the unrectified imagery. The GCP coordinates are used to compute the coefficients necessary to warp the unrectified image to a planimetric map projection. The accuracy of the image-to-map rectification or image-to-image registration is speci-fied in Root-Mean-Square-Error (RMSE) units, e.g. the pixels in the image are within ± 10 m of their true planimetric location. The user must also be able to specify the geoid and datum (e.g. NAD83 refers to the North American Datum 1983 to which all US digital orthophoto quarterquads must be referenced). Change detection projects are especially dependent upon accurate geometric rectification (or registration) of multiple date images. Therefore, it is imperative that the image processing system software allow accurate geometric rectification.

Automated Global Positioning System (GPS) data collection has been used for data collection for GISs for many years. Currently, several new remote sensing systems automatically collect GPS information onboard the aircraft at the exact moment that the radiant energy within each IFOV (pixel) is recorded. Future digital image proces-sing software must be able to use these onboard GPS data to perform near-real time geometric correction of the remote sensor data. Few digital image processing systems currently have this capability.

Table 20.4 Image processing functions required to analyze remote sensor data for earth resource management and planning applications

Preprocessing	1	Radiometric correction of error introduced by the sensor system electronics and/or environmental effects (includes relative image-to-image normalization and absolute radiometric correction of atmospheric attenuation)
	2	Geometric correction (image-to-map rectification or image-to-image registration)
Display and enhancement	3	Black and white (8-bit)
	4	Color-composite display (24-bit)
	5	Black and white or color density slice
	6	Magnification, reduction, roaming
	7	Contrast manipulation (linear, non-linear)
	8	Color space transformations (e.g. RGB to intensity-hue-saturation)
	9	Image algebra (e.g. band ratioing, image differencing)
	10	Linear combinations (e.g. NDVI, SAVI, Kauth transform)
	11	Spatial filtering (e.g. high-pass, low-pass, band-pass)
	12	Edge enhancement (e.g. Sobel, Robert's, Kirsch, Laplacian)
	13	Principal components (e.g. standardized, unstandardized)
	14	Texture transforms (e.g. min-max, texture spectrum, fractal dimension)
	15	Frequency transformations (e.g. Fourier, Walsh)
	16	DEMs (e.g. analytical hill shading, compute slope, aspect)
	17	Animation, e.g. movies depicting change
Remote sensing information extraction	18	Pixel brightness value (digital number value – DN)
	19	Transects
	20	Univariate and multivariate statistical analysis (e.g. mean, covariance)
	21	Feature (band) selection (graphical and statistical)
	22	Supervised classification, e.g. minimum distance, maximum likelihood
	23	Unsupervised classification, e.g. ISODATA
	24	Contextual classification
	25	Incorporation of ancillary data during classification
	26	Expert system image analysis
	27	Neural network image analysis
	28	Fuzzy logic classification
	29	Hyperspectral data analysis
	30	Radar image processing
	31	Accuracy assessment (descriptive and analytical)
Photogrammetric information extraction	32	Soft-copy extraction of digital elevation models
	33	Soft-copy production of orthoimages
Metadata and image/map lineage documentation	34	Metadata
	35	Complete image and GIS file processing history (lineage)
Image/map cartography	36	Scaled postscript level II output of images and maps
GIS	37	Raster (image) based GIS
	38	Vector (polygon) based GIS (must allow polygon overlay)
Integrated image processing and GIS	39	Complete image processing systems (functions 1–36 plus utilities)
	40	Complete image processing systems and GIS (functions 1–43)
Utilities	41	Network, e.g. local area network, Internet
	42	Image compression (single image, video)
	43	Import and export of various file formats

20.3.2. Display and enhancement

An analyst must be able to display individual black and white images (usually 8-bit) and color composites of three bands at one time (24-bit). The true nature of the remote sensor data must be visible without any dithering (a method of display based on the combined use of a smaller number of colors in conjunction with texture). The analyst can then density slice, magnify, roam, or manipulate the contrast of the individual bands or color composites. To this end, algebraic and linear combinations of bands of remote sensor data have proven useful. Therefore, the system must be able to build simple algebraic statements, e.g. ratio bands 2 and 3, and more useful transformations of the remote sensor data.

Spatial and frequency filtering algorithms are used to enhance and display subtle high and low frequency features and edges of these features in the remote sensor data. Texture algorithms enhance areas of uniform texture (e.g. coarse, smooth, rough). Some bands of remote sensor data are highly correlated with other bands, which signifies redundant information. Principal components analysis is often applied to reduce the dimensionality (number of bands) used in the analysis while still maintaining the bands that contain most of the variance.

Digital Elevation Models (DEMs) are critical to successful modeling and understanding of many landscape processes. The analyst must be able to display a DEM in a planimetric (vertical) view using analytical hill shading as well as in a pseudo three-dimensional perspective view. Ideally, it is possible to drape thematic information such as a hydrologic or road network on top of the hill-shaded DEM. The DEMs and orthophotos are produced using photogrammetric principles as discussed in the soft-copy photogrammetry information extraction section. Finally, it is important to be able to monitor change in the landscape by displaying multiple dates of imagery in an animated fashion. Change information can be used to gather information about the processes at work.

20.3.3. Remote sensing information extraction

Analysts must be able to extract the brightness value (z) at any x, y location in the raster image and along user-specified linear and non-linear transects. The user must also be able to draw rubberband-like polygons around objects or areas of interest using *heads-up* or *on-screen* digitization. The software must then be able to extract polygon area, perimeter, and/or volume information. Ideally, the polygon and its attribute information are saved in a standard format for subsequent processing, e.g. in ERDAS or ArcInfo coverage formats. Heads-up on-screen image interpretation and digitization is becoming more important as very high spatial resolution satellite remote sensor data becomes available and fundamental photo-interpretation techniques are merged with automated feature extraction (Jensen 1995a, 2000; Firestone *et al.* 1996).

Most popular digital image processing systems allow the user to classify multispectral remote sensor data using supervised or unsupervised techniques (refer to Chapter 19). In a supervised classification the analyst supervises the training of the algorithm. In an unsupervised classification the analyst relies on the computer to identify pixels in the terrain that have approximately the same multispectral characteristics. These 'clusters' are then labeled by the analyst to produce a thematic map. Most image processing systems still do not easily allow the incorporation of contextual or other types of ancillary data in the classification. Such information could greatly increase classification accuracy.

Several digital image processing systems now include software that allows an analyst to build a knowledge base and use an expert system to classify and analyze remote sensor data, e.g. ERDAS Imagine. Relatively few systems have incorporated an artificial neural network analysis capability although it has shown great potential and has the capability of 'learning' as new training cases are evaluated by the network (Jensen and Qiu 1998). However, stand-alone artificial neural network software may be integrated into mage classification with moderate difficulty. Finally, the terrain usually grades from one land cover into another without 'hard' partitions. In fact, the 'fuzzy transition interface' between homogeneous terrain elements is often where the greatest species diversity of plants and animals exists. Therefore, 'fuzzy' classification algorithms will likely see even more utility in the future.

The launch of Japan's JERS-1, Canada's RADARSAT, the European Space Agency's ERS-1 and 2, NASA's Shuttle Imaging Radar (SIR) missions, and NASA's Shuttle Radar Topography Mission (SRTM) (which collected interferometric synthetic aperture radar data to derive a worldwide DEM) have stimulated the use of radar data. Unfortunately, only a few image processing systems provide the software necessary to remove or adjust the speckle associated with the raw radar data and to geometrically warp it into a ground-range map projection (as opposed to the original slant-range geometry).

20.3.4. Photogrammetric information extraction

Photogrammetric information extraction from stereoscopic high spatial resolution imagery is one of the most exciting areas of digital image processing. For 50 years it has been necessary to use expensive and time consuming stereoplotters to (a) extract DEMs from stereoscopic vertical aerial photography, and (b) produce orthophotographs. With advances in desktop computers and soft-copy photogrammetry software, it is now possible to extract DEMs from both aerial photography and satellite digital data and then produce accurate orthoimages, (Greve *et al.* 1992; Jensen 1995b, 1997). Some of the new software can produce true orthophotos where the building footprint is in its proper planimetric location over the foundation and all relief displacement has been removed.

Planners, civil engineers, tax mapping agencies, natural resource managers, and water management groups can now produce accurate DEMs and orthophotos on demand for their local modeling purposes rather than being dependent on the dreadfully slow cycle of government DEM and orthophoto production. Table 20.5 identifies several vendors that provide photogrammetric image processing software.

20.3.5. Metadata and image/map lineage documentation

Many countries have adopted rigorous national standards concerning the content, accuracy, and transmission characteristics of map and image spatial data. Such information is called *metadata* – data about data. For example, the US Federal Geographic Data Committee (FGDC) developed stringent metadata standards for all image and

Table 20.5 Selected commercial and public image processing systems used for earth resource mapping and their functions (Operating systems: W = Microsoft Windows 98; U = Unix; NT = Microsoft Windows NT (2000), M = Macintosh; Functions: ● = significant capability; ○ = moderate capability; no symbol = little or no capability) (updated from Jensen 1996, 2000a)[a]

System	Operating system	Preprocessing	Display and enhancement	Info extraction	Soft-copy photo	Metadata lineage	Image/map cartography	GIS	IP/GIS
Commercial									
AGIS	W NT	○						●	○
Applied Analysis	W U NT		●	○					
CORE HardCore	W U		●	○					
CORE ImageNet	W U		●	●					
Dragon	W	●	●	●			●		
EarthView	W	●	●	●			●		
EIDETIC	W	○	●	○	○	○			○
ESRI ArcInfo	NT U	○	●	○			●	●	●
ESRI Image Analyst	M W U NT	●	●	○	●	●	●	●	●
ENVI	M W U NT	●	●	●	●	●	●	○	●
ERDAS Imagine	W U NT	●	●	●	●		●	●	●
ERIM	U	●	●	●			●	●	●
ER Mapper	U NT	●	●	●	●	●	●	●	●
GAIA	Mac	○	○	○				●	
GENASYS	W U	●	●	●					
GenIsis	W	○	●	○	○		●		
Global Lab Image	U		●	○					
GRASS	W U	●	●	●	●	●	●	●	●
Helava Associates	W NT	●	●	●	●		●	●	●
IDRISI	W NT U	●	●	●	●	●	●	●	●
Intergraph	W U	●	●	●	●		●	●	●
Leica	M W NT U	●	●	●	●	●	●	●	●
PCI	M W NT U	●	●	●	●		●	●	●
Photoshop	M W NT U	○	●	○					
MapInfo	Mac		●		●	●	●	●	●
MacSadie	W	●	●	●					
Jandel MOCHA	U	●	●	●					
OrthoView		●	●		●			○	

Table 20.5 (continued)

System	Operating system	Preprocessing	Display and enhancement	Info extraction	Soft-copy photo	Metadata lineage	Image/map cartography	GIS	IP/GIS
R2V	W NT	●	●				○	●	●
R-WEL	W	●	●	●	●		●	●	●
GDE Socet Set	NT U	●	●	●	●		●	●	●
Vision Softplotter	W U	●	●	●	●		●	○	○
Microimage TNTmips	W NT U	●	●	●	●	●	●	●	●
Trifid TruVue	W U	●	●	●	●		●	●	●
Vexel IDAS	U	●	●	●	●		●	●	●
VISILOG	U	●	●	●	●				
Public Systems									
C-Coast	W	●	●	●			●	●	●
Cosmic VICAR	U	○	●	●			○	●	○
EPPL7	W	●	●	○					
MultiSpec	Mac		●	●					
NIH-Image	U	○	○						
NOAA	U		○						
XV	U		●						

[a] Sources of information:
AGIS Software, PO Box 441, Franklin, Tasmania, Australia 7113.
R2V-Able Software Corp. 5 Appletree Lane. Lexington, MA 02420-2406.
Applied Analysis Inc., 46 Manning Road, Suite 201, Billerica, MA 01821; subpixel processing.
ArcInfo/ArcView/ArcGrid/ArcInfo/Image Analyst - ESRI, 380 New York St., Redlands, CA 92373; http://www.esri.com/.
C-Coast, JA20 Building 1000, Stennis Space Center, MS 29519.
CORE, Box 50845, Pasadena, CA 91115.
Cosmic, University of Georgia, Athens, GA 30602.
Dragon, Goldin-Rudahl Systems, Six University Dr. Suite 213, Amherst, MA 01002.
EarthView, Atlantis Scientific Systems Group, 1827 Woodward Dr. Ottawa, Canada K2C 0P9; www.atlsci.com.
ENVI, Research Systems, Inc., 2995 Wilderness Place, Boulder CO 80301; www.rsinc.com.
EPPL7, Land Management Information Center, 300 Centennial Building, 638 Cedar St., St. Paul, MN 55155.
ERIM, Box 134001, Ann Arbor, MI 48113-4001.
ERDAS Imagine, 2801 Buford Hwy., NE, Suite 300, Atlanta, GA 30329.

ER Mapper, 4370 La Jolla Village Dr., San Diego, CA 92122.

GAIA, 235 W. 56th St., 20N, New York, NY, 10019.

Global Lab, Data Translation, 100 Locke Dr., Marlboro, MA 01752-1192.

GRASS – http://www.cecer.army.mil/announcements/grass.html.

Helava Associates, Inc., 10965 Via Frontera, #100, San Diego, CA 92127–1703.

IDAS – Vexcel Imaging, Inc., 3131 Indian Road, Boulder, CO 80301.

IDRISI, Graduate School of Geography, Clarke Univ. 950 Main, Worcester, MA 01610.

Intergraph, Huntsville, AL, 35894.

Leica, Inc., 2 Inverness Drive East, #108, Englewood, CO 80112.

MapInfo Corporate Headquarters, One Global View, Troy, New York 12180.

MOCHA Jandel Scientific, 2591 Kerner Blvd., San Rafael, CA 94901.

MultiSpec, Dr. David Landgrebe, Purdue Research Foundation, W. Lafayette, IN 47907.

NIH-Image, National Institutes of Health, Washington, D.C.

OrthoView, Hammon-Jensen-Wallen, 8407 Edgewater Dr., Oakland, CA 94621.

PCI, 50 W. Wilmot, Richmond Hill, Ontario Canada L4B 1M5; www.pci.on.ca.

PHOTOSHOP, Adobe Systems Inc., 1585 Charleston Road, Mountain View, CA 94039.

R-WEL Inc., Box 6206, Athens, GA 30604.

SOCET SET, GDE Systems, Inc., Sand Diego, CA and Helava Associates, Inc., 10965 Via Frontera, #100, San Diego, CA 92127-1703.

SPHIGS (Simple Programmer's Hierarchial Interactive Graphics System) and SRGP (Simple Raster Graphics Package).

Terra-Mar Resource Information Services, Inc., 1937 Landings Dr., Mountain View, CA 94043.

TNTmips, MicroImages, 201 N. 8th St., Lincoln, NB 68508; info@microimages.com.

TruVue, Trifid Corp., 680 Craig Rd., Suite 308, St. Louis, MO 63141.

VISILOG, NOESIS Vision, Inc., 6800 Cote de Liesse, Suite 200, St. Laurent, Quebec, H4T 2A7.

Vision International Inc., Division of Autometric, Inc., 81 Park Street, Bangor, ME 04401.

VI2STA, International Imaging Systems, Inc., 1500 Buckeye Drive, Milpitas, CA 95035.

XV, image viewer program written by John Bradley: http://phoenix.csc.calpoly.edu/CSL/cobra/xv.html

map data produced for government use (Chapter 36). All digital image processing systems and GIS in the future will eventually provide detailed metadata and image lineage (genealogy) information. The lineage information summarizes the different processes that were sequentially applied to each image or map (Lunetta *et al.* 1991; Lanter *et al.* 1992; Jensen *et al.* 1992; FGDC 1998). The image lineage information becomes very important when the products derived from the analysis of remotely sensed data are involved in peer-reviewed scientific publication, litigation and/or intense public scrutiny.

20.3.6. Image and map cartographic composition

Users have two options when producing image (photo) maps or thematic maps: rectified or unrectified. Popular image processing software (e.g. Adobe Photoshop) and graphics programs (e.g. Adobe Illustrator, Macromedia Freehand) can be used to produce useful unrectified images and diagrams. However, if cartographically accurate images and/or thematic maps are required, then a fully functional digital image processing system or GIS must be utilized that produces quality cartographic products, e.g. IDRISI, ERDAS, ER Mapper, ESRI Arc-View. Ideally, the software supports Postscript level II output devices. Several of the most popular digital image processing systems have 'wizard' applications that walk the analyst through a standardized map production process thus insuring that the output product conforms to a standard map template. For example, ERDAS and ER Mapper contain a template for the production of scaled 1:24,000 7.5-minute map products that mimic the look and feel of US Geological Survey 7.5-minute map and image products.

The selection of printers and plotters is an important consideration for remote sensing laboratories. Printers range in quality from dot-matrix, which creates characters by striking pins against an ink ribbon, to electrostatic printers that produce high quality output through electrostatic technology that uses a print bar to produce an electrical charge to apply 'dots' linearly across the width of specially treated media. After the charge is placed, the media are run across an inking station where the correct colors are applied. While electrostatic printers produce high-quality output, they are also expensive. Depending on the purpose and audience, ink jet and laser jet printers are usually adequate for map production. Ink jet printers spray ink at paper and produce good text and graphics. Laser printers are based on copy machine technology, and also produce high quality text and graphics. Large output (or E-sized) plotters are sometimes necessary for remote sensing laboratories because they allow for large poster-sized output. Most large output plotters use ink-jet technology. As with all computer peripherals, the cost of printers and plotters continues to decrease.

20.3.7. GIS

Information derived from remote sensor data is most valuable when used in conjunction with other ancillary data (e.g. soils, elevation, slope, aspect, depth-to-ground-water) stored in a GIS (Lunetta *et al.* 1991). Therefore, the ideal integrated system performs both digital image processing and GIS spatial modeling and considers map data as image data (and vice-versa) (Cowen *et al.* 1995). The GIS analytical capabilities are hopefully based on 'map algebra' logic that can easily perform linear combinations of

GIS operations to model the desired process. The GIS must also be able to perform raster-to-vector and vector-to-raster conversion accurately and efficiently. GISs provide relatively easy map output that integrates cartographic tools and layout techniques. Output using a GIS seamlessly integrates both raster and vector data sources.

20.3.8. Utilities

The digital image processing system should have the ability to network not only with colleagues and computer databases in close proximity but with those throughout the world. Therefore, efficient transmission lines and communication software (protocol) must be available. Much information is now routinely served on the World-Wide-Web (WWW) (Ubois 1993). Some have suggested that the Internet is the sales channel of the future for imagery (Thorpe 1996).

The type of *data compression* algorithm used to store the image data can have a serious impact on the amount of mass storage required. The basic idea of image compression is to remove redundancy from the image data, hopefully, without sacrificing valuable information. This is usually done by mapping the image to a set of coefficients. The resulting set is then quantized to a number of possible values that are recorded by an appropriate coding method. Most commonly used image compression methods are based on the discrete cosine transform such as the Joint Photographic Experts Group (JPEG) algorithm, on vector quantization, on differential pulse code modulation, and on the use of image pyramids (Lammi *et al.* 1995).

One must decide on whether to use a *loss-less* or *lossy* data compression algorithm (Sayood 2000). When an image is compressed using lossy logic and then uncompressed, it may appear similar to the original image but it does not contain all of the subtle multispectral brightness value differences present in the original (Bryan 1995; Nelson 1996). Imagery that has been compressed using a lossy algorithm may be suitable as an illustrative image where cursory visual photo-interpretation is all that is required. If lossy compression is absolutely necessary, the algorithm of choice at the present time appears to be JPEG which has a compression ratio of about 1:10 for color photos without considerable degradation in the visual or geometric quality of the image for photogrammetric applications (Lammi *et al.* 1995). Unfortunately, such lossy data may be unsuitable for scientists performing quantitative analysis of the data. Therefore, it is good practice to store the archive image using a loss-less data compression algorithm based on (a) JPEG differential pulse code modulation – DPCM or (b) run-length encoding (e.g. the simple UNIX compress command) so that both novice users and scientists have access to the best reproduction of the original data. The optimum lossy and/or loss-less image compression algorithm(s) are still being debated, e.g. fractal, wavelet, quadtree, run-length encoding (Russ 1992; Pennebaker *et al.* 1993; Jensen 1996; Nelson 1996; Sayood 2000). Robust multiple frame video data compression algorithms now exist that are of benefit for remote sensing image animation projects, e.g. MPEG.

The image processing system must have the capability to import and export remote sensing and GIS data files stored in a variety of standard formats. The system must be able to read at least the following file formats: encapsulated postscript file – EPSF, tagged interchange file format – TIFF, Macintosh PICT, ERDAS, ESRI coverages, GeoTIFF, and CompuServe GIF.

20.4. Commercial and publicly available digital image processing systems

The development and marketing of digital image processing systems is a multi-million dollar industry. The image processing software/hardware may be used for non-destructive evaluation of items on an assembly line, medical image diagnosis, and/or analysis of remote sensor data. Several of the most widely used systems that are used to analyze remotely sensed data are summarized in Table 20.5. Their capabilities are cross-referenced to the general image processing functions summarized in Table 20.4.

Universities and public government agencies have developed digital image processing software. Several of the most widely used and publicly available digital image processing systems are summarized in Table 20.5. COSMIC at the University of Georgia is a clearinghouse for obtaining NASA sponsored digital image processing software for the cost of media duplication.

20.5. Summary

Analysis of remotely sensed data and GIS information requires access to sophisticated digital image processing system and GIS software. This chapter identified typical (a) computer hardware/software characteristics, (b) image processing functions required to analyze remote sensor data for a variety of applications, and (c) selected commercial and public digital image processing systems and their generic functions available for earth resource mapping. The Earth Observing System (EOS) and several commercial enterprises now provide dramatically improved spatial and spectral resolution remote sensor data, e.g. NASA's Moderate Resolution Imaging Spectrometer – MODIS, France's improved Spot Image, and EOSAT Space Imaging's IKONOS (Asrar and Greenstone 1996; Jensen and Qiu 1998). Continually improving digital image processing software will hopefully be developed to extract the required land use/cover and biophysical information needed for many applications.

References

ASPRS, *Manual of Photographic Interpretation*, Philipson, W. (ed.), American Society for Photogrammetry & Remote Sensing, Bethesda, MD, 1997.

Asrar, G. and Greenstone, R., *Mission to Planet Earth EOS Reference Handbook*, NASA, Washington, DC, 1996.

Bishop, M. P., 'Scale-dependent analysis of satellite imagery for characterization of glaciers in the Karakoram Himalaya', *Geomorphology*, 21, 1998, 217–232.

Bryan, J., 'Compression scorecard', *Byte*, 20, 1995, 107–111.

Cowen, D. J. et al., 'The design and implementation of an integrated geographic information system for environmental applications', *Photogrammetric Engineering & Remote Sensing*, 61, 1995, 1393–1404.

EDC, *Correspondence with EROS Data Center personnel*, Sioux Falls, SD, 1995.

Faust, N. L. et al., 'Geographic information systems and remote sensing future computing environment', *Photogrammetric Engineering and Remote Sensing*, 57, 1991, 655–668.

Firestone, L. et al., 'Automated feature extraction: the key to future productivity', *Photogrammetric Engineering & Remote Sensing*, 62, 1996, 671–674.

FGDC, *The Value of METADATA*, Federal Geographic Data Committee, Reston, VA, 1998.

Foley, J. D. et al., *Introduction to Computer Graphics*, Addison-Wesley, New York.

Freedman, A., *The Computer Glossary*, American Management Association, New York.

Greve, C. W. et al., 'Image processing on open systems', *Photogrammetric Engineering & Remote Sensing*, 58, 1992, 85–89.

Jensen, J. R., 'President's inaugural address', *Photogrammetric Engineering & Remote Sensing*, 10, 1995a, 835–840.

Jensen, J. R., 'Issues involving the creation of digital elevation models and terrain corrected orthoimagery using soft-copy photogrammetry', *Geocarto International*, 10, 1995b, 5–21.

Jensen, J. R., *Introductory Digital Image Processing: a Remote Sensing Perspective* (2nd ed.), Prentice Hall, Saddle River, NJ, 1996, 318 pp.

Jensen, J. R., 'Issues involving the creation of digital elevation models and terrain corrected orthoimagery using soft-copy photogrammetry', *Manual of Photogrammetry Addendum*, Greve, C. (ed.), American Society for Photogrammetry & Remote Sensing, Bethesda, MD, 1996.

Jensen, J. R., 'Processing remotely sensed data: hardware and software', *Remote Sensing in Hydrology and Water Management*, Schultz, G. A. and Engman, E. T. (eds.), Springer, Berlin, 2000a, pp. 41–63.

Jensen, J. R., *Remote Sensing of the Environment: an Earth Resource Perspective*, Prentice Hall, Saddle River, NJ, 2000b, 544 pp.

Jensen, J. R. and Qiu, F., 'A neural network based system for visual landscape interpretation using high resolution remotely sensed imagery', *Proceedings, Annual Meeting of the American Society for Photogrammetry & Remote Sensing*, Tampa, FL, CD:15, 1998.

Jensen, J. R. et al., 'Improved remote sensing and GIS reliability diagrams, image genealogy diagrams, and thematic map legends to enhance communication', *International Archives of Photogrammetry and Remote Sensing*, 6, 1992, 125–132.

Jensen, J. R. et al., 'Remote sensing image browse and archival systems', *Geocarto International*, 11, 1996, 33–42.

Kodak, 1995, Correspondence with Kodak Aerial Systems Division, Rochester, New York.

Lanter, D. P. et al., 'A research paradigm for propagating error in layer-based GIS', *Photogrammetric Engineering & Remote Sensing*, 58, 1992, 825–833.

Lammi, J. et al., 'Image compression by the JPEG algorithm', *Photogrammetric Engineering & Remote Sensing*, 61, 1995, 1261–1266.

Lillesand, T.M. and Kiefer, R.W. *Remote Sensing and Image Interpretation*, (4th Ed.), John Wiley & Sons, New York, 2000, 724 pp.

Lunetta, R. S. et al., Remote sensing and geographic information system data integration: error sources and research issues, *Photogrammetric Engineering & Remote Sensing*, 57, 1991, 677–687.

Mazlish, B., *The Fourth Discontinuity: the Co-Evolution of Humans and Machines*, Yale University Press, New Haven, CN, 1993.

Miller, R. L. et al., 'C-coast: a PC-based program for the analysis of coastal processes using NOAA CoastWatch data', *Photogrammetric Engineering & Remote Sensing*, 60, 1994, 155–159.

Nelson, L. J., 'Wavelet-based image compression: commercializing the capabilities', *Advanced Imaging*, 11, 1996, 16–18.

Normile, D., 'Get set for the super disc', *Popular Science*, 96, 1996, 55–58.

Pennebaker, W. B. et al., *JPEG: Still Image Data Compression Standard*, Van Nostrand Reinhold, New York, 1993.

Rencz, A. N., *Remote Sensing for the Earth Sciences*, John Wiley & Sons, New York, 1999, 707 pp.

Rothenberg, J., 'Ensuring the longevity of digital documents', *Scientific American*, 272, 1995, 42–47.

Russ, J. C., *The Image Processing Handbook*, CRC Press, Boca Raton, FL, 1992.

Sayood, K., *Introduction to Data Compression*, (2nd Ed.), Morgan Kaufmann Publishers, San Francisco, CA, 2000, 600 pp.

Spooner, J. G., 'Next Celeron to merge graphics, audio', *PC Week*, April 26, 1999.

Thorpe, J., 'Aerial photography and satellite imagery: competing or complementary?' *Earth Observation Magazine*, June, 1996, 35–39.

Ubois, J., 'The internet today', *SunWorld*, April, 1996, 90–95.

Chapter 21

Quality assurance and accuracy assessment of information derived from remotely sensed data

Russell G. Congalton and Lucie C. Plourde

21.1. Introduction

The spatial and thematic information extracted from remotely sensed data is usually summarized in cartographic map format. It is essential that lay persons, professionals and scientists know the accuracy of the information summarized in the map. Therefore, a critical step in analyzing any map created from the analysis of remotely sensed data is quality assurance and accuracy assessment. Quantitative accuracy assessment is performed to identify and measure map errors. There are two primary motivations for conducting a quantitative accuracy assessment:

- To understand the errors in the map (so they can be corrected), and
- To provide an overall assessment of the reliability of the map (Gopal and Woodcock 1994).

The history of accuracy assessment of cartographic products derived from digital remotely sensed data is relatively short. Before the mid 1970s, maps derived from analog remotely sensed data using visual photo interpretation techniques were rarely subjected to any kind of quantitative accuracy assessment. Field checking was performed as part of the interpretation process, but no overall map accuracy or other quantitative measure of quality was generally produced. It was only after photo interpretation began to be used as the reference data to compare to digital remotely sensed data that issues concerning the accuracy of the photo interpretation arose.

This history of accuracy assessment can be effectively divided into four developmental stages or epochs. Initially, no real accuracy assessment was performed but rather an 'it looks good' mentality prevailed. This approach is typical of any new, emerging technology in which everything is changing rapidly and there is little time available for review. The second stage is called the epoch of non-site specific assessment. During this period, overall acreages were compared between ground estimates and the remote sensing derived map without regard for location, e.g. Meyer *et al.* (1975). The second epoch was relatively short-lived and quickly led to the age of site specific assessments. In a site specific assessment, actual places on the ground were compared to the same place on the map and a measure of overall accuracy (i.e. per cent correct) was presented. Finally, the fourth and current age of accuracy assessment may be called the age of the error matrix. This epoch includes a significant number of analysis techniques, most importantly the Kappa analysis. A review of the techniques and considerations of the error matrix age can be found in Congalton (1991) and in more detail in Congalton and Green (1999).

21.2. Sources of error

It is important to realize that some error is introduced into every component of a mapping project. Failure to consider these errors does not minimize their impact. A number of papers have been published in the literature with various representations of possible error. Congalton and Green (1993) looked specifically at the error or sources of confusion between the remotely sensed derived classification map and the reference data, including:

1   Registration differences between the reference data and the remotely sensed map classification.
2   Delineation error encountered when the sites chosen for accuracy assessment are manually digitized.
3   Data entry error when the reference data are entered into the accuracy assessment data base.
4   Error in interpretation and delineation of the reference data, e.g. photo interpretation error.
5   Changes in land cover between the date of the remotely sensed data and the date of the reference data (temporal error). For example, changes due to fires or urban development or harvesting.
6   Variation in classification and delineation of the reference data due to inconsistencies in human interpretation of heterogeneous vegetation.
7   Errors in the remotely sensed map classification.
8   Errors in the remotely sensed map delineation.

Lunetta *et al.* (1991) took a much broader approach and looked at a more inclusive list of errors and how they accumulate throughout a remotely sensed mapping project. The various sources and cumulative nature of these errors are summarized in Figure 21.1.

Given the many potential sources of error and the relative magnitude of each, it is critical that accuracy assessment be an integral component of any mapping project. Failure to incorporate the process into a mapping project dooms the map to misuse and misinterpretation, and the map user to potential trouble.

21.3. Map accuracy assessment considerations

Assessing the accuracy of any map generated from remotely sensed data includes many factors. These factors are highly correlated and are separated here only for purposes of being able to describe each one. In most cases, the issue is some kind of thematic accuracy. However, failure to consider radiometric and geometric corrections to the imagery can greatly influence the thematic accuracy. Therefore, before we address thematic accuracy, both radiometric and geometric accuracy are addressed. Methods of radiometric and geometric correction of remotely sensed data are summarized in Chapter 19.

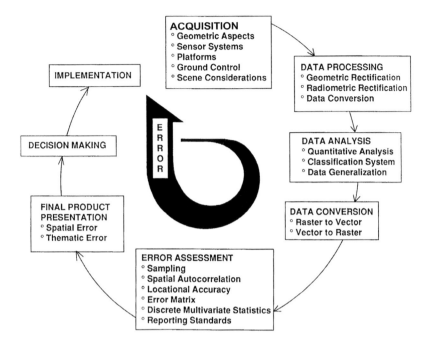

Figure 21.1 Sources of error in information extracted from remotely sensed data. Reproduced with permission of the American Society for Photogrammetry and Remote Sensing, from Lunetta, R. et al. (1991).

21.3.1. Radiometric accuracy

Radiometric accuracy is a measure of the effect of various factors on the Digital Number (DN) recorded by a digital remote sensing system. Remember that most digital remote sensor data contains 1 byte or 8 bits of information and therefore the DN ranges from 0 to 255, depending on the amount of light reflected in that particular band or wavelength by the object of interest. In addition, factors such as scene illumination, sensor characteristics, the atmosphere, and the viewing angle (bi-directional reflectance distribution function) can alter the DN values (Jensen 1996). When dealing with a single scene or when only relative differences are important, radiometric accuracy is less critical. When using multiple scenes to create an image mosaic or when comparing multiple dates for change detection, however, radiometric accuracy becomes very important.

21.3.2. Positional or geometric accuracy

Positional or geometric accuracy is a measure of how close the imagery fits the ground. In other words, it is the accuracy of the location of a point in the imagery with reference to its physical x, y, z location on the ground. It is imperative for any accuracy assessment that the same exact location can be determined both on the image and on the ground. The major factor influencing geometric accuracy is topography, while sensor characteristics and viewing angles can also have some effect. It is commonly accepted that a positional accuracy of one half a pixel is sufficient for sensors such as Landsat Thematic Mapper (15×15 m^2 panchromatic and 30×30 m^2 multispectral) and SPOT (10×10 m^2 panchromatic and 20×20 m^2 multispectral). As sensors increase in spatial resolution, such as the 1×1 m^2 panchromatic and 4×4 m^2 multi-spectral IKONOS data, positional accuracy increases in importance and new standards will need to be established.

Like radiometric accuracy, geometric accuracy is an important component of thematic accuracy. If an image is registered to the ground to within half a pixel and a Global Positioning System (GPS) unit is used to locate the place on the ground to within about 3–10 m, then it is impossible to use a single pixel as the sampling unit for assessing the thematic accuracy of the map. If positional accuracy is not up to the standard or GPS is not used to precisely locate the point on the ground, then these factors increase in importance and can significantly affect the thematic accuracy assessment.

21.4. Single-date thematic map accuracy assessment

Single-date thematic accuracy assessment refers to the accuracy of a mapped land cover category at a particular time compared to what was actually on the ground at that time. To perform a meaningful assessment of map accuracy, land cover classifications must be assessed with reference to data that are believed to be correct. Thus, it is vital to have at least some knowledge of the accuracy of the reference data before using them for comparison against the remotely sensed map. Congalton (1991) summarized as follows: 'Although no reference data set may be completely accurate, it is important that the reference data have high accuracy or else it is not a fair assessment. Therefore, it is critical that the ground or reference data collection be carefully considered in any accuracy assessment.'

21.4.1. The error matrix

Thematic map accuracy assessment begins with the generation of an error matrix (Table 21.1). An error matrix consists of a square array of numbers, or cells, set out in rows and columns, that expresses the number of sample units assigned to each land cover type as compared to what is on the ground. The columns in the matrix represent the reference data (actual land cover) and the rows represent assigned (remote sensing mapped) land cover types. The major diagonal of the matrix indicates agreement between the reference data and the interpreted land cover types. The process of generating an error matrix is summarized in Figure 21.2.

The error matrix is useful for both visualizing image classification results and, perhaps more importantly, for statistically measuring the results. Indeed, an error matrix is the only way to effectively compare two maps *quantitatively*. A measure of overall accuracy can be calculated by dividing the sum of all the entries in the major diagonal of the matrix by the total number of sample units in the matrix (Story and Congalton 1986). In the ideal situation, all the non-major diagonal elements of the error matrix would be zero, indicating that no area had been misclassified (Congalton *et al.* 1983). The error matrix also provides accuracies for each land cover category as well as both errors of exclusion (omission errors) and errors of inclusion (commission

Table 21.1 An example of an error matrix

		Reference Data					**Land Cover Categories**
		D	C	AG	SB	row total	
Classified Data	D	63	3	18	22	106	D = deciduous
	C	2	76	3	8	89	C = conifer
	AG	0	6	85	19	110	AG = agriculture
	SB	1	5	2	87	95	SB = shrub
	column total	66	90	108	136	400	

OVERALL ACCURACY = 311 / 400 = 78%

PRODUCER'S ACCURACY
D = 63 / 66 = 95%
C = 76 / 90 = 84%
AG = 85 / 108 = 79%
SB = 87 / 136 = 64%

USER'S ACCURACY
D = 63 / 106 = 59%
C = 76 / 89 = 85%
AG = 85 / 110 = 77%
SB = 87 / 95 = 92%

errors) present in the classification (Card 1982; Congalton 1991; Congalton and Green 1999).

Omission errors can be calculated by dividing the total number of correctly classified samples, or cells, in a category by the total number of samples in that category from the reference data (the column total) (Story and Congalton 1986; Congalton 1991). This measure of omission error is sometimes referred to as 'producer's accuracy,' because from this measurement the producer of the classification will know how well a certain area was classified (Congalton 1991). For example, the producer may be interested in knowing how many times 'deciduous' forest was in fact classified as deciduous (and not, say, conifer). To determine this, the 63 correctly classified deciduous samples (Table 21.1) would be divided by the total 66 units of deciduous from the reference data, for a producer's accuracy of 95 per cent. In other words, deciduous forest was correctly identified as deciduous 95 per cent of the time.

Commission errors, on the other hand, are calculated by dividing the number of correctly classified samples for a category by the total number of samples that were classified in that category (Story and Congalton 1986; Congalton 1991; Congalton and Green 1999). This is often referred to as the 'user's accuracy,' indicating for the user of the map the probability that a pixel classified on the map actually represents that category on the ground (Story and Congalton 1986; Congalton and Green 1999). So, while the producer's accuracy for the deciduous forest category is 95 per cent, the user's accuracy is only 59 per cent. That is, only 59 per cent of the areas mapped as deciduous are actually deciduous on the ground. However, because each omission from the correct category is a commission to the wrong category, it is critical that both producer's and user's accuracies are considered, since reporting only one value can be misleading (Story and Congalton 1986; Congalton and Green 1999).

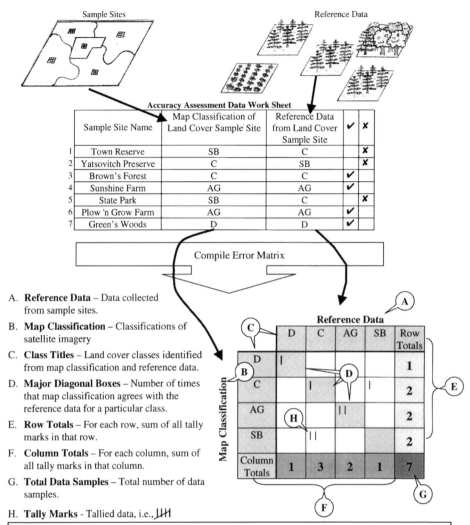

Figure 21.2 Diagram of the error matrix generation process.

An accuracy assessment based on an error matrix that was not properly generated can be meaningless, since the improperly generated error matrix may not be truly representative of the entire classification. To properly generate an error matrix, the following factors must be considered (Congalton 1991):

1   Ground data collection
2   Classification scheme
3   Sampling scheme (Congalton 1988b; Stehman 1992)
4   Spatial autocorrelation (Campbell 1981; Congalton 1988a)
5   Sample size and sample unit (van Genderen and Lock 1977; Congalton and Green 1999).

Congalton (1991) warned that failure to consider even one of these factors could lead to significant shortcomings in the accuracy assessment process.

Ground data collection. Ground data collection is the first step in any assessment procedure, and may be the single most important factor in accuracy assessment, since an assessment will be meaningless if the reference data cannot be trusted.

Ground reference data have been collected in many ways, including photo inter-pretation; aerial reconnaissance with a helicopter or airplane; videography; drive-by surveys; and visiting the area of interest on the ground (Congalton and Biging 1992). Ground visits, themselves, have ranged from visual inspection by walking through the area to detailed measurements of parameters such as species, size class, and crown closure (Congalton and Biging 1992). Despite these varied approaches, reference data have historically been assumed to be 100 per cent accurate, or at least substantially more accurate than the map (Card 1982). Congalton and Green (1993), however, have shown that differences between photo interpreted reference data and mapped data are often caused by factors *other* than map/classification error, including photo interpretation error. While photo interpretation remains a widely used means of collecting reference data, Congalton and Biging (1992) demonstrated that at least some ground data should be collected using field measurements. Indeed, they found that field measurements coupled with other visual estimates provided the most efficient ground reference data.

In addition to verifying a remotely sensed classification, discovering confusion between cover types, and perhaps improving the classification, reference data are also used initially to 'train' the classification algorithm. In the recent past, most assessments of remotely sensed maps were conducted using the same data set used to train the classifier (Congalton 1991). For example, in a supervised classification spectral characteristics of known areas were used to train the algorithm that then classified the rest of the image, and these known areas were also used to test the accuracy of the map. This training and testing on the same data set resulted in an improperly generated error matrix that clearly overestimated classification accuracy (Congalton 1991). In order for accuracy assessment procedures to be valid and truly representative of the classified map, data used to train the image processing system should not be used for accuracy assessment. These data sets must be inde-pendent.

Finally, the information used to assess the accuracy of remotely sensed maps should be of the same general vintage as those originally used in map classification. The greater the time period between the media used in map classification and that used in assessing map accuracy, the greater the likelihood that differences are due to change in vegetation (from harvesting, land use changes, etc.) rather than misclassification. Therefore, ground data collection should occur as close as possible to the date of the remotely sensed data.

Classification scheme. Classification schemes or systems categorize remotely sensed map information into a meaningful and useful format. The rules used to label the map must therefore be rigorous and well defined. One way to assure this precision is to define a classification system that is totally exhaustive, mutually exclusive, and hierarchical (Congalton and Green 1999).

A totally exhaustive classification scheme ensures that *everything* in the image fits into a category, i.e. nothing is left unclassified. A mutually exclusive classification scheme further assures that everything in the image fits into one *and only one* category, i.e. one object in an image can be labeled only once. Total exhaustion and mutual exclusivity rely on two critical components: (1) a set of labels (e.g. white pine forest, oak forest, non-forest, etc.), and (2) a set of rules (e.g. white pine forest must comprise at least 70 per cent of the stand). Without these components, the image classification would be arbitrary and inconsistent.

Finally, hierarchical classification schemes – those that can be collapsed from specific categories into more general categories – can be advantageous (Congalton and Green 1999). For example, if it is discovered that white pine, red pine, and hemlock forest cannot be reliably mapped, these three categories could be collapsed into one general category called coniferous forest.

Sampling scheme. An accuracy assessment very rarely involves a complete census of the classified image (i.e. every pixel in the image), since this is too large a data set to be practical (van Genderen and Lock 1977; Hay 1979; Stehman 1996). Creating an error matrix to evaluate the accuracy of a remotely sensed map therefore requires sampling to determine if the mapped categories agree with field-identified categories (Rosenfield *et al.* 1982).

In order to select an appropriate sampling scheme for accuracy assessment, some knowledge of the distribution of the vegetation/land cover classes should be known. For example, if there are cover types that comprise only a small area of the map, simple random sampling and systematic sampling may undersample this land cover class, or completely omit it. In this instance, simple random sampling or systematic sampling may create an error matrix that is not truly representative of the map. Stratified random sampling, by which a minimum number of samples is selected from each land cover category, may be more appropriate.

Ginevan (1979) suggested general guidelines for selecting a sampling scheme for accuracy assessment: (1) the sampling scheme should have a low probability of accepting a map of low accuracy; (2) the sampling scheme should have a high probability of accepting a map of high accuracy; and (3) the sampling scheme should require a minimum number of 'ground reference' samples.

Stratified random sampling has historically prevailed for assessing the accuracy of remotely sensed maps (van Genderen and Lock 1977; Jensen 1996). Stratified sampling has been shown to be useful for adequately sampling important minor categories, whereas simple random sampling or systematic sampling tended to oversample categories of high frequency and undersample categories of low frequency (van Genderen *et al.* 1978; Card 1982). Systematic sampling, however, has been less widely agreed upon. Cochran (1977) found that the properties of the mapped data greatly affected the performance of systematic sampling in relation to that of stratified or simple random sampling. He found that systematic sampling could be extremely precise for some data and much less precise than, say, simple random sampling for other data.

No one sampling method can be recommended globally for accuracy assessment. Rather, the best choice of sampling scheme requires a knowledge of the structure and distribution of characteristics in the map so that cover types will be sampled proportionately and a meaningful error matrix can be generated.

Spatial autocorrelation. Because of sensor resolution, landscape variability, and other factors, remotely sensed data are often spatially autocorrelated (Congalton 1988a). Spatial autocorrelation involves a dependency between neighboring pixels such that a certain quality or characteristic at one location has an effect on that same quality or characteristic at neighboring locations (Cliff and Ord 1973; Congalton 1988a). Spatial autocorrelation can affect the result of an accuracy assessment if an error in a certain location can be found to positively or negatively influence errors in surrounding locations.

Spatial autocorrelation is closely linked with sampling scheme. For example, if each sample carries some information about its neighborhood, that information may be duplicated in random sampling where some samples are inevitably close, subsequently violating the assumption of sample independence (Curran and Williamson 1986). For this reason, Curran and Williamson (1986) found that this duplication of information could be minimized by using systematic sampling, 'which ensures that neighboring sample points are as far from one another as is possible for a fixed sample size and site area.' However, spatial autocorrelation could also be responsible for periodicity in the data, which could dramatically affect the results of a systematic sample (Congalton and Green 1999). Indeed, Pugh and Congalton's (1997) spatial autocorrelation analysis indicated that systematic sampling should not be used when assessing error in Landsat TM data for New England. Therefore, while spatial autocorrelation may not be avoidable, its effects should be considered when selecting an appropriate sample size and sampling scheme.

Sample size. An appropriate sample size is essential in order to derive any meaningful estimates from the error matrix. In particular, small sample sizes can produce misleading results. That is, small samples that imply that there are no errors can be deceptive since the error-free result may have occurred by chance, when in fact a large portion of the classification was in error (van Genderen and Lock 1977). For this reason, sample sizes of at least 30 per category have been recommended (van Genderen *et al.* 1978), while others have concluded that fewer than 50 samples per category was not appropriate (Hay 1979). Sample sizes can be calculated from equations from the multinomial distribution, ensuring that a sample of appropriate size is obtained (Tortora 1978). In general, however, sample sizes of 50–100 for each cover type are recommended, so that each category can be assessed individually (Congalton and Green 1999).

In addition to determining appropriate sample size, an appropriate sample unit must be chosen. The sample unit may be a single pixel, a cluster of pixels, a polygon, or a cluster of polygons. For raster data, a single pixel has historically been a poor choice of sample unit (Congalton and Green 1999), since it is an arbitrary delineation of the land cover and may have little relation to the actual land cover delineation. Further, it is nearly impossible to align 1 pixel in an image to the exact same area in the reference data. A cluster of pixels (e.g. a 3 × 3 pixel square) is thus often a better choice for the sample unit, since it minimizes registration problems. A good rule of thumb is to choose a sample unit whose area most closely matches the minimum mapping unit of the reference data. For example, if the reference data have been collected in 2-hectare minimum mapping units, then an appropriate sample unit may be a 2-hectare polygon.

21.4.2. Single date classification map error matrix analysis techniques

Once an error matrix has been properly generated, it can be used as a starting point to calculate various measures of accuracy in addition to overall, producer's, and user's accuracy. For example, a discrete multivariate technique called Kappa (Bishop *et al.* 1975) can be used to statistically determine if: (1) the remotely sensed classification is better than a random classification and (2) two or more error matrices are significantly different from each other. Kappa calculates a KHAT value (Cohen 1960), which is an estimate of Kappa, as follows (Congalton and Mead 1983):

$$\mathrm{KHAT} = \frac{p_o - p_c}{1 - p} \tag{21.1}$$

where p_o is the actual agreement, or the overall proportion of the map correctly classified, and p_c is the 'chance' or random agreement. Actual agreement is the total number of correctly mapped samples (i.e. the summation of the major diagonal in the error matrix). Chance agreement is calculated by the summation over all categories of the proportion of samples in each category from the map data multiplied by the proportion of samples in each category from the reference data. In this way, it incorporates all off-diagonal cell values in the error matrix.

KHAT is thus a measure of the *actual* agreement of the cell values minus the *chance* (i.e. random) agreement (Congalton and Mead 1983; Rosenfield and Fitzpatrick-Lins 1986). The KHAT value can therefore be used to determine whether the results in the error matrix are significantly better than a random result (Congalton 1991). The KHAT value inherently includes more information than the overall accuracy measure since it indirectly incorporates the error (off-diagonal elements) from the error matrix (Rosenfield and Fitzpatrick-Lins 1986; Chuvieco and Congalton 1988).

In addition, confidence limits can be calculated for the KHAT statistic, which allows for an evaluation of significant differences between KHAT values (Aronoff 1982; Congalton and Green 1999). Because the Kappa analysis is based on the standard normal deviate and the fact that although remotely sensed data are discrete, the KHAT statistic is asymptotically normally distributed, confidence intervals can be calculated using the approximate large sample variance (Congalton and Green 1999). To test if a classification is better than a random classification, the Z test is performed, as follows:

$$Z = \frac{\mathrm{KHAT}}{\sqrt{\mathrm{var(KHAT)}}} \tag{21.2}$$

where Z is standardized and normally distributed. To test if two error matrices are significantly different from one another, the following Z test is used:

$$Z = \frac{|\mathrm{KHAT}_1 - \mathrm{KHAT}_2|}{\sqrt{\mathrm{var(KHAT}_1) + \mathrm{var(KHAT}_2)}} \tag{21.3}$$

where KHAT_1 represents the KHAT statistic from one error matrix, and KHAT_2 is the statistic from the other matrix. Kappa analysis thus brings to accuracy assessment the

Table 21.2 An example of a normalized error matrix

Reference Data

		D	C	AG	SB
Classified Data	D	0.84	0.02	0.09	0.06
	C	0.08	0.83	0.04	0.05
	AG	0.01	0.06	0.83	0.10
	SB	0.07	0.09	0.04	0.79

Land Cover Categories

D = deciduous

C = conifer

AG = agriculture

SB = shrub

NORMALIZED ACCURACY = (0.84 + 0.83 + 0.83 + 0.79) / 4 = 82%

power and efficiency of a parametric model with the ability to detect smaller differences than a non-parametric alternative (Rosenfield 1982). Further, kappa can be used to measure accuracy for individual categories by summing respective numerators and denominators of KHAT separately over all categories (Rosenfield and Fitzpatrick-Lins 1986; Congalton and Green 1999). This results in conditional agreement, or conditional Kappa.

Analysis of the error matrix can be taken yet another step further by normalizing the cell values. For example, in an iterative process called *margfit*, the matrix is normalized to one (Table 21.2). Because the cell values in each row and column in the matrix are forced to sum to one, each cell value becomes a proportion of one, which can easily be multiplied by 100 to obtain percentages. Consequently, there is no need for producer's and user's accuracy's since the cell values along the major diagonal represent the proportions correctly mapped. Congalton *et al.* (1983) argued that the normalized accuracy is a more inclusive measure of accuracy than either KHAT or overall accuracy because it *directly* includes the information in the off-diagonal element of the error matrix. Because each row and column sums to the same value, different cell values (e.g. different forest cover classes) within an error matrix and among different error matrices can be compared despite differences in sampling scheme (Aronoff 1982). Aronoff (1982) cautioned, however, that normalizing can tend to hide the data on sample size. This can be important, since, for example, an accuracy of four out of five sample points correct has less confidence than 40 out of 50 correct.

21.5. Change detection map thematic accuracy assessment

Change detection thematic accuracy assessment presents even more difficulties and challenges than single-date thematic accuracy assessment. Many questions arise such as how does one obtain information on reference data for images that were taken in the past? Likewise, how can one sample enough areas that will change in the future to have a statistically valid assessment? Until recently, most studies on change detection did not present any quantitative results with their work.

Table 21.3 A comparison between a single-date error matrix and a change detection error matrix for the same land cover categories

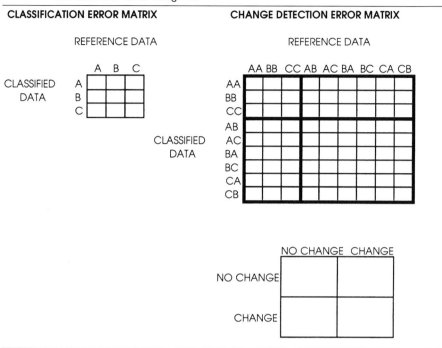

Table 21.3 presents a comparison between a single-date error matrix and the corresponding error matrix for change detection (Congalton and Green 1999). Table 21.3 shows a single date error matrix for three vegetation/land cover categories (A, B, and C). The matrix is of dimension 3 × 3. As we saw in the previous section, one axis of the matrix represents the three categories as derived from the remotely sensed classification, the other axis shows the three categories identified from the reference data, and the major diagonal of this matrix indicates correct classification.

Table 21.3 also shows a change detection error matrix generated for the same three vegetation/land cover categories (A, B, and C). Note, however, that the matrix is no longer of dimension 3 × 3 but rather 9 × 9. This is because we are no longer looking at a single classification but rather a change between two different classifications generated at different times. Note that in a typical single-date error matrix there is only one row and column for each map category. However, in change detection analysis the error matrix is the size of the number of categories squared. Therefore, the question of interest is, 'What category was this area at time one and what is it at time two?'. The answer has nine possible outcomes for each dimension of the matrix (A at time one and A at time two, A at time one and B at time two, A at time one and C at time two,..., C at time one and C at time two) all of which are indicated in the error matrix. It is then important to note what the remotely sensed data said about the change and compare it to what the reference data indicate. This comparison uses the exact same logic as for the single classification error matrix, it is just complicated by the two time periods (i.e. the change). Again, the major diagonal indicates correct classification while the off-diagonal elements indicate the errors or confusion.

The change detection error matrix can also be simplified into a no-change/change error matrix. The no-change/change error matrix can be formulated by summing the cells in the four appropriate sections of the change detection error matrix (Table 21.3). For example, to get the number of areas that both the classification and reference data correctly determined that no change had occurred between two dates, you would simply add together all the areas in the upper left box (the areas that did not change in either the classification or reference data). You would proceed to the upper right box to find the areas that the classification detected no change and the reference data considered change. From this no-change/change error matrix, analysts can easily determine if a low accuracy was due to a poor change detection technique, misclassification, or both.

Recently, National Oceanic and Atmospheric Administration (NOAA); as part of its Coastal Change Analysis Program, funded a task force that investigated and proposed methods for assessing the accuracy of change detection products generated by analysis of remote sensor data. This task force wrote a monograph to recommend appropriate techniques and suggest future work (Khorram *et al.* 1999). The change detection error matrix (Table 21.3) was recommended and sampling techniques using unequal probability sampling were developed. One of the biggest problems with conducting a change detection accuracy assessment is to insure sampling in the areas that have actually changed, since change is typically such a rare event (i.e. most of the mapped area has not changed between the two dates). Another issue is finding accurate reference data for some date in the past so that the map could be assessed for that date. The interested reader should review the monograph which discusses many of these complex issues.

21.6. Conclusions

Remotely sensed data are used to make decisions that may have a global impact – decisions regarding land use/tenure, climate change effects, resource treatments, water quality, ecosystem health considerations, wildlife habitat, sustainability, and a multi-tude of other applications and issues. In order to make any effective, intelligent de-cisions, however, the data must be accurate and reliable. For this reason, accuracy assessment will continue to be an integral component of any map generated from remotely sensed data.

References

Aronoff, S., 'Classification accuracy: a user approach', *Photogrammetric Engineering and Remote Sensing*, 48(8), 1982, 1299–1307.

Bishop, Y., Fienberg, S. and Holland, P., *Discrete Multivariate Analysis – Theory and Practice*, MIT Press, Cambridge, MA, 1975, 575 pp.

Campbell, J., 'Spatial autocorrelation effects upon the accuracy of supervised classification of land cover', *Photogrammetric Engineering and Remote Sensing*, 47, 1981, 555–563.

Card, D., 'Using known map category marginal frequencies to improve estimates of thematic map accuracy', *Photogrammetric Engineering and Remote Sensing*, 48(3), 1982, 431–439.

Chuvieco, E. and Congalton, R. G., 'Using cluster analysis to improve the selection of training statistics in classifying remotely sensed data', *Photogrammetric Engineering and Remote Sensing*, 54(9), 1988, 1275–1281.

Cliff, A. D. and Ord, J. K., *Spatial Autocorrelation*, Pion Limited, London, 1973, 178 pp.

Cochran, W. G., *Sampling Techniques* (3rd ed.), John Wiley and Sons, New York, 1977, 428 pp.

Cohen, J., 'A coefficient of agreement for nominal scale', *Educational and Psychological Measurement*, 20, 1960, 37–46.

Congalton, R. G., 'Using spatial autocorrelation analysis to explore errors in maps generated from remotely sensed data', *Photogrammetric Engineering and Remote Sensing*, 54, 1988a, 587–592.

Congalton, R. G., 'A comparison of sampling schemes used in generating error matrices for assessing the accuracy of maps generated from remotely sensed data', *Photogrammetric Engineering and Remote Sensing*, 54, 1988b, 593–600.

Congalton, R. G., 'A review of assessing the accuracy of classifications of remotely sensed data', *Remote Sensing of Environment*, 37, 1991, 35–46.

Congalton, R. G. and Biging, G. S., 'A pilot study evaluating ground reference data collection efforts for use in forest inventory', *Photogrammetric Engineering and Remote Sensing*, 58(12), 1992, 1669–1671.

Congalton, R. and Green, K., 'A practical look at the sources of confusion in error matrix generation', *Photogrammetric Engineering and Remote Sensing*, 59, 1993, 641–644.

Congalton, R. G. and Green, K., *Assessing the Accuracy of Remotely Sensed Data: Principles and Practices*, Lewis Publishers, Boca Raton, FL, 1999, 137 pp.

Congalton, R. G. and Mead, R. A., 'A quantitative method to test for consistency and correctness in photo-interpretation', *Photogrammetric Engineering and Remote Sensing*, 49, 1983, 69–74.

Congalton, R. G., Oderwald, R. G. and Mead, R. A., 'Assessing Landsat classification accuracy using discrete multivariate statistical techniques', *Photogrammetric Engineering and Remote Sensing*, 49, 1983, 1671–1678.

Curran, P. J. and Williamson, H. D., 'Sample size for ground and remotely sensed data', *Remote Sensing of Environment*, 20, 1986, 31–41.

van Genderen, J. L. and Lock, B. F., 'Testing land use map accuracy', *Photogrammetric Engineering and Remote Sensing*, 43, 1977, 1135–1137.

van Genderen, J. L, Lock, B. F. and Vass, P. A., 'Remote sensing: statistical testing of thematic map accuracy', *Remote Sensing of Environment*, 7, 1978, 3–14.

Ginevan, M. E., 'Testing land-use map accuracy: another look', *Photogrammetric Engineering and Remote Sensing*, 45, 1979, 1371–1377.

Gopal, S. and Woodcock, C., 'Theory and methods for accuracy assessment of thematic maps using fuzzy sets', *Photogrammetric Engineering and Remote Sensing*, 60, 1994, 181–188.

Hay, A. M., 'Sampling designs to test land-use map accuracy', *Photogrammetric Engineering and Remote Sensing*, 45, 1979, 529–533.

Jensen, J. R., *Introductory Digital Image Processing: a Remote Sensing Perspective*, Prentice Hall, Upper Saddle River, NJ, 1996, 318 pp.

Khorram, S., Biging, G., Chrisman, N., Colby, D., Congalton, R., Dobson, J., Ferguson, R., Goochild, M., Jensen, J. and Mace, T., *Accuracy Assessment of Remote Sensing-Derived Change Detection*, American Society for Photogrammetry and Remote Sensing, Bethesda, MD, 1999, 64 pp.

Lunetta, R., Congalton, R., Fenstermaker, L., Jensen, J., McGwire, K. and Tinney, L., 'Remote sensing and geographic information system data integration: error sources and research issues', *Photogrammetric Engineering and Remote Sensing*, 57(6), 1991, 677–687.

Meyer, M., Brass, J., Gerbig, B. and Batson, F., 'ERTS data applications to surface resource surveys of potential coal production lands in southeast Montana', *IARSL Research Report 75-1 Final Report*, University of Minnesota, 1975, 24 pp.

Pugh, S. and Congalton, R. G., 'Applying spatial autocorrelation analysis to evaluate error in New England forest cover type maps derived from Thematic Mapper data', *Proceedings of the 63rd ASPRS Annual Meeting*, Vol. 3, American Society of Photogrammetry and Remote Sensing, Seattle, WA, 1997, pp. 648–657.

Rosenfield, G. H., 'The analysis of areal data in thematic mapping experiments', *Photogrammetric Engineering and Remote Sensing*, 48(9), 1982, 1455–1462.

Rosenfield, G. and Fitzpatrick-Lins, K., 'A coefficient of agreement as a measure of thematic classification accuracy', *Photogrammetric Engineering and Remote Sensing*, 52, 1986, 223–227.

Rosenfield, G. H., Fitzpatrick-Lins, K. and Ling, H. 'Sampling for thematic map accuracy testing', *Photogrammetric Engineering and Remote Sensing*, 48, 1982, 131–137.

Stehman, S., 'Comparison of systematic and random sampling for estimating the accuracy of maps generated from remotely sensed data', *Photogrammetric Engineering and Remote Sensing*, 58, 1992, 1343–1350.

Stehman, S., 'Estimating the kappa coefficient and its variance under stratified random sampling', *Photogrammetric Engineering and Remote Sensing*, 62(4), 1996, 401-402.

Story, M. and Congalton, R. G., 'Accuracy assessment: a user's perspective', *Photogrammetric Engineering and Remote Sensing*, 52, 1986, 397–399.

Tortora, R., 'A note on sample size estimation for multinomial populations', *The American Statistician*, 32, 1978, 100–102.

Chapter 22

Selected examples of remote sensing projects

Donald C. Rundquist, John R. Jensen, Maurice Nyquist and Thomas W. Owens

Remote sensing technology is used routinely in numerous practical applications. This chapter provides four examples of county-wide or regional remote sensing data acquisition and analysis projects, including:

1   Agricultural crop classification using Landsat Thematic Mapper (TM) imagery;
2   Natural vegetation mapping using aerial photography;
3   Mapping and assessing the water quality of lakes, reservoirs and ponds using Landsat TM imagery, and
4   Updating tax maps using aerial photography and soft-copy photogrammetry techniques.

22.1. Agricultural crop classification using satellite remote sensing

Resource managers or agriculturists must often record and/or map the distribution of crop types within a certain study area, for example, one county or a group of counties. Remote sensing may be used for general crop inventory and analysis. What are the keys to identifying cropland by means of remotely sensed data? Experienced air-photo interpreters know that the color of vegetation, the size and shape of individual fields, the spatial resolution of the photography and the texture of crop canopies can all be important considerations when making decisions about the type of crop growing in a particular location. The use of multi-date images and knowledge of crop calendars is also very important. The following discussion focuses on the spectral and temporal components as they relate to crop type discrimination, which is achieved by digital processing of multispectral satellite data.

22.1.1. Spectral properties of vegetation

A simple way to begin a discussion of the spectral properties of growing vegetation is to consider the colors as we see them with the human eye. We know that tree leaves are green in summer and (at least some) are yellow in fall. Of course, the reason vegetation is green in summer is that the leaves, in which chlorophyll is the principal photosynthetic pigment, reflect some energy at green wavelengths. In fall after a frost, the chlorophyll and the photosynthetic process are diminished, and other pigments, for example the carotenoids, result in often spectacular displays of fall color. Our experience shows that this green-to-yellow (or green-to-red) color pattern for tree leaves is generally repeatable and true. Thus, these spectral signatures, as developed by means of our own remote sensors (our eyes), seem quite logical to us.

At a more conceptual level, we can state that when solar radiation reaches a field containing live, photosynthetically active vegetation, predictable interactions occur. In general, all vegetation tends to react to sunlight in a similar fashion, although some specific differences may be caused by variations in plant morphology and physiology, local soil type, and the climate of the growing area. Known biophysical interactions include the fact that most incoming visible light is absorbed by the chloroplasts (i.e. only a small amount of incoming light is reflected from the outer surface of plant leaves), while a greater amount of near-infrared energy is transmitted into the spongy mesophyll tissue in the leaf where the light rays are reflected back by the cell walls toward the light source. The spectral character of the interactions varies according to the phenology of the plant (i.e. the relation between climate and periodic biological phenomena, such as flowering and fruiting). As a typical plant begins its life cycle and is exposed to sunlight, the amount of chlorophyll in the leaf increases for a time. Blue, green, and red light tends to be absorbed by chlorophyll, while a small amount of green light is reflected.

To elaborate upon and illustrate the typical spectral signature for healthy, actively photosynthesizing vegetation, we can make use of data collected by a spectroradiometer. Spectroradiometers acquire data in the visible spectral region, as well as other invisible regions, such as the near and/or middle infrared. Basically, a spectroradiometer provides the user with both the intensity (or magnitude) of reflectance from the target and the spectral distribution of the energy being reflected. Collecting data involves simply pointing the instrument at a target. The scanning is activated and controlled by an attached microcomputer, which also stores the gathered spectral data.

The principal output product from a spectroradiometer is not an image, but rather numerical data summarizing reflectance acquired in different spectral channels (bands). These responses can be used to compile a graph (or spectral profile). The profiles generated from numerical spectroradiometer data are generally displayed as a simple plot with wavelength on the x-axis and reflectivity on the y-axis. However, in order to be able to construct a near-continuous curve, one needs many individual data points. The most meaningful spectral response profiles are constructed with data collected by instruments that are more than multispectral, in fact, they are called 'hyperspectral.' Such an instrument has high spectral resolution, i.e. more than a hundred spectral channels (or bands), each of which constitutes a very small part of the total electromagnetic spectrum (as little as 1 nanometer (nm)). The 100+ bands may include parts of two or three spectral regions, for example, the ultraviolet, the visible, and/or the near-infrared. The plotted spectral curves are often considered diagnostic because the distinct reflection and/or absorption maxima and minima provide useful information to the skilled analyst.

In Figure 22.1, consider the curve's characteristics in light of the previous discussion. For example, compare the reflectance from the vegetation in the visible region of the spectrum (400–700 nm) with those in the near-infrared (700–1,300 nm). In the visible portion of the spectrum, most of the incoming solar is not reflected, it is absorbed (even in the green range!). Note that the reflectance in the visible region is only 10–15 per cent. For dense canopies of growing

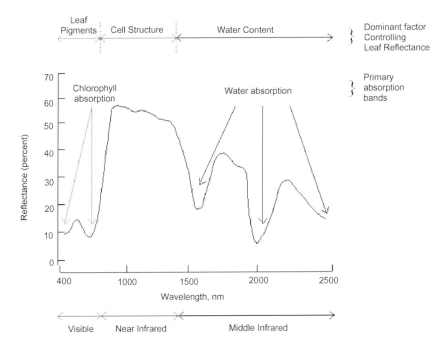

Figure 22.1   Typical spectral profile for healthy, actively photosynthesizing vegetation (adapted from Jensen 1983; Gausman 1985).

vegetation, visible reflectance can be even lower than 10 per cent. In the near infrared region, on the other hand, reflectance is approximately 55 per cent, and the highest levels of reflectance (i.e. lowest levels of absorption) occur in the (invisible) near-infrared spectral region.

The curve in Figure 22.1 contains prominent absorption features in the blue and red spectral regions. All living green vegetation makes use of blue and red light as part of the photosynthetic process, and reflects some green (which is obvious because we see living vegetation in a green color). Thus, in the visible region, it is the pigments that control leaf reflectance. The factors controlling reflectance from leaves in the near infrared and middle infrared regions are cell structure and leaf-water content, respectively. The major spectral location for assessing the water content in leaves is at 1,650 nm.

22.1.2. Digital crop classification

Dawson County, Nebraska, was chosen to demonstrate how satellite data could be used to classify and map the crops growing in a large Midwestern county. The 1997 growing season was chosen due to the availability of quality ground reference data (field maps with crop types) acquired by a federal agency.

The Landsat TM remote sensing system was selected for classifying and mapping the crop types. The 30 × 30 m spatial resolution of the Landsat TM was satisfactory for mapping at the county level. The spectral characteristics of the TM are as follows:

Band 1 = blue; Band 2 = green; Band 3 = red; Band 4 = near-infrared; Band 5 = middle-infrared; Band 6 = thermal-infrared; and Band 7 = middle-infrared (refer to Chapter 18 for wavelength intervals).

The goal was to identify and map nine crop types plus three other land-cover classes, including: (1) corn; (2) soybeans; (3) sorghum; (4) alfalfa (5) small grains (oats, wheat); (6) range, pasture, grass; (7) urban; (8) open water; (9) riparian forest and woodlands; (10) wetlands and subirrigated grasslands; (11) other agricultural land (farmsteads, feedlots, etc.) and (12) summer fallow. Crop calendar information suggested that from mid-May to September would be optimal for acquiring TM imagery. Therefore, TM images were acquired over Dawson County on 4 May, 8 August, and 25 September. The three images were georeferenced, co-registered, and the brightness values were converted to per cent reflectance in preparation for the digital classification.

A false-color composite of the 8 August 1997 image of Dawson County is shown in Figure 22.2 (see plate 15). Four cities, Gothenburg, Cozad, Lexington, and Overton, can be seen. The Platte River is clearly shown, as are the crops (red tones) growing in the heavily irrigated river valley. Interstate 80 is immediately adjacent to the river, and Highway 30 connects the towns. Notice the sharp break between the lowland and the upland areas, and also the actively growing riparian vegetation (red tones) along the adjacent tributary streams.

Remember from the earlier discussion concerning the spectral properties of growing vegetation that the visible region (blue, green, and red) provides information about pigments in the leaves, the near infrared region provides information about cell structure, and the middle infrared region provides information about leaf water content. For classifying crop types, one obvious procedural choice is to use all the TM bands as input to the classification algorithm, but it is unlikely that the thermal infrared band, which provides data on the temperature of the vegetation and land surfaces, will prove useful. Another consideration is that the spatial resolution of TM band 6 on Landsat-4 and -5 has a spatial resolution of 120 m, making the registration and overlay process more difficult. Also, it is well known that blue light is susceptible to intense atmospheric attenuation and scattering. Consequently, TM band 1 will probably contribute little to the final classification. Therefore, TM bands 2, 3, 4, 5 and 7 may be useful in the classification.

Another procedural option is to convert the raw data into a vegetation index, such as the Normalized Difference Vegetation Index (NDVI), for each image date. In this calculation, only the near infrared and red band data are used to compute a new image that provides useful information about the density and vigor of vegetation. If a multi-date approach is to be used, and this is generally very important for analyzing agricultural crops using remote sensing, then an NDVI image from each date (i.e. a 'spectral-temporal' dataset consisting of three derived NDVI images) might be input to the classification software (Jensen 1996).

The strategy for the crop classification example was to exploit not only spectral responses in the visible, near infrared, and middle infrared portions of the electromagnetic spectrum, but also the multi-date aspect of the data. Therefore, five spectral bands for each of the three image-acquisition dates were combined to form a 15-channel spectral-temporal dataset. The blue and thermal-infrared bands were omitted from the digital analysis for the reasons discussed above.

Because an abundance of useful and reliable ground reference data were available for the 1997 growing season, which allows for training fields to be accurately delineated, a decision was made to use a supervised classification strategy (see Chapter 19 for details of this procedure). A statistical-sampling strategy led to the selection of training fields for each of the 12 land cover classes. After the locations of the training fields and the statistics derived from them were input to the classification algorithm, the process was begun. The result of the digital classification is shown in Figure 22.3 (see plate 16). It is apparent that the bulk of the cropland in Dawson County is corn and soybeans, with a significant amount of alfalfa. It is relatively easy to calculate the actual area planted to each crop (by simply counting pixels). There were 219,741 acres of corn, 66,162 acres of alfalfa, and 23,823 acres of soybeans grown during 1997 in Dawson County.

Accuracy assessment is an important consideration with any digital classification of remote sensing image data (Chapter 21 has details). The classification accuracy of this map was calculated based on a point-by-point comparison between the remote sensing derived crop classification and the actual crop as determined by the ground reference data. The overall crop classification map accuracy of Dawson County was 85.49 per cent correct.

Classifying crop types using satellite sensor data is sometimes relatively easy, and sometimes difficult. The level of difficulty varies from place to place and depends, in large part on the crop mix of the area under investigation. For example, crop identification and mapping using remote sensing in central Iowa, where corn is grown to the near-exclusion of other crops, is generally less difficult than crop classification in the Central Valley of California where diverse types of specialty crops and truck farming are prevalent. In addition, the more crops that one is trying to classify by means of digital processing of spectral data, the more difficult it is to obtain satellite data at precisely the right times of the year to allow maximum discrimination. Remote sensing of crop types can become very difficult when one is dealing with both irrigated and dryland (non-irrigated) versions of the same crop. For example, in parts of the Central Plains, one often finds fields of irrigated corn interspersed with fields of dryland corn, and irrigated soybeans interspersed with dryland soybeans. The problem, from the standpoint of remote sensing is that two fields, one irrigated and one dryland, may have been planted on the same day, but the amount of water (or lack of) used by the plants in the two fields may vary greatly. This, of course, can lead to differences in growth rates or levels of stress, which often causes different spectral signatures for the same crop.

Yet another problem in crop classification is that two different crops may exhibit virtually the same spectral signature. For example, an observer standing in the field sometimes mistakes corn and sorghum. Therefore, it is not surprising that the spectral differences are minimal. Varietal differences can lead to phenological variations among one crop, which in turn, can lead to digital classification problems. Alfalfa and similar plants cut for hay present special problems because several crops are often harvested from a single field in one growing season. Therefore, the image analyst may find two adjacent fields of alfalfa, one with a full canopy and one with no canopy because the latter was, by chance, cut the day prior to a satellite overpass.

In summary, it is possible to conduct accurate agricultural crop type inventories if: (1) accurate ground reference information is available to train the classifier and assess the accuracy of the remote sensing derived classification map, and (2) the remotely sensed data are collected at optimum times in the phenological cycle of the respective crops, the imagery are processed correctly, and an error evaluation is performed.

22.2. Vegetation mapping program using aerial photography

Agencies such as the National Park Service, the US Forest Service, the Bureau of Land Management, and state and local organizations have land and resource management responsibilities. They are required to manage their lands to meet national and agency mandates. Managers require information about their resources to understand and properly manage their lands. These resources include (but are not limited to) vegetation, soils, geology, water, wildlife, and air. It is critical that resource managers have a complete inventory and description of each theme to make informed decisions about resource issues.

For example, consider the giant sequoia. This tree is the largest living organism on the planet and one of the oldest living things on earth with individuals over 2,700 years old. The largest groves of this majestic tree are in Sequoia-Kings Canyon National Park in California. Ecologists have determined that sequoias have not been reproducing at a rate high enough to sustain the species. They found that sequoias require light burns in the understory to properly prepare the forest floor for regeneration. This creates the need for prescribed fire burns in the national park. To properly run a prescribed fire program to sustain the regeneration of sequoias, park managers need to know where sequoias are located, the age of the trees, the stand structure, what vegetation is in the understory, and a whole host of other information. To meet information needs for National Park managers for this issue and other complex natural resource issues, the US Geological Survey and the National Park Service (USGS-NPS) Vegetation Mapping Program was created.

The mission of the cooperative program is to classify, describe, and map vegetation communities of all National Park units that have a natural resource component. In addition, the NPS manages many units that have primarily cultural and historical interest, such as the Washington Monument and the Statue of Liberty, which are not included in the program. For detailed information on the program, visit the web site at http://biology.usgs.gov/npsveg.

The USGS-NPS Vegetation Mapping Program is a high priority requirement of the NPS Inventory and Monitoring (I&M) Program and is managed by the USGS Center for Biological Informatics. The Vegetation Mapping Program conducts long-term vegetation monitoring with many short-term, immediate applications for park managers. Examples of immediate applications are monitoring of invasive plants, development of wildlife habitat databases, fire management, scenic management, and visitor management (e.g. control of park visitors walking across fragile alpine tundra).

This program is noteworthy because it attempts to classify and map vegetation for the Park Service on a local basis, while creating standard products that have common quality attributes on a national scale. It meets the very real needs of the Park Service with scientifically valid, peer-reviewed methods. The program uses an international vegetation classification system – the National Vegetation Classification System (NVCS) – and uses standardized inventory and mapping protocols that are adapted to local conditions. The program creates products that meet the Federal Geographic

Data Committee (FGDC) standards for metadata, data transfer (the Spatial Data Transfer Standard (SDTS)), and classification (the NVCS). The program also meets USGS National Map Accuracy standards for data at a scale of 1:12,000 (a well-defined object on a map must be within 33.3 ft of its actual location). The program's classification accuracy standard is that each class must be at least 80 per cent accurate at the 90 per cent confidence level. There is a minimum mapping unit of 0.5 ha, although a smaller minimum mapping unit may be used in special situations.

The NVCS was originally developed for the United Nations Educational, Scientific, and Cultural Organization (UNESCO) as an international standard in the 1970s. The Nature Conservancy has further refined and developed the NVCS, in partnership with the USGS, the FGDC, and the Ecological Society of America. The NVCS is based on standard field and data analysis methods. The classification standard classifies existing biological associations that repeat across the landscape and are mappable from imagery. The NVCS is hierarchically organized so it can be applied at multiple scales and identifies classification units that are appropriately scaled to meet objectives for biodiversity conservation, as well as resource and ecosystem management needs. The NVCS is flexible and open ended and allows additions, modifications and continuous refinement, is well documented, and can be cross-walked between other frequently used systems.

The NVCS is organized into two main levels – physiognomy at higher levels and floristics at lower levels – to classify vegetation communities (Table 22.1). The primary emphasis in the classification system at the park level is with floristics. Often alliances and associations are not defined locally and extensive fieldwork is required. The floristic level is most useful to resource managers because it provides them with specific information on the vegetation communities in their park. This information is critical to resource issues, such as wildlife management, protection of rare species, fire management, and invasive exotic management. For detailed information on the NVCS system, visit the website at http://www.tnc.org and look under 'Conservation Science.' An example of a typical output product for a national park is shown in Figure 22.4.

Table 22.1 Organization of the NVCS (source: USGS-NPS vegetation mapping program 1994a; Grossman *et al.* 1998a,b)

	Example
Physiognomy	
Division/order	Tree dominant (dominant life form)
Class	Woodland (spacing and height of dominant form)
Subclass	Evergreen woodland (morphological and phenological similarity)
Group	Temperate evergreen needle-leafed (climate, latitude, growth form, leaf form)
Subgroup	Natural and semi-natural vs. cultivated communities
Formation	Evergreen needle-leafed woodland with rounded crowns (mappable units)
Floristics	
Alliance (cover type)	Douglas fir woodland (dominant species)
Association (community)	Douglas fir/snowberry woodland (subdominant or associated species)

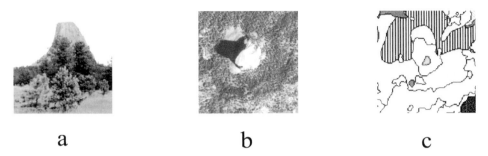

a   b   c

Figure 22.4   Project elements from classifying and mapping the vegetation communities at Devils Tower National Monument. (a) Ground photo of a ponderosa pine vegetation community with Devils Tower in background. (b) Aerial photo of Devils Tower with surrounding ponderosa pine communities. (c) Final map of Devils Tower derived from aerial photography with vegetation communities delineated.

22.2.1. In situ measurement

Classification of the vegetation communities of each National Park requires extensive field work. At least three *in situ* plots per vegetation association are identified in the park to sample possible variation in the communities. Plots vary in size and shape, but are commonly square with sides of 20 m for forest sampling and 10 m for shrubland and herbaceous sampling (Figure 22.5). The plot sampling effort uses the 'Gradsect' (*Grad*ient Directed Tran*sect*) method that was developed by Gillison and Brewer (1985). This method identifies major environmental gradients in an area that affect vegetation composition, such as elevation, aspect, soils, or hydrography. These factors are classified into significant integrated classes through geospatial analysis. Areas with high diversity of these classes are likely to contain high vegetation diversity and

Figure 22.5   Field crew collecting vegetation data in a 10 × 10 m plot at Agate Fossil Beds National Monument.

placing samples in these areas is an efficient way to sample diverse vegetation communities within a relatively small area.

When field data are collected, The Nature Conservancy places them in the PLOTS database, which is maintained as a national database. Experienced ecologists use quantitative and qualitative analysis to classify the vegetation into communities. Once the communities are classified, a dichotomous field key is produced so that anyone who is generally familiar with the plants of an area can place the vegetation into communities. An extensive description is developed for each class that describes many community characteristics, including which species are found in the community, the environmental conditions, references pertaining to further characterization and distribution of this community, its conservation status, and whether it was previously undescribed (USGS-NPS Vegetation Mapping Program 1994b).

22.2.2. Interpretation of aerial photography

Once the vegetation communities of interest are determined, aerial photography is used to map their geographic location. In addition to mapping the vegetation communities, plant height, density, and pattern are also identified for each vegetation polygon mapped from the aerial photography. This structural information is valuable to resource managers for fire management or wildlife habitat assessment.

Aerial photography is used because it is still the best remote sensing tool to reliably identify and map complex vegetation communities. The resolution of the photography is much higher than other imagery, the shape of individual trees and shrubs can be discerned, and the texture and pattern of communities can also be discerned. Experienced aerial photo interpreters identify the communities using stereoscopes by considering color, texture, pattern, size, shape, and placement in the landscape. An experienced photo interpreter is often the best image processing system available, because he or she can interpret the complex and often contradictory modeling algorithms necessary to properly identify vegetation communities and still maintain a high level of accuracy and consistency (Figure 22.6).

The photo interpreter also performs extensive ground reconnaissance and confers closely with plant ecologists to understand the vegetation communities. Occasionally it is not possible to identify all the associations from aerial photographs, because the plants that define those associations may be in the understory of a forest. For example, in Devils Tower National Monument in Wyoming, there are five ponderosa pine associations that are defined by understory plants. It is only possible to reliably identify two ponderosa pine map classes on the aerial photography; one map class contains three associations and the other map class contains two associations.

The map class polygons are drawn on overlays registered to the aerial photographs. As stated above, the minimum mapping unit is 0.5 ha (a little more than 1 acre) although important vegetation types smaller than 0.5 ha (e.g. wetland or riparian zones) are also mapped. Park managers may also request to have additional features mapped that are not part of the national standards, such as prairie dog towns, beaver houses, islands smaller than 0.5 ha, or other features. These added features are also identified and delineated on the overlays.

A quality control/quality assurance step follows the completion of the air photo interpretation. The interpreters revisit the field to ensure that the mapping was correct

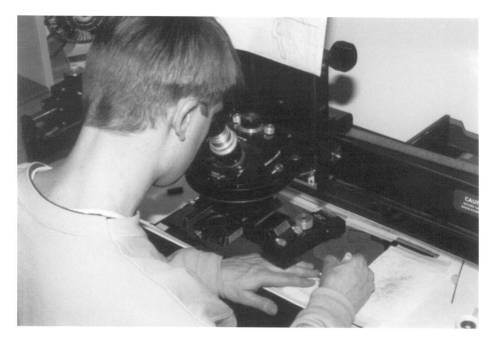

Figure 22.6 Aerial photo interpretation using a prism stereoscope.

and complete. A technical review of the interpretation methodology is also completed to ensure that polygons are closed and properly labeled.

The photo interpreted delineations are rectified and registered to spatial coordinates by several different methods. One method involves the use of a zoom transfer scope that allows the operator to visually superimpose features on the delineated photos with features on ortho images. The data are traced onto an overlay registered to the ortho images and subsequently scanned and placed into a spatial database (Figure 22.7). Another method used 'heads-up' on-screen digitizing in which the delineated features on the photos are visually traced using a computer screen that has an ortho image as a background. A third method uses rubber sheet rectification. In this method the photo overlays are scanned and registration points found on the overlays. The base ortho image is matched with the scanned overlays through a rubber sheet rectification process.

22.2.3. Accuracy assessment

Once the mapped vegetation data are placed into a digital spatial database, an accuracy assessment is performed. For common classes at least 30 points are located for accuracy assessment. For less common classes, fewer points are required. The points are randomly placed in the polygons. An independent field crew travels to the points using Global Positioning System (GPS) receivers for navigation. The crew has a map showing the polygons overlain on a digital ortho image along with the accuracy assessment points, which allows them to identify where they are on the image and where they are in relation to the polygons. They do not know the polygon's classific-

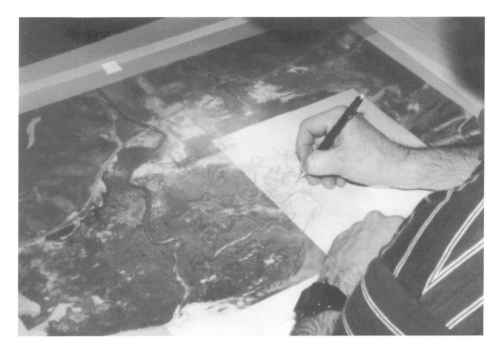

Figure 22.7 Transfer of delineated aerial photo polygonal data to base map for rectification.

ation. They classify the vegetation in the area around the point using the dichotomous field key. These field data are automated and compared to the map data in a contingency table that allows the analysis of the errors of omission and commission. Each map class must be at least 80 per cent accurate at the 90 per cent confidence level. The accuracy assessment process also provides an excellent opportunity to test the validity of the field key (USGS-NPS Vegetation Mapping Program 1996).

When the accuracy assessment is complete, final products are delivered to the park managers. For access to park datasets, visit the program's website at http://biology.usgs.gov/npsveg/products/parkname.html. A partial list of products is included in Table 22.2.

22.2.4. Uses of remote sensing-derived vegetation information

Park managers may use the data in many different ways. An interesting example is at Scotts Bluff National Monument in western Nebraska in the US. The park had recently acquired some property that was formerly a golf course and wanted to restore the area to a more natural state – short grass prairie. The park natural resource manager used the species composition information from the community descriptions as a recipe to create a seed mixture. The area was sown with this seed mixture and the area is on its way to becoming a short grass prairie.

Another example of using the classification system is at Assateague National Seashore in Maryland. The park ecologist was recently transferred from an eastern park further inland and was not familiar with barrier island vegetation. Information

Table 22.2 A partial list of vegetation mapping products delivered to national park managers

Classification
  Methods report
  Vegetation classification
  Vegetation community descriptions
  Dichotomous vegetation field key
  Ground photos of vegetation classes
Aerial photography
  Photos
  Interpreted overlays
  Aerial photo index
Spatial data
  ARC/INFO export file of vegetation communities with density, height, pattern (*.e00 file)
  SDTS format directory of vegetation communities with density, height, pattern
  Graphic file of map composition
  AML files creating map composition
  Hard copies (2) of map composition
  Mapping methods report
  Photo interpretation key
  Mapping class list
  Metadata
Field data (plots format)
  Export file of plot locations
  Physical description of plots
  Species listing of plots
  Strata description of plots
  Metadata
  Hard copies of field data
Accuracy assessment
  Export file of plot location
  Methods report
  Accuracy assessment data
  Contingency table
  Metadata
  Hard copies of field data
Web site
  http://biology.usgs.gov/npsveg
  All data posted on web site

contained in the classification description allowed him to familiarize himself with the plants and vegetation communities of the island. This highlights the advantages of standard data and information formats; as park personnel move from park to park, they are able to quickly master the information in a dataset because it is in a familiar format.

Finally, the NPS Fire Program is a cooperator with the USGS-NPS Vegetation Mapping Program. The Fire Program requires much of the data that the Vegetation Mapping Program produces for its activities. For example, fire managers need to know the structure, composition, and distribution of vegetation communities to predict fire behavior in an emergency situation. The vegetation community descriptions, the spatial data, along with some additional information on dead and down fuels, fulfill this data requirement. Fire managers also need to know the structure, composition,

Figure 22.8 Prescribed fire at Zion National Park.

and distribution of vegetation communities to manage the vegetation through prescribed fire. A park's general management plan may require that the vegetation distribution and composition resemble its state when Europeans first settled the area. The best tool to accomplish this is prescribed fire and an essential data set is the vegetation data (Figure 22.8).

22.3. Mapping and assessing water quality using remote sensing

Suppose you have been given the task of using image data collected by the Landsat TM to: (1) identify and map the spatial distribution of all lakes, reservoirs, and farm ponds in your home county; and (2) to examine the quality of water in those basins. Several questions have probably already occurred to you. How does one recognize the surface water on the image? Do water bodies have a distinctive shape, size, or color that facilitates identification? What wavelengths of the electromagnetic spectrum are important for this purpose? Is one satellite overpass enough, or is it necessary to obtain multi-date coverage for locating surface water? How does one analyze water quality using remote sensing?

Most people realize that although lakes and ponds vary considerably in shape and size, there are certain colors (e.g. shades of blue, green, and brown) that are characteristic of inland waters. It also seems intuitively obvious that the turbidity levels in a lake or reservoir are often the cause of the color, but what is the precise nature of the relationship between color and water clarity?

Regardless of the true color of an individual lake, reservoir, or pond, however, success in identifying these water bodies when using an aerial image depends, to a large extent, on the overall contrast between the color of the water feature and the surrounding terrain. For example, a reservoir built in the California desert to provide water for irrigation might be easier to identify or locate than a lake surrounded by freshly plowed farmland in Illinois, or a frozen lake located among the snow-covered peaks of Colorado. In these latter two examples, one would probably need to use other 'clues' to assist in the search for surface water. These clues might include the irregular shape of a shoreline compared to the straight lines associated with field patterns of Illinois farmland, or the smooth texture of a lake surface as compared to the rough texture of a surrounding stand of pine trees in the mountains of Colorado. Thus, it seems that there is more to delineating surface waters on aerial imagery than just the color, and the human brain is capable of incorporating some of these other consider-ations in its decision-making process. Nevertheless, the spectral characteristics of water, especially color, are certainly fundamental and important. Here we will focus only on using the spectral information to identify and analyze water features in Garden County, in the western portion of the Sandhills region of Nebraska. Before addressing practical considerations in using remote sensing to locate and map water on the landscape, it is appropriate to briefly consider the spectral reflectance charac-teristics of water.

22.3.1. Spectral properties of water

If the intention of a particular project is to conduct a remote sensing survey of surface waters, one must, first of all, be generally familiar with the basic reflectance properties, i.e. spectral signatures associated with both clear and turbid waters. Most people would describe deep, clear water as being distinctly blue in color, which is certainly a reasonable response. While we may not fully understand that the clear water is blue because of the scattering of (short-wavelength) blue light by the water molecules themselves, most would agree that a lake that is blue in color is generally cleaner and more aesthetically pleasing than one that appears brown. This perception is no doubt due to our associating brown-colored water with a lake or stream carrying large amounts of eroded soil in suspension (high turbidity). Similarly, algae-laden water appears green to us, and is also perceived by most of us as being undesirable, often with an unpleasant odor. Again, the water is turbid, but for a quite different reason. Therefore, the color of water can lead to some useful, if only general and qualitative, inferences about the water quality.

Figure 22.9 depicts two spectral profiles derived using a field spectroradiometer for water under various experimental conditions. Figure 22.9a depicts the result of a controlled experiment involving data collection over a large tank of clear water, which was then scanned as suspended sediment (soil in solution) was added incremen-tally. Notice that the overall level of reflectance of clear water (0 mg/L) is low, less than 2 per cent (compare it to the reflectance from growing vegetation shown in Figure 22.1). Therefore, remote sensing of clear surface water is a challenge in that the signal upwelling from the water body is relatively weak. Also note that as wavelength increases, reflectance decreases (or put another way, absorption increases with wave-length). Therefore, from Figure 22.9a, we conclude not only that clear water absorbs most of the visible light incident upon it, but also that the water absorbs virtually all near-infrared energy.

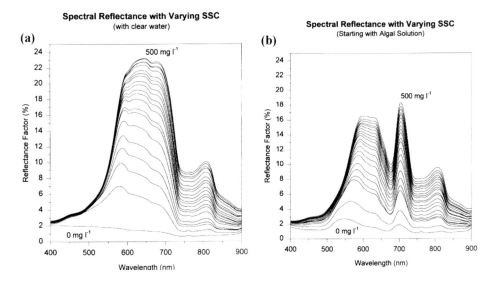

Figure 22.9   (a) Spectral profiles depicting changes in reflectivity with incremental additions of suspended soil material to clear water. (b) Spectral profiles depicting changes in reflectivity with incremental additions of suspended soil material to water containing green algae in solution (adapted from Han 1997).

The graph in Figure 22.9a makes it apparent that overall reflectivity increases as suspended sediment is added to clear water. In the experiment summarized here, reflectance reached about 23 per cent in the visible spectrum when the sediment load was 500 mg/L. Also, notice that the reflectance peak in the visible portion of the spectrum 'broadens' and shifts to longer wavelengths. This should not be surprising, if one considers the color of water that is clear vs. water heavily laden with suspended soil material.

Figure 22.9b summarizes an experiment involving the addition of suspended sediments to water, but this time to a solution containing a concentration of photosynthetically active algal material. Once again, the overall reflectivity increases with increased suspended load. The difference between Figures 22.9a, and 22.9b, however, is that the latter profile contains a distinctive algal signature, and that signature remains distinct even as suspended load increases. Green algae are photosynthetic organisms, i.e. the algal phytoplankton make use of light in the process of photosynthesis. The primary light-harvesting pigment in many algal types is chlorophyll a, which absorbs blue and red light. Therefore, the spectral profile for green algae is characterized by important absorption features near 443 and 670 nm, much like terrestrial vegetation (compare this to Figure 22.1). Reflectance spectra can also result in the presence of narrow peaks at spectral locations where absorption is low. A spectrum for green algae, and therefore, chlorophyll a, is characterized by prominent reflectance features near 550 and 700 nm.

As previously noted, surface water absorbs most of the light incident upon it. Despite the relatively weak signal from surface waters, we can utilize the inherent spectral characteristics of water in locating lakes, reservoirs, and ponds within remotely sensed images.

22.3.2. Locating water on the landscape using satellite remote sensing

Suppose that one has been assigned the task of mapping all surface waters within a county-sized area of the Central Great Plains of the US by means of remote sensing. From the standpoint of identifying typical land cover found in the Midwest, the identification of surface water is a relatively simple task, and there is most likely no overriding need for real-time (or even near-real-time) data. Also, since the typical county in the Central Plains is relatively large, it is probably logical to use satellite data of moderate spatial resolution (20–30 m). For purposes of illustrating how to inventory lakes in a county, let us assume that the choice of sensors is the Landsat TM.

The satellite images shown in Figures 22.10 and 22.11 (see plate 17) cover the northern half of Garden County in the Western Sandhills of Nebraska. Figure 22.10 depicts the landscape, as imaged in TM band 2 (visible green), TM band 3 (visible red), and TM band 4 (near infrared). We know from the previous discussion that much of the near-infrared energy incident upon a lake is absorbed by the water; therefore, the resulting image depicts lakes in dark, homogeneous tones (as in Figure 22.10c). Thus, the land-water contrast is maximized in the near infrared image and open surficial water can easily be identified and mapped. All of the areas appearing black in Figure 22.10c are indeed lakes of various sizes. It is relatively easy using digital-processing techniques to not only identify and map the location of the lakes, but to also measure the surface area of either one particular lake or all lakes combined. In fact, with multi-date satellite data, one can monitor changes in lake areas over time and relate inferred variability to climate change (Gosselin *et al.* 2000).

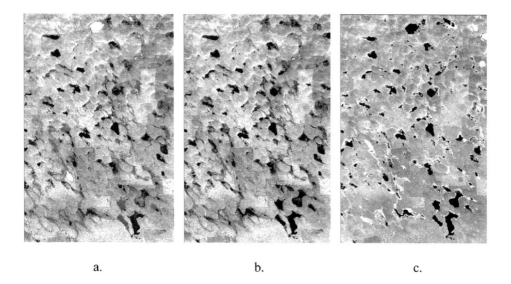

a.   b.   c.

Figure 22.10   Landsat TM images of the northern half of Garden County in the Western Sandhills of Nebraska. (a) TM band 2 (visible green). (b) TM band 3 (visible red). (c) TM band 4 (near-infrared). The areas in dark tones shown in (c) are lakes.

22.3.3. Assessing water quality using satellite remote sensing

Assessing the quality (condition) of surface waters is a more difficult problem than identifying and mapping their aerial extent. To examine water quality, one must refer to parts of the electromagnetic spectrum other than merely the near infrared. For example, compare the spectral characteristics of three lakes found in Figure 22.10. A good example of how a lake's spectral characteristics vary can be appreciated by looking at Schoonover Lake, the large circular water body near the top edge of each of the images. Schoonover Lake has the characteristic dark, homogeneous tone on the near infrared image (Figure 22.10c), but is very light in tone on the visible green image (Figure 22.10a). Notice, in fact, that the signatures of the many lakes vary from white to black when using TM 2. In the visible red spectral region (Figure 22.10b), there is also a great amount of variability in lake signatures. In fact, Schoonover Lake is essentially invisible to a sensor operating in the visible red! Of course, it is not really invisible. There are certainly some small amounts of visible red light being reflected from the lake, but the level of reflectance is approximately the same as that from the surrounding sparsely vegetated terrain. Therefore, there is no contrast between land and water, and the lake is not easily seen.

To further underscore the differences in the water quality in the lakes of the Western Sandhills, consider Figure 22.11 (see plate 17). This Landsat TM false-color composite was constructed by combining TM bands 2, 3, and 4 and using the blue, green, and red color guns of the computer display system. Again, notice the variability in lake signatures (colors) from place to place. On the false-color composite, Schoonover lake exhibits a light-blue tone, while Crescent Lake (near the bottom of the image) is black in tone. Most of the lakes in the county are very dark in tone on the false-color composite, but there are a few exceptions.

The spectral profiles shown in Figure 22.11b–d (see plate 17), were obtained with a field spectroradiometer for three of the lakes in the region – Schoonover Lake, Island Lake, and Crescent Lake (Fraser 1996). Compare the spectral curves obtained from each of the lakes and then compare the color tones in the TM image (Figure 22.11a, see plate 17). In the case of Schoonover Lake, the level of reflectance is the highest of the three graphs (Figure 22.11b, see plate 17). In fact, peak reflectivity (which occurs near 550 nm) reaches 18 per cent, a very high reflectance for a water body. Island Lake, on the other hand has a peak reflectance of only 4 per cent (also near 550 nm), while the peak for Crescent Lake is about 7 per cent (but notice that the peak for Crescent occurs near 705 nm). The lakes in this region are diverse in composition, ranging from fresh to brine. Some lakes, such as Schoonover, obviously have very high levels of suspended inorganics.

What can be inferred about the quality of the water in Schoonover, Island, and Crescent Lakes based on only the signatures shown by the Landsat TM false-color composite (Figure 22.11a, see plate 17) and the corresponding field-acquired spectral profiles? As noted above, Schoonover is highly reflective in all visible wavelengths, but contains almost no photosynthetically active algal material. This is obvious due to the existence of only a very slight absorption feature near 670 nm (visible red). Therefore, it is possible to infer that there are substantial amounts of suspended inorganics (which are effective scatterers of light) in Schoonover Lake, but there seem to be no other 'spectral indicators' (i.e. absorption features in the spectral profile). Crescent Lake, on the other hand, has not only a very pronounced absorption feature near 670 nm, but also a distinct reflectance feature near 705 nm; both are indicative of a significant suspended algal concentration. In addition, the spectral profile for Crescent Lake contains a minor absorption feature near 625 nm, which indicates the presence of a particular type of algae (blue-green). Island Lake (the 'hour-glass shaped' lake immediately above Crescent) had the lowest overall reflectance, with some relatively minor amounts of photosynthesizing algae. Notice that some of the inferences discussed here cannot be made using the broad-band image data alone; rather, they are the result of reference to the field-acquired data (hyperspectral data). Unfortunately, there are, at present, no hyperspectral sensors on orbital platforms. Systems such as the newly launched Hyperion sensor on the EO-1 platform, with 220 spectral bands, should prove useful for such work.

In summary, it is possible to make some inferences about the characteristics of the water column in a lake or reservoir by means of spectral data. Remote sensing techniques for surface water have been applied to the evaluation of water quality in both the oceans and inland waters (e.g. Khorram 1981; Lathrop and Lillesand 1986). The goal is to be able to analyze a body of water from aircraft and/or satellite altitudes and characterize it with respect to level of suspended sediments, chlorophyll content, chemistry, temperature, or other pertinent parameters. To identify specific components of the water column, it is necessary to use a hyperspectral remote sensing system, because biologically, there are literally hundreds of chemicals that absorb selectively, often resulting in narrow spectral peaks. Unfortunately, the broad-band sensors, such as Landsat-TM, SPOT-4, and even IKONOS-2 allow only for assessments of suspended- inorganic loads and algal productivity, but not for detailed interpretations of water-quality parameters.

22.4. Updating tax maps using aerial photography and soft-copy photogrammetry

Many of the most important tasks that are often undertaken by local and county government municipalities in urban/suburban environments are summarized in Table 22.3. Table 22.3 also includes the spatial, spectral, and temporal resolution of the remote sensing data that is typically used to complete the task (Jensen and Cowen 1999). It is well-known that most of these tasks are best accomplished using high spatial resolution remote sensor data, rather than by using high spectral resolution data (Jensen 2000). This is demonstrated convincingly in Figure 22.12, where the spatial and temporal resolution required to perform the functions in Table 22.3 are displayed in relation to the major remote sensing systems in operation today and projected for deployment in the near future. It is clear from Table 22.3 and Figure 22.12 that the vast majority of urban/suburban applications require remotely sensed data with spatial resolutions of ≤1 m. High spatial resolution aerial photography or satellite imagery, such as Space Imaging's IKONOS data, are the only data sets that satisfy most of these requirements.

One of the most important tasks that a county typically has to perform is tax mapping. In fact, many of the counties in the US are mandated to use the latest geographic information science technology to systematically update their tax base maps. This is a very difficult and often a costly endeavor. The application that follows

Table 22.3 Relationship between urban/suburban attributes and the minimum remote sensing resolutions required to provide such data (adapted from Jensen and Cowen 1999)

Attributes	Minimum resolution requirements		
	Temporal	*Spatial*	*Spectral*
Land-use/land-cover			
L1 – USGS Level I	5–10 years	20–100 m	V – NIR – MIR – Radar
L2 – USGS Level II	5–10 years	5–20 m	V – NIR – MIR – Radar
L3 – USGS Level III	3–5 years	1–5 m	Pan – V – NIR – MIR
L4 – USGS Level IV	1–3 years	0.25–1 m	Panchromatic – V – NIR
Building and cadastral (property-line) infrastructure			
B1 – building perimeter, area, volume, height, and property lines	1–5 years	0.25–0.5 m	Panchromatic – V – NIR
Transportation infrastructure			
T1 – general road centerline	1–5 years	0.25–0.5 m	Pan – V – NIR
T2 – precise road width	1–2 years	0.25–0.5 m	Pan – Visible
T3 – traffic count studies	5–10 min	1–30 m	Pan–Visible
T4 – parking studies	10–60 min	0.25–0.5 m	Pan – Visible
Utility infrastructure			
U1 – general utility line mapping	1–5 years	1–30 m	Pan – V – NIR
U2 – precise utility line width	1–2 years	0.25–0.6 m	Pan – Visible
U3 – location of poles, manholes	1–2 years	0.25–0.6 m	Panchromatic
DEM creation			
D1 – large-scale DEM	5–10 years	0.25–0.5 m	Pan – Visible
D2 – large-scale slope map	5–10 years	0.25–0.5 m	Pan – Visible
Socioeconomic characteristics			
S1 – local population estimates	5–7 years	0.25–5 m	Pan – V – NIR
S2 – national population estimates	5–15 years	5–20 m	Pan – V – NIR
S3 – quality of life indicators	5–10 years	0.25–30 m	Pan – V – NIR
Energy demand and conservation			
E1 – energy demand and production	1–5 years	0.25–1 m	Pan – V – NIR
E2 – building-insulation surveys	1–5 years	1–5 m	Thermal infrared
Meteorological data			
M1 – weather prediction	3–25 min	1–8 km	V – NIR – MI
M2 – current temperature	3–25 min	1–8 km	Thermal infrared
M3 – clear air and precipitation	6–10 min	1 km	WSR-88D Radar
M4 – severe weather mode	5 min	1 km	WSR-88D Radar
M5 – monitoring urban heat island	12–24 h	5–30 m	Thermal infrared
Critical area assessment			
C1 – stable sensitive environments	1–2 years	1–10 m	V – NIR – MIR
C2 – dynamic sensitive environments	1–6 months	0.25–2 m	V – NIR – MIR – TIR
Disaster emergency response			
DE1 – pre-emergency imagery	1–5 years	1–5 m	Pan – V – NIR
DE2 – post-emergency imagery	0.5–2 days	0.25–2 m	Pan – NIR – Radar
DE3 – damaged housing stock	1–2 days	0.25–1 m	Pan – NIR
DE4 – damaged transportation	1–2 days	0.25–1 m	Pan – NIR
DE5 – damaged utilities, services	1–2 days	0.25–1 m	Pan – NIR

demonstrates how one county in Mississippi used high spatial resolution aerial photography and soft-copy photogrammetric techniques to develop an ortho-corrected basemap for tax mapping.

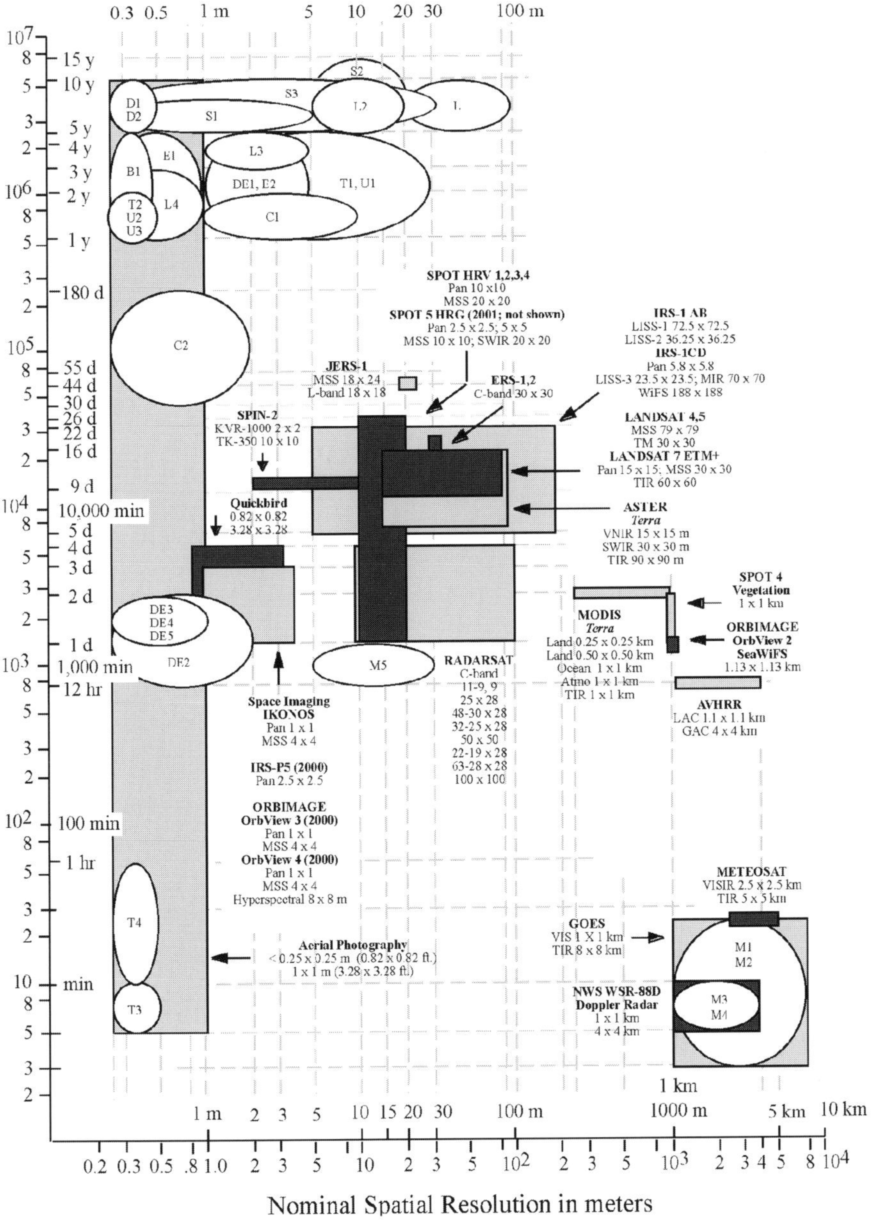

Figure 22.12 Remote sensing nominal spatial and temporal resolution requirements for the urban/suburban tasks summarized in Table 22.3 (Jensen and Cowen 1999).

22.4.1. Rankin county, MS, tax mapping

Mississippi's 1997 Tax Commission regulations stipulate that any county with over 20,000 tax parcels must update its aerial photo inventory every 10 years. Rankin County is one of the top three fastest growing counties in Mississippi (Rochelle 2000). The county's population has grown from 80,000 to 110,000 people since

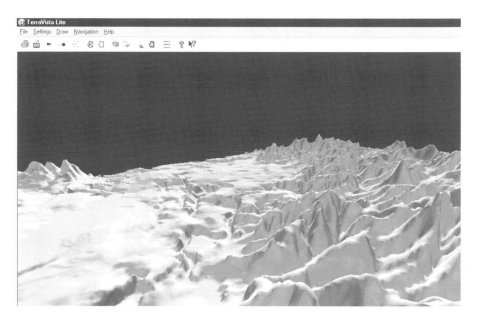

Figure 22.13 A 10:1 vertical exaggeration of a DEM of a portion of Rankin County used to model the orthophoto (courtesy of GeoGraphix, Inc., ESRI Inc., ERDAS, Inc.).

1990. The number of tax parcels now exceeds 60,000. Therefore, Rankin County had to update its 1980 tax assessment aerial photography database by the spring of 2000 to comply with the 1997 state law.

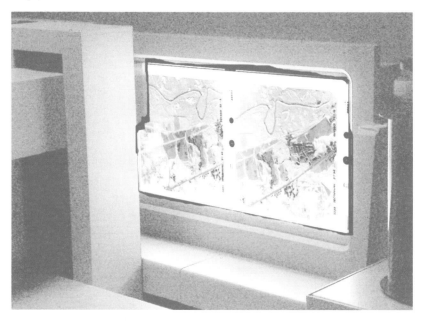

Figure 22.14 Digitization of individual negative frames of aerial photography in a roll using a Vexcel VX4000 scanner (courtesy of GeoGraphix, Inc., ESRI Inc., ERDAS, Inc.).

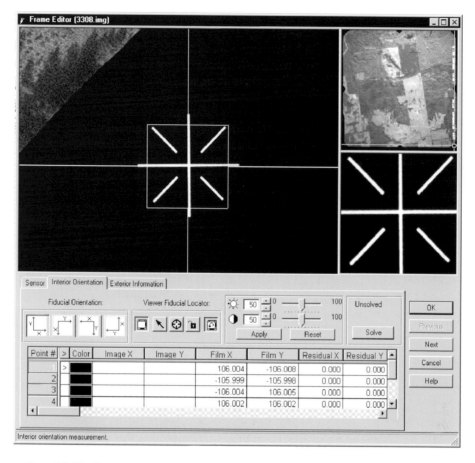

Figure 22.15 The interior orientation takes the camera calibration information and relates it to the image based upon the fiducials and other inputs. This establishes a 'pixel space', a coordinate system based upon the geometry of the aerial image (courtesy of GeoGraphix, Inc., ESRI Inc., and ERDAS, Inc).

In 1998, the Rankin County Tax Assessor initiated an extensive effort to update the aerial photo archive and create an up-to-date parcel map base that could be combined with existing vector data from the county's ArcInfo-based Geographic Information System and be compatible with AutoCAD. The County determined that the project could be outsourced. GeoGraphix Inc. (Huntsville, Alabama) was awarded the contract via a competitive bidding process.

22.4.2. Aerial photography data collection and soft-copy photogrammetric processing

Thirty flight lines of metric, vertical aerial photography containing more than 2,600 stereoscopic aerial photographs at a contact scale of $1'' = 1,000'$ (1:12,000) were obtained during clear weather. The study area is not flat and contains substantial local relief, as shown in the Digital Elevation Model (DEM) in Figure 22.13. There-

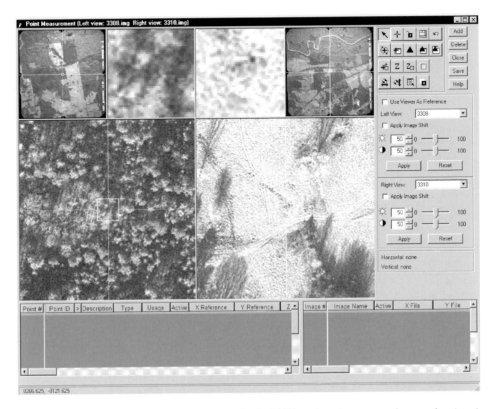

Figure 22.16 This shows how IMAGINE OrthoBASE easily selects ground control points in multiple images. Note the simultaneous multiple zoom levels attainable for rapid photo interpretation (courtesy of GeoGraphix, Inc., ESRI Inc., and ERDAS, Inc).

fore, it was not possible to simply perform first- or second-order rubber-sheeting rectification on the aerial photography and hope that it would remove all of the radial and relief distortion discussed in Chapters 4 and 19. Therefore, the company used ERDAS OrthoBASE soft-copy digital photogrammetry software to turn the unrectified aerial photography into an ortho-rectified image map base suitable for accurately overlying vector parcel cadastral information.

Standard soft-copy photogrammetric procedures were used. First, each frame of the 1:12,000-scale analog aerial photography was digitized using a very high resolution area-array scanner. Negative diapositives are shown being scanned in Figure 22.14. This resulted in imagery with a ground resolution of 1 × 1 ft. Next, soft-copy photogrammetric software was used to perform: (a) interior orientation which makes use of camera calibration information and very precise measurements of the distances between fiducial marks in the aerial photography (Figure 22.15), and (b) exterior orientation where known x, y, z Ground Control Points (GCP) surveyed in the real-world are related to the same points recorded in the aerial photography (Figure 22.16). The subsequent OrthoBASE aerial triangulation allowed the company to obtain low residuals across large areas. The final residuals were kept below a pre-determined Root Mean Square Error (RMSE), meaning that the points used for anchor points

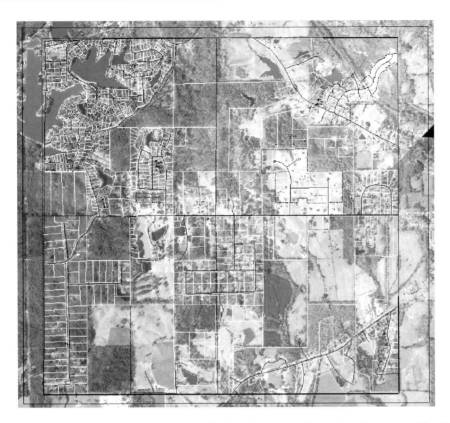

Figure 22.17   Final mosaic overlaid with a Rankin County parcel map sheet (courtesy of GeoGraphix, Inc., ESRI Inc., ERDAS, Inc.).

were highly accurate relative to each other and the geospatial data provided by the
DEM surface.

In January, 2000, approximately 214 map sheets at $1'' = 400'$ scale, each containing
four Public Land Survey sections, were delivered. One of the map sheets is shown in
Figure 22.17. All of the digital map files are available for use in ArcInfo by the
County's GIS users.

22.4.3. Benefits

Rankin County now provides better service to its taxpayers by producing plots of any
area, at any scale, with an ortho-rectified digital image backdrop. The 1×1 ft. pixel
resolution allows the Mapping Department to plot image maps up to $1'' = 100'$ scale
anywhere within the county, which will be helpful to the county's appraisers as they
conduct a physical inventory of parcels. The county's E-911 road maps are now more
accurate. One Rankin County official said, 'The new map not only helps us to accurately
assess improvements among parcels, but we can also identify high growth areas and if
necessary in the next 2–5 years, focus on a particular area that might need updating
without having to fly (aerial photos) for the entire county. This will save us thousands of
dollars and thousands of hours.' Other case studies are found in Part 5 of this manual.

References

Fraser, R., *Multiscale remote sensing of biogeochemical conditions in Nebraska Sand Hills lakes*, unpublished Ph.D. dissertation, Department of Geography, University of Nebraska, Lincoln, NB, 1996.

Gausman, H., *Plant Leaf Optical Properties in Visible and Near-Infrared Light*, Texas Tech Press, 1985, 78 pp.

Gillison, A. N. and Brewer, K. R. W., 'The use of gradient directed transects or gradsects in natural resource survey', *Journal of Environmental Management*, 20, 1985, 103–127.

Gosselin, D., Rundquist, D. and McFeeters, S., 'Remote monitoring of selected groundwater-dominated lakes in the Nebraska Sand Hills. *Journal of the Amrican Water Resources Association*, 36:5, 1039–1051.

Grossman, D. H., Faber-Langendoen, D., Weakley, A. S., Anderson, M., Bourgeron, P., Crawford, R., Goodin, K., Landaal, S., Metzler, K., Patterson, K., Pyne, M., Reid, M. and Sneddon, L., *International Classification of Ecological Communities: Terrestrial Vegetation of the United States, Volume I. The National Vegetation Classification System: Development, Status, and Applications*, The Nature Conservancy, Arlington, VA, 1998a.

Grossman, D. H., Faber-Langendoen, D., Weakley, A. S., Anderson, M., Bourgeron, P., Crawford, R., Goodin, K., Landaal, S., Metzler, K., Patterson, K., Pyne, M., Reid, M. and Sneddon, L., *International Classification of Ecological Communities: Terrestrial Vegetation of the United States, Volume II. The National Vegetation Classification System: List of Types*, The Nature Conservancy, Arlington, VA, 1998b.

Han, L., 'Spectral reflectance with varying suspended sediment concentrations in pure and algae-laden waters', *Photogrammetric Engineering & Remote Sensing*, 63(6), 1997, 701–705.

Jensen, J. R., *Introductory Digital Image Processing*, Prentice Hall, Saddle River, NJ, 1996, 318 pp.

Jensen, J. R., *Remote Sensing of the Environment: an Earth Resource Perspective*, Prentice Hall, Saddle River, NJ, 2000, 544 pp.

Jensen, J. R. and Cowen, D. C., 'Remote sensing of urban/suburban infrastructure and socioeconomic attributes', *Photogrammetric Engineering & Remote Sensing*, 65(6), 1999, 611–622.

Khorram, S., 'Use of ocean color scanner data in water quality mapping', *Photogrammetric Engineering & Remote Sensing*, 47(5), 1981, 667–676.

Lathrop, R., Jr. and Lillesand, T., 'Use of Thematic Mapper data to assess water quality in Green Bay and Central Lake Michigan', *Photogrammetric Engineering & Remote Sensing*, 52(5), 1986, 671–680.

Rochelle, S., 'Updating tax maps in Mississippi', *ESRI ArcNews*, 22(2), 2000, 26.

USGS-NPS Vegetation Mapping Program, *Developing and Documenting a National Vegetation Classification Standard*, prepared under contract for the U.S. Geological Survey Center for Biological Informatics by The Nature Conservancy and Environmental Systems Research Institute, Denver, CO, 1994a.

USGS-NPS Vegetation Mapping Program, *Establishing Standards for Field Methods and Mapping Procedures*, prepared under contract for the U.S. Geological Survey Center for Biological Informatics by The Nature Conservancy and Environmental Systems Research Institute, Denver, CO, 1994b.

USGS-NPS Vegetation Mapping Program, *Producing Rigorous and Consistent Accuracy Assessment Procedure* , prepared under contract for the U.S. Geological Survey Center for Biological Informatics by The Nature Conservancy and Environmental Systems Research Institute, Denver, CO, 1996.

Chapter 23

Remote sensing – future considerations

John R. Jensen

> Since the first satellites had orbited, almost 50 years earlier, trillions and quad-rillions of pulses of information had been pouring down from space, to be stored against the day when they might contribute to the advance of knowledge. Only a minute fraction of all this raw material would ever be processed; but there was no way of telling what observation some scientist might wish to consult 10, or 50, or 100 years from now. So, everything had to be kept on file, stacked in endless air-conditioned galleries, triplicated at the three centers against the possibility of accidental loss. It was part of the real treasure of mankind, more valuable than all the gold locked uselessly away in bank vaults.
>
> Arthur C. Clarke *2001*

23.1. Introduction

This chapter identifies some of the major issues associated with using remote sensing data for monitoring and modeling natural ecosystems and cultural landscapes to make informed decisions, including:

- Data continuity
- Lack of certain remote sensing instruments, e.g. orbital hyperspectral
- The necessity of improved information extraction algorithms
- Data integration (conflation), standards, and interoperability

The chapter begins with a discussion of data continuity. It then identifies some of the major issues that must be addressed if users want to use remote sensing technology operationally in their day-to-day responsibilities.

23.2. Remote sensing data continuity

An agency, individual, or commercial firm must be able to count on having access to timely remotely sensed data if they plan on using it in a dependable, operational manner in their daily activities or in a marketable product line (Potestio 2000). Hope-fully, the required remote sensor data is available into the foreseeable future without the significant risk of having the entire data stream interrupted or terminated (Jensen 1992, 2000). This is generally referred to as the *data continuity* problem.

23.2.1. Sub-orbital remote sensing data continuity

For more than 80 years it has been possible to contract with a photogrammetric engineering commercial firm to collect metric stereoscopic aerial photography on demand, e.g. GEONEX, Inc.; AeroMetric, Inc. Every state in the US and most countries have photogrammetric companies. Weather permitting, the aerial photography collected by these companies are extremely accurate and may be used to produce land use/cover thematic maps, planimetric and topographic maps, and derived Digital Elevation Models (DEM). Many commercial photogrammetric engineering companies and special purpose remote sensing firms are increasing their capability by collecting: (1) digital camera imagery, e.g. Litton Emerge Spatial, Inc.; Positive Systems, Inc., (2) hyperspectral remote sensor data using a variety of instruments, e.g. CASI, HYMAP, (3) radar imagery, e.g. Intermap Star 3i x-band, and (4) LIDAR imagery (Davis 1999). It is likely that an individual or agency can count on obtaining the sub-orbital imagery of choice from these firms, if the price is right. It is unlikely that the data-stream will dry-up completely (Schill et al. 1999).

23.2.2. Orbital remote sensing data continuity

Government agencies such as the National Aeronautics and Space Administration (NASA) and private commercial firms such as Space Imaging, Inc. and government subsidized companies (e.g. the French SPOT Image, Inc.) continue to place remote sensing systems in orbit. Let us briefly discuss the data continuity problems associated with each of these three types of data providers.

23.2.2.1. Government agencies

Government agencies are not the most reliable source of remotely sensed data. Government agencies that place remote sensing systems in orbit are subject to:

1 Political pressure from: (a) within the government (e.g. the military intelligence community may not want the public to have access to high spatial resolution hyperspectral data), (b) private special interest groups (e.g. organizations representing photogrammetric engineering firms may lobby to ensure that the government does not collect the same type of remote sensing data that the commercial firms can provide), and/or (c) international special interest groups or governments (e.g. it is common for government agencies to request or initiate 'shutter control', i.e. they turn off the flow of remote sensing data to the public, especially commercial news agencies, during heightened national or international incidents).
2 An Office of Management and Budget (OMB) and/or a general accounting office that constantly monitor the expenditure of public funds (a good idea in principle if they are not affected by special interest groups both within and outside the government).

The series of US Landsat satellite remote sensing systems is a good example of how data continuity can be severely hampered by government bureaucracy and indecision. Landsat 1 (the Earth Resource Technology Satellite (ERTS)) was launched in 1972. Landsat 7 was launched in 1999. Unfortunately, there is no guarantee that there will be a Landsat follow-on due to the following reasons succinctly described by the Committee on Earth Studies, Space Studies Board of the National Research Council (Space Studies Board 1995):

> In summary, the more than two decades of experience with Landsat show missteps, wasteful starts and stops, conflicting government policies, and a budget process that at each step of the way threatened public and private users with a lack of data continuity. The user community can scarcely be blamed for wariness at committing resources to the use of Landsat-like data when the government publicly announces annually that the current satellite in orbit is to be the last.

As a result,

- The nation that pioneered multispectral sensing of the land has no operational system upon which it can rely...
- The nation has no effective applied R&D arm to develop new sensor technology or to explore new applications of space remote sensing systems to meet immediate needs. The R&D program principally addresses research on climate change, while the nearer-term benefits are not being addressed.
- The single element of continuity in government action regarding Landsat is the unwavering attempts of the OMB and its predecessor, the Bureau of the Budget, to block the use of this technology in an operational context. At every step of the way, short-sighted decisions have been taken that initially prevented the effective use of the technology and now threaten to keep the US from even remaining a player in this strategically and, in the longer term, economically important technology.
- ...No long-range plan exists for the exploitation of advances in Landsat, and even the rudiments of a short-range plan to secure continued data for science and applications users are absent.
- Under current policies and practices, the US will shortly have ceded to France, Japan, the nations who are members of the European Space Agency, Russia, and potentially India and China the hard-won technology leadership in land remote sensing it once possessed. While international cooperation is expected in earth observations, decision to cede responsibility must only be taken with assurances of data availability.

The bottom line is that the government can place in space some excellent remote sensing systems such as Landsat 7 and the new NASA Earth Science Enterprise *Terra* satellite with its package of instruments geared toward global climate change analysis. Unfortunately, one cannot count on data continuity if the remote sensing systems are under government control due to the reasons provided above.

23.2.2.2. Private commercial remote sensing firms

US commercial remote sensing firms such as Space Imaging, Inc., ORBIMAGE, Inc., and EarthWatch, Inc. have obtained licenses from the US government to launch remote sensing systems into orbit and sell the imagery at commercial competitive prices to anyone who will buy it. All three of these vendors propose obtaining 1×1 m panchromatic data and approximately 4×4 m multispectral data. Significant improvements in the spatial resolution of satellite remote sensing is expected to introduce many of the Geographic Information System (GIS) practitioners to the value of remote sensor data (Baker 2000; Dehqanzada and Florini 2000). ORBIMAGE Inc. is also planning to launch an 8 × 8 m hyperspectral remote sensing instrument. Hopefully, these commercial suppliers of remote sensor data will solve the data continuity problem that has plagued orbital remote sensing science since its inception.

Unfortunately, there are also problems associated with the commercial remote sensing data providers. First, it takes a tremendous amount of money to build the satellite and remote sensing instrument, launch it into space, build ground receiving stations, maintain constant communication with the satellite, and format and market the remote sensor data. It is often assumed that this tremendous dollar investment will be returned with a profit by selling the imagery to the federal government, to state and local agencies, and to private individuals. This is a bold assumption. Can the private firms actually sell enough imagery to re-coup the cost of their significant investment and generate sufficient revenue to keep their shareholders pleased? Many believe this is possible while others suggest that the market is still not mature enough to generate the number of sales necessary. In any case, the market may not be sufficient to insure that all the new private remote sensing companies survive. Hopefully, additional private companies will enter the marketplace using small-satellite technology and extremely cost-effective data management and marketing techniques. If this happens, then we may have remote sensing data continuity from a variety of commercial vendors.

23.2.2.3. Government/private industry

Some of the most successful remote sensing data providers are commercial firms that have been subsidized by the federal government to collect and market the remote sensor data. A good example is SPOT Image, Inc. which is heavily subsidized by the French government. It has been in existence since 1986 (SPOT satellites 1–4) with plans for additional SPOT satellites to be launched into the foreseeable future. A similar successful subsidized relationship was initiated by the Canadian government in the creation of RADARSAT, Inc. for the collection of radar imagery. Federal government/private industry relationships such as these are very important because:

1   The government has the financial resources to fund substantial research and development so that the sensors represent the state of the art. For example, the French SPOT 1 satellite was the first to use linear push-broom detector technology as early as 1986. The Landsat 7 Enhanced Thematic Mapper plus launched in 1999 still uses scanning mirror technology.

2   If the government is committed to the relationship, it has a very high probability of providing data continuity. The French SPOT satellites have been around since 1986 with several more planned for launch during 2001–2010. Several other countries (e.g. Japan, India) see this government/private industry relationship as one of the major ways they can assure remote sensing data continuity of the homeland as well as of other worldwide markets.

3   Because the business relationship is subsidized by the federal government, the subsidized remote sensing data provider company may not have to charge as

much for each scene as purely private companies who must bear the full financial burden.

The government/private industry model of remote sensing data collection and marketing may eventually be the most successful model. Conversely, perhaps the commercial remote sensing data market is mature enough to support the private vendors without government subsidization. Time will tell.

23.3. Lack of certain remote sensing instruments

It is possible to fly almost any sophisticated remote sensing instrument onboard an aircraft if one has sufficient financial resources. However, sometimes time is of the essence (as with agricultural applications) or it is not possible to fly sub-orbital through the air-space of another country. In such cases, it may be better to have access to a dedicated satellite in orbit with the required remote sensing system that can be pointed off-nadir if necessary to collect the desired information. It is clear from many investigations that the user community needs several types of improved orbital satellite remote sensing systems, including:

1   An orbital hyperspectral remote sensing instrument. This sensor would look very much like the NASA JPL AVIRIS instrument (224 bands at 10 nm bandwidths from 0.4–2.5 µm). This need may be satisfied by ORBIMAGE Inc. if and when it is allowed to launch OrbView 4 (Warfighter).
2   An orbital satellite active microwave (radar) system optimized for earth resource investigations. The system should be a multi-frequency, multi-polarization radar that is optimized for agriculture, wetland, disaster, and forestry applications. Until such a system is available, it is necessary to use Canadian RADARSAT imagery that is optimized for sea ice and snow monitoring.
3   A multi-frequency, multi-polarization passive microwave radiometer should be placed in orbit with sufficient spatial and radiometric resolution to allow county and regional soil and vegetation moisture maps to be generated.
4   Landsat follow-on remote sensing systems. These sensor systems are required for continuity of inventorying land use/land cover and other earth resources. As usual, the future of Landsat-related remote sensing systems is unclear making it difficult for user agencies to commit long-term financial resources to these data. Ideally, the *Terra* satellite and its configuration of sensors would fulfill this need. Unfortunately, the future of *Terra* and its payload are also uncertain.
5   An orbital LIDAR instrument. The system should be capable of obtaining extremely accurate *x*, *y*, *z* measurements so that bald-earth DEM can be derived. LIDAR sensors optimized for forestry applications are of value but may not be sufficient for the generation of accurate DEM.

23.4. The necessity of improved information extraction algorithms

Chapter 19 briefly summarized the state-of-the-art of digital image processing of remotely sensed data. It is beyond the scope of this chapter to identify all of the basic research areas that should be addressed to make the collection and analysis of remotely sensed data easier and more accurate. However, some general observations can be made.

23.4.1. Ease of use of digital image processing software

There are numerous excellent digital image processing systems that one can purchase from private commercial vendors, including: ERDAS Imagine, Research Systems' Environment for Visualizing Images (ENVI), ER Mapper, Clark University's IDRISI, and PCI (Jensen 1996; Estes and Jensen 1998). Unfortunately, many new users routinely say that these digital image processing systems are difficult to use. While some systems have initiated a 'wizard' approach to certain functions, most have not. The biggest problem in the use of most of these systems is that the user does not know what to do next in a complex chain of procedures. Hyperspectral image analysis is the most difficult. Whenever possible the software should include 'wizards' to walk the novice analyst through the correct procedures. Advanced users must be able to easily by-pass the wizards.

23.4.2. Improved atmospheric correction/image normalization

It is very difficult to remove the deleterious effects of atmospheric attenuation in a remotely sensed image. It requires a good background in atmospheric physics, knowledge of existing atmospheric constituent models, and how to use relatively sophisticated atmospheric correction algorithms, e.g. MODTRAN, ATREM, EFFORT. Thus, most users are very hesitant to atmospherically correct a remote sensing image or they do it poorly. Unfortunately, to really extract the information of value in a remote sensing image it is absolutely essential to perform a quality atmospheric correction and convert the remote sensor data from radiance to actual per cent reflectance. This is especially the case when more than one image is going to be used to assess change. Significant research is needed to develop simple, user-friendly atmospheric correction algorithms. Hopefully, these require relatively little *in situ* atmospheric information and can rely on the characteristics of the spectral data themselves for calibration, i.e. certain bands contain information on atmospheric water vapor content, aerosol scattering, etc. Ideally, the process is conducted using a 'wizard'.

23.4.3. Improved expert system and neural network analysis

Few current image processing systems have easy to use expert system and/or neural network image analysis capabilities. Many require the analyst to pass data back and forth between the digital image processing system and an external expert system or neural network. This is very inefficient. More holistic systems are needed that contain the digital image processing system, and/or expert system shell or neural network. Developing the knowledge base and formulating the expert system rules is a very complex process. Ideally, machine learning technology can use the examples provided automatically to develop 'rules' (Huang and Jensen 1998). More adaptive algorithms are necessary that learn from the examples provided by the user.

23.4.4. Change detection error analysis

Chapter 21 reviewed how to assess the accuracy of land use/cover maps derived from remote sensor data. Khorram *et al.* (1999) suggest that improved sampling and statistical methods are required to assess the accuracy of change detection maps. Additional research is required so that we can place accurate confidence limits on the change detection products derived from remotely sensed data.

23.5. Data integration (conflation), standards, and interoperability

Geographic information provides the basis for many types of decisions, in areas ranging from simple way finding to management of complex networks of facilities and the sustainable management of natural resources (Jensen and Cowen 1999). In all of these areas, better data should lead to better conclusions and better decisions. According to several standards groups and users groups, better data would include greater positional accuracy and logical consistency and completeness. Technological advances are making it possible to capture positional information with ever improving accuracy and precision. Commercial remotely sensed images from space now offer a spatial resolution of less than 1 m (Jaffe 2000; Miller 2000). Satellite telemetry using the Global Positioning System (GPS) can now achieve accuracy within 1 cm. But each new data set, each new data item that is collected, however, accurate it may be, can be fully utilized only if it can be placed correctly into the context of other available data and information.

Geographic data sets show two clear trends: (1) they are becoming increasingly abundant, and (2) they are growing ever more precise. Remote sensing technology alone generates a vast amount of raw spatial data daily, much of which is redundant or too complex for adequate analysis with current technology before the next batch of raw data arrives. Remote sensing technology also promises and delivers continuous precision improvements in image resolution and data capture methods. Unfortunately, data assimilation strategies and methodologies have not kept pace with these advances.

One approach to facilitating the integration of spatial data is to mandate uniformity through standardization and agreed-upon formats and requirements. While this may work in the short term for a temporal cross-section of similar data, it cannot fully address the ever-evolving character of the captured spatial data. Historical databases from a pre-standards era must coexist with current standardized products; and future fully three-dimensional spatial datasets must be reconciled with other data within a context of current two-dimensional spatial data standards.

While remotely sensed and imaged data are becoming available in greater and greater quantities and at higher resolutions, integrating data from different sources is not yet easy because of variations in resolution, registration, and sensor characteristics. Without the ability to integrate data from different sources, we are faced with extensive duplication of effort and unnecessary cost. Imagery can play a very valuable role in updating old data sets, but this process is similarly impeded by the problems of integration.

23.5.1. Conflation (data integration)

The term *conflation* is often used to refer to the integration of data from different sources (Saalfeld 1988). It may apply to the transfer of attributes from old versions of feature geometry to new, more accurate versions; to the detection of changes by comparing images of an area from two different dates; or to the automatic registration of one data set to another through the recognition of common features. Too often, however, methods of conflation and integration have been *ad hoc*, designed for a single, very specific purposes and of no general value. For example, much effort in the past was directed toward updating the relatively low geometric accuracy TIGER database from the US Bureau of the Census with more accurate topographic data. Fortunately, technological advances including vastly greater computing speeds, larger storage volumes, better human-computer interaction, better algorithms, and better database tools are making conflation and integration more feasible than ever before.

A general theoretical and conceptual framework should address at least five distinct forms of conflation (integration) associated with a common database:

1   Map to map (different scales, different coverages, etc.)
2   Image(s) to map (elevation mapping, map revision, etc.)
3   Image to image (different resolutions, wavelengths, etc.)
4   Map to measurement (verification, registration, etc.), and
5   Measurement to measurement (adjustment, variance, etc.)

23.5.2. Interoperability and standards

The capture and integration of spatial data requires the collaboration of many participating disciplines, including cartography, computer science, photogrammetry, geodesy, mathematics, remote sensing, statistics, modeling, geography, and various physical, social, and behavioral sciences with spatial analysis applications. We must solve key problems of capturing the right data and relating diverse data sources to each other by involving participants from all specialty areas, including the traditional data collectors, the applications users, and the computer scientists and statisticians who optimize data management and analysis for all types of data sets (Abel and Wilson 1990). We must develop mathematical and statistical models for integrating spatial data at different scales and different resolutions. Tools must be developed for identifying, quantifying, and dealing with imperfections and imprecision in the data throughout every phase of building a comprehensive spatial database.

Many organizations and data users have developed and promoted *standards* for spatial data collection and representation (e.g. Department of Commerce 1992; Federal Geographic Data Committee (FGDC) 2000). By adhering to these standards, data collectors and data integrators will improve the consistency and overall quality of their products (National Research Council 1993). The standards alone, while facilitating the sound construction of multifaceted spatial (and spatio-temporal) databases, do not, in and of themselves, offer the means by which to integrate fully all types of spatial data efficiently and consistently. Different standards exist for imagery at different scales, for maps at different scales, and for adjustment of measurements taken with instruments of different precisions. A single common framework is needed for the diverse types of spatial data. Spatial data integration permits the coexistence of multiple spatially coherent inputs. Spatial data integration must include:

- Horizontal integration (merging adjacent data sets) and vertical data integration (operations involving the overlaying of maps);
- Handling differences in spatial data content, scales, data acquisition methods, standards, definitions, and practices;
- Managing uncertainty and representational differences; and
- Detecting and removing redundancy and ambiguity of representation.

Careful attention to these data integration (conflation), interoperability, and standards issues will make remote sensing derived information much more valuable.

References

Abel, D. J. and Wilson, M. A., 'A systems approach to integration of raster and vector data and operations', *Proceedings, 4th International Symposium on Spatial Data Handling*, vol. 2, Brassel, K. and Kishimoto, H. (eds.), Zurich, 1990, pp. 559–566.

Baker, J. C., 'New users and established experts: bridging the knowledge gap in interpreting commercial satellite imagery,' *Commercial Observation Satellites: at the Leading Edge of Global Transparency*, RAND Corporation and the American Society for Photogrammetry & Remote Sensing, Washington, DC, 2000.

Department of Commerce, *Spatial Data Transfer Standard (SDTS)*, Federal Information Processing Standard 173, Department of Commerce, National Institute of Standards and Technology, Washington, DC, 1992.

Davis, B. A., 'An overview of NASA's commercial remote sensing program', *Earth Observation Magazine*, 8(3), 1999, 58–60.

Dehqanzada, Y. A. and Florini, A. M., *Secrets for Sale: How Commercial Satellite Imagery Will Change the World*, Carnegie Endowment for International Peace, Washington, DC, 2000, 45 pp.

Estes, J. E. and Jensen, J. R., 'Development of remote sensing digital image processing systems and raster GIS', *The History of Geographic Information Systems*, Foresman, T. (ed.), Longman, New York, 1998, pp. 163–180.

Federal Geographic Data Committee, *Development of a National Digital Geospatial Data Framework*, Federal Geographic Data Committee, Department of Interior, Washington, DC, 2000, ftp://www.fgdc.gov/pub/standards

Huang, X. and Jensen, J.R., 'A machine learning approach to automated construction of knowledge bases for image analysis expert systems that incorporate GIS data'; *Photogrammetric Engineering and Remote Sensing*, 63(10), 1998, 1185–1194.

Jaffe, V., 'Bozeman banks on 1-meter imagery,' *Imaging Notes*, 15(4), 2000, 18–20.

Jensen, J. R., 'Testimony on S. 2297, The Land Remote Sensing Policy Act of 1992,' Senate Committee on Commerce, Science, and Transportation, *Congressional Record*, May 6, 1992, pp. 55–69.

Jensen, J. R., *Introductory Digital Image Processing: a Remote Sensing Perspective*, Prentice Hall, Saddle River, NJ, 1996, 318 pp.

Jensen, J. R., *Remote Sensing of the Environment: an Earth Resource Perspective*, Prentice Hall, Saddle River, NJ, 2000, 544 pp.

Jensen, J. R. and Cowen, D. C., 'Remote sensing of urban/suburban infrastructure and socioeconomic attributes,' *Photogrammetric Engineering & Remote Sensing*, 65(5), 1999, 611–622.

Khorram, S. *et al.*, *Accuracy Assessment of Remote Sensing-Derived Change Detection*, ASPRS Monograph Series, Bethesda, MD., American Society for Photogrammetry and Remote Sensing, 1999, 64 pp.

Miller, B., 'Utah County anticipates better data, lower costs,' *Imaging Notes*, 15(4), 2000, 21–22.

National Research Council, Mapping Science Committee, *Towards a Spatial Data Infrastructure for the Nation*, National Academy Press Washington, DC, 1993.

Potestio, D. S., *An Introduction to Geographic Information Technologies and Their Applications*, National Conference of State Legislatures, Washington, DC, 2000, 92 pp.

Saalfeld, A., 'Conflation: automated map compilation', *International Journal of Geographical Information Systems*, 2(3), 1988, 217–228.

Schill, S. R., Jensen, J. R. and Cowen, D. C., 'Bridging the gap between government and industry,' *Geo Info Systems*, 9(9), 1999, 26–33.

Space Studies Board, *Earth Observations from Space: History, Promise, and Reality*, Space Studies Board, Committee on Earth Studies, National Research Council, Washington, DC, 1995, 310 pp.

Part 4

Geographic information systems

Chapter 24

Geographic Information Systems (GIS) and science

Robert B. McMaster and William J. Craig

24.1. Introduction

Geographic information science or GSci is a relatively new discipline, having emerged from the quantitative revolution and evolution of a digital cartography, and rapid improvements in computer technology over the past 20 years. However, the term most commonly used until the mid 1990s was GIS, which focused on the software itself. More recently, the term GSci, has been applied to describe the theoretical underpinnings of the technology, including database theory, methods of analysis, and visualization techniques whereas GIS is the term that still refers to the hardware and software component. Unfortunately, most individuals use the term GIS to refer to the systems themselves, and are unaware of the rapidly evolving science.

From its incipient roots in digital mapping technology, GIS has now become a billion dollar industry in the US, being utilized by the private sector (including environmental consulting and engineering companies), the public sector (including local, state, and the federal government), and in academia (nearly all universities offer a series of courses, and often a minor/major, in GIS.

GIS is becoming a routine analysis and display tool for spatial data, and used extensively in applications such as:

- Land use mapping (for urban planning purposes)
- Transportation mapping and analysis (for determining efficient transportation routes for deliveries and emergency response)
- Geodemographic analysis (for store location)
- Utilities infrastructure mapping (for precise gas, water, and electric line mapping and maintenance)
- And multiple applications in natural resource assessment (including water quality assessment and wildlife habitat studies)

GIS allows efficient and flexible storage, display, and exchange of spatial data, as well as use in models of all kinds. More recently, the term GSci has emerged as representing 'the science of spatial data processing' – including conceptual problems in spatial data acquisition, storage, analysis, and display – while GIS is reserved for the actual hardware/software component of the technology.

24.2. What is a GIS?

The term GIS is often applied to any package that involves mapping capability or spatial data. However, as Kraak and Ormeling (1997) point out, there are actually several different types of systems – *spatial* information systems – that may be categorized based on their functionality. Figure 24.1 arrays these systems based on both their cartographic and spatial analytic capabilities (Kraak and Ormeling 1997). The simplest of the systems involves Computer-Aided Design (CAD). CAD systems are used by engineers, architects, and designers to assist with automated drawing, and normally provide powerful design tools. A typical application would be in the architectural design of a house. There is no real cartographic or spatial-analytic capability with CAD systems. Similarly, facilities management software allows for the organization of complex 'spatial' databases such as those used by utilities companies for the maintenance of customer accounts, yet allow for no analysis or mapping. More sophisticated software involves computer mapping, where spatial databases may be displayed and complex symbolization types portrayed. Many computer mapping

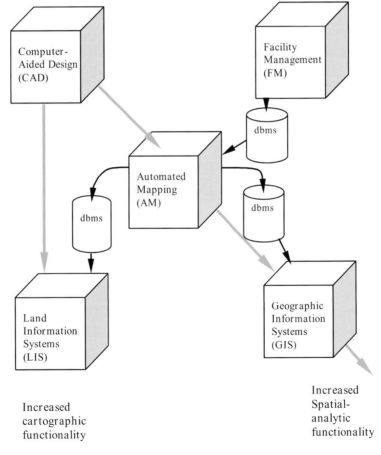

Figure 24.1 The relationship between several types of information systems (redrawn from Kraak and Ormeling 1997).

systems allow for limited analyses, such as non-topological overlay and address matching, but are still not considered full GISs. Land Information Systems (LISs) are designed for the storage and cartographic display of large-scale property – cadastral – databases. Normally urban LISs are utilized for maintaining the parcel-level data needed for city management – taxation, infrastructure repair, and the mapping of crime. Finally a true GIS allows for the powerful spatial analysis and cartographic display of spatial databases. A working definition of a GIS is *a computer-based set of methods for the acquisition, storage, analysis, and display of spatially-addressable data.*

24.3. Basic core knowledge in GSci

There are a core set of basic topics that are considered fundamental to GIS. Some of these include:

- Geometric aspects of coordinate systems and map projections, including construction methods and geometric distortions produced;
- Basic cartometric techniques including measurement of distance and area from maps and digital databases;
- The notion of map scale as the mathematical relationship between map and earth distance, as well as mathematical transformations of scale in computer-based map display operations;
- The derivation of multiple scale databases and automated generalization;
- Statistical classification and analysis of attribute data for effective thematic map display;
- Geographical data structures including vector and raster-based methods;
- Spatial analysis and geographical problem solving using a map algebra;
- Data quality, error propagation through databases, and error assessment;
- Principles of cartographic representation, including four-dimensional cartographies and multimedia cartography.

24.4. A brief history of GIS

A comprehensive history of GIS/GSci is not possible in this short chapter, but the reader may refer to Foresman's recent edited volume on, *A History of GIS* (Foresman 1998). Several key events will be reviewed in the chapter, including the State of Minnesota's MLMIS, the Harvard Laboratory for Spatial Analysis and Computer Graphics, the National Center for Geographic Information and Analysis, and the recently created University Consortium for GSci.

24.4.1. The Minnesota land management information system (MLMIS)

The MLMIS started in the late 1960s as a joint research project between the State Planning Agency and the Center for Urban and Regional Affairs (CURA) at the University of Minnesota. The seeds for MLMIS were planted in 1967, when the Minnesota State Legislature became interested in how Minnesota's lakeshore lands were being developed and perhaps overused. The goal was to attempt to influence decision making to improve land-use planning along Minnesota's thousands of miles of shoreline, and answer questions such as, 'Display all the government lots with wet soil plus lowland bush vegetation plus flat to gentle shore with a weedy bottom offshore' (Niemann and Niemann 1996). The Statewide Lakeshore Development Study itself led to the creation of the Statewide Land Use Map, the basis of MLMIS. This early statewide database utilized the US Public Land Survey System (PLSS) quarter-quarter section unit (40 acres) as the data storage and graphic link for the study. Landuse/landcover was determined from aerial photography, geocoded at the quarter-quarter section, and represented with line-printer characters (3 × 5 representing a square of '40'). The actual computer production and printing of this first ever statewide landuse map represents this country's first true GIS. Additionally, much of Environmental Planning Programming Language (EPPL), the first real software designed for mapping and analysis, was developed from these early efforts. Thus, using the maps and analysis, the researchers pressed for better land-use planning. In large part this was because of the maps and supporting analysis the legislation passed. 'Environmentally-based land-use planning in Minnesota had become a reality thanks to the computer-based data and analysis' (Niemann and Niemann 1996).

24.4.2. Harvard laboratory

A second major activity in the history of GIS involved Harvard University's Laboratory for Computer Graphics and Spatial Analysis, developed by Howard Fisher. The Harvard Laboratory was responsible for creating the first commercially-available computer-mapping package, SYMAP, for introducing a set of new analysis and mapping methods, and for sponsoring a series of computer mapping workshops.

SYMAP, which stood for Synagraphic mapping, or 'acting together graphically' was a computer mapping program that allowed users to build simple spatial databases and created three types of maps: Conformant (choropleth), Isarithmic, and Proximal, based on Thiessen polygons. As with the MLMIS project, the output from the SYMAP package was grid-based, and represented with line-printer characters (Figure 24.2). SYMAP's structure included a series of modules, or packages, including those for the creation of the database. The major components were the:

A-Outline	(bounding polygon of the study area, e.g. US border)
A-Conformolines	(bounding polygons of any subunits, e.g. states' borders) used for choropleth mapping
B-Data points	(centroids of subunits) used for Isarithmic mapping
E-values	statistical values used for mapping
F-map package	a set of electives, including those to manipulate the size, number of data classes, value range minimum and maximum, and other variables)

The Harvard Laboratory, which also developed a vector-based package called Odessy – that allowed for complex topological data structures – was dismantled during the 1970s with many of its key researchers continuing their work at other institutions.

24.4.3. NCGIA

During the mid-1980s the National Science Foundation quickly realized that some type of coordinated effort in GIS was needed, and a call for proposals to establish a national center was issued in 1987. Eight academic institutions or consortia submitted proposals, and in 1988 a winning group of the University of California, Santa Barbara, State University of New York at Buffalo, and the University of Maine was selected. For the first 10 years of NCGIA, the center focused on basic research and a series of expert meetings on fundamental topics were held. Table 24.1 shows listings of these initiatives, along with dates of activity

The reader should be aware that the National Center for Geographic Information and Analysis (NCGIA) had a prolific publications program, including books, reports of the specialists meetings, and scientific papers. The full listing of NCGIA publications may be found at: http://www.ncgia.ucsb.edu/pubs/pubslist.html. In addition to scientific research, the NCGIA also developed specialized software that can be found at: http://www.ncgia.ucsb.edu/pubs/pubslist.html#software.

During the mid 1990s, the NCGIA established a new research program, called Project Varenius. Where the initial set of basic research topics attempted to cover all of GSci, this new project focused on three specific areas: Cognitive models of geographic space, Computational Implementations of Geographic Concepts, and Geographies of the Information Society. As with the previous effort, Varenius was organized around a set of specialist meetings, as listed below. One can see from the topics addressed by the NCGIA the initial set of core research areas considered significant by the GIS research community (Table 24.2).

In addition to forwarding the basic research agenda, the NCGIA also worked on issues of education in the emerging area of GSci. In the late 1980s, the Center produced a print version of a model curriculum, which contained modules on introductory material, technical issues, and applications. Later, this project migrated into

Table 24.1 Specialist meetings of the NCGIA core research program

Initiative 1	Accuracy of spatial databases (1989–90)
Initiative 2	Languages of spatial relations (1989–91)
Initiative 3	Multiple representations (1989–91)
Initiative 4	Use and value of geographic information (1989–92)
Initiative 5	Large spatial databases (1989–92)
Initiative 6	Spatial decision support systems (1990–2)
Initiative 7	Visualization of spatial data quality (1991–3)
Initiative 8	Formalizing cartographic knowledge (1993–)
Initiative 9	Institutions sharing geographic information (1992–)
Initiative 10	Spatio-temporal reasoning in GIS (1993–)
Initiative 12	Integration of remote sensing and GIS (1990–3)
Initiative 13	User interfaces for GIS (1991–3)
Initiative 14	GIS and spatial analysis (1992–4)
Initiative 15	Multiple roles for GIS in US global change research (1994–)
Initiative 16	Law, information policy and spatial databases (1994–)
Initiative 17	Collaborative spatial decision-making (1994–)
Initiative 19	(The social implications of how people, space, and environment are represented in GIS)
Initiative 20	(Interoperating GISs)

Table 24.2 Specialist meetings of the Varenius project

Discovering geographic knowledge in data-rich environments
Multiple modalities and multiple frames of reference for spatial knowledge
Measuring and representing accessibility in the information age
Cognitive models of dynamic geographic phenomena and representations
Empowerment, marginalization and public participation GIS
Place and identity in an age of technologically regulated movement
Ontology of fields
International conference and workshop on interoperating GIS

the virtual domain, where an on-line version may now be found at: http://www.ncgia.ucsb.edu/education/curricula/giscc/cc_outline.html. NCGIA also developed a K-12 school program in order to bubble the technology down into the pre-collegiate levels.

24.4.4. University Consortium for GIS (UCGIS)

During the 1990s, it became clear that the GIS community required a common voice that was broader than the three-member NCGIA. After several years of discussion, the UCGIS was established in December of 1994. As established at the founding meeting, the UCGIS has several primary purposes:

- To serve as an effective, unified voice for the GSci research community
- To foster multidisciplinary research and education
- To promote the informed and responsible use of GSci and geographic analysis for the benefit of society

One of the major contributions of the UCGIS has been the establishment of a series of research, education, and application priorities, or what are now called challenges. The initial series of research challenges established in 1996, and updated in 1998, include:

- Spatial data acquisition and integration
- Distributed computing
- Extensions to geographic representation
- Cognition of geographic information
- Interoperability of geographic information
- Scale
- Spatial analysis in a GIS environment
- The future of the spatial information infrastructure
- Uncertainty in spatial data and GIS-based analyses
- GIS and society

The goal of creating such 'challenges' was to integrate the various disciplinary expertise that exists with UCGIS, to identify those true impediments to critical research areas, and to push for increased federal funding for these areas. Readers can find detailed 'white papers' on each of these topics on the UCGIS Web page. UCGIS has also established a series of education challenges (1997), and application challenges (1999). Full white papers on these and other topics made be found at: www.ucgis.org.

From its incipient roots, the UCGIS has matured into an organization consisting of nearly 50 universities, with an annual assembly and significant national-level activity. For instance, each January the UCGIS sponsors a Congressional Breakfast, where members of Congress learn about the potential for GIS in solving basic societal and environmental problems.

24.5. Societal implications of GIS

An area of increasing research and concern in the discipline of GIS involves work at the interface of GSci and the societal implications of these technologies. Without a firm understanding of the consequences of GIS use, much effort may be wasted or lost on more technology and good intentions with little benefit and possible misunderstanding. Several conferences, and a growing number of publications, represent the growing activity in GIS-Society. One of the first major efforts was the NCGIA Initiative 19, The social implications of how people, space, and environment are represented in GIS (http://www.geo.wvu.edu/i19/default.htm) that focused attention on the social contexts of GIS production and use and addressed a series of conceptual issues, including:

- In what ways have particular logic and visualization techniques, value systems, forms of reasoning, and ways of understanding the world been incorporated into existing GIS techniques, and in what ways have alternative forms of representation been filtered out?
- How has the proliferation and dissemination of databases associated with GIS, as well as differentiatial access to spatial databases, influenced the ability of different social groups to utilize information for their own empowerment?
- How can the knowledge, needs, desires, and hopes of marginalized social groups be adequately represented in GIS-based decision-making processes?
- What possibilities and limitations are associated with using GIS as a participatory tool for more democratic resolution of social and environmental conflicts?
- What ethical and regulatory issues are raised in the context of GIS and Society research and debate?

It is clear that in the discipline of GIS, as with all information technologies, investigators must be both aware of, and concerned with, the potential impact of their analyses and visualizations. As an example, GISs are increasingly being utilized in assessing 'environmental justice', or whether certain marginalized populations (the poor, minorities, children) are disproportionately affected by toxic and waste sites and various forms of pollution. Although GISs can indeed show relationships between, for instance, toxic sites and low-income populations, might the results of such analyses and the resultant visualizations be used by government and/or insurance companies to argue for altered zoning or higher insurance premiums? Likewise, could the results of an analysis of car theft locations result in an insurance 'redlining' of such areas? As with all such technologies, the best solution is to have well-trained, knowledgeable and ethical persons working with the technologies, and policy maker who clearly understand the limitations.

24.6. The literature of GIS/GSci

The literature in the discipline of GIS has grown rapidly over the past 15 years, with a growing number of conferences, books, and journals. Some of the literature tends to be very basic, while increasingly we see more applied publications. Some of the resources available for both researcher and practitioners include:

Scholarly journals

Cartography and GIS
International journal of GIS
Geographical analysis
Geographical systems
Transactions in GIS
Geoinformatica
The URISA journal

Popular magazines

GeoWorld (GIS World)
GeoSpatial Solutions (GeoInfo Systems)
ARC news
ARC user
Business GEOgraphics

Proceedings

Auto-Carto-xx, 1972–97
Spatial data handling, started in 1984
GIS/LIS, 1988–98
GI Science 2000

Major textbooks

Aronoff, S., *GIS: A Management Perspective*, 1989.
Bernhardsen, T., *Geographic Information Systems*, 1992, 1999.
Burrough, P.A. and McDonald, R.A., *Principles of Geographic Information Science*, 1988.
Chrisman, N., *Exploring Geographic Information Systems*, 1997.
Clarke, K., *Getting Started with Geographic Information Systems*, 1999.
DeMers, M., *Fundamentals of Geographic Information Systems*, 2000.
Heywood, I., Cornelius, S. and Carver, S., *An Introduction to Geographical Information Systems*, 1998.
Huxhold, W.E., *An Introduction to Urban Geographic Information Systems*, 1991.
Korte, G.B. *The GIS Book*, 1997.
Laurini, R. and Thompson, D., *Fundamentals of Spatial Information Systems*, 1992.
Longler, P.A., *et al.*, *Geographic Information Systems*, 1999.
Obermeyer, N.J. and Pinto, J.K., *Managing Geographic Information Systems*, 1994.

Peuquet, D. and Duane M.F., *Introductory Readings in GIS*, 1990.
Pickles, J. *Ground Truth: The Societal Implications of Geographic Information Systems*, 1995.
Tomlin, C.D. *Geographic Information Systems and Cartographic Modeling*, 1990.

Conferences and workshops

GIS/LIS conferences
ACSM/ASPRS conferences

24.7. Conclusions

The discipline of geographic systems/science has witnessed remarkable growth and maturation during its first 20 years. From its incipient roots in rudimentary computer mapping software, GIS is now a multi-billion dollar business, with its tentacles in all aspects of society, including local, state, and federal government as well as extensively in the private sector. Nearly all universities now have established coursework in GIS, with several professional masters degrees now appearing. GIS software is finally being used for more than simple mapping, and to assist with complex spatial analysis and modeling. The remainder of Part 4 of this manual covers the basics of GIS, including fundamentals, transformations, geographic data structures, hardware/software issues, spatial analysis, basics of cartography, and implementing a GIS. Chapter 25 covers some of the fundamental issues in GIS, including types of data, sources of data, and provides an explanation of GIS modeling. Chapter 26 reviews geographic data structures, and provides examples of both vector and raster approaches. Chapter 27 delves into the fundamental transformations of GIS, or how spatial data are converted from one form to another. Chapter 28 details both GIS software and hardware, and also provides a series of practical recommendations on selection. Chapter 29 will cover principles of spatial analysis and modeling, while Chapter 30 reviews considerations with spatial data quality. Finally, Chapter 31 addresses basic principles of cartography, symbolization and visualization while Chapter 32 closes this section by discussing issues when implementing a GIS.

References

Aronoff, S., *Geographic Information Systems: a Management Perspective*, WDL Publications, Ottawa, 1989.
Bernhardsen, Tor., *Geographic Informatiion Systems: An Introduction* (2nd Edition), John Wiley and Sons, New York, 1999.
Burrough, P.A., *Principles of GIS for Land Resources Assessment*, Oxford University Press, Oxford, 1996.
Burrough, P. A. and McDonnell, R. A., *Principles of Geographical Information Systems*, Oxford University Press, Oxford, 1998.
Clarke, K. C., *Getting Started with Geographic Information Systems* (2nd ed.), Prentice Hall, Upper Saddle River, NJ, 1999.
Chrisman, N. R., *Exploring Geographic Information Systems*, John Wiley & Sons, New York, 1997.
DeMers, M. N., *Fundamentals of Geographic Information Systems* (2nd ed.), John Wiley and

Sons, New York, 2000.

Foresman, T. (ed.), *The History of Geographic Information Systems*, Prentice Hall, Upper Saddle River, NJ, 1998.

Heywood, I., Cornelius, S. and Carver, S., *An Introduction to Geographical Information Systems*, Prentice Hall, Upper Saddle River, NJ, 1998.

Huxhold, W. E., *An Introduction to Urban Geographic Information Systems*, Oxford University Press, New York, 1991.

Korte, G. B., *The GIS Book* (4th ed.), Onword Press, Sante Fe, NM, 1997.

Kraak, M. J. and Ormeling, F., *Cartography and Visualization*, 1997.

Laurini, R. and Thompson, D., *Fundamentals of Spatial Information Systems*, Academic Press, London, 1992.

Longley, P. A., Goodchild, M. F., Maguire, D. J. and Rhind, R. W. (eds.), *Geographical Information Systems*, Vol. 1, John Wiley and Sons, New York, pp. 1–20, 1999.

Obermeyer, N. J. and Pinto, J. K., *Managing Geographic Information Systems*, The Guilford Press, New York, 1994.

Pickles, J., *Ground Truth: the Societal Implications of Geographic Information Systems*, The Guilford Press, New York, 1995.

Peuquet, D. J. and Marble, D. F. (eds.), *Introductory Readings in Geographic Information Systems*, Taylor and Francis, London, 1990.

Niemann, B. J. and Niemann, S. S., LMIC: pioneers in wilderness protection, *Geo Info Systems*, September, 1996, 16–19.

Tomlin, C. D., *Geographic Information Systems and Cartographic Modeling*, Prentice Hall, Englewood Cliffs, NJ, 1990.

Chapter 25

Fundamentals of Geographic Information Systems (GIS)

David A. Bennett and Marc P. Armstrong

25.1. Introduction

Record keeping systems that help people collect, manage, analyze and display geographic information can be traced back thousands of years (Dale and McLaughlin 1988: 46; Thrower 1972). During the past several decades these systems have had to evolve rapidly to keep pace with changes in computer technologies and in the practice of business and government. Despite these advances and the attention that such systems have received in recent years, the concept of a 'Geographic Information System' (GIS) remains elusive since it is subject to a broad range of context-dependent interpretations. In this chapter we will define commonly used terms, sketch out a framework designed to help practitioners conceptualize, organize and implement GIS applications, and demonstrate, using a prototype application, how GIS software can be used to address an environmental problem.

The previous chapter discussed the concepts of GIS and science. To help place this discussion in context we will draw a further distinction between a 'GIS' and 'GIS software' (we wish to discourage the use of the term 'GIS system' because of its obvious redundancy). A GIS refers to technology embedded in a particular social structure (e.g. the city governance of Iowa City, IA, US) in which information about places and their characteristics is recorded and transformed. The combination of a particular technology and a particular social setting creates a unique set of constraints and opportunities for the design, implementation, and maintenance of a GIS (Chrisman 1997). The success or failure of a specific GIS implementation often depends on an appropriate match between technology and the institutional culture of a particular agency. GIS software, on the other hand, is a set of stored computer instructions that transforms geographic information. Though the basic concepts that underlie the specification of these instructions may have long histories, the functions available in GIS software, and their applications have evolved rapidly during the past two decades. In the following sections, we first describe what we mean by GIS, and then describe the kinds of functional transformations that are embedded in GIS software.

25.2. Background

A GIS must provide some measure of value (e.g. economic, social, political) to ensure its survival. Some researchers have attempted to attach an economic value to implemented systems while others have described methods for conducting cost-benefit analyses prior to system implementation (Obermeyer 1999; Gillespie 2000). Value can accrue to businesses, for example, through increased efficiency in the execution of routine business functions (e.g. the delivery of goods and services), or through more effective methods for selecting from among competing plans for infrastructure development (e.g. retail site selection). Within government, the assessment of value is less straightforward but equally important. Many of the early failures of GIS can be traced to the use of data that lacked appropriate resolution and to the lack of application-specific functions that were needed to support the activities routinely performed by government agencies (Dueker 1979; Tomlinson *et al.* 1976; see Chapter 33 for a summary of applications). Consequently, continued maintenance and support for the flawed systems could not be easily rationalized within existing bureaucratic structures, funding withered, and the implemented, but not particularly useful, systems were doomed.

Lessons were learned from early failures. Commercial firms engaged in the development and sales of GIS software have paid careful attention to the needs of their users and this has, in turn, shaped the form of available software functions and the rate at which new functionality is developed. Initial concerns with the efficient management of geographic information, while still critical given the explosion in volume of available data (Armstrong 2000), have begun to be displaced by an increased focus on analytical capabilities. This has led to a compartmentalization of software functions that are tailored to specific application domains. This trend has allowed vendors to market products designed to support the specific work tasks of their clientele (e.g. toolboxes designed explicitly to support location decisions or watershed management activities). Improvements to methods of analysis, either as fundamental shifts in routine practices, or in improvements to the performance of existing approaches, can have demonstrable payoffs in time-savings and in the quality of analyses provided. Greater access to digital geographical data and advances in data handling technologies do not, however, guarantee a successful GIS implementation. The development of such systems requires a sound understanding of GIS fundamentals.

25.2.1. A transformational view of GIS software

The purpose of GIS software is to transform geographically referenced data using a set of software tools that facilitate the capture, storage, manipulation, analysis, and display of geographical data. Each of these processes can be expressed as a transformation of format, attributes, or geometry. Consequently, we can view GIS software as a kind of 'transformational engine' that uses data inputs and produces information that has some value-added component (e.g. decision support). This transformational view has a lineage traced back to early work by Tobler (1979) that has persisted through cartography, analytical cartography (Clarke 1995) and GIS (Chrisman 1999). Clarke (1995) identifies four types of cartographic transformations that can be placed directly into the context of GIS projects:

1 Geometric (e.g. changes in coordinate systems)
2 Dimension (e.g. from a two-dimensional polygon to a one-dimensional point)
3 Scale (e.g. from 1:24,000–1:250,000 scale map), and
4 Map symbolization (e.g. the selection, classification and presentation of elements on a map)

While it is true that the functional gap between digital cartographic software and GIS software has essentially disappeared, users of GIS software are, in general, more interested in the analysis and management of geographical data and the phenomena that these data represent (e.g. streets networks, land use, watersheds). GIS software, therefore, typically includes transformational functions that:

1　Change the form of geographical data (e.g. capture spatial data by transforming paper maps into digital form, or change digital data (bits) into scientific visualizations (pixels));
2　Change the digital representation of geographical data (e.g. a raster-to-vector data transformation (Flanagan *et al.* 1994), transforming data stored in Spatial Data Transfer Standard (SDTS) format (USGS 2000) into a proprietary format); and
3　Change the content of geographical data (e.g. eliminate unneeded data, add information content to raw data through analysis).

These three basic categories can be further decomposed into more specific functional classes. For example, changes in content can be classified into transformations that assist in the:

1　Modification of data (e.g. update cadastral maps to reflect new subdivisions or new owners). The transformations traditionally associated with automated cartography belong in this category.
2　Extraction of data (e.g. create a new dataset as a subset of a larger dataset, proximity operators, the extraction of topographic features from Digital Elevation Models (DEM), the extraction of the shortest path between two points in a street network).
3　Analysis of data (e.g. synthesize or abstract data to make it compatible with other datasets, provide analytical value to raw data, integrate multiple datasets into a single dataset).

The above lists are only illustrative of what can be accomplished using GIS software. For a more formal review of proposed taxonomies for GIS transformations the reader is referred to Chrisman (1999).

GIS applications can be described as a sequenced series of transformations that begins with the 'capture' of one or more datasets and concludes with the synthesis of these data into a form that helps users to answer questions (Figure 25.1). It is important to note that the way in which geographical data are stored places real limits on the kinds of transformations that can be performed and, thus, the kinds of analyses that a GIS user can conduct. Furthermore, though each GIS software package implements a particular set of transformational tools, no single software package implements all available tools. The selection of a particular software package will, therefore, affect the kinds of transformations that can be accomplished (see Chapter 28). It is, therefore, good practice to begin each GIS project with a careful design process.

During the GIS design process the application developer must carefully define the problem. This definition can begin by seeking to delimit the scope of inquiry:

• What are the key questions that must be answered to solve the problem under consideration?

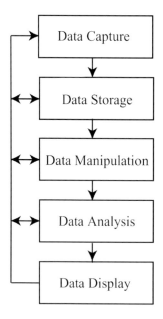

Figure 25.1 The basic five steps associated with GIS applications. Note that as the process proceeds it is often necessary to return to a previous step to, for example, gather

- What are the essential geographical features that must be captured in the digital domain to answer these questions?
- How do these features vary across space or through time?
- What tools are required to quantify and analyze these spatio-temporal patterns?

The insight gained from this initial problem definition stage will guide key decisions regarding software, geographical representation, database content, scale, resolution, spatial extent, and geographical reference systems. For further information on GIS implementation, see Chapter 32.

25.3. The language of GIS

As with many technological fields, GIS has a unique language that must be mastered before one can take full advantage of the capabilities of associated software. Historically, there were many different 'dialects' of this language as the various loci of GIS development generated software to meet the needs of specific application domains. For example, many of the concepts that led Tomlin to develop map algebra (Tomlin 1990; Tomlin and Tomlin 1981) were apparently independently developed in other places by researchers with different disciplinary backgrounds (Alsberg *et al.* 1975; Fabos *et al.* 1978). While multiple definitions of technical terms still cause some ambiguity, for example, alternative usage's of the term *arc*, much progress has been made in the standardization of a GIS language. Here we introduce the reader to terminology that has become somewhat standard within the GIS community. Many of the issues introduced here are discussed in greater detail in subsequent chapters.

25.3.1. Spatial representation

One of the first challenges that a user of digital geographic data must confront is how to best represent the geographical systems of interest within the digital domain (Peuquet 1988). Real world geographical systems are complex, dynamic and inter-related. While seemingly complex to the new user, geographical datasets are normally simplified, static models of reality (Peuquet and Duan 1995). The choice of a particular model, or representation, will affect almost every aspect of a GIS project, from the cost of data acquisition to the types of conclusions that can be drawn from associated analyses.

In 1982 the US National Committee for Digital Cartographic Data Standards set out to standardize the terminology associated with the digital representation of geographical data. This work evolved into the SDTS, which was first ratified as US Federal Information Processing Standard number 173 in 1994 (FIPS 1994). The current version of this standard is known as ANSI NCITS 320-1998, which was ratified by the American National Standards Institute in 1998 (ANSI 2000). These standards provide a conceptual foundation on which application developers can build digital representations of geographical phenomena and a language that allows them to communicate this representation, either verbally or in digital form, to others in an unambiguous manner.

25.3.1.1. The conceptual model

GIS datasets store a digital representation of real world geographical phenomena (Figure 25.2a). Such phenomena can be tangible features (e.g. the segment of Main Street between 1st and 2nd Avenue) or they can be more abstract, like a measurement of elevation at a particular point on the earth's surface. GIS datasets typically represent a specific class of geographical phenomena (e.g. the phenomenon *Main Street between 1st and 2nd Avenue* is a kind of *street*, Figure 25.2b). These classes are referred to as *entity types*. While the representation of entity types as independent datasets (often referred to as a *layer* or *theme*) is useful from a data management perspective, the choice of a particular data classification scheme can have a significant effect on the outcome of any subsequent analyses (Anderson 1980: 104; Bowker and Star 1999).

Each entity type is associated with a set of attributes that are used to uniquely identify specific phenomena (key attributes) and provide information of value to users (e.g. street name, length, and ownership). A geographical phenomenon (e.g. the segment of Main Street between 1st and 2nd Avenue) is known as an *entity instance* and the digital representation of this instance is an *entity object* (Figure 25.2c). Entity objects often contain data on location, attribute, and topology (relative location).

25.3.1.2. Dimensionality and entity objects

Entity objects are often classified by their spatial dimension. This carry over from

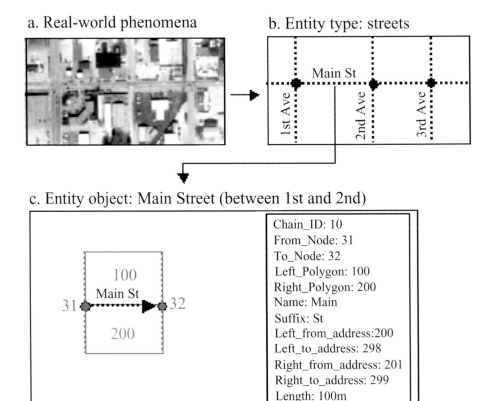

Figure 25.2   Geographical datasets store real world phenomena (a street network) as entity types (a street dataset) that, in turn, are comprised of one or more entity objects (a street segment). Attribute information uniquely identifies each object (e.g. Chain_ID) and provides information useful for the management of the entity type.

traditional cartographic symbolization has been translated into the conceptual conventions of GIS and serves partly as a convenience for programmers and their data structures. Entity objects that represent geographical phenomena are formed from zero, one, two, and three-dimensional spatial objects (these same objects are referred to as zero, one, two, and three cells in some of the technical literature). For each dimension a set of well-defined spatial objects has been defined (see examples in Figure 25.3). For example, a *point* is defined as 'a zero-dimensional object that specifies geometric location', while a *node* is 'a zero-dimensional object that is a topological junction of two or more links or chains, or an end point of a link or chain.' (USGS 2000). As this definition suggests, complex objects are often constructed from a collection of simple objects. For example, a *complete chain* is 'a directed non-branching sequence of non-intersecting line segments and (or) arcs...' that '...references left and right polygons and start and end nodes' (USGS 2000). The construction of complex

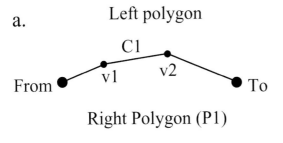

a. Left polygon

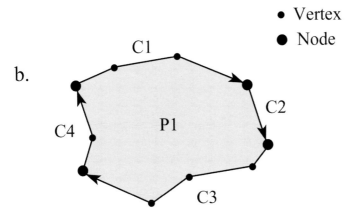

Figure 25.3   ANSI NCITS 320-1998 provides an established set of definitions for a variety of spatial objects. For example, C1 in (a) is a representation of a complete chain that bounds the GT-polygon P1 in (b) (a polygon defined by geometry and topology).

objects from simple objects and the explicit representation of topology (e.g. the polygon to the left of Main Street from 1st to 2nd Street represents census block *y*) provides advantages for data management and spatial analysis (Peucker and Chrisman 1975).

Two-dimensional entities can be placed into three categories: polygons, grids, and images. As suggested above, polygons may be defined as a set of one-dimensional spatial objects or as a set of points. Entity types built from points, lines, or polygons are collectively referred to as vector-based representations. Topological vector-based representations explicitly store relative spatial location (Figures 25.2 and 25.3; Peucker and Chrisman 1975). Grids are formed from a tessellation of the landscape into cells (typically square). Images are comprised of pixels (picture elements) and are often captured from some form of remote sensing platform (e.g. aerial photographs or satellite images). Collectively, grids and images are classed as raster-based representations.

Though the advantages and apparent flexibility associated with the use of these techniques for the representation of geographical phenomena is real, this approach also can prove to be vexing to novice and expert alike (Buttenfield and McMaster 1991). For example, a city might be encoded and represented as a point at one scale and as an area at another. The way in which such dimensional shifts take place with respect to scale and purpose remains an area of active research (Leitner and Buttenfield

1995) and can have important implications when selecting applicable analytical methods.

25.3.1.3. From entity objects to data structures

The ANSI NCITS 320-1998 standard provides a common language that can be used to communicate information about the structure and content of geographical datasets. It does not, however, specify how these datasets are to be implemented. The specific implementation of an entity object, referred to as a data structure, is left to software producers and, thus, is often viewed as proprietary knowledge. Various tools are being developed to facilitate access to such propriety data structures. For example: (1) the SDTS profile formats in the US (http://mcmcweb.er.usgs.gov/sdts/) and the National Transfer Format (NTF) in Britain (http://www.bsi-global.com/group.html) provide well defined data structures that vendors can export to and import from; and (2) the Open GIS Consortium promotes access to proprietary data through the specification of a common communication protocol (McKee 1999; http://www.opengis.org).

25.3.1.4. The role of metadata

To assess the utility of a particular geographic dataset the GIS analyst must know: (1) its spatial extent; (2) how it was developed (e.g. its lineage); (3) the set of stored attributes; (4) the resolution of the stored data; and (5) the accuracy of the stored data. Resolution is a statement about the level of detail recorded in the dataset (e.g. how many land use classes are used, what is the minimum mapping unit recorded on the map). Accuracy, on the other hand, is a measure of how closely the data reflects a real location or attribute value. Such issues are critical to the assessment of utility. For example, the data must cover the geographical extent of an area of interest and the spatial and attribute data must be sufficiently detailed, current, and accurate. Other issues are largely a matter of convenience. For example, coordinate systems and data structures can be transformed and databases can be joined.

The information required to assess the fitness of a dataset in the context of a particular problem is stored as *metadata*. The development and maintenance of metadata is becoming increasingly important as large volumes of data are being distributed across computer networks and, as a consequence, there is no 'local' knowledge about the conditions under which the data were produced. From a thorough reading of the metadata, an individual should be able to determine what types of transformations are needed to prepare the data for analysis.

25.3.2. Finding the necessary data

Historically, the cost of data acquisition has been a significant impediment to the development of GIS applications. Publicly accessible repositories of digital geographical data were rare and when such data were available, they often were stored in proprietary formats or lacked the accuracy and resolution needed to be widely applicable to geographical problem solving. In recent years the volume of data that can be downloaded from network accessible data repositories has increased rapidly. In the US, the Federal Geographic Data Committee (FGDC) has acted as a catalyst for the development of online geospatial data repositories through its efforts to coordinate the National Spatial Data Infrastructure (NSDI).

The NSDI is being developed to promote the distribution of geographical data produced by governments, industry, non-profit organizations, and the academic community (http://www.fgdc.gov). At the heart of the NSDI effort is the geospatial data clearinghouse. Each node on the clearinghouse network of data repositories is managed by a sponsoring agency that is responsible for the maintenance of the geographical databases within its jurisdiction. While the clearinghouse network is focused on data repositories within the US, there are several international sites linked into this network. There were 208 clearinghouse nodes as of August 2000 and this number continues to grow (see for example, http://gisdasc.kgs.ukans.edu/dasc_ie.html).

Much of the data that is stored on nodes within the US can trace their lineage back to data collection efforts initiated by different federal agencies. The following discussion provides a brief introduction to a sample of these datasets. Please keep in mind that this is not meant to be an exhaustive list of datasets or data providers.

25.3.2.1. US Geological Survey (USGS) datasets

The USGS provides four commonly used datasets: Digital Line Graphs (DLG), Digital Raster Graphics (DRG), DEM, and Digital Orthophoto Quarter-quadrangles (DOQ). A DLG file can (but does not always) contain vector representations of the following entity types: hypsography, hydrology, vegetative surface cover, non-vegetative features, jurisdictional boundaries, survey control markers, transportation, man-made features, and the public land survey systems (for details see http://rockyweb.cr.usgs.gov/nmpstds/acrodocs/dlg-3/1dlg0798.pdf). The USGS has produced DLG files from their 1:24,000, 1:100,000, and 250,000 scale topographic maps. DRG files are raster representations of a USGS topographic quadrangle maps. These images have been geographically referenced and rectified to facilitate their use within GIS software packages. DEMs provide topographic data in raster form (USGS 1993). These files are available at three resolutions: (1) 30 × 30 m datasets derived from 1:24,000 USGS topographic quadrangles; (2) 2 × 2 arc second datasets derived from 1:100,000 USGS topographic quadrangles; and (3) by 3 arc second datasets derived from 1:250,000 USGS topographic quadrangles (USGS 1993). DOQs are digital, geographically referenced and geometrically corrected, aerial photographs with a 1 m resolution (sometimes provided in resampled form to reduce storage requirements).

The USGS distributes these data in SDTS format. However, it is often possible to find these data in proprietary data formats (e.g. Environmental Systems Research Institute's (ESRI's) vector or raster formats) at geospatial data clearinghouse nodes. Note that not all USGS datasets are available for all areas in the US. The status of these datasets is documented at http://mapping.usgs.gov/www/product s/status.html and much of the data can be downloaded from the EROS Data Center (http://edc.usgs.gov/).

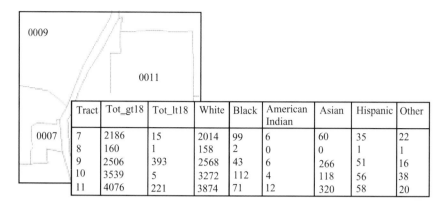

Tract	Tot_gt18	Tot_lt18	White	Black	American Indian	Asian	Hispanic	Other
7	2186	15	2014	99	6	60	35	22
8	160	1	158	2	0	0	1	1
9	2506	393	2568	43	6	266	51	16
10	3539	5	3272	112	4	118	56	38
11	4076	221	3874	71	12	320	58	20

Figure 25.4 Census geography can be linked to a wealth of socio-demographic data generated as part of the US decennial census.

25.3.2.2. US Census Bureau

The US Census Bureau maintains geographically referenced datasets derived from a variety of data collection activities. Among the most commonly used datasets are the Decennial Census of Population and the associated Topologically Integrated Geographic Encoding and Referencing (TIGER) files. A key to the utility of TIGER files is their rich topological and attribute content. Features (i.e. chains) in a TIGER line file contain information about multiple levels of census geography, and, if appropriate, the street name and associated address range for each side of the street (Broome and Meixler 1990). By storing census geography and street address information, census surveys returned by the general population can be easily linked to census geography. Through this link spatially aggregated data can easily be computed and used to produce a wide variety of socio-demographic maps and analyses (Figure 25.4). This same functionality can also be used to geographically locate any observation that has been assigned a street address. For example, retail establishments can quickly map customers based on their home address.

25.3.2.3. Natural Resource Conservation Service

Increasingly, soils data are being made available in digital form. These vector datasets represent soil series polygons and can be linked to a wealth of descriptive data that are stored in the Soil Survey Geographic Database (SSURGO). These datasets include information on the physical and chemical properties of the soil (e.g. salinity and depth to water table) and interpretive data that suggests appropriate uses for the land. These data can be found at the following sites: http://www.ftw.nrcs.usda.gov/ssur_ data.html, http://www.statlab.iastate.edu/soils/muir/.

25.4. Putting it all together through data transformations

In this section we illustrate how geographical problem solving can be represented as a well-defined set of transformations. A set of transformations that is designed to address a particular geographical problem is often referred to as a cartographic model and is expressed in a dialect of map algebra. Several alternative taxonomies for GIS-based transformations have been put forth in the literature (e.g. Chrisman 1997, 1999; Maguire and Dangermond 1991; Tomlin 1990). In section 2 we enumerated seven generic functions that are recurrent themes in most of these taxonomies. These transformations change the geometry, dimension, scale, symbolization, form, structure, or content of geographical datasets. To motivate this discussion we will develop a simple cartographic model designed to facilitate the selection of a sanitary landfill within Johnson County, IA, US. The following is a subset of the locational requirements for a landfill site set forth in the Iowa Administrative Code (http://www.legis. state.ia.us/IAC.html), though these requirements have been slightly modified for the purpose of this illustration:

1 The proposed site must be so situated that the base of the proposed site is at least 5 ft above the high water table.
2 The proposed site must be outside a floodplain.
3 The proposed site must be so situated to ensure no adverse effect on any well within 1,000 ft of the site.
4 The proposed site must be so situated to ensure no adverse effect on the source of any community water system within 1 mile of the site or at least 1 mile from the source of any community water system in existence at the time of application for the original permit.
5 The proposed site must be at least 50 ft from the adjacent property.
6 The proposed site must be beyond 500 ft from any existing habitable residence.

Some of these regulatory statements provide hard-and-fast rules. These *binary* variables are unambiguous and, in theory, are easy to document. For example, it is a straightforward GIS operation to determine the spatial relation between a proposed landfill site and a well if both are positioned accurately. Others rules are less well defined. When, for example, does the impact of a landfill become 'adverse' and how do we evaluate *a priori* what the likely impact of a proposed landfill will be on a water supply? Such an evaluation would require advanced ground water models. While interfaces are being constructed that link such models to GIS software, the representation of dynamic spatial processes is largely beyond the capability of 'out-of-the-box' GIS software.

Still other regulations are well defined but require data that are not generally available at the resolution needed to address this particular problem. For example, surveyors can say with some certainty that a point is within or beyond 50 ft of an adjacent property line. Digital datasets that accurately document parcel location, however, are not always available.

Nevertheless, each of these site requirements refers to a geographical phenomenon of interest that can, in theory, be rendered as a separate theme in a GIS database. Not surprisingly, the entity types of interest in this application include land use, public water supplies, private wells, water table, and floodplains (Figure 25.5, see plate 18). To instantiate an entity type as digital entity objects the GIS analyst must select a spatial object that captures the form and spatial pattern of the associated phenomenon. For example, a GIS analyst may choose to represent land use, depth to water table, and floodplains as polygons, and water supplies and well locations as point objects.

Table 25.1 An abbreviated set of metadata for datasets used in the landfill example. Note that while many of these datasets exist, the metadata presented below have been modified from their original form for the purpose of illustration

Soils	
Projection	UTM, Zone 15, NAD83
Format	Polygon
Attributes	Mapping unit ID
Original source	USDA Soil Conservation Service, 1:15,840 scale mylar map sheets
Floodplain	
Projection	UTM, Zone 15, NAD83
Format	Polygon
Attributes	Flood frequency (FREQ)
Original source	Federal Emergency Management Agency, 1:1,2000–1:24,000 scale paper maps
Private water wells	
Projection	UTM, Zone 15, NAD27
Format	Point
Attributes	Permit number, owner, date, address,...
Original source	Iowa Department Of Natural Resources, well reports
Private water wells	
Format	Comma separated text file
Attributes	Permit number, owner, date, address,...
Original source	Iowa Department of Natural Resources, well reports
Community water sources	
Projection	UTM, Zone 15, NAD83
Format	Comma separated text file
Attributes	Permit number, public water supply name, depth,...
Original source	Iowa Department of Natural Resources, 1:100,000 scale maps
Parcel maps	
Projection	UTM, Zone 15, NAD83
Format	Mylar maps
Attributes	
Original source	Johnson County Assessor's Office
Land use	
Projection	UTM, Zone 15, NAD83
Format	Grid
Attributes	Land cover (LC)...
Original source	Iowa Department of Natural Resources, supervised classification of Landsat imagery
Soils database	
Format	Comma separated text file
Attributes	Mapping unit ID, average depth to water table
Original source	National Resource Conservation Service, Map Unit Interpretation Database (MUIR)

25.4.1. Preparing the data

Table 25.1 presents an abbreviated set of metadata for the geographical data that are needed to address this problem. Note that not all of these data are available in a digital, geographically referenced form. For example, data for land parcels are in hard copy form and private wells developed after 1995 are referenced only by street address. These gaps in the availability of digital data raise three additional important issues.

25.4.1.1. Digitizing and scale

In the example developed here, parcel data stored as analog maps must be transformed to digital form (i.e. digitized). Several strategies can be explored to convert paper maps to digital form, but one commonly employed approach is referred to as manual digitizing (Marble *et al.* 1984). The process works by placing a map on a digitizing tablet and tracing selected geographical features with an electronic device called a puck. When the user activates the puck, perhaps to signify a street intersection or a point along a curve in a stream, its location is sensed using the time delay between an emitted electromagnetic impulse and a fine mesh embedded in the tablet. The resolution of a digitizer can be very high (e.g. 0.01 mm). Despite this apparent accuracy, however, the user must be aware of the impact of error on the resulting database.

Consider, for example, the process of digitizing features from a 1:24,000 USGS topographic map. First, the original map document will contain error. Assuming that the map meets National Map Accuracy Standards, the best that we can assume is that roughly 90 per cent of all 'well-defined' points are within 12.2 m of their actual location. Second, a digitizing mistake of, for example, a millimeter produces a positional shift of 24 m at a scale of 1:24,000. Third, users must approximate continuous features (e.g. a sinuous stream or the curve of a cul de sac) as a series of discrete line segments (for a review of error attributable to the digitizing process see Jenks 1981; Bolstad *et al.* 1990).

A review of the metadata associated with the datasets used in our prototypical project (Table 25.1) illustrates why we must be concerned about error resulting from the digitizing or cartographic process even when we download digital datasets from network accessible repositories. All of the datasets used here can trace their lineage to a hard copy map and each has an error component. Positional errors resulting from scale, map generalization, policies and practices, human error, and the discretization of continuous features are all faithfully copied into a digital database by the software and such errors will propagate through any analyses performed using these data (Lanter and Veregin 1992; Hunter and Goodchild 1996; Heuvelink 1998). Thus, while it is true that digital geographic information has no real scale and it can be enlarged or reduced to the limits of precision of a computer system, the raw material for a GIS dataset is often taken from analog maps. The scale of the source material is, therefore, integral to the dataset and can influence the quality of solutions obtained from GIS-based analyses. This is now widely recognized and consequently scale is an important part of most metadata schemes that have been devised.

25.4.1.2. The impact of map projections

The co-registration of the various geographic themes requires a consistent coordinate system (see Chapter 3). Here all themes will be *projected* to Universal Transverse Mercator (UTM) zone 15 using the North American Datum of 1983 (NAD83). A map projection is the transformation of a three-dimensional surface (the earth) to a plane (a flat map) (Snyder 1987; see also Chapters 2 and 4 of this manual). The projection of a dataset to a planar surface simplifies geometrical analyses but introduces distortion into the dataset. There are, of course, many ways to perform such transformations and each projection will have its own unique pattern of error and distortion. GIS users should understand the impact of this error on their projects.

25.4.1.3. The need to address match

Note that not all of the data needed for our landfill location problem possess geographical coordinates. For example, our only knowledge about wells drilled after 1995 is through permit references to street addresses. The question becomes: how are addresses converted to plane coordinates? The answer lies in address matching (Dueker 1974; Drummond 1995). The process of address matching relies on an up-to-date street centerline file that contains attribute data for street names and address ranges, for the left and right side of the road, for each chain in the database. In the US, TIGER files provide a common source of such information (Broome and Meixler 1990). However, steps must be taken to ensure that these datasets are sufficiently current and accurate to support a particular application. A number of private firms provide centerline datasets that are continually updated.

Given an appropriate street file, address matching the street addresses associated with the well records becomes, in part, a data extraction (query) procedure. For example, to place a well at 250 Main Street onto a map we would find a instance in the *streets* database where street name = 'Main', suffix = 'Street', left_from_address > = 200 and left_to_address <300 (see Figure 25.2c). In the address matching process, it is often assumed that location can be inferred through linear interpolation between the coordinates of the points of the associated street segment. In the absence of additional information, the best estimate for the location of 250 Main Street, for example, is halfway along the selected chain.

25.4.2. Data extraction and integration

The transformations presented in Figure 25.6a will generate input data that share a common data structure and coordinate system and, thus, provide raw material for data extraction and analysis (Figure 25.6b). Data extraction, in this situation, is used to subset the dataset and, thus, focus attention on the area of interest. Here we perform two common data extraction procedures: (1) a *select* operator is used to extract relevant polygons from the *soils* and *land use* themes; and (2) a *proximity* operator (in this case a buffer function) is used to find all land that may not be appropriate for the development of a landfill because of its proximity to features identified by the state's administrative code (e.g. within 1 mile of public water supply, see Figure 25.7).

Finally, the various themes produced through the above transformations must be merged into a single theme that represents those areas that are unsuitable for the development of a landfill. This transformation can be implemented using an overlay function (Figure 25.8). Care must be taken, however, in the interpretation of this analysis. As suggested above each of the individual datasets contains error and the error is propagated into this composite dataset. The accumulated error can be significant (MacDougall 1975; Chrisman 1987; Lanter and Veregin 1992) and decisions derived from overlay analysis must take into consideration the uncertainty generated by less than perfect data. The analysis presented here, for example, is best viewed as an inexpensive screening exercise that allows decision makers to focus more detailed

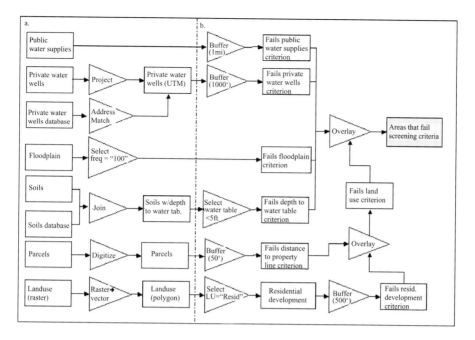

Figure 25.6 A cartographic model representing a solution to the example landfill location problem.

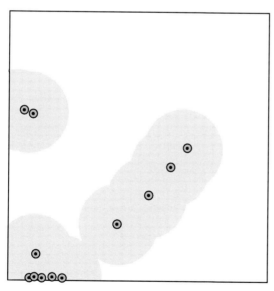

⊙ Public water supplies

    1 mile buffer

Figure 25.7 The results of a proximity operator (buffer) applied to public water supply locations.

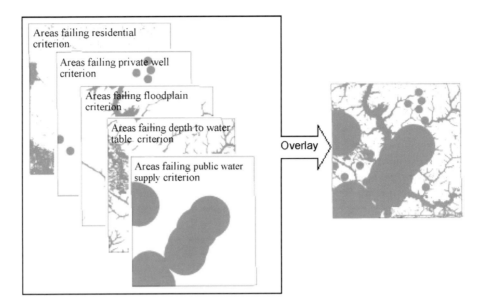

Figure 25.8 The results of an overlay operator (union) applied to the example landfill location problem.

analyses on promising sites. Users must also be cognizant that GIS software will allow them to perform analyses that are conceptually flawed (Hopkins 1977; Malczewski 2000).

25.5. Limitations and opportunities

The conceptual advancements made by GIS researchers over the past 20 years have been impressive and the utility of the resulting software is undeniable. However, it is equally safe to say that much work remains to be done in the field of Geographic Information Science (GISci) before these software packages reach their full potential. Here we select two issues that, on the one hand, limit the utility of current GIS software packages and, on the other, offer research opportunities for those interested in GISci. These two issues were selected because they are particularly relevant to the example presented in Section 25.4. For a more thorough accounting of current GISci research themes in the US the reader is directed to the University Consortium for GIS (http://www.ncgis.org) and The National Center for Geographic Information and Analysis (http://www.ncgia.ucsb.edu).

25.5.1. Geographic representation

Geographical systems are dynamic (change through time), interconnected (energy and material flows through space), and three-dimensional. However, the lineage of GIS software has its roots in automated cartography and image processing. This heritage has left what has seemed to be an indelible mark on GIS software (although recent developments are promising). This is both good and bad. The ability to communicate the results of geographical analyses in cartographic form has obvious advantages and image-processing techniques provide an invaluable source of derived data for geographical analysis. Thus, the tools of cartography and imaging have found a natural home in the GIS toolbox. Unfortunately, the forms of geographic representation that are well suited to cartographic representation and digital image processing do not necessarily transfer well when one is interested in the dynamic nature of geographical processes (Peuquet and Duan, 1995; Langran 1993; Yuan 1999). Research into the digital representation of dynamic geographical systems has been active over the past several years (Bennett 1997; Wesseling et al. 1996; Westervelt and Hopkins 1999). Much of this work has been experimental and is not readily portable to mainstream GIS software packages. As software development companies adopt more open and extensible architectures, however, four-dimensional GIS software (three dimensions in space, one in time) becomes more feasible.

25.5.2. The multifaceted nature of geographical problems

Geographical problem solving is often complex and multifaceted. The search for solutions to such problems often requires users to synthesize data from disparate sources and to find common ground among multiple competing stakeholders (Armstrong 1994). One of the most powerful features of GIS software is its ability to integrate information. The GIS application presented above is driven by simple exclusionary rules derived, for the most part, from mandatory state regulations. The problem, however, becomes more complicated when multiple stakeholders are considered. For example, landfill operators might be concerned about finding an economical solution, such as one that minimizes land acquisition and transportation costs. The general public may be concerned about such issues as the impact of the landfill on property values, agricultural land, local aesthetics, or environmental justice. Balancing issues related to acquisition costs, transportation costs, the politics of the 'Not In My Back Yard' problem, and environmental justice is a complicated problem. How does one weigh the relative importance of property values and transportation costs? Can you define and calculate an appropriate index for 'local aesthetics' or 'environmental justice'?

The integration of multicriteria evaluation techniques with GIS software has become increasingly common (Carver 1991; Malczewski 1999; Jankowski 1995). However, there are serious theoretical issues that must be addressed when using any technique that integrates data that are based on different metrics (Chrisman 1997). Spatial decision support systems are developed to acknowledge this complexity and provide a means for developing alternatives that can be used to foster discussion about facets of a problem that may not be included in the digital domain (Armstrong et al. 1986; Densham 1991).

25.6. Conclusions

GIS software provides an engine that, when in the hands of a trained user, can transform spatial data into usable information. The successful application of GIS software and geographical data to a highly diversified set of problems provides solid evidence that the transformational framework on which this software is built is both robust and general. Yet it is equally true that much work remains to done before GIS reaches its full potential. GIS software, for example, is limited by its inability to fully represent the spatio-temporal dynamics of geographical systems. Paradoxically, when and if the potential of GIS is fully realized, it is possible that many of the resulting products will not be recognizable as GIS software. Rather the capabilities of GIS software may become deeply embedded into the functionality of a wide variety of application specific software and, thus, geoprocessing software will become both transparent and ubiquitous (Armstrong 1997).

However, regardless of the form taken by the next generation of spatially enabled software, the proper use of such technology will require an understanding of the conceptual framework on which geoprocessing tools are built. Our intent here was to introduce the reader to such a framework. Subsequent chapters will flesh out this framework through discussions that are both theoretical and practical in nature.

References

ANSI, 2000, http: //webstore.ansi.org/ansidocstore/product.asp?sku=ANSI+NCITS+320%2D1998

Alsberg, P. A., McTeer, W. D. and Schuster, S. A., *IRIS/NARIS: A Geographic Information System for Planners*, CAC Document No. 188, Center for Advanced Computing, Urbana, IL, 1975.

Anderson, P. F., *Regional Landscape Analysis*, Environmental Design Press, Reston, VA, 1980.

Armstrong, M. P., Densham, P. J. and Rushton, G., 'Architecture for a microcomputer based spatial decision support system', *Proceedings of the Second International Symposium on Spatial Data Handling*, IGU Commission on Geographical Data Sensing and Processing, Williamsville, NY, 1986, pp. 120–131.

Armstrong, M. P., 'Requirements for the development of GIS-based group decision support systems', *Journal of the American Society for Information Science*, 45 (9), 1994, 669–677.

Armstrong, M. P., 'Emerging technologies and the changing nature of work in GIS', *Proceedings of GIS/LIS '97*, unpaged CD-ROM, American Congress on Surveying and Mapping, Bethesda, MD, 1997.

Armstrong, M. P., 'Geography and computational science', *Annals of the Association of American Geographers*, 90(1), 2000, 146–156.

Bennett, D. A., 'A framework for the integration of geographic information systems and model-base management', *International Journal of Geographical Information Science*, 11(4), 1997, 337–357.

Bolstad, P. V., Gessler, P. and Lillesand, T. M., 'Positional uncertainty in manually digitized map data', *International Journal of Geographical Information Systems*, 4, 1990, 399–412.

Bowker, G. C. and Star, S. L., *Sorting Things Out: Classification and its Consequences*, MIT Press, Cambridge, MA, 1999.

Broome, F. R. and Meixler, D. B., 'The TIGER data base structure', *Cartography and Geographic Information Systems*, 17(1), 1990, 39–47.

Buttenfield, B. P. and McMaster, R. B., *Map Generalization: Making Rules for Knowledge Representation*, John Wiley and Sons, New York, 1991.

Carver, S. J., 'Integrating multicriteria evaluation with geographical information systems', *International Journal of Geographic Information Systems*, 5(3), 1991, 321–339.

Chrisman, N. R., 'The accuracy of map overlays: a reassessment', *Landscape and Urban Planning*, 14, 1987, 427–439.

Chrisman, N. R., *Exploring Geographic Information Systems*, John Wiley and Sons, New York, 1997.

Chrisman, N. R., 'A transformational approach to GIS operations', *International Journal of Geographic Information Science*, 13(7), 1999, 617–638.

Clarke, K. C., *Analytical and Computer Cartography*, Prentice Hall, Englewood Cliffs, NJ, 1995.

Dale, P. F. and McLaughlin, J. D., *Land Information Management*, Oxford University Press, New York, 1988.

Densham, P. J., 'Spatial decision support systems', *Geographical Information Systems: Principles and Applications*, Maguire, D. J., Goodchild, M. F. and Rhind, D. W. (eds.), John Wiley and Sons, New York, 1991, pp. 403–412.

Drummond, W. J., 'Address matching: GIS technology for mapping human activity patterns', *APA Journal*, Spring, 1995, 240–251.

Dueker, K. J., 'Urban geocoding', *Annals of the Association of American Geographers*, 64(2), 1974, 318–325.

Dueker, K. J., Land resource information systems: a review of fifteen years experience', *Geo-Processing* 1, 1979, 105–128.

Fabos, J. G., Greene, C. M. and Joyer, S. A., *The Metland Landscape Planning Process: Composite Landscape Assessment, Alternative Plans Formulation and Plan Evaluation*, Part 3 of the Metropolitan Landscape Planning Model. Research Bulletin 653, Mass AG Experimental Station, Amherst, MA, 1978.

FIPS, *Spatial Data Transfer Standard (SDTS) Federal Information Processing Standards*, National Institute of Standards and Technology, Washington, DC, 1994.

Flanagan, N., Jennings, C. and Flanagan, C., 'Automatic GIS data capture and conversion', *Innovations in GIS 1*, Worboys, M.F. (ed.), Taylor and Francis, Bristol, PA, 1994, pp. 25–38.

Gillespie, S. R., 'An empirical approach to estimating GIS benefits', *Journal of the Urban and Regional Information Systems Association*, 12(1), 2000, 7–14.

Heuvelink, G. B. M., *Error Propagation in Environmental Modelling with GIS*, Taylor and Francis, Bristol, PA, 1998.

Hopkins L. D., 'Methods for generating land suitability maps: a comparative evaluation', *Journal for American Institute of Planners*, 34(1), 1977, 19–29.

Hunter, G. J. and Goodchild, M. F., 'A new model for handling vector data uncertainty in geographic information systems', *URISA Journal*, 8, 1996, 51–57.

Jankowski, P., 'Integrating geographical information systems and multiple criteria decision-making methods', *International Journal of Geographical Information Systems*, 9(3), 1995, 251–273.

Jenks, G. F., 'Lines, computers and human frailties', *Annals of the Association of American Geographers*, 71(1), 1981, 1–10.

Langran, G., *Time in Geographic Information Systems*, Taylor and Francis, Washington, DC, 1993..

Lanter, D. P. and Veregin, H., 'A research paradigm for propagating error in layer-based GIS', *Photogrammetric Engineering and Remote Sensing*, 58, 1992, 825–833.

Leitner, M. and Buttenfield, B. P., 'Multi-scale knowledge acquisition: Inventory of European topographic maps', *Cartography and GIS*, 22(3), 1995, 232–241.

Maguire, D.J., and Dangermond, J., 'The functionality of GIS', *Geographical Information Systems: Principles and Applications*, Maguire, D. J., Goodchild, M. F. and Rhind, D. W. (eds.), John Wiley and Sons, New York, 1991, pp. 319–335.

Marble, D. F., Lauzon, J. P. and McGranaghan, M., 'Development of a conceptual model of the manual digitizing process', *Proceedings of the International Symposium on Spatial Data Handling*, vol. 1, Geographisches Institut, Zürich, Switzerland, 1984, pp. 146–171.

MacDougall, E. B., 'The accuracy of map overlays', *Landscape Planning*, 2, 1975, 23–30.

Malczewski, J., *GIS and Multicriteria Decision Analysis*, John Wiley and Sons, New York, 1999.

Malczewski, J., 'On the use of weighted linear combination method in GIS: common and best practice approaches', *Transactions in GIS*, 4(1), 2000, 5–22.

McKee, L., 'The impact of interoperable geoprocessing', *Photogrammetric Engineering and Remote Sensing*, 65, 1999, 564–566.

Obermeyer, N. J., 'Measuring the benefits and costs of GIS', *Geographical Information Systems: Principles and Applications* (2nd ed.), Longley, P. A., Goodchild, M. F., Maguire, D. J. and Rhind, D. W. (eds.), John Wiley and Sons, New York, 1999, pp. 601–610.

Peucker, T. K. and Chrisman, N. R., 'Cartographic data structures', *American Cartographer*, 2, 1975, 55–69.

Peuquet, D. J., 'Representation of geographic space: toward a conceptual synthesis', *Annals of the Association of American Geographers*, 78(3), 1988, 375–394.

Peuquet, D. J. and Duan, N., 'An event-based spatiotemporal data model (ESTDM) for temporal analysis of geographical data', *International Journal of Geographical Information Systems*, 9(1), 1995, 7–24.

Snyder, J. P., *Map Projections – A Working Manual*, USGS Professional Paper 1395, U.S. Government Printing Office, Washington, DC, 1987.

Thrower, N. J. W., *Maps & Man*, Prentice Hall, Englewood Cliffs, NJ, 1972.

Tobler, W. R., 'A transformational view of cartography', *American Cartographer*, 6(2), 1979, 101–106.

Tomlinson, R. F., Calkins, H. W. and Marble, D. F., *Computer Handling of Geographical Data*, The UNESCO Press, Paris, France, 1976.

Tomlin, C. D. and Tomlin, S. M., 'An overlay mapping language', *Regional Landscape Planning*, American Society of Landscape Architects, Washington, DC, 1981.

Tomlin, C. D., *Geographic Information Systems and Cartographic Modeling*, Prentice Hall, Englewood Cliffs, NJ, 1990.

USGS, *Data Users Guide 5: Digital Elevation Model*, United States Department of Interior, Reston, VA, 1993.

USGS, 2000, http://mcmcweb.er.usgs.gov/sdts/standard.html.

Wesseling, C. G., Karssenberg, D., Burrough, P. A. and van Deursen, W. P., 'Integrating dynamic environmental models in GIS: the development of a dynamic modelling language', *Transactions in GIS*, 1(1), 1996, 40–48.

Westervelt, J. D. and Hopkins, L. D., 'Modeling mobile individuals in dynamic landscapes', *International Journal of Geographic Information Systems*, 13, 1999, 191–208.

Yuan, M., 'Representing geographic information to enhance GIS support for complex spatio-temporal queries', *Transactions in GIS*, 3(2), 1999, 137–160.

Chapter 26

Geographic data structures

May Yuan

26.1. Introduction

Geographic data structures are methods for organizing the spatial information (points, lines, areas, or cells) inside the computer. Identifying the best geographic data structure is a critical step in any problem-solving task. A good data structure, or what is often called a computer representation, will 'make important things explicit,' while exposing 'the natural constraints inherent in the problem' (Winston 1984). In a computing world, data structures are thus implementations of representations; they specify the conceptual views of reality and the frameworks in which data should be organized. In particular, a geographic data structure provides both conceptual and computational foundations for processing, integrating, analyzing, and visualizing Geographic Information Systems (GIS) data. Most importantly, in GIS the kinds of geographic information that can be encoded, computed, analyzed and visualized largely depend upon the embedded data structures (Fotheringham and Wegener 1999).

There are many levels of spatial data structures in a GIS, ranging from computer-encrypted codes of zeros and ones to models that are closer to human conceptualizations of reality. While detailed implementation of these representations is only significant to technical professionals, all GIS users should understand spatial data objects, data structures, and data models to have a good grasp of how a GIS works, its applications, and its limitations. What are the main differences among spatial data objects, spatial data structures, and spatial data models? *Spatial data objects* are digital representations of real-world entities, and they are the basic data unit that users can manipulate and analyze in a GIS database. They carry data about both the geometric and thematic properties of represented. *Spatial data structures* refer to methods for organizing spatial data with emphases on efficiency on storage and performance (Franklin 1991). Furthermore, *spatial data models* are high-level data structures that focus on 'formalization of the concepts humans use to conceptualize space' (Egenhofer and Herring 1991). The following sections in this chapter provide essential discussions on representing spatial data in terms of spatial data models, spatial data structures, and in incorporating time into GIS. There are many other excellent references on these subjects. Puequet (1984) provides analytical comparisons and synthesis of spatial data models. Samet (1989a,b) offers a comprehensive introduction to spatial data structures and their applications. Interested readers should also consult standard work in information storage and retrieval, such as Date (1995), for in-depth discussions on various general data structures and relational data models that are commonly applied to handle attribute data in GIS databases.

26.2. Spatial data models: object and field representations

Before discussing the specific data structures used in GISs, a distinction between the terms object and field is provided. In general, humans perceive the geographic world as a set of discrete entities and continuous fields. Discrete entities, often simply referred to as objects, are distinguishable by their independent and localized existence; they are identified as individuals before any attributes they may possess (Coucleclis 1992). Examples of discrete entities are buildings, streets, and water towers. On the other hand, a continuous field describes the spatial variation of a geographic variable in a space-time frame. Most fields are scalar fields in which we only have one value at a given location (such as elevation). Some fields are vector fields (such as wind) in which directions are important (Goodchild 1997a). Both objects and fields are represented well in topographic maps (Figure 26.1, see plate 19) where discrete entities like towns, roads, and lakes are marked by points, lines, and polygons, and a continuous field of topography is portrayed by contours and levels of shades.

However, current spatial data models cannot accommodate both object and field perspectives in a single model. It is because object- and field-based conceptualizations require distinct sets of data objects to portray geographic space. Spatial data models designed to represent discrete entities use points, lines, and polygons as geometric primitives to form spatial data objects that depict spatial characteristics of the entities. Such an object-view of the world describes reality as an empty space containing a combination of conceptual objects, primitive objects, and compound objects. An object representation is most appropriate for geographic entities where boundaries can be well defined. Such boundaries, often represented with points, lines, and polygons, do not normally exist in nature but are practical for engineering works or administrative and property lines, such as well locations, highways and streets, states and counties, and land parcels.

Alternatively, since a field is spatially continuous, a representation of the field must have a value or be able to imply a value at every location. Unlike the object representation, there are no identified things in a field representation. Rather, geographic things (or features) emerge through spatial and temporal aggregation of spatial units. Although it is impossible to fully represent a continuous space in a digital world, several spatial data models are designed to provide various degrees of approximation. Commonly used models include regular spaced points (lattice), irregular spaced points, contours, regular cells (raster or grid), Triangular Irregular Networks (TIN), and polygons (Goodchild 1997a). Regular or irregular points are most popular for field surveys and weather observatory networks (Figure 26.2a,b). Isolines apply a set of lines of equal values to show the spatial pattern of a geographic variable, such as elevation (contours, Figure 26.1, see plate 19), temperature (isotherms), or chemical concentration (see Chapter 31). Isolines are very effective representation for visualization, but they are ineffective in spatial analysis and digital computation compared to other field-based models.

Point-based (e.g. regular or irregular points) or line-based (e.g. isolines) models cannot represent a field completely because values of the focused geographic variable

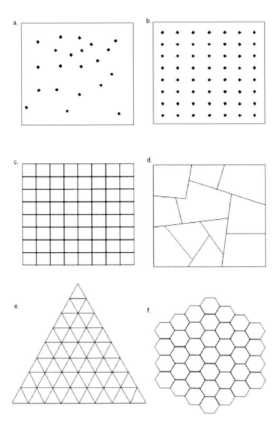

Figure 26.2 Examples of field representations ((a) irregular points, (b) regular points, a grid, (d) polygons, (e) a triangular mesh, (f) a hexagonal mesh).

are only available at the specific locations of points or lines. In order to create a continuous surface that covers the entire field, spatial interpolation is necessary to transform point- or line-based models to area-based models, i.e. grid, TIN, or polygons (Burrough and McDonnell 1998). There are many spatial interpolation methods, and each of them has different assumptions on spatial distributions. Lam (1983) provides an excellent discussion on commonly used spatial interpolation methods.

Area-based field data models divide a space into a finite number of smaller areas (Figure 26.2c–f). While area-based models cover the entire space, the value within each small area is set to a constant, assuming that there is no spatial variation within each of the small areas. Consequently, the size of these small areas, which determines the actual resolution or granularity of the data structure, determines the amount of information and the degree of detail in a field that an area-based model can represent. A field representation is regular if its smaller areas are of the same area and geometry; otherwise, it is irregular. A square is the most commonly used geometric unit to construct a regular field model, named a *grid*. The size of the grid is often directly related to data acquired by remote sensing technology, which is especially useful to acquire large-scale geographic data. Other geometries, especially the triangular or hexagon mesh, will provide better coverage for the spheroidal Earth, but converting squared cell-based data captured by satellites to triangular or hexagonal meshes will inevitably introduce uncertainty.

Nevertheless, irregular triangular meshes, known as *TIN*, are particular effective to represent surfaces that are highly variable and contain discontinuities and breaklines (Peucker *et al.* 1978, Figure 26.3). It is effective because a TIN connects a set of irregularly spaced significant locations, each of which defines a point where there is a change in the surface. For example, all neighboring points of the peak of a mountain are downhill; all neighbors of a point along a stream, except for the downstream point, are uphill. These significant points form nodes of triangles, while linear features (such as streams or ridges) and boundaries of area features with constant elevation (such as shorelines) frame triangle edges. Alternatively, a field can be represented by a set of irregular polygons, also known as *irregular tessellation* (Frank and Mark 1991, Figure 26.2d.). This approach is similar to thematic mapping where the value in each area

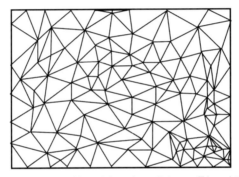

(a) Triangular Irregular Network based on a Delauney Triangulation

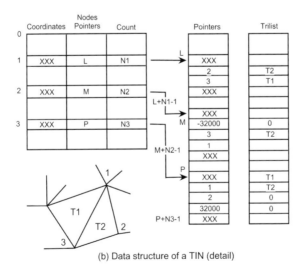

(b) Data structure of a TIN (detail)

Figure 26.3 The data structure of Triangulated Irregular Networks (from Burrough and McDonnell (1998), reproduced with permission from Oxford University Press).

represents an average, total, or some other aggregate property of the field within the area. An example here is the aggregation of individual census records to the block, block-group, or tract level. Every point in the space-time frame of interest must lie in exactly one polygon, except for the ones on boundaries. Irregular tessellation allows no overlapping polygons and requires all polygons together must completely cover space. Typical geographic fields represented by irregular polygons include vegetation cover classes, soil types, and climate zones.

Another field model of irregular polygons are *Thiessen polygons*, also known as *Dirichlet* or *Voronoi polygons*. Thiessen polygons are derived by first connecting data points to triangles via Delauney triangulation (the same procedure to form a TIN) and then linking perpendicular bisectors of these triangles (Figure 26.4). Thiessen polygons are often used in meteorology and hydrology as a quick method for relating point-based data to space under the assumption that data for a given location can be taken from the nearest observatory station. However, the size of Thiessen polygons is strongly related to the spatial distribution of data points, and an observatory point may represent unreasonably large area within which all locations are given the same value. Consequently, the model may represent the field well in areas where sample points are appropriately dense to the variable under study, but it may overlook spatial variation in sparsely sampled area.

Together, the object and field representations of spatial data reflect two complementary views of the geographic world. There are geographic things that can be easily identified as individuals as objects, and there are properties of the land that can only be described through continuous fields of attributes. These fields serve as bases upon which objects are built through human activities (such as roads or ranges) or natural

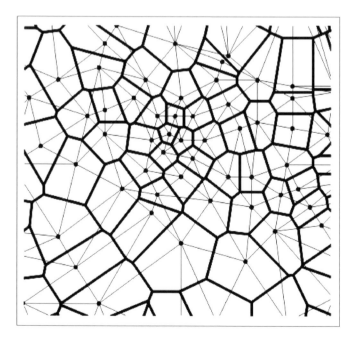

Figure 26.4 An example of Thiessen polygons (thick lines) and Delauney triangles (thin lines).

processes (such as rivers or islands). While current spatial data models are object-based or field-based, most commercial GISs provide functions to convert data from one model to another with a certain degree of compromising accuracy. Nevertheless, there are geographic phenomena (such as wildfire and precipitation), that have both object-like and field-like properties. Object or field representations alone cannot fully capture the intrinsic properties of these phenomena (Yuan 2001). An integrated model that supports both object and field perspectives will facilitate a better modeling of reality.

26.3. Spatial data structures: vector and raster data

The object and field representations of spatial data outline the conceptual models of reality. Implementation of the models relies upon appropriate data structures. Vector and raster are two major types of spatial data structures. Vector data are based on Euclidean geometry with *points*, *lines*, and *polygons* as primitive zero-, one-, and two-dimensional objects, respectively. Raster data consist of only one spatial object type: *cells*. Since geographic things are clearly identified in an object representation, they are better modeled by vector objects to describe their location and geometry. As to field representation, spatial data models may use vector data (for regular or irregular points, contour lines, TIN, and Thiessen or irregular polygons), or raster data (for grids to represent spatial distributions of geographic properties).

26.3.1. Vector-based data structures

In a vector GIS, primitive spatial data objects are points, lines, and polygons located by Cartesian coordinates in a spatial referencing system. These simple geometric primitives indicate static locations and spatial extents of geographic phenomena in terms of XY coordinates. A point object only marks the location of a geographic entity, such as a well, by a pair of XY coordinates. A line object shows the location and linear extent of a geographic entity, such as a river, by a series of XY coordinates. Furthermore, a polygon object depicts the location and two-dimensional extent of a geographic area, such as a county, also by a series of XY coordinates along the boundary of the area. However, a polygon has its first XY pair the same as its last to ensure a closure. These simple geometric primitives constitute many early vector data sets in computer cartography, in which data entry was the primary consideration in structuring spatial data, and little data manipulation was performed after the data had been entered into the system from maps (Peucker and Chrisman 1975). *The spaghetti data structure* is the representative of spatial data structures with simple points, lines, and polygons (Figure 26.5). In this data structure, every data object is independent of the others without regard for connectivity or adjacency, or what allows for *topology*. A point shared by two lines or a boundary shared by two adjacent polygons is to be encoded twice in a spaghetti data structure. This not only results in data redundancy, but impedes error checking and data analysis.

The spaghetti data structure was quickly replaced by spatial data structures that incorporate connectivity among spatial objects, i.e. *topology* (Dangermond 1982). Topological data structures distinguish two types of points: (1) *nodes* that are endpoints on a line (often referred to as *from-node* and *to-node*) and (2) *points* that mark

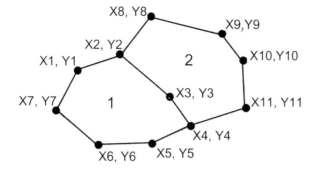

Polygon 1	Polygon 2
X1, Y1	X2, Y2
X2, Y2	X8, Y8
X3, Y3	X9,Y9
X4, Y4	X10,Y10
X6, Y6	X11, Y11
X7, Y7	X4, Y4
X1, Y1	X3, Y3
	X2, Y2

Figure 26.5 An example of the spaghetti data structure. Note that the points along the common boundary are recorded twice.

the geometric locations along a line (between nodes). Thus points located on a line also describe the shape of that line. For polygon data, topological data structures encode *polygon adjacency* by noting the polygons on both sides of a boundary line as the left-polygon and right-polygon in relation to the direction of the line. With these arrangements, it is easy to discern connected lines and adjacent polygons. If two lines have a common node, then they are connected. Likewise, if two polygons are the left- and right-polygons of a line, then they are adjacent.

There are many topological data structures, such as the Graphic Base File/Dual Independent Map Encoding (GBF/DIME), POLYVERT (POLYgon ConVERTer), and Topologically Integrated Geographic Encoding and Referencing System (TIGER). Both the GBF/DIME and TIGER systems were developed by the US Census Bureau to automatically encode street networks and census units for 1970 and 1990 , respectively. The GBF/DIME system is centered on linear geographic entities, such as streets and rivers, and uses straight-line segments as the basic data objects to represent these linear entities. For each line segment, it encodes from-node, to-node, left-polygon (left-block), right-polygon (right-block), left-address ranges, and right-address ranges. Figure 26.6 illustrates the basic format of the DIME file, which would allow for both nodes and polygons to be encoded twice. Such redundancy allowed for automated checking of data consistency (Peuquet 1984), and was later applied to

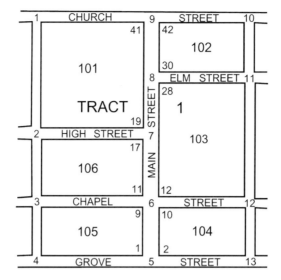

CENSUS ADDRESS CODING GUIDE RECORDS

STREET	TRACT	BLOCK	LOW ADDR	HIGH ADDR
Main	1	102	30	42
Main	1	103	12	28
Main	1	104	2	10
Main	1	105	1	9
Main	1	106	11	11
Main	1	101	19	41

DIME STREET SEGMENT RECORDS

STREET	NODE START	NODE END	TRACT LEFT	BLOCK LEFT	TRACT RIGHT	BLOCK RIGHT	LOW ADDR	HIGH ADDR
Main	5	6	1	105	1	104	1	10
Main	6	7	1	106	1	103	11	17
Main	7	8	1	101	1	103	19	28
Main	8	9	1	101	1	102	30	42

Figure 26.6 An example of the DIME file structure. (Reproduced from Peucker and Chrisman (1975), *The American Cartographer*, Vol. 2, No. 1, p. 59 with permission from the American Congress on Surveying and Mapping.)

the 'chains' or 'arcs' structure underlying POLYVERT and many modern vector GIS data models such as USGS Digital Line Graphs (DLG), Spatial Data Transfer Standard (SDTS), and the polygon layers in commercial systems such as ARC/INFO.

Although DIME incorporates topology through line segments, it is ineffective to assemble the outline of a polygon. Moreover, it is laborious to update line segments and retrieve polygons (Peucker and Chrisman 1975). Alternatively, POLYVERT, developed at the Harvard Laboratory (see Chapter 29), uses the *chain* as its basic spatial object. A chain, like a line segment, has two nodes at its ends, but, unlike a line segment, it may consist of many shape points. Thus, a single chain can sufficiently represent the boundary between two polygons in POLYVERT, rather than a series of

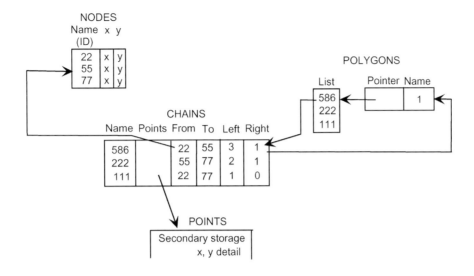

Figure 26.7 An example of the internal representation of the POLYVERT chain file. (Reproduced from Peucker and Chrisman (1975), *The American Cartographer*, Vol. 2, No. 1, p. 59 with permission from the American Congress on Surveying and Mapping).

straight-line segments as in DIME. POLYVERT organizes geometric and topological information into three tables of points, nodes, and chains, and a polygon list and polygon-chain list (Figure 26.7). Since chains are the basic spatial objects, the chain table references all other tables and the polygon list. Each chain has an identifier, references to the point table to obtain coordinates for its shape points, references to from- and to-node identifiers in the node table for node coordinates, and references to the left- and right-polygon. The polygon list has built-in pointers, each of which directs a polygon to a list of all chains that make up that polygon. Such a data structure enables effective retrieval of polygons in two ways: (1) by adjacency from left- and right-polygons in the chain table; and (2) by polygon identifiers from the polygon-chain list.

The data structure used in POLYVERT is similar to the design of the TIGER system. While the GBF/DIME encodes only straight lines and requires assembling of lines to retrieve a polygon, the TIGER system improves the census encoding by allowing the storage of curve lines (i.e. *chains*) and directly retrieving polygons of census units (Marx 1986). Additional data objects are incorporated in the TIGER system to further

Table 26.1 A summary of spatial data objects used in the TIGER system (from TIGER/Line 1999 Technical Documentation) http://www.census.gov/geo/www/tiger/tiger99.pdf)

	Point (0-cell)	Line (1-cell)	Polygon (2-cell)
Topology	Node	Complete chains or Network chains	GT-polygons
Non-topology Attribute	Entity point Shape point		

differentiate topological significance (Table 26.1). An *entity point* is used to identify the locations of point features, such as wells and towers, or to provide generalized locations for areal features, such as buildings and parks. A *complete chain* has both topology (marked by its end nodes) and geometry (characterized by its shape points). Complete chains form polygon boundaries and intersect other chains only at nodes. When complete chains form a closure, they form a Geometry and Topology polygon (*GT-polygon*). Chains that are not associated with polygons (i.e. have no left and right polygons) are *network chains*.

In total, the TIGER system consists of 17 record types (each of which forms a table) to encode geometric coordinates and attribute information for spatial data objects. Among the 17 record types, Type seven has coordinates of entity points for landmark features, Types one, two, three, and five record coordinates and geographic attributes of complete chains, Type I provides the linkages among complete chains and GT-polygons, and Type P contains coordinates of entity points for polygons. Common data items, such as TLID, FEAT, CENID, and POLYID, can be used to relate tables of the 17 record types. Although the TIGER system is structurally complex, it provides a rich spatial data framework for census and socio-economic mapping. Many more complex spatial data objects have been introduced to account for diverse geometric and topological properties of geographic phenomena. For example, the SDTS developed in 1992 by the US Geological Survey (USGS) contains many additional spatial objects, such as rings, universal polygons, and void polygons (SDTS document, http://mcmcweb.er.usgs.gov/sdts/standard.html). When more and more complex spatial objects become necessary to satisfy the need for representing diverse geographic phenomena, topological structures that associate objects from various tables and lists create significant overhead for data management and retrieval. Non-topological data structures have recently re-gained popularity, for they can provide faster drawing speed and editing ability, allow overlapping and non-contiguous polygons, and require less disk storage.

Non-topological data structures are considerably simple compared to topological data structures. Every data object has its own stand-alone table or list to encode identifiers and coordinates. Since there is no topology embedded, there is no need to link tables in any way. However, topology is critical for spatial data manipulation and analysis. In order to fulfill the need for topology, advancement in computing power has made it possible to incorporate appropriate functions to build topology on the fly (i.e. build it when it becomes necessary). ArcView shapefiles and Spatial Database Engine (SDE), both developed by Environmental Systems and Research Institute (ESRI, Redlands, California) are examples of the modern non-topological data structures (Figure 26.8).

26.3.2. Raster-based data structures

Unlike vector data, raster data are regular arrays of cells without explicitly associations with XY coordinates except for one of its corner cells (usually the southwest corner). While a cell is easier for programming purposes to be identified by row and column number, it can also be accessed by XY coordinates. In simple raster structures, there is a one-to-one relation between data value, cell, and location. Therefore, all data values can be stored as a simple array with a meta file indicating the

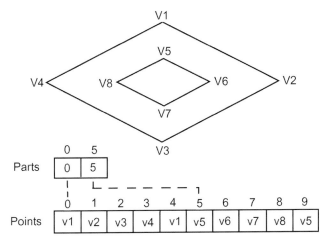

Polygon Record Contents

Position	Field	Value	Type	Number	Byte Order
Byte 0	Shape Type	5	Integer	1	Little
Byte 4	Box	Box	Double	4	Little
Byte 36	NumParts	NumParts	Integer	1	Little
Byte 40	NumPoints	NumPoints	Integer	1	Little
Byte 44	Parts	Parts	Integer	NumParts	Little
Byte X	Points	Points	Point	NumPoints	Little

Note: X = 44 + 4 + NumParts

Figure 26.8 An example of the shape file structure. Here is a polygon with a hole and eight vertices. Parts and points indicate the number of parts and the vertices (shape points) that constitute each part, linked by the address numbers or identifiers of the vertices. Polygon record contents detail x and y coordinates and relevant structural information for the object (graphic image supplied courtesy of Environmental Systems Research Institute, Inc.).

numbers of rows and columns, cell size, projection and coordinate system, and minimum values for X and Y coordinates (Figure 26.9). However, in terms of data storage this method is inefficient because it uses the disk space for the entire array regardless of the data distribution. Demand for large data storage can also degrade data processing performance. To effectively increase data processing performance and reduce the demand for data storage, two issues involved in raster data structures need to be addressed: (1) compression methods: how to more efficiently store the data, and (2) scan order: how to scan the data in an array. Since geographic phenomena often show a certain degree of spatial autocorrelation, or similar values near each other, it is common to have blocks of cells in a raster array with the same data value. For example, when raster structures are used to represent an area – such as a homogeneous stand of oak, all cells of the area will have the same value. Such properties are the basis

```
ncols        270
nrows        476
xllcorner    391253.1875
yllcorner    4064188.25
cellsize     3
NODATA_value -9999

-9999 -9999 -9999 -9999 -9999 -9999 2321.5 2321.295 2320.653 2319.938 2319.385 .....
-9999 -9999 -9999 -9999 2321.5 2321.5 2321.5 2321.093 2320.492 2319.851 2319.341 .....
-9999 -9999 2321.5 2321.5 2321.5 2321.5 2321.421 2320.977 2320.449 2319.905 2319.438 .....
-9999 -9999 2321.5 2321.5 2321.5 2321.327 2320.94 2320.492 2320.024 2319.595 .....
-9999 -9999 2321.5 2321.5 2321.5 2321.281 2320.964 2320.588 2320.179 2319.777 .....
```

Figure 26.9   An example of simple raster data structures.

for many compress and scan-order methods. Commonly used compression methods include chain codes, run-length codes, block codes, and quadtrees. Commonly used scan orders include row, row-prime, Morton and Peano-Hilbert. Quadtrees can be used as both compression and scan-order methods (Goodchild 1997b). Examples of some of these will be provided below.

Compression methods aim to reduce the demand for data storage. The *run-length coding* method is a simple yet effective data compression technique. It groups cells of the same value row by row and encodes these cells by a beginning cell, an end cell, and an attribute value (Figure 26.10a). A general format for the code is (Row, (Min_Columm, Max_Columm, Attribute)). The run-length coding method is particularly useful when there are just a few classes (attribute values), but is ineffective when there is a high degree of spatial variability in the data.

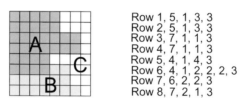

```
Row 1, 5, 1, 3, 3
Row 2, 5, 1, 3, 3
Row 3, 7, 1, 1, 3
Row 4, 7, 1, 1, 3
Row 5, 4, 1, 4, 3
Row 6, 4, 1, 2, 2, 2, 3
Row 7, 6, 2, 2, 3
Row 8, 7, 2, 1, 3
```

(a) Run-length codes: A=1, B=2, and C=3

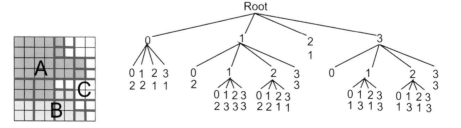

(b) A quadtree: A=1, B=2, C=3

Figure 26.10   Examples of (a) run-length codes and (b) quadtree codes.

Another very popular coding method is known as *quadtree*, which is also a scan-order method. Distinguished from all the other coding methods above, a quadtree is a hierarchical data structure in that it is based on the principle of recursive decomposition of space (Samet 1989a). The quadtree data structure divides a grid into four quadrants (NW, NE, SW, and SE), and will further divide a quadrant to four subquadrants if the quadrant is not homogeneous (i.e. contains only one attribute value, Figure 26.10b). As a result, the quadtree method is only applicable to grids with both the numbers of rows and columns equal to a power of 2 (i.e. 2^n). A quadtree has a *root node*, which corresponds to the entire grid, and *leaf nodes*, which identify attribute values and quadrants without further divisions. It provides effective data access and spatial operations at multiple levels of resolutions (Samet 1989b; Mark and Lauzon 1984).

Scan-order methods are mainly concerned with performance in terms of data processing. *Row* and *row-prime* methods scan one row at a time (Figure 26.10a,b),

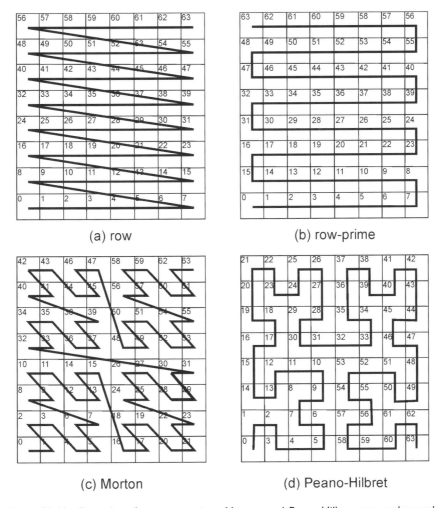

(a) row

(b) row-prime

(c) Morton

(d) Peano-Hilbret

Figure 26.11   Examples of row, row-prime, Morton, and Peano-Hilbert scan order methods. Numbers represent the scan order in each method.

but the row-prime method reverses every other row. Two-dimensional scan methods, such as *Morton* and *Peano-Hilbert*, put neighboring cells in a cluster by spatially recursive shapes (Figure 26.11c,d). The Morton method repeats a Z-like shape with four neighboring cells as a unit and repeats the shape at all levels. Unfortunately, the complexity of a Morton scan does not guarantee a better compression. Goodchild and Grandfield (1983) show that the row-prime method in general produces better results than a Morton scan. Alternatively, the Peano-Hilbert method, also known as *Pi-order scan* or *Peano-curve*, has a basic U-like shape that repeats at all levels, and this method generally gives the best results (Goodchild and Grandfield 1983).

26.4. Representing time in geographic data structures

The above discussion on spatial data modeling and data structures is designed to represent static and planar geographic space. As a result, they overlook the need to handle dynamic or volumetric geographic phenomena. Incorporation of temporal components into a spatial representation is not a trivial task because time has distinct properties from space (Langran 1992). Time-stamping is the most popular approach of incorporating temporal components into Relational Database Management Systems (RDBMS) and GIS (Yuan 1999). Correspondingly, GIS data are time-stamped layers (the snapshot model in Armstrong 1988, Figure 26.12a), attributes (the space-time composite model in Langran and Chrisman 1988, Figure 26.12b), or spatial objects (the spatiotemporal objects model in Worboys 1992, Figure 26.12c).

The snapshot model is the simplest way to incorporate time into spatial data using a set of independent states. Since the method stores a complete state for every given snapshot, it encounters problems of data redundancy and possible data inconsistency, especially in dealing with large data sets. Alternatively, space-time composites and spatiotemporal object models explicitly represent information about changes. Both eliminate the problems of data redundancy and inconsistency to a degree. While the spatiotemporal object model is able to maintain spatial object identifiers, it, as in all time-stamping approaches, has difficulty representing dynamic information, such as transition, motion, and processes.

Recent development in GIS representation has emphasized dynamic processes. These models include Smith *et al.* (1993) Modeling and Database System (MDBS), Peuquet and Duan's (1995) Event-Based Spatiotemporal Data Model (ESTDM), Raper and Livingstone's (1995) geomorphologic spatial model (OOgeomorph) and Yuan's (1996) three-domain model. MDBS takes a domain-oriented approach to support high-level modeling of spatiotemporal phenomena by incorporating semantics into data modeling, specifically for hydrological applications. However, MDBS is designed for hydrological data processing and modeling; it is not designed for representing or managing temporal information in GIS. On the other hand, ESTDM is conceptually simple and easily adaptable to other raster-based systems to represent information about locational changes at pre-defined cells along the passage of an event. While the model has shown its efficiency and capability to support spatial and temporal queries in raster systems, it will require a substantial redesign for use in a vector-based system (Peuquet and Duan 1995). On the other hand OOgeomorph, which is a vector-based system, is designed to handle point data of time-stamped locations. Its ability to handle spatial objects of higher dimensions and its applicability

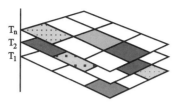

a. Time-stamped layers (Armstrong 1988).

Poly id	T_1	T_2	T_3	T_4
1	Rural	Rural	Rural	Rural
2	Rural	Urban	Urban	Urban
3	Rural	Rural	Urban	Urban
4	Rural	Rural	Urban	Urban
5	Rural	Rural	Rural	Urban

b. Time-stamped attributes (columns): Space-Time Composites
(Langran and Chrisman 1988).

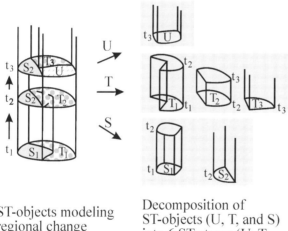

ST-objects modeling
regional change

Decomposition of
ST-objects (U, T, and S)
into 6 ST-atoms (U, T_1,
T_2, T_3, S_1, and S_2).

☐ Agriculture ◌ Urban ▨ Industry

c. Time-stamped space-time objects: the spatiotemporal object model
(Worboys 1992).

Figure 26.12 Time-stamp methods used to incorporate time into GIS databases. (From Yuan
(1999), reproduced with permission from Blackwell Publishers.)

to systems other than geomorphology are not yet evident. The three-domain model provides a framework that extends both the space-time composite model and ESTDM model to enable representing histories at locations as well as occurrences of events in space and time, but it needs to be refined and tested. In addition to these approaches, the object-oriented design is gaining recognition in the development of temporal GIS (Wachowicz 1999).

26.5. Challenges in spatial data structures

Harvey (1969) identifies that 'The whole practice and philosophy of geography depends upon the development of a conceptual framework for handling the distribution of objects and events in space.' He further cites from Nystuen (1963) that the conceptual framework should be appropriate for '(1) stating spatial distributions and the morphometric laws governing such distributions, and (2) examining the operation of processes and process laws in a spatial context.' (Harvey 1969).

The developments in GIS and GIScience have brought out renewed calls for such a conceptual framework through enforcing powerful and robust spatial data structures that can embrace diverse geographic objects and events and support analysis of various spatial relations. The debate on field versus object representations evolves to the need for a better understanding of human cognition for representing spatial data (Coucleclis 1992), and recently, the idea has been further explored (Peuquet 1994; Yuan 1996, 1997; Mennis et al. 2000). The return of non-topological data structures from topological data structures allows a better integration of geometry and attribute data to better match how human perceive the world. Although topology is still critical for spatial analysis, rapid advancements in computer processing speed and random access memory enable quick construction of topological relations when necessary. All these advancements have not only enriched the conceptual and theoretical basis in geographic information science but also improved the usefulness of geographic information systems. However, as the world grows to be more and more integrated – yet complex – through development of technologies, the need for better data structures becomes more and more challenging. Emerging research issues reflect the needs for heterogeneous and massive, and distributed geospatial data for example.

Heterogeneous data can include data of multiple formats, dimensions, and resolution. In parallel to the development of multi-media technology, geographic data have recently embraced a variety of large images and video and audio data. Integration of these new forms of data within a GIS database is non-trivial, for the current spatial data models are incapable of handling data of multiple forms. Consequently, query support is limited, if any, to file retrieval (i.e. access individual video files by keywords or some indexing scheme), rather than retrieve information based on an object. For example, a full integration should enable retrieval of all related video or audio clips when a user clicks a lake object on a GIS data layer, such as video clips about landuse change around the lake or audio explanation of the lake's history. These clips are created dynamically from larger video and audio files through the selection of data objects.

Data of multiple dimensions are critical for 3D and temporal applications. Currently, GIS data are two-dimensional. Three-dimensional visualization techniques cannot fully support 3D GIS because a true 3D application requires information that can only be derived from analyzing 3D topological relationships beyond simple visualizing of the data volume. This is because topological integrity forms the basic operations to manipulate and analyze data in either a 2D, 3D, or 4D GIS (Egenhofer and Herring 1991; Hazelton 1998). For example, a GIS must have capabilities to compute information about adjacency in the vertical dimension to answer a 3D query for areas where sandstone lies upon shale layers to identify areas of landslide potential.

Moreover, geospatial data have grown at a phenomenal rate as a result of advanced remote sensing and survey technology over the last decade. Yet, despite the massive amount of data coming on a daily basis, the utility of GIS technology in scientific research is considerably limited because information implicit in GIS data is not easy to discern. This generates an urgent need for new methods and tools that can intelligently and automatically transform geographic data into information and, furthermore, knowledge. The need is in part of a broader information technology in Knowledge Discovery in Databases (KDD) that aims to extract useful information from massive amounts of data in support of decision making (Fayyad 1997). Robust spatial representation with effective data structures is the key to facilitate geospatial knowledge discovery because a GIS cannot support computation on information that it cannot represent. Geospatial knowledge cannot be synthesized when geographic phenomena cannot be fully embraced in a GIS.

Finally, the development of internet technology presents another challenge to spatial data structures because the trend promotes the use of distributed data, and furthermore, distributed computing. When data are distributed at multiple sites, the integration and usability of these data in an application depends upon both structural and semantic interoperability (how the systems can interact). A common representation for spatial data object specifications is critical to ensure effective communications among different data sets and systems. The Open GIS consortium with members from industry and the academy has proposed an object-oriented spatial data model to serve the need for interoperability. The model provides a common base for developing open GISs, i.e. data are interoperable at multiple GIS platforms. Its use for integrating distributed geospatial data on the internet is yet to be investigated. While the term 'distributed data' signifies that data reside at multiple sites, the term 'distributed computing' emphasizes sending functions to multiple data sites for computation and returning results from individual sites to compile the final result. Distributed computing is best suited for intranet applications. With large data sets residing at multiple computers getting all data to a single computer is inefficient and, sometimes, impractical. Similar to issues related to distributed data, data interoperability is critical to the success of distributed computing. A data model that can provide a common structural and semantic basis for data communication will significantly facilitate both distributed data and distributed computing applications.

References

Armstrong, M. P., 'Temporality in spatial databases', *Proceedings: GIS/LIS'88*, 2, 1988, 880–889.

Burrough, P. A. and McDonnell, R. A., *Principles of Geographical Information Systems*, Oxford, New York, 1998.

Coucleclis, H., 'People manipulate objects (but cultivate fields): beyond the raster-vector debate in GIS', *Theories and Methods of Spatio-temporal Reasoning in Geographic Space*, Frank, A. U., Campari, I. and Formentini, U. (eds.), Springer Verlag, Berlin, 1992, pp. 65–77.

Dangermond, J., 'A classification of software components commonly used in geographic information systems', *Proceedings of the U.S.-Australia Workshop on the Design and Implementation of Computer-based Geographic Information Systems*. Honolulu, HI, 1982, pp. 70–91.

Date, C. J., *An Introduction to Database Systems* (6th ed.), Addison-Wesley, Reading, MA, 1995.

Fotheringham, A. S. and Wegener, M., *Spatial Models and GIS: New Potential and New Models*, Taylor & Francis, London, 1999.

Egenhofer, M. J. and Herring, J. R., 'High-level spatial data structures for GIS, *Geographical Information Systems, Volume 1: Principles*, Maguire, D. J., Goodchild, M. F. and Rhind, D. W. (eds.), Longman Scientific & Technical, Essex, 1991, pp. 227–237.

Fayyad, U., 'Editorial. *Data Mining and Knowledge Discovery*, Vol. 1, 1997, pp. 5–10.

Franklin, W., 'Computer systems and low-level data structures for GIS', *Geographical Information Systems, Volume 1: Principles*, Maguire, D. J., Goodchild, M. F. and Rhind, D. W. (eds.), Longman Scientific & Technical, Essex, 1991, pp. 215–225.

Frank, A. U. and Mark, D. M., 'Language issues for GIS', *Geographical Information Systems, Volume 1: Principles*, Maguire, D. J., Goodchild, M. F. and Rhind, D. W. (eds.), Longman Scientific & Technical, Essex, 1991, pp. 147–163.

Goodchild, M. F., 'Representing fields', *NCGIA Core Curriculum in GIScience*, 1997a, (http://www.ncgia.ucsb.edu/giscc/units/u054/u054.html, last revised August 12, 2000).

Goodchild, M. F., 'Quadtrees and scan orders', *NCGIA Core Curriculum in GIScience*, 1997b (Unit 057, http://www.ncgia.ucsb.edu/giscc/units/u057/u057.html, last revised October 23, 1997).

Goodchild, M. F. and Grandfield, A. W., 'Optimizing raster storage: an examination of four alternatives', *Proceedings, AutoCarto 6, Ottawa* 1, 1983, 400–407.

Harvey, D., *Explanation in Geography*, Edward Arnold, London, 1969.

Hazelton, N. W. J., 'Some operational requirements for a multi-temporal 4D GIS', *Spatial and Temporal Reasoning in Geographic Information Systems*, Egenhofer, M. J. and Golledge, R. G. (eds.), Oxford, New York, 1998, pp. 63–73.

Lam, N. S., 'Spatial interpolation methods: a review', *American Cartographer*, 10, 1983, 129–149.

Langran, G., *Time in Geography*, Taylor & Francis, London, 1992.

Langran, G. and Chrisman, N. R., 'A framework for temporal geographic information', *Cartographica*, 25(3), 1988, 1–14.

Mark, D. M. and Lauzon, J. P., 'Linear quadtrees for geographic information systems', *Proceedings: IGU Symposium on Spatial Data Handling*, Zurich, 1984, pp. 412–431.

Marx, R. W., 'The TIGER system: automating the geographic structure of the United States census', *Government Publications Review*, 13, 1986, 181–201.

Mennis, J. L., Peuquet, D. and Qian, L., 'A conceptual framework for incorporating cognitive principles into geographical database representation', *International Journal of Geographical Information Science*, 14(6), 2000, 501–520.

Nystuen, J. D., 'Identification of some fundamental spatial concepts', *Papers in Michigan Academia of Science, Arts, and Letters*, 48, 1963, 373–384.

Peucker (now Poiker), T. K. and Chrisman, N. Cartographic data structure', *The American Cartographer*, 2(1), 1975, 55–69.

Peucker (now Poiker), T. K., Flower, R. J., Little, J. J. and Mark, D. M., 'The triangulated irregular network', *Proceedings of the DTM Symposium*, American Society of Photogrammetry and American Congress on Survey and Mapping, St. Louis, MO, 1978, pp. 24–31.

Puequet, D. J., 'A conceptual framework and comparison of spatial data models', *Cartographica*, 2(2), 1984, 55–69.

Peuquet, D. J., 'It's about time: a conceptual framework for the representation of temporal dynamics in geographic information systems', *Annals of the Association of American Geographers*, 84(3), 1994, 441–462.

Peuquet, D. J. and Duan, N., 'An event-based spatiotemporal data model (ESTDM) for temporal analysis of geographical data', *International Journal of Geographical Information Systems*, 9(1), 1995, 7–24.

Raper, J. and Livingstone, D., 'Development of a geomorphological spatial model using object-oriented design', *International Journal of Geographical Information Systems*, 9(4), 1995, 359–384.

Samet, H., *The Design and Analysis of Spatial Data Structures*, Addison Wesley, Reading, MA, 1989a.

Samet, H., *Applications of Spatial Data Structures*, Addison Wesley, Reading, MA, 1989b.

Smith, T. R., Su, J., Agrawal, D. and El Abbadi, A., 'Database and modeling systems for the earth sciences', *IEEE (Special Issue on Scientific Databases)*, 16(1), 1993.

Wachowicz, M., *Object-Oriented Design for Temporal GIS*, Taylor & Francis, London, 1999.

Winston, P. H., *Artificial Intelligence* (2nd ed.), Addison-Wesley, Reading, MA, 1984.

Worboys, M. F., 'A unified model of spatial and temporal information', *The Computer Journal*, 37(1), 1994, 26–34.

Yuan, M., 'Modeling semantic, temporal, and spatial information in geographic information systems', *Progress in Trans-Atlantic Geographic Information Research*, Craglia, M. and Couclelis, H. (eds.), Taylor & Francis, Bristol, PA, 1996, pp. 334–347.

Yuan, M., 'Knowledge acquisition for building wildfire representation in geographic information systems', *The International Journal of Geographic Information Systems*, 11(8), 1997, 723–745.

Yuan, M., 'Representing geographic information to enhance GIS support for complex spatio-temporal queries', *Transactions in GIS*, 3(2), 1999, 137–160.

Yuan, M., 'Representing complex geographic phenomena with both object- and field-like properties', *Cartography and Geographic Information Science*, 28(2), 2001, 83–96.

Chapter 27

Processing spatial data

Francis J. Harvey

27.1. Materials and tools

In many ways, processing spatial data is analogous to processing wood. Both data and wood are raw materials for a vast variety of products. Processing transforms materials into products. In the carpenter's case, the raw material is the diverse types of wood available and the tools are the sanders, saws, drills, and planers. Spatial data processing begins with data material and uses a vast array of Geographic Information System (GIS) tools to transform data into various products. In this chapter, we examine the processing of spatial data in terms of materials and transformations.

Quality is always an important aspect for considering processing and evaluating materials and tools. We need to remember that materials vary in terms of quality, quality stands in a relationship to expense and tools, and quality considerations circumscribe processing. In processing wood to make chairs, a carpenter will consider various quality characteristics of wood when considering a project: grain, hardness, shrinkage, warp. Cherry wood for fine furniture is finely grained, but it is expensive. Pine is much cheaper, but is very soft. If a carpenter is making a rocking chair for an elegant hotel lobby, cherry is the better choice, but if the hotel needs 100 chairs for a conference room, pine is probably better because of the cost. A carpenter will also need to use different tools for processing according to the quality of the wood. Similar concerns with quality characteristics apply to geographic data. A GIS analyst assesses data quality in terms of lineage, currency, positional accuracy, and attribute accuracy (See Chapter 30 on data quality). A lot line survey for a home sale is done with a theodolite or accurate Global Positioning System (GPS)-based surveying equipment to ensure that the survey has a high degree of positional accuracy. This level of quality costs much more per measurement than a land cover study of, for instance, 100 square miles of ranch land in Montana using 30 m resolution land cover data from the US Geological Survey (USGS).

The quality of the product in each case is related to the raw material used, tools available, and ultimate use. Too much quality for a particular use is usually acceptable (if not overly expensive), but may lead to problems. If the survey defines lot lines of a parcel with an accuracy of ± 10 ft, then any disputes about the ownership of a fence that straddles the assumed boundary will not be settled with these data. The issues in choosing data for spatial processing are, in comparison to the selection of wood for a carpentry project, just as complicated, but obviously different. Spatial data is intangible and the possible modifications are limitless. Wood is wood; a tangible material can be altered only once and never returned thereafter to its original form. Cut a chair leg too short and the chair will be unstable. Data, however, are infinitely malleable and can be copied an unlimited amount of times. In most cases, processing produces copies. The original is usually available. In other ways, data is much more flexible. If the available data misses part of the area you are working in, you can obtain more. Data has other constraints that limit the types of processing related to projections, scale, and accuracy. Learning these limits is a critical component of working with spatial data.

All GIS tools perform transformations. Transformations are processing operations that convert data to other forms of data, combine and integrate data, and classify data to make spatial data products, usually maps or images. Spatial data varies in a thousand ways, and there are ways to transform it in almost every way imaginable. When it uses different attribute measures, you can rescale the measures, when it includes a bigger area, you can remove the unnecessary part. You can also add more data; change raster data to vector data; and change vector data to raster data. While the possibilities are endless, there are vexing problems related to projections, scale, and accuracy that effectively limit transformations. The apprentice sorcerer can take a difficult situation and make things much worse, destroying the data and maybe the laboratory in the process. Recovery from data chaos is much more difficult than learning the limitations of different types of data and how to use transformations.

In the following section we cover the main types of transformations and examine how they transform data into information products. Knowing the diversity of materials and the key issues to pay attention to in processing spatial data is fundamental to the effective and efficient use of geographic information technologies. In this chapter, one will learn key types of data, key transformations, and how data and transformations are used in processing spatial data.

27.2. Types of data to choose from

In Chapter 26, key data structures used by GIS were presented. Spatial data processing uses these data structures. We should think of them as types of material used by GIS. Cherry wood is used for fine furniture, but only the wood without knots can be used for turning a delicate bowl. Likewise each of the four data structures is useful for particular uses, but not always all the data of that type. In the following four sections we discuss the main types of spatial data and limitations to be aware of for processing.

27.2.1. Raster

Raster data (Section 26.3.2) is mainly used in applications that differentiate continuous attributes such as elevation, air pollution, and approximated values, such as soil pH (Figure 27.1).

Raster data structures are derived from the technology used in television. In a television set a 'sweep gun' tracks back and forth in a sweeping motion. The term raster came to refer to the rows and columns of the particles on the screen that are activated by the gun. In a GIS these rows and columns are usually square in format. These squares subdivide space without any gaps. Although for larger areas, the number of squares and therefore the memory requirements can get quite large,

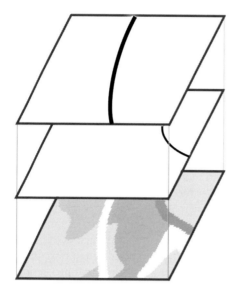

Figure 27.1   Raster data represents various types of phenomena; a road, trail, and land use.

compression techniques are commonly used to reduce the memory required. Another key issue with a raster system is the size of the cells. Smaller cells allow a more precise representation at the cost of increased numbers of cells, larger cells need less storage. Each cell is used to store one attribute and data are grouped into layers.

When the cells are very small, objects can be represented in more detail compared to larger cell sizes. This is, of course, connected to the size of the object. A $40'$ by $50'$ structure would encompass 20 $10'$ square raster cells, but only one $50'$ cell, or possibly two $50'$ square cells. The quality concerns for processing are obvious: if the $10'$ cells are used than the structure is clearly represented, but uses much memory. The $50'$ cells use significantly less memory, but the structure may be coded in the data using up to two cells, thus over representing the structure by 100 per cent. The actual size of the structure may be underreported if only one cell is labeled structure, but grossly exaggerated when two cells are labeled.

Additionally, quality concerns arise when the objects are irregular in shape or at an angle to the north-south, east-west orientation of the raster cells. To judge the quality of raster data it is very important to know the resolution of cells, rules used to assign cell values, and any previous processing. It is extremely important to keep the same projection system throughout processing. As discussed in Chapter 2, projection systems uses different models for representing features on the round earth in the flat coordinate system of a GIS. Changing from one projection system to another can introduce major positional accuracy errors since the orientation of features changes from one system to another.

Processing raster data usually follows concepts presented by Dana Tomlin (1990), called map algebra. By registering raster data by cell location, the values saved for each cell in every layer can be mathematically processed in an infinite number of ways.

27.2.2. Vector data

Vector data structures are best in representing objects ranging from buildings, roads, pipelines, cables, etc. as they are measured. Any object that should be represented as a point, line, or polygon can use a vector data structure.

The previous section (27.2.1) covered raster data, therefore we can compare vector data processing to raster data. Vector data can more accurately represent objects than raster. It is only limited by the quality of measurements, the numerical processing precision, and the degree of generalization. Whereas the amount of detail shown in raster data is determined by the cell size, or resolution; vector data shows different amounts of detail according to its generalization. Vector data is very useful for most types of processing. The results produce objects that correspond to the objects people are used to seeing and data from different sources that are in the same projection and coordinate system can easily be combined. Splinter polygons are significant problems that result because the objects from different data sets do not align completely. All GIS's have tolerances to join vectors that are close but do not exactly align.

27.2.3. Hybrid

This type of data structure is exceedingly rare, and is rarely used in practice. Although hybrid data structure are rare, increases in raw computer processing power have made working jointly with raster and vector formats very common. While it is easy to transform data from one format to another and back, the transformation loses information. For example going from vector to raster will usually lose the detailed attribute information stored for vector features. This transformation will also reduce the resolution of the vector data considerably. A transformation from raster to vector leads to polygons with very jagged edges that will not align with other vector features.

27.2.4. Object orientated data

Along with the increased use of object orientated programming techniques, databases, and programming languages such as Java, object data structures for GIS are becoming more common. Object orientated data makes it possible to link different data types. The success of this type of data structure goes hand-in-hand with a change from top-down information processing, to application specific approaches. For example, an existing database system may store all information in relational tables. Every application must conform to this structure. An object orientated data structure opens up possibilities to customize data structures for a specific application. This is a great advance for large-scale applications that integrate company information management processes. For project work, the scale-ability may be a boon, but also a problem if different activities requires the adaptation of the object model grows too complex (Figure 27.2).

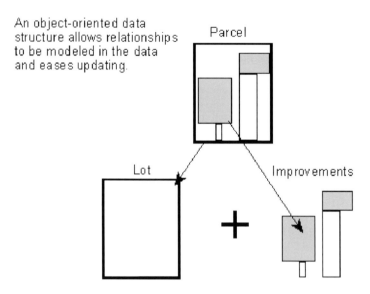

An object-oriented data structure allows relationships to be modeled in the data and eases updating.

Figure 27.2 Object-oriented data.

Since objects can include both vector and raster forms of data structures, they share the same limits and potentials.

27.3. Tools as transformations

This chapter was introduced by showing the relevance of thinking about GIS processing as transformations involving materials and tools. In the previous section, we discussed the materials used in GIS processing. In this section we turn to the tools, which are all transformations in GIS processing. These tools are also part of many operations, or special collections of tools to help with certain processing. All the transformations in this section are critical tools, but the selection reflects a partial list those available. Processing spatial data involves a multitude of tools and new ones are created daily to help with special kinds of processing. The selection here provides insights into the most common types of transformation.

27.3.1. What are transformations?

Transformations change data from one format to another. Converting raster data to vector data, or vice versa, is a type of transformation. Combining data from different sources is another. Classifying data to show less variety is yet another type of transformation. Some transformations will change the data structure, some change the attributes, other transformation combine attributes and data structures. Because transformations are always only relevant in terms of an operation, a discussion only about transformation is an abstraction from actual use. In light of this, the following sections that discuss specific types of transformations stress usage and practical limitations. Section 27.4 examines the use of some transformations in the context of common GIS operations.

27.3.2. Data structure changes

The most visible transformation is changing data from one data structure to another. These changes are very versatile, making data available in the format needed. There are, however, numerous constraints that should be considered. Some of these constraints can be found in Section 27.2 and in Chapter 26.

With regard to the most common data structure changes – from raster to vector and from vector to data – several issues need to be kept in mind. Raster data is always projected and it is very difficult to reproject. Even then, projected as vector data it will usually not align with other vector data. When vector data are converted to raster data, a common problem is assigning attribute values to cells that cover parts of two polygons. Usually the attribute is assigned that makes up the largest proportion of the cells area, but different rules are often used. It is also possible to consider neighboring cells. Knowing what rules were used in transforming data structures is very important when the data will be analyzed.

27.3.3. Scale

Scale presents a never-ending list of issues to be aware of in spatial data processing. Because every observation of geographic phenomena must abstract a part of reality to make a coherent representation, scale plays an important role in determining the level of abstraction. In the past, and for large organizations today, scale has been the most significant criteria differentiating mapping products. These were (and are) usually national mapping organizations that have legislated mandates to produce maps for an entire country at a variety of scales. Usually these scales have become standardized ways of representing the world for state agencies and the large number of private enterprises that either rely on the state data or do work for it. For instance, the USGS maintains topographic series at 1:24,000, 1:100,000, and 1:250,000.

Scale usually indicates a well-defined representation in traditional mapping projects. Due to common usage of term scale in cartography to refer to a map series and its characteristics, people continue to refer to scale when processing geographic information with a GIS. However, scale is only marginally useful for working with geographic information because a GIS relies on coordinates that are the same for the different scales, although the level of abstraction can be very different. It is very easy to use data produced at one scale for another, completely inappropriate, scale. Using data at a scale provided by an agency required to create state-level data, for a local project is an example of using scale data inappropriately. Finally, there are no automatic checks to ensure that similar, or the same scales, are used in data processing. In a way, knowing the scale at which data were prepared will help avoid some large problems. Beyond scale, it is important to know the resolution at which data were collected and the accuracy at which various units were measured.

Modifying features as a result of scale change is called generalization in GIS. Generalization involves reducing the quantity of data within some constraints (McMaster and Shea 1992). An example of a GIS generalization strategy is the Douglas and Peucker (1973) algorithm that thins out these points from lines by deleting vertices that lie outside a defined distance from a line constructed between the two end points of the line (called the anchor and floats) in question (Douglas and Peucker 1973).

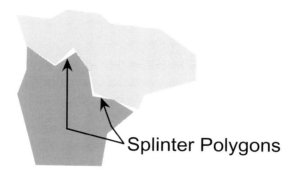

Splinter Polygons

Figure 27.3   Sliver (splinter) polygons between two polygons.

Removing numerous vertices can lead to erroneous overlaps with other lines in the same data set (Figure 27.3).

27.3.4. Projections

Along with scale, an issue facing most people working with GIS is choosing and using the correct projection. Projections are the way cartographers and other people working with geographic information transform measurements from the round surface of the earth to flat (or planar) representations. There are many issues to consider in choosing a projection. Most people find that these decisions have been made already and possibly even legislated by government entities, or defined by agency rules (See Chapters 2 and 4 on projections).

The key issue to remember when processing geographic information is that data from different projections will not align or coincide. Sometimes it may look very similar on a CRT screen, but the apparent alignment may only occur in a very small area. Unfortunately, the problems that arise from different projections can be very complicated and difficult to straighten out. Great care must be paid to ensure that geographic information is in the same projection.

27.4. In the workshop: operations

Operations are the practical ways in which tools are applied to actually transform geographic information. Going back to the wood processing analogy, operations are what the carpenter does in the workshop. There are many different ways to categorize the work carpenter's do and probably many more ways to classify geographic information processing (Chrisman 1999; Longley *et al.* 1999). This section relies on categories that are established by numerous authors writing about spatial information processing. Other categories may be useful for specific settings or fields, but these categories are purposefully very broad. Although covered more fully in the chapter on spatial analysis (Chapter 29), the following sections on overlay and buffers demonstrates that these operations are also fundamental transformations.

27.4.1. Overlay

The operation most associated with GIS, and the operation that long distinguished GIS from other types of software, is overlay. GIS owes a great deal of its success to overlay's ability to take any geographic information registered to a common coordinate system and projection and merge them into one data set.

The boon for GIS makes it possible for spatial information processing to start from diverse data sets and produce a single product. There are many concerns, however. The first issue is whether to overlay the data in vector or raster format (assuming you have a choice). The vector overlay operation provides more precise positional results, but requires more preprocessing and postprocessing. The main problem vector overlay faces are splinter polygons (see Figure 27.3). These are usually very small polygons that can result from generalization (see Section 27.3.3) or even a large difference in the precision with which coordinates are stored and processed in the computer (usually 16 or 32 digits) and the actual accuracy of the coordinates. To deal with this issue, vector overlay uses a tolerance to 'search' for features in the data sets. This tolerance was originally known as the 'epsilon' tolerance after the Polish mathematician who introduced it, but is more commonly known as 'fuzzy' distance (Wiggins and French 1991).

This tolerance solves many situations that would result in splinter polygons, but at the cost of introducing 'fuzzy creep.' 'Fuzzy creep' is an arbitrary shift of coordinates during overlay processing. The best advice is to keep the fuzzy tolerance low and limit the number of overlay operations to the absolute minimum. Some research has been done to address this problem, but it is not yet commercially available (Harvey and Vauglin 1997).

Vector overlay begins with two input data sets and produces a single output data set. It is possible to combine lines, points, and polygons in any combination. It is also possible to use polygons to remove or isolate features from a different data set, combine features for only particular areas, or just merge all data from both data sets (Figure 27.4, see plate 20).

Raster overlay operations face few problems in contrast, but are limited in their accuracy. While smaller cell sizes help alleviate this issue, it increases the requirements on processing power and time. For many purposes, it remains much faster, making it a strong contender when evaluating which type of processing is best suited. Raster data are predominantly used for continuous phenomenon, but can readily be used for discrete data also stored in raster format. The most significant problem for raster overlay is getting the information into the raster cells. If the raster data is in different cell resolutions, the data in the smaller cells can be coarsened to correspond to the data with a lower resolution. If the raster data are in different projections or coordinate systems, the numerical conversions are often unable to rectify nuances between projections used to represent the spherical surface of the earth on a plane.

For both vector and raster data the next issue, interpretation, is often the most involved part of processing. While spatial information can be easily combined with overlay, making sense of the results can be a time consuming activity. As in most information processing, garbage in equals garbage out. Judicious preprocessing before overlay can help enormously in interpreting the results. Removing extraneous attributes, simplifying polygons, and generalizing line shapes can help enormously with the interpretation of results.

27.4.2. Buffers

The second ubiquitous GIS operation is the buffer operation. Buffers are used as much as overlay for spatial processing, but depending on the type of processing one may used more regularly than the other. Usually, a systematic difference is noticeable between application types. In particularly, the implementation of regulations relies heavily on buffers. Buffers are not only extremely useful in implementing regulations, but are commonly used to calculate proximity.

The buffer operation basically is a static representation of a process or relationship. For example, factory aerosol discharges will travel a certain distance with the wind before their weight or chemical reactions causes them to become heavy and fall to earth. With knowledge about the dispersion process, this process can be represented through a buffer operation, in either raster or vector data structures. Buffers can also be used to indicate relationships. For instance, the distance a flying squirrel travels to obtain food for its young can be used to delineate areas that need to be protected from tree or underbrush removal (Figure 27.5, see plate 21).

Buffers are commonly processed using both raster and vector data structures, however, vector results are more accurate. The geometry of the raster cell resolution precludes highly accurate buffer calculations, unless the cells are very small. The actual distance of a buffer will be approximated to the raster cells. Any raster cell on the edge of the buffered area will actually be partially in or outside of the buffered area (Figure 27.6).

Buffers generated using vector data can be calculated for point, line, and polygon features and for multiple distance values. For example, noise emitted by airplanes taking and landing at an airport can be calculated using formulas that compute the complex factors involved in noise dispersion. The distances at which particular noise levels occur are calculated, and represented with buffers following the flight paths that airplanes take.

Raster buffer operations can be used to indicate processes and relationships as well. The advantages of raster buffer operations are clearer when used for representing continuous as opposed to discrete features. The dispersion of gases in the air and substances in water is usually modeled as a continuous surface. In a surface, the raster cells represent a range of values. Zones of values are indicated symbolically, but the data values are continuous.

Concern is often warranted when working with buffers or analyzing the results of buffer operations. Buffers are highly simplified representations of relationships and processes that may be very complicated. Airplane noise, mentioned before, is a good example. Many factors can influence the spread of noise: wind direction, speed, stagnant air layers, local reflection, and engine condition. Calculations can take into account many factors, but important micro-climatic factors may play a decisive role in the dispersion of noise. A buffer that computes a simplified 500 m distance from plane paths is unable to account for such complexities.

27.4.3. Conflation

While not as commonly used as overlay or buffers, conflation is a critical operation for integrating information from different sources. In the map conflation operation two

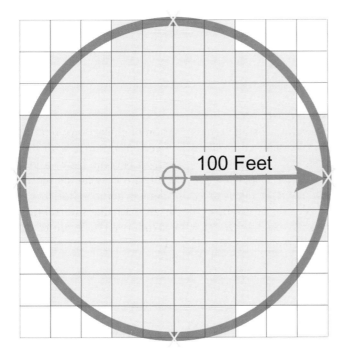

Figure 27.6 Raster buffer operations approximate distances.

data sources that share features, but have different attributes, are systematically or individually joined to merge the attributes. A good example are roads. If a street centerline data set has attributes about pavement quality and another data set has attributes with address ranges, a conflation operation joins the address attributes to the street centerline data with both attributes. Conflation was developed by the US census to merge two different data sets with different positional quality and different attributes (Lupien and Moreland 1987; Saalfeld 1988).

This operation has been used with great success in a number of settings to convert data with lesser accuracy to greater accuracy. Conflation reduces or eliminates the need to go out and collect field data. It also makes it possible to integrate older data with newer data if the data history is important (Figure 27.7).

The key issue with conflation is aligning the correct features. While theoretically the same road in two data sets is easy to identify, practically there can be numerous problems, either geometric or semantical in nature. Imagine, a new road that ends in a cul-de-sac (round-about). The engineers who designed the road may separate the curved part of the road from the straight component in order to cleanly construct the circle and provide specifications for the builder. A city planner will probably have digitized the road as a single feature since it has the same name. When the data sets are conflated it can be easy to overlook the circular end. A more intricate problem can arise with streets that are divided into small segments, which is even more common for streams. Hydrologists and civil engineers will divide streams into very different segments according to their rationale: flood risk and riparian suitability produce extremely different stream networks. A stream can be divided into extremely small segments that be overlooked during conflation.

Because of the complexity, conflation is usually an operation that relies heavily on the human to ensure high quality results. If the amount of data are large, newer research has developed ways to automate the conflation operation. Automation relies on identifying robust semantic rules (based on a definition of common meaning for features) that can successfully merge features from different data sets (Uitermark *et al.* 1999). The need for human verification of the results remains, but automation is important for improving the efficiency of conflation.

27.4.4. Generalization

Generalization is the key to the process of rendering meaning in geographic information (Goodchild *et al.* 1998). It is a very important spatial processing operation that reduces information content to emphasize the most relevant components of geographic information. Further, effective communication of geographic information requires generalization. Scale and accuracy concerns usually make the use of generalization necessary in every kind of spatial information processing.

Following Robinson (1960), generalization involves elements and controls. The elements are simplification, classification, symbolization, and induction. The controls are objective, scale, quality of data, and graphic limits. Although work on automated

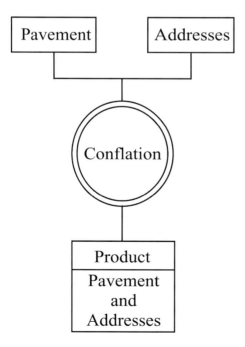

Figure 27.7 Conflation merges attributes from multiple data sets.

generalization has generated numerous numerical techniques, inductive reasoning on a case by case basis remains the rule rather than the exception for generalization. Generalization involves abstraction of both spatial and attribute information.

In practice generalization is rarely a single operation. It usually involves several operations that are used in sequence to reduce information content. Generalization operations can be classified in various ways, but always involve transformations of spatial and attribute components (See Chapter 31). The Douglas-Peucker algorithm discussed above (Section 27.2.2) effects the spatial component of geographic information. While it thins out points which are less important to the shape of a line, attributes, that may be scale dependent, are not included. Following simplification of a line shape, a simplification of attributes should follow. A reclassification of attributes to reduce 72 land use classifications to11 is an example of a generalization operation on the attribute components(see the following section on aggregation for more discussion).

Abstracting the infinitely complicated world to a rescaled, comprehensible, and useful symbolic representations alters meanings. It may distort, delete, or enhance symbolic representations to place relevant aspects of a place in the foreground and diminish the symbolic relevance of other aspects. Through manipulation of the map symbols geographic relationships may be symbolically emphasized to clarify or connote connections for a particular audience. For example, a road map may indicate a dead-end side street ending in a residential area that is impassable from a main road. In actuality, however, this area is a connection between the side street and main road for pedestrians and cyclists. Further, the area in question was once a part of said road, but has now been closed to motor vehicles in order to reduce noise pollution. This reduction of information is pertinent to the automobile or truck driver, but distorts and falsifies the situation to cyclists, pedestrians and others. A benefit for some is a loss of geographic information for others. For the town's transportation department this road map's distortion, although benefiting motorists is considered blatantly incorrect.

Problems commonly arise through feature displacement, another component of generalization. In some cases the size of symbols are clear legibility require that features be displaced. For example, moving a shoreline to properly indicate a railroad next to a road on a highway map can lead to significant problems when this data is later used with riparian habitat data collected at a much more detailed scale (e.g. 1:5,000). Recognizing these problems can be a very involved process, and often any correction is impossible. Most importantly, it is always important to know how geographic information has been processed before using it for ones own processing.

27.4.5. Rotation, translation and scaling (refer to chapter 4 for mathematical detail)

All too frequently, map materials that are digitized come at different scales or orientations. GIS packages provide a number of useful operations based on similarity or affine transformations for rotating, translating and scaling. Rotation operations will rotate coordinates by an angle; translation operations move the coordinates to a new origin; scaling operations can change the scale on one x or y axis, or on both. These operations rely on two transformations. The similarity transformation maintains the scale on both axes. It is regularly used to transform digitizer table coordinates to map coordinates by rotating, translating and scaling.

$$X = A + Cx + Dy \tag{27.1}$$

$$Y = B - Dx + Cy$$

where

$C = $ (scale factor) $\times \cos$(rotational angle)

$D = $ (scale factor) $\times \sin$(rotational angle)

A and $B = $ offsets for the center of rotation in output coordinates

The affine transformation uses different scales for each axis. Situations arise, e.g. paper shrinkage that require the use of different scaling factors on each axis.

$$x' = x \times S_x \tag{27.2}$$

$$y' = y \times S_y$$

where

$S_x = $ (scale factor for the x axis)

$S_y = $ (scale factor for the y axis)

27.4.6. Attribute

Attribute operations reduce or increase the information content of attributes. While the processing involved is more mundane than the other operations, these transformations are used in conjunction with spatial operations to add or simplify geographic information. Some of these operations transform existing values into new ones. Reducing land use categories to a smaller number involves applying some grouping criteria. USGS Land Use Land Cover (LULC) data has a hierarchical classification system that can be used to group land use categories. For example, strongly differentiated residential categories can be grouped together under the single residential category to produce simpler geographic information. The number system of this data makes this easy by grouping the lowest level of differentiation into groups of ten. For instance, all forested areas are classified by a value beginning with 40.

Other operations to reduce attribute information isolate, order, or scale attributes by a constant value. For example, an isolation operation would keep only single family residential areas and delete areas with other attributes. An ordering operation could be used to rank soil pH values by their known toxicity to plants and scaling could be done to convert elevations from feet to meters and rounding the resulting values to integer values only. Since a meter is equal to 39.37 inches and a foot has 12 inches, rounding off the resulting meter values could result in differences of up to 1.5 ft.

Attribute operations can also combine values through mathematical operations, or cross-tabulation (Chrisman 1997). These operations create new values which leave the original values intact. The reduction of attributes often deletes the original values making it impossible to return to these values.

It is also possible to increase attribute information by ranking, evaluating, or rescaling attribute values (Chrisman 1997). Usually nominal measurements of geographic features, for example land use, are upgraded by classifying them in terms of suitability or another characteristic to ordinal values. Continuous measurements can be rescaled using non-linear equations that convert a measurement into an indicator. An example of this is transforming soil pH into a crop suitability index. A complete discussion of measurement scales is provided in Chapter 31.

27.5. Theories for spatial information processing: spatial transformations

In spatial data processing transformations are the fundamental activities. The material, data, comes in various forms and must be considered first. Tools are considered according to the material available and the processing required to make the product. Knowing data, tools, and operations is key to the successful spatial data processing. Since all processing of geographic information involves transformations, it seems logical that processing is thought of in those terms.

Waldo Tobler, a famous geographer, is the first person to describe the processing of information in terms of tools. In 1979 he wrote an important paper that presents a concept for organizing and analyzing geographic information in terms of cartographic transformations between points, lines, and areas (locative) and interpolation, filtering, and generalization (substantive) (Tobler 1979). Transformations also involve conversions from one type of conceptual and representational element to another. For example, changing the representation of a town from a dot, its size indicating the population, to an area representing its actual shape is a transformation from point to area. Represented as an area, the information about the size of its population is lost, unless another graphical variable, such as shading, is used. This is not only a transformation of a graphical variable, it is just as well a transformation of the geographic concepts presented. Transformations can also involve converting a particular geographic conceptualization to a cartographic representation. Interpolating the demographic characteristics of census blocks to census tracts involves aggregation.

These are all the kinds of transformations used daily in processing spatial data and discussed in this chapter. By understanding the material, tools, and operations in terms of transformations, we can develop a better understanding of the spatial processing skills involved. Much of the work with GIS involves sequences of operations to transform data into various intermediate products that are ultimately combined into the final product. A flashy map may involve hundreds of operations. With all these operations it is very easy to make a mistake that may not be noticed until much later, and can therefore become costly. Here is where spatial information processing stands heads above wood processing. If a copy of your data is kept at key junctions in processing spatial information, it becomes easy to locate errors and start over with the least loss of time and effort.

The key to the processing of spatial data is transforming data. By keeping track of these transformations and understanding what spatial processing changes in data, one will be able to better understand the results of GIS, and improve their own use of GIS. Geographic information is different than wood, but by thinking about processing in terms of transformations, GIS becomes less sorcery and more practical.

References

Chrisman, N. R., *Exploring Geographic Information Systems*, John Wiley & Sons, New York, 1997.

Chrisman, N. R., 'A transformational approach to GIS operations', *International Journal of Geographical Information Systems*, 13(7), 1999, 617–637.

Douglas, D. H. and Peucker, T. K., 'Algorithms for the reduction of the number of points required to represent a digitized line or its caricature', *Canadian Cartographer*, 10(2), 1973, 112–122.

Goodchild, M. F., Egenhofer, M. J. and Fegas, R., *Interoperating GISs: Report of the Specialist Meeting*, NCGIA: Varenius Project, Santa Barbara, CA, 1998.

Harvey, F. and Vauglin, F., 'No Fuzzy Creep! A clustering algorithm for controlling arbitrary node movement', *Proceedings, AutoCarto 13*, Seattle, WA, 1997.

Longley, P. A., Goodchild, M. F., Maguire, D. J. and Rhind, D. W. (eds.), 'Introduction', *Geographical Information Systems*, vol. 1, John Wiley and Sons, New York, 1999, pp. 1–20.

Lupien, A. E. and Moreland, W. H., ' approach to map conflation', *Proceedings, AutoCarto 8*, Baltimore, MD, 1987.

McMaster, R. and Shea, K. S., *Generalization in Digital Cartography*, The American Association of Geographers, Washington, DC, 1992.

Robinson, A., *Elements of Cartography* (2nd ed.), Wiley, New York, 1960.

Saalfeld, A., 'Conflation, automated map compilation', *International Journal of Geographic Information Systems*, 2(3), 1988, 217–228.

Tobler, W. R., 'A transformational view of cartography', *The American Cartographer*, 6(2), 1979, 101–106.

Tomlin, C. D., *Geographic Information Systems and Cartographic Modeling*, Prentice Hall, Englewood Cliffs, NJ, 1990.

Uitermark, H., Vogels, A. and van Oosterom, P., 'Semantic and geometric aspects of integrating road networks', *Interoperating Geographic Information Systems*, Vol. 1580, Vckovski, A., Brassel, K. E. and Schek, H.-J. (eds.), Springer, Berlin, 1999, pp. 177–188.

Wiggins, L. L. and French, S. P., *GIS: Assessing Your Needs and Choosing a System*, (Planning Advisory Service 433), American Planning Association, 1991.

Chapter 28

Hardware and software

Mark Lindberg

Deciding what hardware and software to purchase for Geographic Information System (GIS) use should be an easy endeavor. One should examine the array of possibilities and select those that meet their functional requirements and are within their budgetary constraints. Deciding what hardware and software to purchase may be a highly individualistic endeavor. Presumably the GIS facility, whether a single consultant or an office with a large staff, has functional requirements and budgetary constraints specific to them. Attempting to provide guidelines for purchasing hardware and software, without the corresponding analysis of user requirements and recognition of budget, might seem foolish, but there are a number of considerations that might prove useful during this decision-making process.

This chapter's intent is providing some degree of guidance about selecting hardware and software in today's GIS computing environment. This is done with an assumption that readers will have their own particular requirements and constraints that will ultimately be the most important influences on their decisions. The discussion here is meant to augment an important decision-making process. It does not provide a framework for the decision-making process, since that too will probably be specific to any given GIS facility.

28.1. Recent history

We seem to be at the end of a somewhat turbulent period for GIS software. There have been four recent changes to GIS software that are important: (1) an increased importance of providing Web-based services; (2) a maturation of software interfaces resulting in easy-to-use menu-driven systems; (3) the movement of data from proprietary formats to major Relational Database Management System (RDBMS) systems; and (4) a strong movement away from UNIX to Windows NT as the dominant operating system for GIS.

There has been a change in the fundamental economic structure of the GIS industry. This structural change is a movement from being primarily a *producer* industry to a *consumer* industry. The driving force behind GIS software development ten years ago was the government/university/utility GIS community that was involved with providing classic *producer goods*. Today, GIS software development is most strongly influenced by the business market and GIS is becoming a consumer industry.

The size of the producer (government/university/utility) market is greatly surpassed by the size of the business (consumer) market. The producer market will remain a niche market, but the consumer market will be central to most GIS software development in the future. As GIS becomes increasingly mainstream it will take on more and more trappings of business computing. This implies GIS will become for all intents and purposes a PC-based enterprise. The producer market will remain important, and will remain a cornerstone of the GIS industry, but producer requirements will not be the driving force behind GIS software development. GIS software vendors see a large relatively untapped business market that runs PCs. That has become, and will continue to be, the driving force behind GIS software development.

A singular hardware fact remains important for GIS: *Moore*'s *Law* – popularized as stating that hardware performance doubles every 18 months – seems to continue unabated. Computer life spans are determined more by operating system and application software requirements than by hardware operability. Moreover, many GIS operations are computationally intensive, and therefore rely heavily on the basic speed and memory capacity of a computer. For example, one is likely to actually receive a four-fold increase in performance when replacing a 3-year-old machine with a new one for much GIS work. On the other hand, few people are likely to report such performance increases for word processing – based solely on a new computer.

Computer hardware has advanced to the stage where consumer-based desktop systems are now powerful enough to run the most computationally intensive GIS software. This is a primary reason why many existing GIS facilities have switched from UNIX workstations to PCs running Microsoft Windows NT. It remains unclear whether NT systems are actually cheaper than UNIX workstations *in the producer environment*. There is no doubt, however, that Windows-based systems are fundamental in the GIS *consumer environment*.

28.2. The basic parts of a GIS system

Dividing computational procedures into input, storage, manipulation, and output has been a traditional way of describing the internal workings of a computer. A variant of this model, shown in Figure 28.1, identifies the basic physical features of a typical GIS. The model is useful because it distinguishes between four types of hardware. (Consider the data manipulation section to be encompassed by a computer.) Data storage in this model refers to off-line storage, for archiving and storing data sets. The data capture and output categories refer to the basic standard GIS input devices (digitizers and scanners), and standard GIS output devices (large-format printers). The first row of

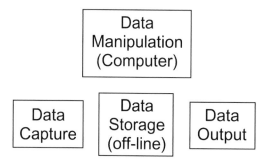

Figure 28.1   Physical components of a GIS system.

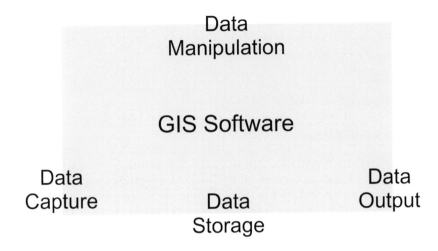

Figure 28.2 Software components of a GIS.

the figure – Data manipulation – consists of a typical computer consisting of a CPU (referring to everything within the main box as opposed to a chip), monitor, keyboard, mouse, and other devices. The second row contains peripheral devices – those that are not normally packaged with a desktop computer and are generally external devices connected to the computer via cabling.

An additional distinction can be drawn from Figure 28.1 based on columns. GIS use tends to require specialized hardware for the first and last columns (data capture and output) whereas GISs make use of highly standardized hardware for the center column. The fundamental reason is size – the large size of maps relative to most other printed materials – and the most notable impact is on price. Whereas there are relatively inexpensive input and output devices, the large-format devices called for in GIS work are considerably more expensive than their smaller-format relatives.

Although Figure 28.1 fundamentally is based on hardware, it is instructive to review how GIS software products fit into this system. Figure 28.2 illustrates how modern GIS software packages fit into this view of GISs. GIS software overlaps substantially with data manipulation, capture, and output. It overlaps to a somewhat lesser extent with data storage. The overlap with manipulation, capture, and output is not complete. Frequently, GIS analysis (manipulation) requires additional work using an external statistical package or database manager. Although data capture from a digitizer is commonly handled by a GIS package alone, scanners are frequently driven by other software. Most GIS packages have rich facilities for data output, but cartographers often choose to use general-purpose drawing packages as part of their standard map production. Different types of off-line data storage generally require specialized software for transferring files to tape, CDs, or other media.

An overriding element of GISs – including both hardware and software components – is that it is obviously and fundamentally *computer technology*. People involved with GIS should have a good understanding of various computer issues when putting a GIS together. GIS users should know much about their computer hardware and operating system – perhaps as much as they do about their GIS application software. This statement appears at odds with the trend of GIS moving from a producer to a consumer type of industry. However, GIS software is complex and GIS data sets can be quite large, and these can put serious strain on any computer system. Additionally, when you integrate specialized (non-consumer) peripherals into your system, you may be confronted with issues such as non-standard serial port cabling or SCSI port problems. The more computer savvy a GIS professional has, the better.

28.3. Software Guidelines

Software considerations should be considered critical because different packages offer different functionality at a very wide range of cost. Learning curves can be quite different, and these have a significant impact on start-up costs. Many vendors offer a range of products and at present there is relatively minor standardization. Selecting an appropriate package need not be difficult, however.

There are some fundamentally different approaches, from large expensive packages designed for the producer markets to small relatively inexpensive packages designed for the consumer markets. Arc/Info from Environmental Systems Research Institute (ESRI) and Geomedia Professional from Intergraph Corporation are examples of large GIS packages that are likely to remain in the producer domain. Software packages in this realm can generally be custom-fitted to some extent by selecting particular module subsets, and different GIS facilities could obtain different functionality mixes. MapInfo Professional from MapInfo Corporation and IDRISI from Clark Labs (Clark University) are examples of smaller and less expensive packages. In general these have feature sets smaller than the large packages, and these will often limit analysis (but not necessarily display) to either vector or raster processing. They are powerful GIS packages, however, and their use is by no means limited to the consumer realm. There are packages that are entering the market geared solely for the consumer and these might not even be advertised as GIS software. MapPoint 2000 from Microsoft Corporation and Street Atlas USA from DeLorme Mapping are examples of these. Some GIS products consist of libraries designed for producers to essentially design and produce their own systems. MapObjects from ESRI and MapInfo MapX from MapInfo Corporation are examples of these. These two selections also provide an example of another common feature of GIS software vendors. Many vendors provide an array of products aimed at different market segments. In addition there are numerous limited-focus packages designed to be used in addition to other GIS software. The range is illustrated by MaPublisher from Avenza Software that allows common GIS data formats to be read into drawing packages such as Adobe Illustrator and MrSID from LizardTech Incorporated that provides mosaicing and compression facilities for different GIS packages. Finally, some vendors provide software enhancements for other vendor's products. For example, Blue Marble Geographics provides Global Positioning System (GPS) support for MapInfo Professional and ERDAS Incorporated provides an image analysis extension to ESRI's ArcView.

Ideally one's software requirements will fall within the range of the software available, and one simply needs to decide from a subset of what is available. A single package or program might meet limited consumer needs. More commonly, GIS facilities will use a range of GIS software products, and producer needs will probably be met only by an array of different packages.

Finding out about different software products is fairly easy because most vendors provide information about their products online. Some online sites, such as Urban and Regional Information Systems Association (URISA) at www.urisa.org, provide links to vendor home pages. Note that obtaining pricing information for some of the packages is not possible without contacting the vendor directly. Some vendors provide considerable detail about their products in the form of *white papers*, and these are often available online. One should certainly peruse the trade magazines such as *GEOWorld* (online at www.geoplace.com) and *Geospatial Solutions* (online at www.geospatial-online.com) for software reviews. Talking to other GIS professionals who have experience with the particular software packages of interest is probably the best source of information. For the beginner, checking introductory-level books on GIS is an appropriate starting point. Many contain sections on the GIS selection process and reading as many of these as possible is a good idea (e.g. Clarke 1989; Korte 1997; Bernhardsen 1999).

28.4. Hardware guidelines

Hardware selections are inseparable from software selections because much GIS software is operating system-specific and therefore by extension hardware-specific. Most GIS software runs under some variant of Microsoft Windows (95/98/NT/2000), some of the larger packages run under UNIX, and few packages have a MacOS presence. In a recent survey of GIS software, 24 per cent of the packages run under UNIX, and only MAPublisher (mentioned above) runs under MacOS (*Point of Beginning* 2000). Most importantly, all vendors that supplied software that runs on UNIX or MacOS have identical or similar versions that run under Microsoft Windows. *For most new users, a computer for GIS means a PC running Microsoft Windows.*

GIS has computer requirements unlike most office or consumer uses and efficient GIS work requires a computer with substantial speed and memory. Some GISs are linked to very large databases, for example tax information for every parcel in a large city – therefore prodigious disk space with extremely fast I/O is often necessary. Fast graphics and high-resolution large monitors are essential. In short, today's GIS environment 'requires' today's best hardware. If processing time is a recurring bottleneck at a GIS facility, the newest and fastest hardware may be appropriate. If however, processing bottlenecks are but an occasional problem then devising a schedule whereby lengthy procedures are done during a break or overnight, or perhaps on an otherwise unused machine, may be a more cost-effective solution. It is the time-tested solution familiar at many GIS facilities.

Computer 'requirements' appear to be a moving target. The purchase of each new computer brings a faster more powerful machine into use. It seems fast...for a while. If one obtains a new computer every third year, the new machine is approximately four times faster than the previous one (given Moore's Law), and it will be perceived as very fast indeed. During the ensuing few years it will become *relatively* slower – it does not slow down, of course, but newer machines are increasingly faster. Notably, it is often impossible to purchase a computer substantially behind the curve. For example, it is currently impossible to purchase a new hard drive that has less capacity than the largest drives commonly available just a few years ago. Hence, much of the discussion on the finer details of hardware is a moot point. The best PCs in the market today will be seriously out of date in a matter of a few years. The best advice is to purchase near the high-end of the presently available computers.

There are some specifics that are important, however. Plan on outfitting a computer with substantial memory (RAM) – with a minimum of 256 MB, and preferably 512 MB or more. Purchase a large monitor – with at least a 19 inch diagonal, and preferably a 20 or 21 inch diagonal. Match it to a fast AGP-format graphics board with at least 16 MB memory, and preferably 32 MB of memory. Disk drive capacity has expanded greatly in recent years and you should be sure to purchase substantial disk space at the outset; you will not regret it. Consider 30 GB to be a minimum, and preferably 45 GB or 60 GB to be a reasonable amount for a desktop GIS. The latter comes with a caveat – large amounts of disk space can spell trouble if you cannot adequately back it up on a regular basis. Do not use ample disk space as a hedge against archiving your data off-line.

GIS uses a range of peripherals that, while not unique to GIS, are still small-market producer products. Digitizers (graphic tablets), large-format scanners, large-format plotters are probably the most commonly used of these. High quality consumer small-format scanners and small-format printers are available inexpensively. Large-format versions are relatively expensive and in today's GIS hardware environment the cost of peripherals will probably greatly exceed the cost of the PC they are attached to. For that reason, they are discussed in somewhat greater detail below.

28.4.1. Data capture devices

Data capture hardware most likely used in GIS includes digitizers and scanners. The digitizer, shown diagrammatically in Figure 28.3 has been the traditional device for encoding digital data from existing map sheets. Since 'digitizing' refers to a generic process of capturing data in digital form, and since a wide variety of devices perform

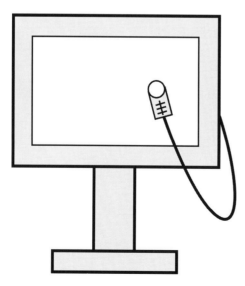

Figure 28.3 A large-format digitizer.

this function (from scanners to digital cameras), these devices are increasingly called *graphics tablets*. By manually moving the cursor (puck) across an image – tracing it – vector data are captured as strings of *x-y* coordinate pairs. Once a primary means of working with existing printed map information, digitizing has been slowly declining in importance in recent years. Increasing amounts of data are already available in digital form and scanning has been making serious inroads into the digitization of printed materials.

There has been considerable consolidation of suppliers of digitizers in recent years. Digitizers by GTCO, CalComp, and Summagraphics – three traditional brands – are now all sold by GTCO CalComp. Large format-digitizers are also available from the ALTEK Corporation. Digitizers are available in sizes ranging from small desktop pads to 44 × 60 inches. They are available in different resolutions, with backlights, and motorized stands allowing easy adjustments. Depending upon size and features one can expect to pay about $2,000 for a 36 × 48 inch tablet and stand.

Scanners are manufactured in several formats – flatbed, continuous-feed, and drum. Although not exclusively, flatbed scanners are typically small devices handling materials up to about 11 × 17 inches. These are now consumer items and are commonplace in home and office computing. Their quality is quite good and their price has plummeted. It is possible to purchase a 1,200 dpi-color scanner for about $100. Drum scanners are currently most frequently used in high-resolution scanning (2000+ dpi) in the printing industry. Continuous-feed scanners move the sheet past a row of optical recording devices – see Figure 28.4. These have proved to be cost-effective for scanning large materials, such as sheet maps.

Scanners rasterize printed maps (and other images). Whereas digitizing is inherently a vector process, scanning is a raster process. If the data are needed in raster form, additional data manipulations are easy. If, however, data are needed in vector format there are two distinct approaches currently available. First, software for raster-to-vector conversion has matured over the past 20 years and continues to improve. Traditionally, there has been a comparison between the time necessary to manually digitize a sheet (including preparation and quality assurance operations) versus the

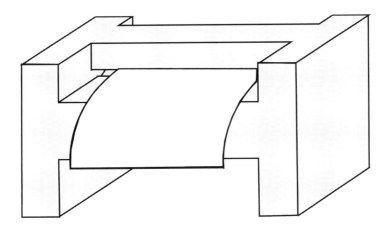

Figure 28.4 A large-format scanner.

time necessary for scanning (and corresponding preparation and editing operations). Software has now improved to the point where scanning and automatically vectorizing a map is viable. A second approach is a variant of manual digitizing. Rather than 'tracing' a map on a digitizer, a person 'traces' a scanned image of the map on-screen. One can therefore manually vectorize a scanned map, or other raster data, using hardware common to most contemporary desktop computers – a mouse and a good monitor.

Scanners work fast. A good scanner can process a sheet map in a matter of a seconds. Therefore, assuming quality scanning materials, a person can scan maps at the rate of several per hour. A GIS facility can keep multiple people working at desktop computers digitizing maps and running a single scanner. Although expensive, many GIS facilities have found a single large-format scanner to be more cost-effective than maintaining multiple digitizers.

Unlike digitizers, which can often be 'driven' by off-the-shelf GIS software, large-format scanners use specialized software for their scanning operations. Large-format scanners suitable for GIS are produced by companies such as Vidar Corporation and Océ Engineering Systems. They are sold in a few sizes – but since they are continuous-feed, their widths alone are the crucial dimension. They are available in black-and-white and color versions in resolutions from 200–1,200 dpi. These machines are relatively expensive. You can expect to pay at least $10,000 for a fast 36 inch black-and-white scanner and $20,000 for a good 36 inch color scanner.

28.4.2. Output devices

The ability to print large-format output is important for most GIS facilities. Producing short-run maps, proofing maps during the design phase, and assisting in quality assurance procedures by reproducing digitized materials at their original sizes is among some of the uses for large-format printers. Like scanners, high quality inexpensive small-format printers are readily available in the consumer market. Also like scanners, the reason GIS uses much more expensive producer devices is simply the need for working with sizes larger than about 11 × 17. A variety of technologies are in use. Pen plotters, once the mainstay of GIS printed output, have all but disappeared. Electrostatic plotters (and closely related technologies) which once held a dominant position for high-end and high-quality GIS printed output have also been rapidly declining. Large-format ink-jet printers have replaced both. Large-format ink-jet plotters are very much like small format color printers in that they work by running paper past a print-head that moves across the paper in swatches. Figure 28.5 shows this diagrammatically.

Unlike pen plotters, ink-jet printers are equally adept at printing raster and vector-based images. Ink-jet printers are cheaper than electrostatic plotters and are significantly easier to maintain. High-quality large-format ink-jet printers are manufactured by Hewlett Packard, Encad, and Colorspan.

Large-format ink-jet printers can be slow; a 24 × 36 inch plot might take a half-hour or more to complete. If speed is important, there are other technologies available. Large format LED printers designed for the reprographics industry are available that are much faster than ink-jet printers, but they are two to three times as expensive. LED printers are available from vendors such as Océ Engineering Systems and Xerox Corporation.

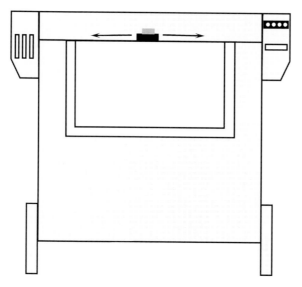

Figure 28.5   A large-format inkjet plotter.

Finding out about different hardware products is a bifurcated process. If it is consumer hardware one has a numerous choices – mainstream computer press, local retailers, and any friend that might be using similar hardware. There are two additional factors that should be considered. First, make sure the hardware meets, or preferably exceeds, requirements specified by your chosen software. Second, if your organization has an Information Technology (IT) or computer-support staff, make sure any selection is done in consultation with them – even in situations where their direct involvement is not required. This is especially important if they will be called on to maintain or service the equipment. Finding out about producer-type hardware peripherals is quite different. There will be much less information readily available because they are niche-market items. Approaches for finding out about GIS software work for specialized hardware as well. This hardware is not GIS-specific and one should expand their search to include trade magazines and online sites for surveying (e.g. *Point of Beginning* online at www.pobonline.com), CAD/CAM (e.g. *CADENCE,* online at www. cadenceweb.com), computer graphics (e.g. *Computer Graphics World* online at cgw.pennnet.com), and the printing industry. For beginners, reading as much about the use of peripheral devices in GIS in introductory texts is a good starting point (e.g. Clarke 1989; Bernhardsen 1999; DeMers 2000).

28.5. Recommendations

The remainder of this chapter is structured around a series of recommendations that are meant to provide food for thought during the selection of hardware and software, rather than highly specific options that are likely to out-of-date within a few months. Although some are sure to be seriously questioned by some readers, they are not meant to be controversial. Hopefully they will spark a recognition of the range considerations that a thorough examination of GIS needs should receive.

28.5.1. Learn all you can about computer technology

Although it might seem a truism that GIS requires high-level computer technology, there are some impacts of this statement that are important. First, GIS practitioners – those who perform the GIS functions in a GIS organization – need to know hardware and software. To be effective, a GIS staff needs to have considerable expertise and experience beyond the running and maintenance of GIS software. Costs associated with this experience need to be considered part of the hardware/software selection process. In large organizations where there is a separate IT staff, impacts of this 'hidden expense' may be minimized, but even then you might not be able to rely on them for all your maintenance and administrative needs.

Why a GIS staff needs substantial computer savvy beyond the running of software and why they might not rely completely on a separate IT staff is due to the complexity of GIS systems. To be most effective, someone on a GIS staff should probably control the installation and maintenance of GIS software. To save time and eliminate frustration in critical situations, someone on the staff should probably control installation and maintenance of GIS hardware peripherals. The same may be true for consumer-based GIS. Should an organization's GIS consist of one person running a desktop GIS system, to be efficient that person should probably be computer savvy and should not have to rely on IT staff for GIS support.

28.5.2. Be wary of one-package solutions

The single software package that handles all your GIS needs might be an ideal, but it is not likely to be found. Even if you are lucky enough to find such a product (assuming your needs are not overly simplistic), it might not last for any significant period of time. Being able to move data in and out of a package is often where problems arise. As products from other vendors change it might impact the ability to easily exchange data. An appropriate software toolbox – utility programs that augment the functionality of your main package – is an important feature of many GIS shops. For some processes you might rely on moving data through a series of programs, relying on their particular strengths.

GIS productivity is impacted by non-GIS software – spreadsheets, text editors, statistical packages, database management systems, and others. Of these, the selection of statistical package and database management system are probably most critical to a GIS endeavor, and for many institutions the selection of a database management system is a *pre-GIS* decision. A GIS package is selected partially based on its interaction with the organization's database manager.

The importance of this statement is twofold. First, putting together a software toolbox should be a concern when establishing a GIS lab. Second, you might seek an 80 or 90 per cent solution – select a vendor's software that accomplishes most of what you need, as long as you have other tools that can do the rest. One often has to rely on auxiliary software and therefore one should recognize this fact when judging different GIS packages.

Maps are primary GIS products. Maps for print – as individual sheets or as illustrations in a manuscript – will at some stage probably go through a series of manipulations while in PostScript format. The PostScript language is the de facto standard in the printing industry, and the Apple Macintosh is the overwhelmingly standard computing platform for the printing industry. How this affects a GIS operation depends upon its orientation. If you provide custom cartographic services, it will impact your hardware and software selections in a fundamental way; if your output is limited to internal memos and reports used for decision-making by your organization, it will have little impact.

Printing and publishing maps adds the specter of expanded software requirements and possibly an additional hardware platform (Macintosh). Initial processing in a GIS software package with subsequent processing in a general purpose drawing package (e.g. Illustrator, FreeHand, or CorelDRAW for vector formats, and PhotoShop for raster formats) has become commonplace in map production. Consideration of how you move maps between different software packages is very important. This may involve third-party software or perhaps software tools produced internally.

Whether for a producer or consumer shop, the GIS software selected must coexist with a wide range of other software systems. Start with the assumption that GIS software is only one, albeit an important, element in the mix – and then focus directly on how potential GIS software selections fit into your particular mix.

28.5.3. Avoid turnkey systems

Turnkey systems are those that bundle hardware and software together in an *integrated* package. As a corollary to avoiding single-software solutions, turnkey systems are problematic because they are single system solutions in the extreme. They have a high probability of creating hardware and software legacy problems. Note that a hardware and software package bundled by a vendor is not necessarily a problem if it is not tightly integrated, and therefore not a turnkey system in the strict sense. If the software runs on other machines without problems, if the hardware accepts mainstream operating system updates, and if one can go to another vendor for hardware upgrades there is probably no cause for concern. Still, the likelihood that a vendor's bundled product meets your particular needs remains cause for concern.

Turnkey systems run counter to consumer GIS, which requires software that runs on existing desktop systems. Moreover, turnkey systems may eliminate advantages in existing producer shops – they may compromise skill sets and experience of the existing staff.

28.5.4. Be wary, but do not automatically avoid, software from start-up GIS vendors

Advertisements in the GIS industry press indicate that entrance into the GIS software market is possible and continues at a fairly rapid rate. There are two highly contradictory conclusions that can be drawn from this. First, one should not overlook software from new start-ups – especially for limited-function software in niche categories. Second, there will be industry shakeouts – most of the new firms will not survive, and you might want to avoid their software. Perhaps the key is limiting interest in new firm's software to where they have a competitive advantage – special purpose programs that provide functions that are not (yet) available from older established firms.

28.5.5. Consider labor costs when selecting hardware and software

Labor is the most expensive part of GIS – and it impacts the GIS enterprise in a variety of ways. Data entry costs, in the form of salaries, far outweigh initial costs of hardware and software for most producer GIS facilities. This is well understood. There are various other 'labor costs' that also impact hardware and software decisions. Costs associated with hardware maintenance and operating system administration must be accounted for when selecting GIS systems. These costs are likely to be highly variable based on the types of expertise already available at the GIS facility. Over time, these costs exceed all others. One can easily imagine a situation where it would be entirely appropriate to select a software package from an array of alternatives because it fits existing personnel best – even though some other package might provide a better match with stated software requirements.

A critical feature of consumer GIS consists of reducing the *apparent* necessity for training costs. The software must be easy-to-use and GIS neophytes must be able to learn it quickly in order to succeed in the consumer market place. This masks an important aspect of training costs. Making GIS software highly accessible to users who might not have a solid understanding of GIS does not necessarily eliminate training costs – it simply allows anyone to avoid such costs. The time and expense of learning to use the software properly – whether formally in classes or informally by reading and talking with GIS professionals – is still a *cost* of the software.

28.5.6. Existing producer GIS facilities should carefully examine their use of UNIX

Although, the movement from UNIX to Windows-based PCs is well known and is a necessary part of the GIS software industry's movement into consumer GIS, UNIX will still be an important operating system for producer GISs for as long as software vendors support it. This statement runs counter to current punditry.

The most commonly cited reason for switching to PCs is the dramatic improvement in *wintel* machines relative to RISC-based UNIX workstations. As discussed above, the movement of GIS software vendors into the consumer market is another important factor. Moore's law has operated regardless of platform and dramatic improvements keep occurring with UNIX workstations as well as PCs. The power/price ratios that favor *wintel* machines do not take all costs into account, and these unmentioned costs will keep UNIX viable for many producer GIS facilities.

First, the major cost associated with GIS is labor. If a GIS staff is already skilled at maintaining UNIX systems there will probably be a cost advantage with UNIX. If a facility struggled with system and network administration, or if it shared IT staff with many offices and was stretched thin, then the cost advantage was clearly with Windows (and they have probably already switched.)

Second, there are hidden costs associated with system and network administration. Perhaps the most important of these are opportunity costs. As an example, at our site (university) there is a notable difference in how our UNIX and NT systems are administered. Our UNIX administrators regularly implement modifications that would be unthinkable by our NT administrators. This is based on fundamental differences between Windows NT and UNIX, and not the skill-levels of the administrators. We have many more locally implemented modifications to our UNIX systems than our NT systems. Importantly, these modifications improve our local computing environment. Although this difference is not based on skill level, there is an additional difference based on experience. We have UNIX administrators with well over 20 years of experience. How much is that experience worth? In a very real sense, our UNIX systems will remain 'cheaper' at least until our lead system administrators retire. Perhaps the death of UNIX as a viable platform for GIS has been somewhat exaggerated.

28.5.7. Plan on purchasing new computers rather than making continuous upgrades

Except for memory and disk drives, planned hardware upgrades probably seldom occur. Selecting a particular hardware product with the intent of upgrading it later has been a frequently recurring theme in computer purchasing over the years. It has become less attractive as systems have become smaller. What was standard procedure for mainframes has become very unusual for PCs.

As hardware prices continue to plummet, it becomes increasingly unlikely to do anything other than memory or disk upgrades. Therefore, an upgrade path is probably not a serious concern when evaluating hardware, moreover it is probably not worth spending extra on a product with a touted upgrade path.

28.5.8. Plan on peripheral costs exceeding other hardware costs

GIS often requires peripheral hardware – especially for producer shops. Large-format digitizers, scanners, and plotters are typically more expensive than substantial PC-based systems and low-end UNIX workstations. Moreover, because these peripherals are niche-market devices their prices will remain relatively high. Unless large-format input and output devices enter the consumer market, the price differential will remain quite large.

Two further considerations derive from this price differential. First, hardware peripherals will probably be on longer replacement cycles. When considering the purchase of a new scanner, one might have to plan for a 5 or 6-year useful life (or longer). Second, unlike the proverbial 'tail wagging the dog' a GIS facility might legitimately find a specific peripheral driving other hardware and software selections.

Expensive peripheral devices often scale well – a shop might require a single large-format scanner regardless of whether they have one or 15 people. Hence, the cost of GIS-peripherals is most problematic for smaller operations.

28.5.9. To reduce costs, try to leverage consumer products

Given that some of the GIS-peripherals are likely to remain very expensive relative to their consumer versions, one should try to leverage costs by employing consumer products as much as possible. For example, if your requirements include a substantial amount of color scanning, yet you rarely need areas covering entire map sheets, you should strongly consider purchasing an inexpensive consumer scanner. Sending your large scans out to a firm that provides scanning services may be less costly than purchasing your own large-format scanner. Some GIS facilities may be able to accommodate the bulk of their printing needs with a consumer printer (especially since their sizes have been creeping up to handle tabloid-sized paper), and perhaps share a large-format plotter with another office.

Leveraging consumer products might appear in unexpected, but not necessarily surprising, ways. This fact might impact your costs if your producer peripherals share components with consumer items. For example, at our site we used to run a large-format plotter that employed the same print head as a commonly available consumer printer line. We had to occasionally replace the print head and, for a period of time, it was actually cheaper to purchase a new printer (and remove its head) than it was to purchase a replacement head for the plotter. (Consumer computer products are often cheaper to replace than maintain.) This price advantage disappeared when the print head was replaced by another model in the consumer printer line.

28.5.10. Do not underestimate the importance of off-line backup systems

Backup systems can become the most important components in your GIS operation when things go bad, but they have a tendency to be considered nuisances when everything is running properly. There are several considerations regarding backups.

First, backup systems must be periodically and perhaps regularly upgraded. This is critical for smaller GIS operations that may not be able to afford redundant systems. Relying on a single backup device is inherently dangerous – although not as dangerous as not regularly backing up your data. There is sufficient anecdotal evidence of small GIS operations that suffered catastrophic disasters because of the failure of a legacy backup system. Consider the unfortunate situation where an old tape drive failed and the owner found out that the read-write head had drifted somewhat – and therefore similar legacy systems still running elsewhere could not read any of the carefully maintained back up tapes. Redundancy is the key to backups, and this includes having multiple devices as well as a highly organized logical system of backups. Backup systems must be rigorously tested. Consider how much work you can afford to lose – an hour's worth, a day's, a week's, etc. Let this consideration guide your decision when selecting a backup system.

28.5.11. Use RAID

RAID, which is an acronym for Redundant Array of Inexpensive Disks, has become widely available. RAID systems have different 'levels' that offer different features. Levels that offer data security in the form of simple redundancy (level 2) or even automatic error-correction and the ability to swap drives without shutting down (level 5) should be considered. In today's computing environment with very inexpensive disk drives, there is little reason not to use raid systems. Disk failure is still a primary cause of catastrophic data loss. A RAID system with hot-swappable drives should provide the data storage cornerstone for any large operation with a concern for data security. Being able to replace a failed drive without loss of data, much less an interruption to users, is a proven technology that should be considered basic. RAID systems have traditionally been in the domain of SCSCI disk drives, however, today there are PC motherboards that implement level 2 RAID using a pair of internal IDE disk drives. All data written to one drive is duplicated on another. It doubles your disk costs, but it greatly diminishes the chance of losing data.

28.5.12. Carefully consider support and life cycle when making hardware decisions

Most computer hardware is sold with a limited warranty, but most allow the purchase of an extended warranty. The value of such a warranty is probably highly individualistic based on the maintenance policies of the GIS facility. Some facilities might do a lot of maintenance internally; some might do little maintenance but purchase hardware more frequently. Both situations may lead to a decision to avoid purchasing extended warranties. Other facilities might choose to purchase extended warranties as a standard procedure.

Although most people pay close attention to support when purchasing hardware, careful attention to life cycle is sometimes avoided. For example, evaluating different maintenance options is probably always done, but linking the maintenance options to computer life span might be avoided. We have gone through a period of short life spans for hardware and some GIS facilities find it difficult to argue that their desktop computers should be replaced every 2 or 3 years.

A limited life cycle for PCs has been enforced by continuous changes in hardware and is therefore a direct result of Moore's law. For example, as PCs age, upgrades become increasingly problematic. Many machines were scrapped in the recent past because their BIOSs limited the size of disk drives they could use, some were scrapped as memory format changed and older formats became more expensive than new. Older machines also typically cause more problems and become increasingly expensive to maintain. Coupled with the fact that a new machine is approximately four times as powerful/fast as a 3-year old machine and GIS operations are typically more computationally intensive than many other desktop operations, there is intense pressure to replace machines quite frequently.

The same pressures have been acting on UNIX workstations, although the life cycles have not been as short as for PCs. Moore's law still holds but the market for these producer machines is not as volatile as for consumer PCs – therefore the extra expense of purchasing and maintaining a UNIX workstation gets a partial payback in a longer useful life. However, an old machine is still a slow machine and GIS computational requirements put substantial pressure for frequent replacements.

Assuming a GIS facility could convince their financial management that they need to replace all of their desktop PCs on a 2-year rotation, it might then be prudent to obtain a 2-year on-site maintenance warranty. Likewise, a 3-year warranty on a 3-year cycle machine might be appropriate. What needs careful consideration is a consideration of how much this is worth. Is the cost of such a warranty worth giving up 100 MHz of CPU speed?

Another hardware life cycle model has new machines being purchased for GIS work, with older machines being switched to other less computationally intensive uses. The newest most powerful computers are reserved for GIS use and older systems are used for other activities. This has a very important benefit – your GIS hardware can be kept clear of a lot of seldom-used software.

28.5.13. Experiment with hardware and software – but for immediate production equipment purchase behind the curve

Another suggestion applicable for all GIS facilities but has been fairly standard in many UNIX-based GIS shops for some time is to integrate new systems slowly. For example, after a new system is purchased and installed, there would be a substantial shakedown period before it is brought online for production purposes. The same would be true for new software versions – new software would be installed and tested for quite some time before shifting production work to it.

The second aspect, purchasing production equipment behind the curve, is the direct result of what appears to be an unavoidable hazard of a fast changing computing environment. The newest hardware and operating systems often cannot be integrated into GIS production immediately. Often the culprit is missing device drivers (for a printer or scanner) and sometimes it is due to outright hardware incompatibilities. In many instances, one must first wait for their GIS software vendor to release a patch or upgrade in order to run a newer version of an operating system. If your operation has the ability to experiment with hardware and software, then purchasing on the cutting edge is perhaps viable. If, however, one must bring new systems online immediately and one has neither the desire nor personnel for experimentation, one should probably purchase hardware somewhat behind the curve and should sit on software upgrades until available evidence suggests there are no major bugs that might adversely impact your operation.

28.5.14. Do not underestimate stability

As an addition to the latter few suggestions, consider stability to be one of the most precious computing commodities. People working on 'flaky' computers become demoralized and their attention wavers from their work. There is probably a corollary to *Murphy*'s *Law* stating that unstable systems will crash at the worst times. For example, computers have an uncanny knack for increasingly acting up as one gets closer to deadlines and computers are almost sure to crash when giving a demonstration to a potential client.

Stability is a fickle commodity. One is not guaranteed that properly maintaining hardware and software will necessarily insure it. However, learning all one can about computer technology – the first of these 'recommendations' – remains the best advice for seeking it.

28.6. Conclusion

The approach to making decisions about hardware and software for GIS systems varies considerably based on level of expertise, whether it is a consumer or producer-based GIS system, and a host of other factors, some of which are unique to particular circumstances. We have presented a series of guidelines that are perhaps most important for inexperienced GIS users, and a series of recommendations intended to provide items to consider or reconsider when facing a new purchase.

References

Bernhardsen, T., *Geographic Information Systems: an Introduction* (2nd ed.), John Wiley & Sons, New York, 1999.

Clarke, K. C., *Getting Started with Geographic Information Systems* (2nd ed.), Prentice Hall, Upper Saddle River, NJ, 1999.

DeMers, M. N., *Fundamentals of Geographic Information Systems* (2nd ed.), John Wiley & Sons, New York, 2000.

Korte, G. B., *The GIS Book* (4th ed.), Onword Press, Sante Fe, NM, 1997.

Point of Beginning, '2000 GIS software survey', *Point of Beginning*, June, 2000, 48–57.

Chapter 29

Spatial analysis and modeling

Michael F. Goodchild

29.1. Introduction

In the previous chapters we have seen how a wide variety of types of geographic data can be created and stored. Methods of digitizing and scanning allow geographic data to be created from paper maps and photographs. Powerful computing hardware makes it possible to store large amounts of data in forms that are readily amenable to manipulation and analysis using the routines stored in powerful software. Thus the stage is set for a discussion in this chapter of the real core of GIS, the methods of analysis and modeling that allow us to solve specific problems, and to support important decisions, using the capabilities of hardware, software, and data.

Chapters 25 and 26 introduced the various ways in which geographic phenomena can be represented in digital form. A road, for example, can be represented by recording an appropriate value in a swath of cells in a raster representation. With a cell size of 1 m, the swath corresponding to a major four-lane highway might be as much as 50 or even 100 cells wide. Alternatively, the road might be represented as a single line, or *centerline*, in a vector database. In this case its location would be recorded by specifying the coordinates of a series of points aligned along the road's center. Such centerline databases are now very commonly used in applications like vehicle routing and scheduling, and by sites such as www.mapquest.com which offer to find the best routes between pairs of places, and are used daily by millions of people. Finally, the road might be represented as an area, encased by its edges (we use the term *cased* to describe this option, and also call it a *double-line* representation). Figure 29.1 shows the three options. Which of these is used in a given case depends on the nature of the application, on the limitations of the available software, on the origins of the data, and on many other factors.

This chapter is about turning data into useful information. Analysis and modeling can make calculations that are too tedious to do by hand, and by doing so provide numbers that are useful for many kinds of applications. They can be used to reveal patterns and trends in data that may not be otherwise apparent. They might also be used by one person or group to draw the attention of another person or group, as might occur in a courtroom where geographic data are used by one side to make a point. Finally, the results of analysis and modeling can be used to provide the information needed to make decisions. In all of these cases GIS is the engine that performs the necessary operations, under the guidance of the user who issues the necessary instructions.

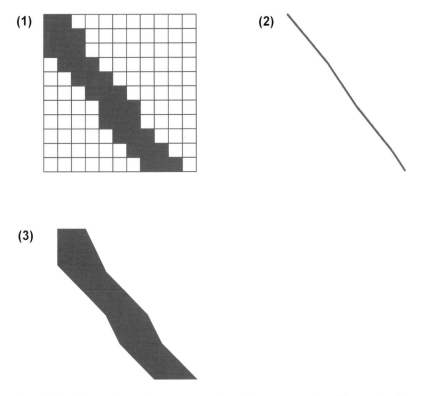

Figure 29.1 Three alternative representations of the same road, as (1) a swath of raster pixels, (2) a vector centerline, and (3) a vector area.

29.2. Organizing the possibilities

A very large number of methods of analysis and modeling have been devised over the years, many dating from well before the advent of GIS, when calculations and measurements had to be performed by hand. The ability to process large amounts of data quickly has led to a rapid explosion in the list, and today it is virtually impossible to know about every form of spatial analysis. The developers of GIS software often provide thousands of methods, and thousands more are added by specialized companies and individuals. So one of the most daunting aspects of analysis and modeling with GIS is simply keeping track of all possible analytic functionality.

One way to organize methods of analysis and modeling is by the data types on which they operate. There are operations that work on discrete objects – points, lines, and areas – and operations that work on phenomena conceptualized as fields. At a different level, and following the concepts introduced in Chapter 25, it is possible to separate methods of analysis of vector data sets from those that operate on raster data sets, since there are very few instances of operations that require input of both kinds. Another possibility is to see every operation as a transformation, taking some kind of input and producing some kind of output, and to organize methods on this basis.

The structure adopted here is a little of all of these, but is based primarily on popularity: since there are so many possibilities, it is most important to understand the ones that are used most often, and to leave the less popular ones to further study. Each method is described in terms of the problem it attempts to solve, the inputs required, and the outputs that it generates. Both raster and vector operations are covered.

The next section deals primarily with points, and the following section with areas. The fourth section looks in detail at rasters, using the framework of *cartographic modeling*, (See Chapter 25) and includes methods of analyzing digital elevation models (DEM). Finally, the last section of the chapter examines methods for optimization and design, where the objective is to use GIS to find the best solution to problems.

29.3. Methods of analysis for points

We begin with the simplest kinds of geographic objects. Suppose we have records of each of the customers of an insurance agent, including each customer's location. Perhaps these originated in an address list, and were subsequently converted to coordinates using the process known as *geocoding* or *address-matching* (Chapter 25). Plotted on a map they might look something like Figure 29.2. Several questions might occur to a market researcher hired by the insurance agent to study the state of the business and recommend strategies. For each there is a straightforward type of GIS analysis that can be used to provide the necessary information.

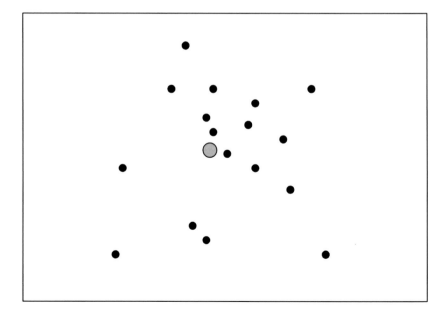

Figure 29.2 The locations of the customers of an insurance agent, shown in relation to the agent's office location, and ready for analysis.

29.3.1. Distance

First, we might ask how far each customer is located from the agent's location. The agent might be interested in knowing how many customers do their business with the agent rather than with some other agent located closer to the customer. What is the average distance between each of the customers and the agent, and are there areas near the agent where advertising might increase the agent's market share? All of these questions require the ability to measure the distance between points, a commonly used function of a GIS.

To measure the distance between two points we need to know their locations in (x, y) coordinates. Coordinate systems were introduced in Chapters 2 and 4, and many of them can be used to measure distances. The Universal Transverse Mercator (UTM) coordinate system, for example, is based on measurements of coordinates in meters, so distances calculated in this coordinate system are easily understood. State plane coordinates are often expressed in feet. In these cases the distance between two points (x_A, y_A) and (x_C, y_C), representing the locations of the agent and a customer, are given by:

$$D = \sqrt{(x_A - x_C)^2 + (y_A - y_C)^2}$$

This equation is derived from Pythagoras's famous theorem, so it is also known as the Pythagorean distance, or the length of a straight line between the two points.

Pythagorean distance applies to points located on a plane, not on the curved surface of the Earth, and so should be used only when the points are close together. A good rule of thumb is that the study area should be no more than 500 km across, because at that distance errors due to ignoring the Earth's curvature begin to approach 0.1 per cent.

It is important to recognize that this problem of the Earth's curvature is not resolved by using latitude and longitude coordinates. Suppose we record location using latitude and longitude, and plug these into the Pythagorean equation as if they were y and x, respectively (this is often called using *unprojected* coordinates, and is also what happens if we specify the so-called Plate Carrée or Cylindrical Equidistant projection) (Snyder 1987). In this case the coordinates are measured in degrees, so the value of D will also be in degrees. But one degree of latitude is not the same distance on the Earth's surface as one degree of longitude, except exactly on the Equator. At 32 degrees North or South, for example, the lines of longitude are only 85 per cent as far apart as the lines of latitude, and that percentage drops all the way to zero at the poles. So the equation cannot be applied to latitude and longitude. Instead, provided the spherical shape of the Earth is an adequate assumption (Chapter 2), we should use the equation for distance over the curved surface of the Earth. This is otherwise known as the Great Circle distance, because the shortest path between two places follows a Great Circle (a Great Circle is defined as the arc formed when the Earth is sliced through the two points and through the center of the Earth). If the point locations are denoted by (ϕ_A, λ_A) and (ϕ_C, λ_C), where ϕ denotes latitude and λ denotes longitude, then the distance between them is given by:

$$D = R\cos^{-1}(\sin\phi_A\sin\phi_C - \cos\phi_A\cos\phi_C\cos(\lambda_A - \lambda_C))$$

where R denotes the radius of the Earth, or approximately 6378 km.

There are many other bases for measuring distance, because it is often necessary to allow for travel that must follow streets, or avoid barriers of one kind or another. So the actual distance traveled between two places may be much more than either of these formulae would predict. Two of these cases are dealt with later in this chapter: when the path followed is the one that minimizes total cost or travel time over a surface (a case of raster analysis, see Section 29.4.2), and where the path follows a network of links with known lengths or travel times (a case of network analysis, see Section 29.4.2).

29.3.2. Buffers

Instead of asking how far one point is from another, we might turn the question around and identify all of the points within a certain distance of a reference point. For example, it might be interesting to outline on a map the area that is within 1 km of the agent's location, and subsequent sections will cover several applications of this concept. In GIS the term *buffer* is used, and we say that the circle created by this operation constitutes the 1 km buffer around the agent. Chapter 27 already provided a view of buffering as a fundamental transformation of spatial data. Buffers can be created for any kind of object – points, lines, or areas – and are very widely used in GIS analysis. Figure 29.3 shows buffers for each of these types of objects.

By finding a buffer, and by combining it with other information using the methods discussed in this chapter, we could answer such questions as: how many customers live within 10 km of the agent's location; what is the total population within 10 km of the agent's location (based on accurate counts of residents obtained from the census); where are the people with the highest incomes within 10 km of the agent's location?

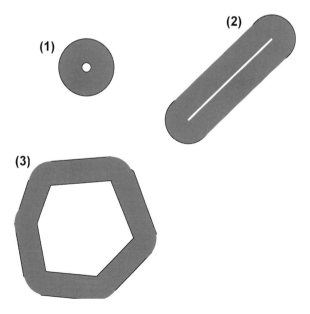

Figure 29.3   Buffers for (1) point, (2) line, and (3) area objects.

29.3.3. Points in polygons

The so-called *point in polygon* operation is another key feature of GIS analysis. Suppose we have a map of the tracts used by the census to publish population statistics, such as the map shown in Figure 29.4. In form, the map looks much like a map of states, or counties, or voting districts, or any of a number of types of maps that divide an area into zones, or what are often called enumeration units, for the purpose of counting and reporting. The point in polygon operation allows us to combine our point map with this map of areas, in order to identify the area that contains each point. For example, we could use it to identify the census tract containing each of the customer locations in our point data set. By counting customers in each tract, and

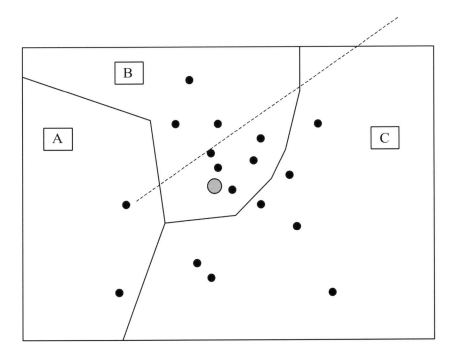

Census Tract	Number of Customers	Total Households
1	2	249
2	8	377
3	7	401

Figure 29.4 By using the point in polygon operation to identify the census tract containing each of the custome locations, it is possible to compare customer counts to other attributes of the tracts, such as the total number of households.

comparing the totals to the known populations of each tract, we could get interesting data on market penetration by tract.

The point in polygon operation is actually very simple to execute, and so the method will be described briefly. One of the points in Figure 29.4 (in tract A) will be used as an example. A line is drawn from the point, in this case diagonally upwards. The number of intersections between this line and the boundaries of each polygon is counted. In the example, there is one intersection with the boundary of Polygon A, two with the boundary of Polygon B, and two with the boundary of Polygon C. The polygon that contains the point is the only one with an odd number of boundary intersections (and it will always be true that exactly one polygon has an odd number – all other polygons will have an even number).

One useful way to think about the result of the point in polygon operation is in terms of tables. The result of the point in polygon operation could be expressed in terms of an additional column for the census tract table shown in Figure 29.4, recording the number of points found to lie in that polygon, as a preliminary calculation to computing the local market penetration. It is applications like these that make the point in polygon operation one of the most valuable in vector GIS.

29.4. Analysis of areas

The previous section looked at one method that combines points and areas (or *polygons*, a term commonly used interchangeably with areas in vector GIS, since areas are normally represented as polygons as discussed in Chapter 26). This section focuses specifically on operations on areas, again from the perspective of vector GIS.

29.4.1. Measurement of area

One of the strongest arguments for the use of GIS, as compared with manual methods of analysis of information shown on maps, is that computers make it easier to take measurements from maps. The measurement of area is in fact the strongest argument of this type. Suppose we need to measure an area from a map, such as the area of a particular class of land use. Perhaps our task is to measure the amount of land being used by industry in a specific city, and we are given a map showing the locations of industrial land use, in the form of appropriately colored areas. In a medium-sized city the number of such areas might be in the thousands, and so there would be thousands of measurements to be made and totaled. But even the measurement of a single area is problematic. Manually, we would have to use one of two traditional methods, known as dot counting and planimetry. Dot counting proceeds by overlaying a prepared transparent sheet covered by dots at a known density, and counting the number of such dots falling within the area of interest. To get a reliable estimate the density of dots needs to be high, and the dots need to be small, and it is easy to see how tedious and inaccurate this task can be. Planimetry proceeds by using a crude mechanical device to trace the outline of the area – the result is read off a dial on the instrument, and again the process is tedious and error-prone. In short, manual methods are frustrating, and expensive in terms of the time required to obtain even a poor level of accuracy.

By contrast, measurement of area is extremely simple once the areas are represented in digital form as polygons. A simple calculation is made for each straight edge of the polygon, using the coordinates of its endpoints, and the calculations are summed around the polygon. Any vector GIS is able to do this quickly and accurately, and the accuracy of the final result is limited only by the accuracy of the original digitizing.

The world's first GIS, the Canada GIS, was developed in the mid 1960s precisely for this reason. A large number of detailed maps had been made of land use and other land properties, with the objective of providing accurate measurements of Canada's land resource. But it would have taken decades, and a large workforce, to produce the promised measurements by hand, and the results would have been of disappointing accuracy. Instead, a GIS was developed, and all of the maps were digitized. In addition to straightforward measurement of area, the system was also used to produce measurements of combined areas from different maps, using the operation of polygon overlay, another important vector operation that is covered in the next section.

29.4.2. Polygon overlay

Figure 29.5 shows a typical application of polygon overlay. The two input maps might represent a map of land use, and a map of county boundaries. A reasonable question to ask of a GIS in a case like this is: 'how much land in County A is in agricultural use?' To answer the question we must somehow compare the two maps, identifying the area that is *both* in County A and in agricultural land use, and measure this area. Polygon overlay allows this to be done. As a method, its history dates back to the 1960s, though efficient software to perform the operation on vector data sets was not developed until the late 1970s.

Figure 29.5 shows the result of the overlay operation, in the form of a new data

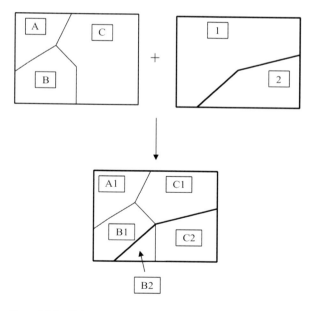

Figure 29.5 Polygon overlay, combining a map with three polygons and a map with two polygons to create a map with five polygons. The attributes of the new map combine the relevant attributes of the input maps.

set. Each of the polygons in the new data set represents a unique combination of the inputs, and the boundaries of the new polygons are combinations of the lines from the two inputs. The operation is actually reversible, since it is possible to recover both of the input data sets by deleting some of the attributes of the new polygons and merging polygons with the same remaining attributes. For example, by deleting land use attributes we recover the county data set. In other words, no information is lost or created in polygon overlay – instead, the information is simply rearranged.

The new data set makes it easy to answer questions like the one we started with. To determine the amount of land in County A in agricultural land use, we simply identify all polygons in the new data set with those attributes, compute their areas, and total. Polygon overlay has a myriad of uses, of which this kind of query is among the simplest. One popular use has to do with population statistics. Suppose a water supply agency wishes to estimate consumption in one of its service areas, and needs to know the total number of households in the area. The census provides statistics on total numbers of households, but for areas defined by the census, not those used by the water supply agency. So a simple way to estimate households is to overlay the service areas on the census areas, and measure the areas of overlap. Then the household counts from the census are apportioned based on these areas, to obtain estimates that are often remarkably accurate. For example, if one census area has 1,000 households, and 50 per cent of its area lies in Service Area 1, and the remaining 50 per cent lies in Service Area 2, then 500 households are allocated to each of these service areas. When all census counts are allocated in this fashion the service area counts are summed.

Polygon overlay is an important operation for vector GIS, but it has two distinct versions. The examples discussed in this section have been of fields (as defined in Chapter 26), such that each point on each map lies in exactly one polygon, and overlay produces a similar map. Overlay in this version is conducted an entire map or *layer* or *coverage* at a time (these terms all have similar meaning in this context). But overlay can also be used for data that represent the *discrete object* view of the world. In this case the input maps can be such that a point can lie in any number of polygons, including zero. This view of the world makes no sense for maps of counties, or land use, but it certainly makes sense for maps of potentially overlapping phenomena like forest fires, or zoning ordinances. In this case, polygon overlay proceeds one polygon at a time, and is used to answer questions about individual polygons, such as 'what zoning ordinances affect Parcel 010067?' Some vector GISs support only the field approach (e.g. versions of ArcInfo up to Version Seven), and some (typically those with strong roots in Computer-Assisted Design (CAD) software) support only the discrete object approach – while others can potentially support both approaches.

Also it is important to recognize the differences between overlay in vector GIS and overlay in raster GIS. The latter is a much simpler operation, and produces very different results, for different purposes. The discussion of raster analysis below includes the essentials of this version of the overlay operation.

29.5. Raster analysis

Raster GIS provides a very powerful basis for analysis that is similar in many respects to the capabilities of other software that also relies on raster representations. For example, some of the raster operations described here will be familiar in concept to people who regularly use software for processing digital photographs or scanned documents, or for processing the images captured by remote sensing satellites. Vector data sets are similar in form to those used in CAD software and in drawing software, but it is unusual to find comparable methods of analysis in these environments.

As discussed in Chapter 26, in raster GIS the world is represented as a series of *layers*, each layer dividing the same project area into the same set of rectangular or square cells or *pixels*. This makes it very easy to compare layers, since the pixels in each layer exactly coincide. But it also means that the ability to represent detail is limited by the pixel size, since any feature on the Earth's surface that is much smaller than a single pixel cannot be represented. Smaller features can be represented by reducing the pixel size, but only at the cost of rapidly increasing data volume. Although data structures that allow for varying pixel sizes (quadtrees) have been developed, they are not commonly available in GIS packages.

Each layer records the values of one variable or attribute, such as land use or county name, for each pixel. The recorded value might be the average value over the pixel, or the value at the exact center of the pixel, or the most common value found in the pixel. If there are many attributes to record (the census, for example, reports hundreds of attributes for each county in the US), then separate layers must be created for each attribute. However, some raster GISs allow a more efficient solution to this particular

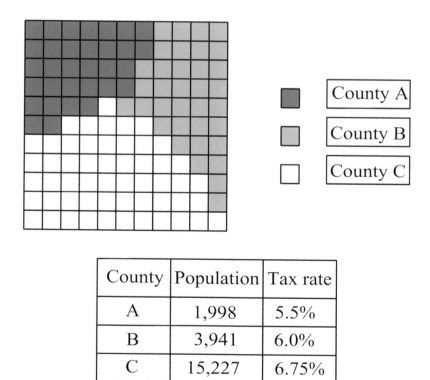

County	Population	Tax rate
A	1,998	5.5%
B	3,941	6.0%
C	15,227	6.75%

Figure 29.6 By using raster cells to store the IDs of polygons with multiple attributes in a related table it is possible to store many attributes in a single raster layer.

problem, in which a single layer is used to record county ID for each pixel, and a related table gives the many attributes corresponding to that ID (Figure 29.6).

An excellent framework for thinking about raster analysis was provided some years ago by Dana Tomlin, who coined the term *cartographic modeling* (Tomlin 1990). In this scheme there are four basic types of operations (there is some variation in the terms in different versions of the scheme):

- *Local* operations, which process the contents of data sets pixel by pixel, performing operations on each pixel, or comparing the contents of the same pixel on each layer;
- *Focal* operations, which compare the contents of a pixel with those of neighboring pixels, using a fixed neighborhood (often the pixel's eight immediate neighbors);
- *Zonal* operations, which perform operations on zones, or contiguous blocks of pixels having the same values;
- *Global* operations, which are performed for all pixels.

Examples of each type of operation are given in the following sections.

29.5.1. Local operations

The simplest kind of local operations occur when a single raster layer is processed. This is often done in order to apply a simple reclassification, such as:

All areas of Soil Classes One, Three, and Five are suitable building sites for residential development; all other areas are not.

This could be operationalized by reclassifying the soil class layer, assigning one to all pixels that currently have Soil Class One, Three, or Five, and zero to all other pixels. Another might be:

Find all areas where the average July temperature is below 25°C.

Again, this could be achieved by reclassifying a layer in which each pixel's attribute is that area's average July temperature.

This kind of reclassification, using simple rules, is very common in GIS. Other local operations on a single layer include the application of simple numerical operations to attributes – for example, the July temperature layer might be converted to Fahrenheit by applying the formula:

$$F = 9C/5 + 32$$

GISs with well-developed raster capabilities, such as Idrisi, allow the user access to a wide range of operations of this type that create a new layer through a local operation on an existing layer.

Other local operations operate on more than one layer. The raster equivalent of polygon overlay is an operation of this type, taking two or more input layers, and applying a rule based on the contents of a given pixel on each layer to create a new layer. The rules can include arithmetic operations, such as adding or subtracting, as well as logical operations such as:

If the average July temperature is above 20°C and soil type is Class One then the pixel is suitable for growing corn, so assign one, otherwise assign zero.

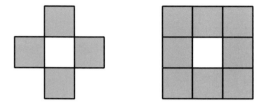

Figure 29.7 Two definitions of a raster cell's neighborhood. In the rook's case only the four cells that share an edge are neighbors, but in the queen's case the neighborhood includes the four diagonal neighbors.

But notice how different this is from the vector equivalent. The operation is not reversible, since the new layer does not preserve all of the information in all of the input layers. Instead of reorganizing the inputs, raster overlay creates new information from them. Fortunately, most often the original layers are not altered, thus allowing the user to recreate an overlay.

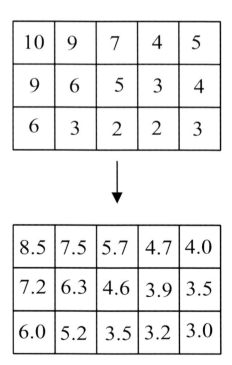

Figure 29.8 Application of a convolution filter. Each pixel's value in the new layer is the average of its queen's case neighborhood in the input layer. The result is smoother, picking up more of the general trend. At the edge, where part of the neighborhood is missing, averages are over fewer than nine pixels.

29.5.2. Focal operations

Focal operations produce results by analyzing each pixel in relation to its immediate neighbors. Some of these operations work on all eight immediate neighbors, but others focus only on the four neighbors that share a common edge, and ignore the four diagonal neighbors. Sometimes we distinguish these options by referring to moves in the game of chess – the eight-neighbor case is called the *queen*'s *case*, and the four-neighbor case is called the *rook*'s *case* (Figure 29.7).

Among the most useful focal operations are the so-called *convolutions*, in which the output is similar to an averaging over the immediate neighborhood. For example, we might produce a new layer in which each pixel's value is the average over the values of the pixel and its eight queen's case neighbors. Figure 29.8 shows a simple instance of this. The result of this operation would be to produce an output layer that is smoother than the input, by reducing gradients, lowering peaks, and filling valleys. In effect, the layer has been *filtered* to remove some of the variation between pixels, and to expose more general trends. Repeated application of the convolution will eventually smooth the data completely. Convolutions are very useful in remote sensing, where they are used to remove noise from images (see Chapter 19). Of course averaging can only be used if the values in each pixel are numeric, and measured on continuous scales, so this operation would make no sense if the values represent classes of land, such as soil classes. But in this case it is possible to filter by selecting the most frequently occurring class in the neighborhood, rather than by averaging. Note also that special rules have to be adopted at the edge, where cells have fewer than the full complement of neighbors.

If the input layer is a DEM, with pixel values equal to terrain elevation, a form of local operation can be used to calculate slope and aspect. This is normally done by comparing each pixel's elevation with those of its eight neighbors, and applying simple formulae (given, for example, by Burrough and McDonnell 1998). The result is not the actual slope and aspect at each pixel's central point, but an average over the neighborhood. Because of this, slope and aspect estimates always depend on the pixel size (often called the distance between adjacent *postings*), and will change if the pixel size changes. For this reason it is always best to quote the pixel size when dealing with slope or aspect in a GIS.

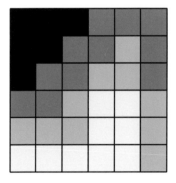 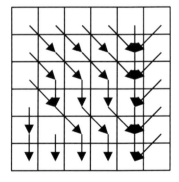

Figure 29.9 A digital elevation model, and the result of inferring drainage directions using the simple rule 'If at least one neighbor is lower, flow goes to the lowest neighbor'.

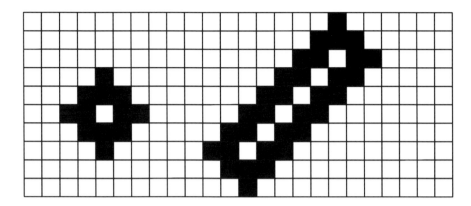

Figure 29.10 The buffer operation in its raster form, for a point and a line (indicated by the white cells).

Slope and aspect can also be used to estimate the pattern of surface water flow over a DEM, a very useful operation in determining watershed boundaries and other aspects of surface hydrology. Each pixel's elevation is compared to those of its eight neighbors. If at least one of the neighbors is lower, then the GIS infers that water will flow to the lowest neighbor. If no neighbor is lower, the pixel is inferred to be a pit, in which a shallow lake will form. If this rule is applied to every pixel in a DEM, the result is a tree-like network of flow directions (Figure 29.9), with associated watersheds. Many more advanced versions of this simple algorithm have been developed, along with a range of sophisticated methods for studying water flow (hydrology) using GIS (see Maidment and Djokic 2000).

Another powerful example of a local operation is the raster equivalent of the buffer operation, and also supports a range of other operations concerned with finding routes across surfaces. To determine a buffer on a raster, it is necessary only to determine which pixels lie within the buffer distance of the object. Figure 29.10 shows how this

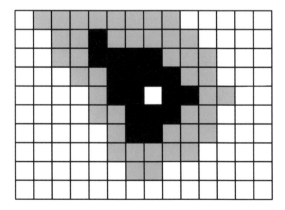

Figure 29.11 Illustration of a spreading operation over a variable friction layer. From the central white cell, all cells in the black area can be reached in 10 min, and all cells in the grey area can be reached in 20 min.

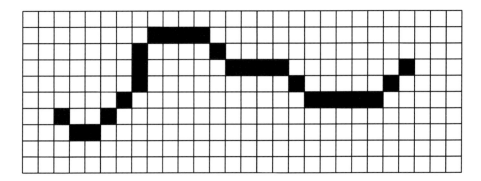

Figure 29.12 Finding the least-cost path across a variable-cost layer from a given origin to a given destination, using queen's case moves between raster cells. This form of analysis is often applied to find routes for power lines, pipelines, or new highways.

works in the cases of a point and a line. But suppose we make the problem a little more complex, by asking for pixels that are within a certain travel *time* of a given point, and allowing travel speed to be determined by a new layer. Figure 29.11 shows such a layer, and the result of determining how far it is possible to travel in given numbers of minutes. The operation is also known as *spreading*. Of course this assumes that travel is possible in all directions, but this would be appropriate in the case of an aircraft looking for the best route across the Atlantic, or a ship in the open ocean.

This method is used very frequently to find the best routes for power transmission lines, highways, pipelines, and military vehicles such as tanks – in all of these cases, the assumption of travel in all directions is reasonably accurate. Origin and destination points for the route are defined. A raster GIS is used to create a layer of travel speed, or *friction*. Because this is not uniform, the best route between the origin and destination is not necessarily a straight line. Instead, the best route is modeled as a series of moves in the raster, from each cell to the most appropriate of its eight neighbors, until a complete route of least total time or least total cost is found (Figure 29.12). There have been many applications of this GIS method over the past three decades by highway departments, power utilities, and pipeline companies.

The final operation considered here is the so-called *viewshed* operation. This also works on a DEM, and is used to compute the area that can be seen from a specified point, by a person at that point. When more than one point is specified, the GIS could be used to determine the area visible from at least one point, or in some military applications it might be useful to determine the area invisible from all points.

The viewshed operation is often used in planning. For example, if a new clearcut is proposed by a forest management agency it might be useful to determine whether the scar will be visible from certain key locations, such as public roads. If a new paper mill is proposed, it might be useful to determine how visible it will be from surrounding areas. One of the complications of the viewshed operation as implemented on a DEM is that it is difficult to incorporate the effects of tree cover on visibility, since a DEM records the elevation of the ground surface, not the top of the tree canopy. Also, it is wise to be careful about the representation of the observer, since far more can often be seen by someone standing on a tower than by someone standing on the ground. Many GIS implementations of the viewshed operation simply assume an observer of average height standing on the ground.

29.5.3. Zonal operations

Rasters represent everything in pixels, so line objects and area objects are modeled as collections of pixel values, with no explicit linkage between them. Because of this, some operations that are very straightforward in a vector GIS, such as the measurement of area described in Section 29.3.1, are quite awkward in raster. If we simply count the pixels assigned to Class One on a soil layer, for example, we get the total area of Class One, not the separate areas of the individual patches of Class One. Zonal operations attempt to address this by focusing on operations that examine a layer as a collection of zones, each zone being defined by contiguous pixels of the same class. Zonal operations can be used to measure the areas of patches, or their perimeter lengths, and the results are returned as new layers in which each pixel is given the appropriate measure, evaluated over the zone of which the pixel is a part.

29.5.4. Global operations

Finally, global operations apply a simple manipulation to an entire layer, usually returning a single value. Examples include operations to determine the average of all pixel values, or the number of pixels having a specific value, or the numbers of pixels having each unique value, or the total area having each unique value. For example, a global measurement of area on a soil layer with six classes might return six areas, plus the total area. Global operations produce summary statistics, so they are often the last operation in a sequence.

29.6. Optimization and design

Many applications of GIS analysis provide information that can be used to make decisions, but some methods come much closer to recommending decisions directly. These are the methods that focus on optimization, by finding the answers that *best* address a problem, and methods that focus on design. Some of the earliest of these were advocated by Ian McHarg, then Professor of Landscape Architecture at the University of Pennsylvania, who developed a method of overlaying maps to find the best locations for new developments (McHarg 1969). Each map to be overlaid represented some specific issue, such as preservation of agricultural land, or cost of construction, so that when all maps were overlaid the best location would be the one with the least impact or cost. McHarg developed this as a manual method, using transparent sheets, but it is easily automated in a raster GIS. The method of optimum route selection discussed in Section 29.4.2 is very similar in concept to McHarg's.

Today, there are many methods of optimization and design available in GIS, in addition to the one already discussed. Many are concerned with optimum routing and scheduling of vehicles operating on a road network, a type of problem that is almost always implemented in vector GIS. In the simplest instance, the GIS is used to find the shortest path through a road and street network between a user-specified origin and destination. Millions of people use this kind of analysis daily when they access Web sites such as www.mapquest.com to request driving directions between two street addresses, and the same kind of analysis is possible in the in-vehicle navigation systems that are becoming increasingly common, especially in vehicles from car rental agencies.

Much more sophisticated versions of the same basic idea are used by school bus authorities to design the best routes to pick up children and take them to school, by parcel delivery companies to design routes for their drivers, and by utility companies that schedule daily work orders for their maintenance staff. The term *logistics* is often used to describe this application of GIS.

Another class of optimization problems is termed *location-allocation*. Here the issue is to find sites for activities that serve geographically distributed demand. Examples are retail stores, which must locate centrally with respect to their customers, schools and hospitals, fire stations and ambulance depots, and a host of other services. In all cases the objective is to find locations that are best able to serve a dispersed population. The problem is known as location-allocation because the best solution involves both the *location* of one or more central facilities, and the *allocation* of demand to it, in the form of service areas (though in some cases this pattern of allocation is controlled by the system designer, as in the case of school districts, but in other cases it is determined by the behavioral choices of customers). In other versions of the problem there is no *location* as such, but instead an area must be divided into the best set of districts, for such purposes as voting, or the assignment of territories to sales staff.

29.7. Summary

It should be clear by now that a vast number of options are available in the form of methods of spatial analysis and modeling implemented in GIS. Any good GIS provides many of these, and more are available in the form of add-ons from other vendors, agencies, or individuals. In the case of one popular GIS, ArcView from the Environmental Systems Research Institute (ESRI), the basic package contains many operations; others are available from ESRI in the form of additional special-purpose modules; more are available from other vendors to build on the power of ArcView in specific applications; and the package provides tools that allow the sophisticated user to add even more. If your favorite GIS appears not to permit a certain kind of operation, it is certainly worth hunting around, probably on the Web, to see if a suitable add-on is available. All of these options have greatly extended the power of GIS to solve problems, and to provide the kinds of information people need to make effective decisions.

Certain aspects should be considered in any application of analysis and modeling using GIS. First, is the analysis more suited to a raster or a vector approach? While many operations are possible in both, this chapter has shown how there can be subtle differences, and how some operations are much more difficult in one form than the other. Second, what level of geographic detail is needed? Most GIS operations give different results depending on the level of geographic detail (data resolution) of the inputs (e.g. the pixel size in a raster operation, or the enumeration unit – block, block-group, or tract in a vector operation), so it is important that the level of detail be adequate. Third, what accuracy is achievable? All results from a GIS will be uncertain to a degree, because all inputs are uncertain, so it is important to determine whether accuracy is sufficient (see Chapter 30 on spatial data quality). Finally, how should the results be presented? GIS is fundamentally a graphic tool, and the images and pictures it produces can have enormous impact if they are well designed. On the other hand, poor graphic design can easily undermine the best presentation (see Chapter 31 on cartography).

References

Burrough P. A. and McDonnell, R. A., *Principles of Geographical Information Systems*, Oxford University Press, New York, 1998.

Maidment D. R. and Djokic, D. (eds.), *Hydrologic and Hydraulic Modeling Support with Geographic Information Systems*, ESRI Press, Redlands, CA, 2000.

McHarg, I., *Design with Nature*, Natural History Press, Garden City, NY, 1969.

Tomlin, C. D., *Geographic Information Systems and Cartographic Modeling*, Prentice Hall, Englewood Cliffs, NJ, 1990.

Snyder, J., *Map Projections - a Working Manual*, US Government Printing Office, Washington, DC, 1987.

Chapter 30

Spatial data quality

Joel Morrison

30.1. Introduction

Data about positions, attributes, and relationships of features in space are termed spatial data. Incidents throughout history can be related where knowledge of spatial data played a key role in the outcome of events. Almost any war, movement of people, vacation travel, exploration, military exercise or campaign, planning activity, or real estate transaction relies on spatial data. Some of these activities provide the corner-stones of our current economic systems.

Prior to the electronic revolution that ushered in the 'Information Age' most spatial data users were in a continual quest for more accurate and up-to-date data. The prevailing mode of operation was always to want more and better data – better data being defined as more positionally accurate and current data. The emphasis was on the x, y position of the data with the quest for attributes (z-values) data being somewhat less imperative.

Spatial data users are currently facing, for perhaps the first time, a situation in which today's technology can and does enable the potential map maker to routinely collect and use data whose quality exceeds that which is needed and/or requested by the user. In the digital world the spatial data situation has changed so that data producers are collecting data and structuring spatial data into databases that may contain topology and enhanced attribute information. The fact that these digital databases may also be distributed allows other data producers to easily add attributes to features already in the database or to even add new features. In some databases relationships among features are also specified. Therefore, any given data file may be the product of a number of data producers. The speeding of the collection of data sets results in many more data sets of varying quality that duplicate earth features at differing resolutions and which compete to satisfy user's needs. These readily available distrib-uted data files are easily accessible electronically at numerous single workstations. Moreover they can be *visualized* in colorful graphics. (Throughout this chapter the term *visualization* will be used to indicate the graphic rendering of spatial data, either on a monitor or in printed map form.)

All of these statements point to the fact that the amount of control exercised by the cartographer over the new end-product is much less than was the case during the preparation of the printed map. In the printed map end-product era, the cartographer controlled exactly what data was shown on the map, how it was shown, and was ultimately responsible for the success of the product. He, or she, through repeated experiences, developed a sense of appropriateness for data use. Relying on that sense, the cartographer carefully evaluated the quality of each data source (or at least was given credit for so doing) and rendered the contents of the map accordingly, putting an invisible 'stamp of approval' of the quality on each created map.

The end result of today's change in technology is that there are many more players involved in creating and using spatial data. In the current scenario the data producer and the data user are often unrelated. The producer collects data, and a pertinent question is; How does the producer describe the quality of the spatial data collected so that it can be assessed against the demands of customers? The data user, however, now must select among competing spatial data sets. What criteria are used to make this decision? Is it a decision based on the known quality of the spatial data, on the availability, or the ease of accessibility of the spatial data? After these decisions are made, the user still must impart to the visualization indications of the quality of the spatial data. Traditionally, cartographers have not done this, and in fact, today we do not know how to do it in the digital environment. Oddly, we are even hampered by the new technology in creating concern about this aspect of a visualization, because most renderings by electronic hardware appear to be very precise and accurate.

30.1.1. Important new questions

Two aspects of the current situation are therefore important. First the user (not the cartographer or the producer) now must decide which, among available competing data sets, to use for a specified visualization or analysis. To aid the user in making this decision, what spatial data quality information should the producer include in a spatial data set? How important is the criterion of spatial data quality in deciding which spatial data sets to use?

Second, since computer renderings of digital data are precise and can easily be made to look accurate, regardless of their data's inherent accuracy, the conscientious user must devise some method to indicate the known quality of the selected data sets in the resulting analyses or visualizations. There are few inherent clues in a well-rendered visualization that indicate the quality of the data. In the digital era, the analog methods cartographers used to select and render data sets are no longer sufficient. Each user, who has the hardware capability to download digital files and to create visualizations, now has the professional obligation to include knowledge of the quality of the data files used. Therefore in the current situation the user of spatial data (both the naive and the sophisticated user) is empowered to fully utilize spatial data sets for personal benefits and, at the same time, required by the technology to make decisions, the consequences of which may be unknown.

To add to the current state of constantly changing affairs, geographic datasets are increasingly being shared, exchanged, and used for purposes other than their producer's intended. The user surfing the Internet can select from among many datasets that were obtained from a variety of sources by a variety of methods by multiple data producers at different scales. Many of these data sets can be directly downloaded to the user's terminal, and in the absence of statements about the quality of the data, the visualization of each data layer looks precise and accurate. Generally, the user will find that data from differing sources when conflated, i.e. merged geometrically, will not perfectly agree, i.e. the visualization of the digital points, lines, and areas representing the same earth features do not exactly match. The user will have no logical way to decide how to resolve the differences in the data sets in the absence of statements containing data quality information. To some extent this situation was present in analog cartography. And in those rare instances of conflated data in analog technology, the cartographer sorely needed information about spatial data accuracy before assuming the responsibility for using the data in the construction of a visualization. In the absence of any other information, often the decision was made based on the reputation of the data producer. Today many data producers are so new to the marketplace that they have not established reputations.

Imagining a perfect world in which all spatial data are available to potential users, we can see that today's empowered user, acting as a cartographer, must make the selection of the most appropriate data set for use in a given situation. How can a spatial data user and the creator of visualizations determine this? What constitutes a statement of spatial data quality? How can the myriad of electronic data producers be required to include statements of spatial data quality with their data sets?

30.1.2. Initial responses to new questions

Society has begun to react to today's confusion, brought about by technological change. One reaction seems to be the adoption of ISO9000, an international quality standard (ISO 1987). If we look at the tenets of ISO9000, we find that the de facto rule is 'give no more quality, or less quality, than the customer requires'. In today's customer oriented society how do spatial data producers produce according to the ISO9000 standard without a thorough knowledge of the spatial data quality of a given data set? What is the responsibility of the spatial data producer in documenting that quality for empowered users? What is the knowledge of data quality in the digital world that is necessary and sufficient?

Whether it is considered important or not, ISO9000 represents a stark contrast to the traditional operating assumptions familiar to most analog cartographers. For most of recorded history, professional cartographers have tried to present the *most* accurate maps that they could produce given the working constraints under which they labored. The reason for this operating assumption was partly due to the fact that printed maps were multipurpose products. Very few widely disseminated printed maps could afford to be single purpose. The cartographer could never be certain of all of the uses for which a given map would be used. Electronic technology now encourages the production of unique products to satisfy single user needs; hence the appropriateness of the ISO9000 philosophy.

The knowledge required today, about the spatial data quality of digital files, goes well beyond simply the positional accuracy and occasional reliability diagram of the analog world. Unfortunately many of the traditional spatial data production institutions have not fully interpreted these changes in user's needs yet. They still produce in the digital environment to the exacting positional standards of the analog data days, ignoring other equally important aspects of spatial data quality.

30.2. The specification of spatial data quality

Three parts are identifiable to the specification and use of spatial data quality information. The definition of the elements of spatial data quality constitutes the first of those parts. The second part of the specification of spatial data quality is the derivation of easily understood indicies for each of the elements of spatial data quality. These metrics may accompany a digital dataset and must have meaningful recognition by the community of spatial data users.

The third part of the specification of spatial data quality is the problem of presenting or rendering the known data quality in visualizations. This third aspect of spatial data quality is receiving some attention by researchers, but much research needs yet to be done. The bulk of the remarks in this chapter will be aimed primarily at the first two parts. There is a parallelism in the developments relating to parts one and two. Research into topics included in part three has been accomplished in a more independent arena and the reader will be directed to the published literature of that arena at the end of this chapter.

30.2.1. Definition of spatial data quality

In analog cartography spatial data quality was synonymous with the quest for positional and attribute measurement accuracy. National Map Accuracy Standards were devised and implemented by most National Mapping Agencies (Thompson 1987). For larger scale work, standards were generated by local governments who were in charge of collection and displaying spatial data (American Society for Photogrammetry and Remote Sensing 1990). With the advent of the Information Age and digital cartography, mapping organizations lost the tight control they enjoyed over spatial data collection and use. The National Map Accuracy Standards are not only of little use but really irrelevant to most mapping in the new age.

30.2.1.1. Initial work on spatial data quality in the US

During the 1980s groups of experts in several parts of the world began research tasks and collaborations in efforts to standardize different aspects of digital spatial data. Most of these efforts included a concern for spatial data quality, and more precisely for a definition of spatial data quality. One of the most comprehensive efforts was undertaken in the US. In 1982, a National Committee on Digital Cartographic Data Standards (NCDCDS) was established under the auspices of the American Congress of Surveying and Mapping (ACSM). Over a 5 year period this committee deliberated and produced a report entitled 'A draft proposed standard for digital cartographic data' (Moellering 1987). One of the four sections of this report was devoted to digital cartographic data quality. This perhaps represents the first comprehensive statement on spatial data quality in the electronic age. Quoting from the report's statement of spatial data quality (Moellering 1987):

> The purpose of the Quality Report is to provide detailed information for a user to evaluate the fitness of the data for a particular use. This style of standard can be characterized as 'truth in labeling', rather than fixing arbitrary numerical thresholds of quality. These specifications therefore provide no fixed levels of quality

because such fixed levels are product dependent. In the places where testing is required, several options for different levels of testing are provided. In this environment the producer provides the quality information about the data and the user makes the decision of whether to use the data for a specific application.

In the Moellering report the specification of the components for reporting data quality is divided into five sections: lineage, positional accuracy, attribute accuracy, logical consistency, and completeness (Moellering 1987). After 1987 and several task force meetings, a modified version of the proposed standard for the exchange of spatial data created by the Moellering committee was accepted by the National Institute of Standards and Technology as the Federal Information Processing Standard – 173 (National Institute of Standards and Technology 1994).

30.2.1.2. International work on spatial data quality

Although it appears that the data quality subcommittee of the NCDCDS in the US may have held the most extensive discussions on identifying the elements of spatial data quality, cartographers in South Africa (Clarke *et al.* 1992), in the UK (Walker 1991) and in Australia, all seem to have adopted aspects of the five elements embodied in the NCDCDS report and the NIST standard.

In the early 1990s the Comite Europeen de Normalisation Technical Committee 287, CEN/TC287, became one of the first international (regional) standards bodies to begin to focus on standards for geographic information to aid portability of geographic data within Europe (Comite Europeen de Normalisation 1992). Spatial data quality was one of its initial work items.

Beginning its work in 1991, The Commission on Spatial Data Quality of the International Cartographic Association accepted the five elements of the NCDCDS as important aspects of spatial data quality. The Commission in its deliberations also considered several other potential aspects of spatial data quality and reached consensus on two additional elements: semantic and temporal. In the Commission's book 'Elements of spatial data quality' (Guptill and Morrison 1995), all seven aspects are explained.

In 1997 the International Organization for Standardization (ISO) established a new technical committee, TC211, Geographic Information/Geomatics, to specifically devise a set of international standards for geographic information. Through an agreement of cooperation, the work of CEN/TC287 was taken into account in the development of the more global ISO standards. Experts working in CEN/TC287, as well as experts from the US that had worked on the NIST standard and experts from the ICA Commission, combined to ensure a continuity of ideas in the work of ISO/TC211.

Within ISO/TC211 five working groups (WG) were established and among the tasks identified for working group three (WG3), geospatial data administration, were work item 13 (NO13), geographic information-quality, work item 14 (NO14), geographic information-quality evaluation, and work item 15 (NO15), geographic information-metadata. Also of relevance to this discussion is work item 4 (NO4), geographic information-terminology included under the tasks assigned to working group 1 (WG1), framework and reference model. ISO TC211/WG3/NO13 has divided the aspects of spatial data quality into two categories: (1) data quality elements, and (2) data quality overview elements. Five data quality elements and three data quality overview elements are included in the current Draft International Standard (ISO 2001).

This work is important to all levels of spatial data users, international, national, regional, and local, for the following reasons. First it defines terminology which enables all users to communicate with one another about spatial data. Second, it establishes standards which ease the exchange and inter use of spatial data, and finally it defines the elements of metadata that should accompany any dataset to enable the wise use of the data. Experience to date suggests that digital spatial datasets are routinely used for purposes that differ from the producers intended use. It is only through the accompanying metadata, and specifically through the metadata about data quality that a potential user can make a rational decision on the use of a dataset.

30.2.1.3. The work of ISO/TC211/WG3/NO13 – quality principles

As of this writing, the work of the ISO/TC211/WG3/NO13 on quality principles represents the current state-of-the-art in defining the elements of spatial data quality. It combines the major threads of previous work done in various parts of the world over the past two decades. The following discussion therefore relies heavily on that work.

The ISO/TC211 project team working on quality principles found it necessary and convenient to define a model of data quality. Figure 30.1 is a modification of the current model created by the ISO/TC211 project team.

'A verbal description of the data quality model begins with a data set, which is an identifiable collection of related data. These data represent real or hypothetical entities of the real-world (which are characterized by having spatial, thematic and temporal aspects.) The process of modeling the potentially infinite characteristics of real-world entities into an ideal form for the purpose of making these entities intelligible and representable is defined as the process of abstracting from the real-world to a universe of discourse. The universe of discourse is described by a product specification, against which the quality content of a data set, or specific parts of it, can be tested. The quality of a data set depends either on the intended use of the data set or the actual use of the data set: both data producers and data users can assess the quality of a data set within the confines of the data quality model. A data producer is given the means for specifying how well the mapping used to create the data set reflects its universe of discourse; data producers validate how well a data set meets the criteria set forth in its product specification. Data users are given the means for assessing a data set derived from a universe of discourse identified as being coincident with (some or all) requirements of the data producer's application. Data users can assess quality to ascertain if a data set can satisfy their specific applications, as specified within their user requirements' (Godwin forthcoming).

This model is compatible with ISO terminology that defines quality as the 'totality of characteristics of a product that bear on its ability to satisfy stated and implied needs' (ISO 2001). The ISO/TC211 project team's model in Figure 30.1 makes an important distinction. It acknowledges that data users commonly have differing needs or requirements from the specifications used by the data producers in creating a data set. Therefore both data producers and data users bear a responsibility for assessing the quality of a data set.

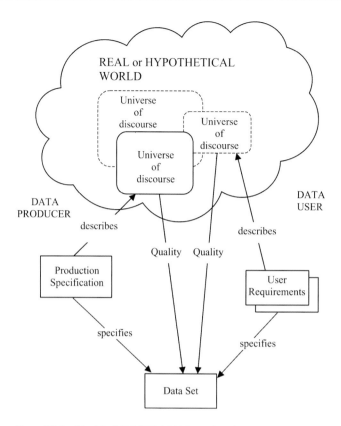

REAL or HYPOTHETICAL
WORLD

Universe
of
discourse

Universe
of
discourse

Universe
of
discourse

DATA
PRODUCER

DATA
USER

describes

describes

Quality   Quality

Production
Specification

User
Requirements

specifies

specifies

Data Set

Figure 30.1   Modified ISO/DIS 19113 model of data quality.

30.2.2. The quantitative elements of spatial data quality

The draft international standard on quality principles produced by the deliberations of the ISO/TC211 project team defines five data quality elements that are subject to quantitative assessment and three data quality overview elements which are non-quantitative in nature. These are listed in Figure 30.2. There is provision for additional elements to be user specified.

30.2.2.1. Completeness

The first recognized element of spatial data quality is *completeness*. The draft international standard states that completeness shall describe the 'presence and absence of features, their attributes and their relationships' (ISO 2001). Completeness is applied to both features and attributes and can be measured by commission (the inclusion of features and/or attributes not found in reality) and by omission (the exclusion of features and/or attributes found in reality).

Completeness has been generally agreed by all working groups to be an element of spatial data quality. The FIPS 173 states: 'The quality report shall include information about selection criteria, definitions used and other relevant mapping rules. For example, geometric thresholds such as minimum area or minimum width must be reported'

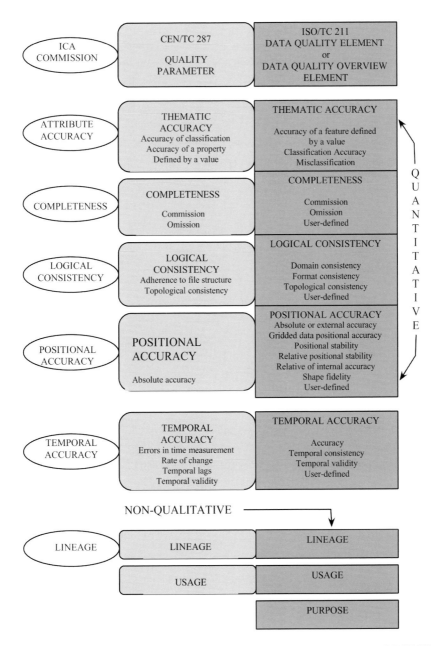

Figure 30.2 Crosswalk of spatial data elements adapted from ICA Commission, CEN/TC287, to ISO/TC211/WG3/NO13.

and 'The report on completeness shall describe the relationship between the objects represented and the abstract universe of all such objects. In particular, the report shall describe the exhaustiveness of a set of features' (National Institute of Standards and Technology 1994).

Professor Kurt Brassel and his associates at the University of Zurich expanded on the FIPS 173 definition of completeness in Chapter 5 of the book *Elements of spatial data quality*. 'Completeness describes whether the entity objects within a data set represent all entity instances of the abstract universe…the degree of completeness describes to what extent the entity objects within a data set represent all entity instances of the abstract universe'(Guptill and Morrison 1995).

In practical terms, many map users never considered completeness when analog maps were the only type available. The tacit assumption was that all instances of existing features were being shown on the map, or at least all instances known to the cartographer were being faithfully rendered. Clearly using digital files, it is no longer necessary for that to be the case and a relevant question about a digital file is to ask what percentage of feature instances in reality are encoded in the dataset. Are all culverts listed, or only those over 4 m in length? What percentage of those culverts over 4 m in length are claimed to be encoded by the metadata of the dataset? Are all farm ponds encoded, or only those over 5 acres in surface extent? Are all water towers listed, or only those in active use over 30 ft in height? Completeness is now of critical concern to digital spatial data users and is an integral part of spatial data quality.

30.2.2.2. Logical consistency

A second data quality element is *logical consistency* that shall describe the 'degree of adherence to logical rules of data structure, attribution and relationships' (ISO 2001). This element is also defined in FIPS 173 as: '…logical consistency shall describe the fidelity of relationships encoded in the data structure of the digital spatial data.' (National Institute of Standards and Technology 1994). Logical consistency describes the number of features, relationships, or attributes that have been correctly encoded in accordance with the integrity constraints of the feature data specification. Tests can be either graphic or visual, and topological. Conceptual consistency, domain consistency, format consistency and topological consistency are all testable for omission and commission.

Kainz states in Chapter 6 of the book *Elements of Spatial Data Quality*, 'According to the structures in use, different methods may be applied to test for logical consistency in a spatial data set.' Kainz 'gives an overview of the current state-of-the-art modeling of spatial data and relationships and their mathematical background', and follows that discussion with 'a section on consistency rules derived from the characteristics of the underlying mathematical structure and a collection of tests that can be performed to test a data set or scene for any logical inconsistencies present' (Guptill and Morrison 1995).

Spatial data in digital form present a different set of concerns for the spatial data user. From the physical properties of entities and from man's use of these entities a set of logical relationships can be deduced. For example, when two highways cross either there is an intersection, an underpass, or an overpass. All drainage flows from higher to lower elevations. A given area can not be simultaneously land and water. If water, it can not have a railroad or a road without a bridge, etc. Since digital data are often stored by feature layer, changes or updates to one digital dataset layer will affect other layers of data. You can not add a dam to a data layer without affecting another data layer. Automated checks for these logical relationships can reveal inconsistencies in a dataset.

The logical consistency element of spatial data quality measures the internal consistency of the data in the various data layers of a dataset.

30.2.2.3. Positional accuracy

Positional accuracy has a longer history of concern by spatial data producers and users, and still represents a major concern in the electronic age. The ISO/TC211 project team states very simply that positional accuracy shall describe the 'accuracy of the position of features' (ISO 2001). However, the aspects of positional accuracy mentioned are numerous. According to the FIPS 173, 'The quality report portion on positional accuracy shall include the degree of compliance to the spatial registration standard' (primarily latitude and longitude) (National Institute of Standards and Technology 1994). It includes measures of the horizontal and vertical accuracy of the features in the data set. It must consider the effects on the quality of all transformations performed on the data, and report the results of any positional accuracy testing performed on the data.

ISO/DIS 19113 refers to absolute, or external accuracy, relative, or internal accuracy, and gridded data position accuracy. It also mentions positional stability and shape fidelity. (Godwin forthcoming) Clearly, even though perhaps more work has been done on positional accuracy than on any other single element, much remains to be researched.

30.2.2.4. Temporal accuracy

During the 1980s as more digital spatial data began to become available, researchers began to pay more attention to temporal aspects of spatial data quality. The ISO/TC211 project team states that *temporal accuracy* shall describe the 'accuracy of the temporal attributes and temporal relationships of features' (ISO 2001). Temporal accuracy includes errors in time measurement, temporal consistency, and temporal validity.

The ICA Commission also accepted temporal information as an element of spatial data quality in its deliberations. Temporal information describes the date of observation, type of update, (creation, modification, deletion, unchanged), and validity periods for spatial data records. Prior to its specification as an element of spatial data quality, aspects of the temporal element were contained in each of the other elements of spatial data quality as defined in FIPS 173. Guptill states that in the use of digital spatial data and geographic information system technology, 'users will collect, combine, modify, and update various components of spatial information. This will occur in a distributed, heterogeneous environment, with many parties participating in the data enterprise. In such an environment, having information about the temporal aspects of the spatial data becomes paramount' (Guptill and Morrison 1995).

To realize the implications of the temporal dimension, spatial data users need only think about producing a map of all the fence lines in a county that have existed through history on the one hand, and a map of the current fence lines in the county on the other. In analog topographic mapping it was accepted that the features shown on a given quadrangle may never have actually existed at the same time in reality (but did exist sometime within the approximate average of 4 years that it took to produce a topographic map). Today when selecting a file of digital spatial data for use in the preparation of a visualization, the data user must have some indication of the dates of validity for a digital feature in order to produce a map. Time tags on all digital features are therefore necessary, and the availability and reliability of those time tags reflect on the spatial data quality of a dataset.

30.2.2.5. Thematic accuracy

The fifth data quality element is *thematic accuracy*. Thematic accuracy shall describe the 'accuracy of quantitative attributes and the correctness of non-quantitative attributes and of the classifications of features and their relationships' (ISO 2001). It is important to note that included in thematic accuracy is the accuracy of the classification and/or misclassification. It can be maintained that the use in analog technology of reliability diagrams to display attribute accuracy is the analog equivalent of this element. However, it is now deemed crucial to the use of digital data that the accuracy of the value assigned for any single attribute tied to an earth position, *as well as*, the accuracy of the classification of that attribute, be specified.

The preceding five elements are relatively consistent among the work of the ICA Commission, the CEN/TC287, and ISO/TC211 (Figure 30.2). The work by Salge on semantic accuracy, included as Chapter 7 in the ICA Commission's work (Guptill and Morrison 1995), and the textual fidelity referring to the accuracy of spelling and the consistency of abbreviations in CEN/TC287, has not been accepted by ISO/TC211/WG3/NO13. According to Godwin, the ISO/TC211 project team does allow for the addition of data quality elements. The team defines this element as describing '…a component of the quantitative quality of a data set not addressed in the standard.' (Godwin forthcoming). Clearly this 'catch-all' category could be used for such items as spelling, abbreviations, and the use of names for feature generics.

30.2.3. The non-quantitative elements of spatial data quality

Three non-quantitative spatial data overview elements are also defined by the ISO/TC211 project team. These are lineage, usage, and purpose. This is an expansion of previous work as the ICA Commission, like the FIPS 173 standard, only defines lineage. The CEN/TC287 defines both lineage and usage.

30.2.3.1. Lineage

Lineage: The ISO/TC211 project team defines lineage as a description of the history of a dataset and, in as much as is known, a recounting of the life cycle of a dataset from collection and acquisition through compilation and derivation to its current form (ISO 2000a). Lineage contains two unique components: (1) source information describing the parentage of a dataset, and (2) process step or history information on the events or transformations in the life of a dataset, including the process used to maintain the dataset whether continuous or periodic, and the lead time (ISO 2000a). The FIPS 173 states: 'The lineage portion of a quality report shall include a description of the source material from which the data were derived, and the methods of derivation, including all transformations involved in producing the final digital files' (National Institute of

Standards and Technology, 1994). The description shall include the dates of the source material and the dates of ancillary information used for update. The lineage portion shall also include reference to the specific control information used, and describe the mathematical transformations of coordinates used in each stage from the source material to the final product.

Clarke and Clark state in Chapter 2 of *Elements of spatial data quality* that, 'Lineage is usually the first component given in a data quality statement. This is probably because all of the other components of data quality are affected by the contents of the lineage and vice-a-versa.' They admit that 'The ultimate purpose of lineage is to preserve for future generations the valuable historical data resource'(Guptill and Morrison 1995). Clarke and Clark list four components that provide the contents of a quality statement on lineage, and a format is suggested for the recording of lineage data for digital data files.

30.2.3.2. Usage

Usage, according to the ISO/TC211 project team shall 'describe the application(s) for which a dataset has been used. Usage describes uses of the dataset by the data producer or by other, distinct, data users' (ISO 2001).

30.2.3.3. Purpose

'Purpose shall describe the rationale for crating a data set and contain information about its intended use.' The project team adds a footnote that a dataset's intended use is not necessarily the same as its actual use. Actual use is described using the data quality overview element usage (ISO 2000a).

Figure 30.2 gives a good summary of Section 30.2, the specification of spatial data quality and the identification of its elements. Three of the more important studies on spatial data quality are easily compared and in general there is much agreement among the three. The world is ready to move beyond this stage of development and emphasis now needs to be placed on increased research on parts two and three of spatial data quality.

30.3. Spatial data quality and its measurement

The second necessary part of spatial data quality, easily understood and implemented methods for measuring and stating spatial data quality in terms of its defined elements, is the scope of work of the ISO/TC211/WG3/NO14. The results of the needed research are vital to both producers and spatial data users.

Much less work has been done on the specification of spatial data quality and its measurement. The ICA Commission presented a descriptive framework by Veregin and Hargitai in Chapter 9 of its book (Guptill and Morrison 1995), and in addition several individual researchers have addressed the problem, so we find in the literature a few beginning attempts to 'objectively' specify and measure data quality.

Veregin and Hargitai attempted to define a metric for the seven elements of spatial data quality as defined by the ICA Commission. It was exploratory by its very nature but offered an outline of a logical schema for data quality assessment in the context of geographical databases. The schema is based on two concepts: 'geographical observations are defined in terms of space, time and theme. Each of these dimensions can be treated separately (but not always independently) in data quality assessment', and 'the quality of geographic databases cannot be adequately described with a single component.' The multidimensional aspect of spatial data quality is readily accommodated in the framework.

More recently Aalders and Morrison in Chapter 34 of *Geographic Information Research: Trans-Atlantic Perspectives* have addressed both parts two and three. They emphasized that in the Information Age three competing possibilities exist for a potential data user when a data set is selected for the delineation of a set of features or a region of the world (Craglia and Onsrud 1999). These are: (1) the quality of the data, (2) the availability of the data, and (3) the ease of accessibility. A conscientious data user needs to know what spatial data quality information is required, and how to impart that quality information in the visualization being prepared. Although their work gives examples, no actual metrics for the specification of spatial data quality are introduced.

In contrast Hunter has devoted an entire article to 'New tools for handling spatial data quality: moving from academic concepts to practical reality' (Hunter 1999). Hunter's work represents an important beginning of the type of research that is needed. Hunter presents five examples: (1) tracking of feature coordinate edits and their reporting in visual data quality statements, (2) testing and reporting the positional accuracy of linear features of unknown lineage, (3) simulating uncertainty in products derived from Digital Elevation Models (DEMs), (4) incorporating uncertainty modeling in vector point, line and polygon files, and (5) reporting data quality information at different levels of Geographic Information Systems (GIS) database structures. While these examples do not relate one-to-one with the elements of spatial data quality, it is a good beginning.

Simley has recently written about the practical application of ISO9000 techniques to cartographic production (Simley forthcoming). Based on the philosophy of ISO9000, Simley has appraised cartographers of tools available in the ISO9000 arena that could prove useful in increasing and maintaining the quality of spatial data. These same tools could be used for the assessment of positional and thematic accuracy and logical consistency. They are compatible with the ISO/CD 19114.

30.3.1. The work of ISO/TC211/WG3/NO14

The most comprehensive work on the derivation of evaluation metrics for the elements of spatial data quality has been accomplished by a project team working within ISO/TC211 under the direction of Dr. Ryosuke Shibasaki. The work exists as a Committee Draft which means that it is not quite as mature as the Draft International Standard on Quality Principles. The two project teams creating ISO/DIS 19113 and ISO/CD 19114 have worked closely together to maintain compatibility

ISO/CD 19114 'recognizes that a data producer and a data user may view data quality from different perspectives' (ISO 2000a). Each may set conformance quality levels. The ISO/CD 19114 project team developed a quality evaluation process which is a sequence of steps that can be followed by both a dataset producer and a dataset user to produce a quality evaluation procedure that yields a specific quality evaluation

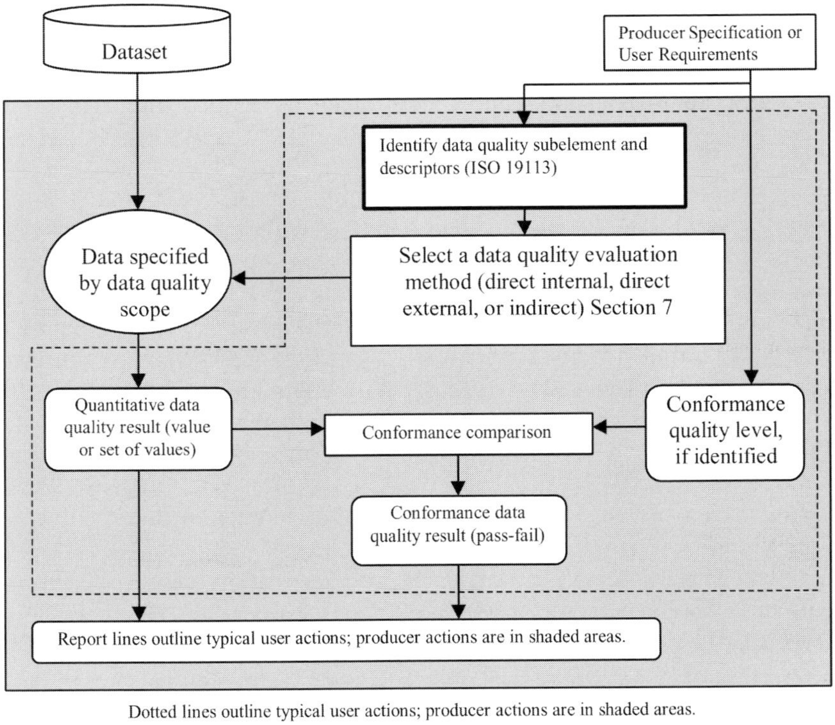

Dotted lines outline typical user actions; producer actions are in shaded areas.

Figure 30.3 Data quality evaluation process flow (per draft dated May 31, 2000).

result. Figure 30.3 illustrates the ISO/CD 19114 project team's data quality evaluation process flow in their May 31, 2000 draft document.

For each different test of a dataset required by a product specification or user requirement, a data quality element, data quality sub-element, and data quality scope, in accordance with ISO/DIS 19113, must be identified. Next a data quality measure is identified for each test, and a data quality evaluation method is selected. A data quality evaluation method can be either direct or indirect. Direct methods are applied to the data specified, while indirect methods use external information to determine a data quality result. Direct evaluation may be accomplished by automated or non-automated means (ISO 2000a). Finally, the data quality results must be reported as metadata compliant with ISO/CD 19115 (ISO 2000b).

ISO/CD 19114 contains 45 'examples of data quality measures' as Annex D (informative) (ISO 2000a). In all ISO/CD 19114 contains 63 pages of Annexes which illustrate rather than describe the proposed standard. Much work remains however, in order to provide data producers and data users with sufficient tools and methods to specify the quality of spatial data.

30.4. Spatial data quality and its visualization

The ability of researchers to focus on the visualization of spatial data quality has changed dramatically with the tools available to cartographers from the electronic revolution. The use of computers has freed the visualization of spatial data from serving both as a database and as a visual display. The database is now separated and the need to retrieve precise information from the visual display is no longer necessary. This makes the need for creators of visualizations from digital spatial databases to be more 'honest' in their displays. Until the Information Age, cartographers have not been able to separate their research energies between database creation and visualization creation. Rather independently of the work reported above in Sections 30.2 and 30.3, several groups of scholars have been working on the graphic presentation of spatial data quality. The intent of this rather abbreviated Section 30.4 is to direct the reader to a sample of the current literature. This on-going work is not as structured as the ISO work reported in Sections 30.2 and 30.3.

Initially the term 'uncertainty' was used in an attempt to convey weaknesses in a dataset. The obverse of 'uncertainty' is the term 'certainty' which with its more positive inference is seeing more use. Most recently Leitner and Buttenfield have addressed some aspects of the problem in an article in *Cartography and Geographic Information Science* (Leitner and Buttenfield 2000). The bibliography at the end of the article points to other researchers identified with the visualization of spatial data quality.

Buttenfield (Buttenfield and Beard 1991; Buttenfield 1993), MacEachren (1991, 1992) and McGranaghan (1993) are researchers with strong interests in the subject. The interested reader is also referred to books by MacEachren and Taylor (1994), MacEachren (1995), and Goodchild and Gopal (1989) and a technical report from the National Center for Geographic Information and Analysis edited by Beard *et al.* (1991).

30.5. Conclusions

Spatial data producers and spatial data users are forced to seriously consider spatial data quality as a result of the electronic revolution. In the Industrial Age of analog maps, where the control of the collection, maintenance, and visualization of spatial data was tightly controlled by a few organizations, it was possible to provide spatial data to users based on the establishment of a reputation built over a lengthy period of providing useful products to the public. The products were static and the user could not alter the information on the printed map. In the Information Age, electronic production and use of spatial data requires a systematization of the specification of spatial data quality so that a potential user may make an educated selection of datasets for use. Every individual is now a potential spatial data user, and as a consumer of spatial data has both a need and a right to know the quality of the data.

In response, over the past 20 years, groups of interested individuals throughout the world have worked on systematizing our knowledge about and processes for evaluating spatial datasets. The culmination of this work is the two parts of a new ISO standard, ISO 15046. In general the international community needed to define the elements of spatial data quality. Agreement has essentially been reached on this first step. Next a series of acceptable metrics to measure each element of spatial data quality needs to be created. The concerned community is only beginning this second step. ISO 15046 has specified a process flow that will enable metrics to be selected and reported to potential data users, but the metrics themselves are still not systematized. Finally in related, yet independent, research, international scholars have begun to seriously research the visualization of spatial data quality. This third step is more important in the Information Age than it was in the Industrial Age due to the separation of the analog function of a spatial display into a digital database and a visualization.

Much work remains to be done, but the last 20 years has seen more systematic progress in the creation of knowledge about spatial data quality than the previous century.

Therefore the spatial data community can expect great progress in this arena in the near future. Morrison and Guptill undertook to outline the future direction of this line of inquiry (Guptill and Morrison 1995). They point to directions in need of further investigation and speculate about the use and transfer of digital spatial data in tomorrows electronic world. It should be an interesting time.

Unfortunately, the fact that this work is so new to the spatial data mapping community means that readily available sources of information about how to collect and use digital spatial data, including the encoding and presentation of the quality of that data, do not exist. The attached bibliography offers some of the highlights of the existing literature.

References

American Society for Photogrammetry and Remote Sensing, Committee for Specifications and Standards, 'ASPRS Accuracy Standards for Large Maps,' *Photogrammetric Engineering and Remote Sensing*, LVI (7), 1990, 1068–1070.

Beard, K.M., Buttenfield, B. P. and Clapham, S. B. (eds.), *Visualization of Spatial Data Quality*, National Center for Geographic Information and Analysis Technical Paper 91-26, Santa Barbara, CA, 1991.

Buttenfield, B. P. and Beard, K. M., 'Visualizing the data quality of spatial information', *Auto-Carto 10*, American Congress on Surveying and Mapping, Bethesda, MD, 1991, pp. 423–427.

Buttenfield, B. P., 'Representing Data Quality' *Cartographica*, 30, 1993, 1–7.

Clarke, D. G., Cooper, A. K., Liebenberg, E. C. and van Rooyen, M. H., 'Exchanging data quality information,' Ch. 2, *A National Standard for the Exchange of Digital Geo-referenced Information*, National Research Institute for Mathematical Sciences, Republic of South Africa, 1992.

Comite Europeen de Normalisation, *Electronic Data Interchange in the field of Geographic Information - EDIGEO*, prepublication of CEN/TC 287, AFNOR, 1992.

Craglia, M. and Onsrud, H. (eds.), *Geographic Information Research: Trans-Atlantic Perspectives*, Taylor and Francis, London, 1999.

Godwin, L., *Elements of Spatial Data Quality* (2nd ed.), Guptill, S. C. and Morrison, J. L. (eds.), Elsevier Science, Oxford, forthcoming.

Goodchild, M. F. and Gopal, S., *Accuracy of Spatial Databases*, Taylor and Francis, New York, 1989.

Guptill, S. C., and Morrison, J. L. (eds.), *Elements of Spatial Data Quality*, Elsevier Science, Oxford, 1995, 202 pp.

Hunter, G. J., 'New Tools for handling spatial data quality: moving from academic concepts to practical reality', *Journal of the Urban and Regional Information Systems Association*, 11, 1999, 25–34.

ISO, *ISO 9000 - 1987 Quality Management and Quality Assurance Standards*, International Standards Organization Copyright Office, Geneva, 1987.

ISO, ISO/Dis 19113, Draft International Standard, *Geographic Information – Quality Principles*, International Standards Organization Copyright Office, Geneva, 2001.

ISO, ISO/TC211/WG3/Editing Committee 19114, Final text of CD 19114, *Geographic Information – Quality Evaluation Procedures*, International Standards Organization Copyright Office, Geneva, 2000a.

ISOm ISO/TC211/WG3/Editing committee 19115, Final text of CD 19115, *Geographic Information – Metadata*, International Standards Organization Copyright Office, Geneva, 2000b.

Leitner, M. and Buttenfield, B. P., 'Guidelines for the display of attribute certainty' *Cartography and Geographic Information Science*, 27, 2000, 3–14.

MacEachren, A. M., 'Visualization of data uncertainty: Representational issues.', Beard, K. M., Buttenfield, B. P. and Clapham, S. B. (eds.), *Visualization of spatial data quality*. National Center for Geographic Information and Analysis Technical Paper 91-26, Santa Barbara, CA, 1991.

MacEachren, A. M., 'Visualizing uncertain information' *Cartographic Perspectives*, 13, 1992, 10–19.

MacEachren, A. M., *How Maps Work: Representation, Visualization, and Design*. The Guilford Press, New York and London, 1995.

MacEachren, A. M. and Taylor, D. R. F. (eds.), *Visualization in Modern Cartography*, Elsevier Applied Science, New York, 1994.

McGranaghan, M., 'a cartographic view of spatial data quality', *Cartographica*, 30, 1993, 8–19.

Moellering, H. (ed.), *A Draft Proposed Standard for Digital Cartographic Data*, National Committee for Digital Cartographic Standards, American Congress on Surveying and Mapping Report #8, 1987.

National Institute of Standards and Technology, *Federal Information Processing Standard Publication 173 (Spatial Data Transfer Standard Part 1, Version 1.1)*, U.S. Department of Commerce, 1994.

Simley, J., 'The application of quality engineering techniques to cartographic production', *Cartography and Geographic Information Science*, forthcoming.

Thompson, M. M., *Maps for America* 3rd ed., U.S. Department of the Interior, U.S. Geological Survey, U.S. Government Printing Office, Washington, DC, 1987.

Walker, R. S., 'Standards for GIS data', *Proceedings, Third National Conference and Exhibition*, The Association for Geographic Information, Birmingham, UK, 1991.

Chapter 31

Cartographic symbolization and visualization

Robert B. McMaster

31.1. Introduction

Cartography is defined as the art and science of making maps. It is a discipline that is thousands of years old, yet has just in the last several decades seen enormous changes as a result of digital technologies. Cartography's history dates back thousands of years, with the concept of projections, a systematic earth graticule, measurement of the Earth, and comprehensive mapping of Earth detail dating back to the Greeks. It is during this period that the orthographic, stereographic, and gnomonic projections were developed. Comprehensive surveys of the history of cartography are provided Bagrow (1964), and Thrower (1996). A basic classification of maps development by Robinson and Petchenik (1976) involves the scale of the map, and its intrinsic purpose (Figure 31.1).

31.1.1. A classification of maps based on scale and purpose

General-purpose maps are those that provide a geographical base. The primary example of this type of map is the topographic map, which depicts the basic location of features on the surface of the earth, including transportation, hydrography, cultural features, and elevation. These types of maps are considered 'geographic dictionaries', where the focus is on position and location, and on the accuracy of the planimetry. For

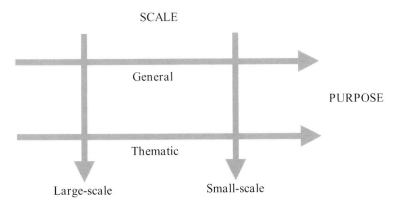

Figure 31.1 A classification of maps based on purpose and scale.

instance, the standard topographic map is designed to accurately locate basic geographical features.

Alternatively, cartographers have increasingly focused their attention on special-purpose, or thematic, maps. These maps are not designed to depict accurate geographical positions, but to show the distribution of a theme, such as population or land use. They are geographical essays, in contrast to dictionaries. The field of thematic cartography focuses on generalization, symbolization, classification, and graphical design. Starting in the nineteenth century, geographers, statisticians, and cartographers have developed a series of symbolization methods for the portrayal of thematic data, including the choropleth, graduated symbol, dot, isarithmic, and dasymetric techniques. These will be covered in more detail later in this chapter. A fascinating history of the development of thematic cartography may be found in Robinson's (1982), *A History of Thematic Cartography*. Maps may also be based on their scale, or the mathematical relationship between the map distance and the commensurate earth distance as represented by the *representative fraction*, or RF value. Map scales are normally broken down into large-scale (e.g. RF = 1:10,000) where the map represents a smaller area in greater detail, or small-scale (e.g. RF = 1:1,000,000) where a greater area is represented in smaller detail. As seen in the illustration, one can find small- and large-scale general or thematic maps. However, the presentation of theme/scale is somewhat problematic, as these really represent a continuum in both directions. An increasing problem within modern GISs is the mixture of disparate scales (as database layers) without accounting for differential generalization, detail, and accuracy between and among these layers. The initial scale from which the data were obtained should be part of the meta-data.

31.2. Cartographic generalization

Generalization is a fundamental human activity where, for instance, in all aspects of science the desire to select, simplify, and enhance certain characteristics is essential. As mentioned in Chapter 27, generalization is a significant topic in GIS. This chapter will focus on the cartographic aspects of scale change, in contrast to the database concerns with generalization that are detailed extensively in the European literature. But it is important to initially differentiate these two terms: cartographic generalization and database generalization. Database generalization involves the filtering of the cartographic information in the database, and does not include any operation to improve graphical clarity. For instance, it might be that a master database is created at a scale of 1:25,000. For efficiency, and to prevent having to regenerate smaller scale databases, this 1:25,000 database is filtered to a scale of 1:50,000, creating a new database. This process is not visual, but occurs through the application of internal database operations. Alternatively, cartographic generalization is a visual process, where the features are modified to maintain the aesthetic quality of the map at a reduced scale. A basic definition of generalization, developed by McMaster and Shea (1992, 1993) is, 'the process of deriving, from a data source, a symbolically or digitally-encoded cartographic data set through the application of spatial and attribute transformations.' Basic objectives of generalization include the reduction in the amount, type, and cartographic portrayal of the mapped or encoded data consistent with the chosen map purpose and intended audience; and the maintenance of clarity of information at the target scale.

31.2.1. Philosophical objectives of generalization

The basic philosophical objectives of generalization include:

- Reducing the complexity of the map/database
- Maintaining the spatial accuracy
- Maintaining the attribute accuracy
- Maintaining the aesthetic quality
- Maintaining a logical hierarchy
- Consistently applying the rules of generalization

Robinson and his colleagues first identified the fundamental components of generalization, including selection (as a preprocessing step) simplification, classification, and symbolization. A key activity over recent years has been the identification of the fundamental operations of generalization in order to automate the process. Figure 31.2 illustrates many of these generalization operations, showing the results of a 50 per cent scale reduction.

31.2.2. The fundamental operations of generalization

A first set of operations are based on line generalization. Simplification involves the elimination of unnecessary information, which normally utilizes coordinate reduction routines. Alternatively, line smoothing shifts the coordinates to improve aesthetic quality, where mathematical splining and averaging methods are classic smoothing operations. Merging is a method for 'zipping' together two parallel lines at a reduced scale, such as two casing of a road symbol, or two banks of a river. Exaggeration and enhancement are operators that actually accentuate certain features in order to preserve their significance under scale reduction. The reopening of a bay that would likely close under reduction is one such example. Related to exaggeration is the critical operation of displacement, where features are purposefully shifted apart to prevent coalescence under scale reduction. Other fundamental generalization operations are designed for point and aerial data. For instance, aggregation fuses multiple point features to create an aerial feature. Amalgamation fuses several smaller polygons into a larger polygon while the collapse operator reduces aerial features to point features. Finally, refinement and typification are operations that create a smaller-scale symbolic representation of a set of features. While cartographers have identified such operations, they have had only minimal success in automating these methods. Perhaps the most commonly applied operator in a GIS is the simplification operation, where coordinate data are weeded out. In Chapter 27 one very commonly applied algorithm, the Douglas-Peucker simplification algorithm, was briefly detailed as an example of a transformation of spatial data.

31.3. Cartographic symbolization

Fundamental to the understanding of cartographic symbolization is the concept of the

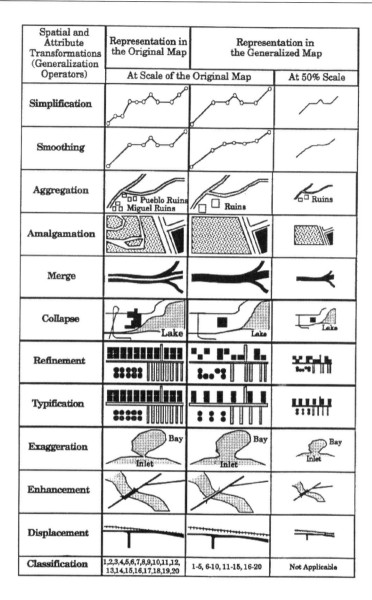

Spatial and Attribute Transformations (Generalization Operators)	Representation in the Original Map	Representation in the Generalized Map	
	At Scale of the Original Map	At 50% Scale	
Simplification			
Smoothing			
Aggregation			
Amalgamation			
Merge			
Collapse			
Refinement			
Typification			
Exaggeration			
Enhancement			
Displacement			
Classification	1,2,3,4,5,6,7,8,9,10,11,12, 13,14,15,16,17,18,19,20	1-5, 6-10, 11-15, 16-20	Not Applicable

Figure 31.2   The fundamental generalization operators (after McMaster and Shea 1992). Permission received from ACSM.

primary visual variables, as presented by Jacque Bertin in his influential book, *The Semiology of Graphics* (Bertin 1983). To fully understand cartographic symbolization, one must realize the relationship among the fundamental types of geographical data, the measurement levels, the forms of cartographic symbolization, and the visual variables, or the graphic elements that make up a specific symbol (e.g. such as the variation of size for graduated symbols). The variables then are used to construct various cartographic symbolization types. Figure 31.3, as adopted from Muehrcke (1983), depicts these relationships.

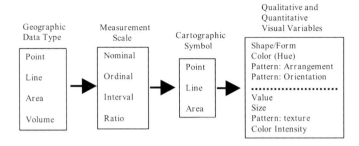

Figure 31.3 The relationship between geographic data type, measurement scale, cartographic symbol and the visual variables (after Muehrcke 1983).

Traditionally, cartographers have identified four primary types of geographical data, including point, linear, aerial, and volumetric. However, increasingly the idea of a fourth dimension, or the temporal component, has been studied. These terms – point, line, and area – have more recently been referred to as zero-dimensional, one-dimensional, and two-dimensional features (see Chapter 26). It should also be noted that the 'dimensionality' of an object changes with scale. For instance, at a large scale a house would be represented with an area, whereas at a smaller scale the same feature would be represented by a point. At a large scale a river would be represented by an area; at a smaller scale by a line. Thus the form of symbolization used (point, line, or area) can change dramatically as a result of the scale of the feature. Modern GISs do not yet have the ability to automatically change the dimensionality of the feature when scales are easily, perhaps all too easily, changed.

31.3.1. The measurement level of data

For symbolization purposes, the measurement level of the data must be considered. Cartographic data falls into the classical four measurement levels of nominal, ordinal, interval, and ratio.

As seen in Table 31.1, the data can be measured at a very basic level (nominal) that involves a simple classification. Examples of this include land-use (residential, transportation, agriculture, or forest) and race (White, African American, or Asian). Nominal-level data are unique in that they are at the same level (different classes or categories) and any form of symbolization must not impart quantity. Ordinal data involve a ranking, and allow for a quantitative differentiation involving higher-lower such as socio-economic status (lower, middle, higher). Visual variables that are quantitative in nature may be used for such data. Both interval and ratio data are consid-

Table 31.1 The four measurement levels of data

Measurement level	Example	Description	Quantitative/qualitative
Nominal	Landuse/landcover categories	Categorical	Qualitative
Ordinal	Socioeconomic status	Ranked	Quantitative
Interval	Elevation	Metric	Quantitative
Ratio	Population density	Metric	Quantitative

ered metric (involve true numbers) and allow for richer types of analyses. Interval-level data do not have a true zero-point, such as temperature, whereas ratio-level data have a true zero point, such as distance, area, and volume. One can find few examples of true interval-level data, although a good example involves elevation. The establishment of a 'mean sea level' for the purposes of mapping is arbitrary. Sea-level changes hourly, daily, and through geologic time; thus the datum (zero-point) here is socially-constructed. One could easily use the top of Mt. Everest or the lowest point in the Marianas Trench as a datum. (Chapter 3). All three of the quantitative measurement scales – ordinal, interval, and ratio – may be represented with a quantitative visual variable.

31.3.2. The fundamental visual variables

Cartographers have traditionally identified three types of cartographic symbols – point, line, and area symbols. Due to the two-dimensionality of paper and computer screens, volumetric data are represented in two dimensions, as an area, although virtual technologies are changing our ability to 'visualize' the third and fourth dimensions. The symbolization types are strongly related to the fundamental visual variables of cartography. Many sets of visual variables have been identified since the original work of Bertin, but the eight variables presented here were identified by Muehrcke (1983).

Figure 31.4 depicts the visual variables as interpreted by DiBiase *et al.* (1992). One can see from this illustration the relationship between the actual visual variable (position, size, value, texture, hue, orientation, and shape) and the dimensionality of the symbol (point, line, and area). The authors also relate the effectiveness of the measurement level to this visual variable-dimensionality matrix.

31.3.3. Geospatial data and symbolization

Another approach to understanding the underpinnings of cartographic symbolization was provided by MacEachren and DiBiase (1991). They argue that mappable phenomena can be arrayed along two axes: continuous-discrete and abrupt-smooth. 'To decide whether a distribution is continuous or discrete,' they assert, 'the designer needs to visualize an abstract surface representing the distribution'. Gaps, tears or overlaps in the surface can be visualized where the distribution is discontinuous. (MacEachren and DiBiase 1991). The second continuum (abrupt-smooth) relates to the character of variation in the phenomenon across space where, 'Some phenomenon (e.g. tax rates) can vary abruptly as political boundaries are crossed, while others (e.g. average farm size in Kansas) can exhibit a relatively smooth variation independent of the units to which data are aggregated' (MacEachren and DiBiase 1991).

An example of a typology of mappable phenomena is provided in Figure 31.5. Each of the boxes provides an indication of the typical data type found along the continuum as well as a specific symbolization type (below the box). For instance, the number of government employees is both abrupt and discrete and would require a graduated symbol. Alternatively the average farm size is both continuous and smooth and would normally utilize the isopleth method.

Those using GISs must be very careful when applying cartographic symbolization.

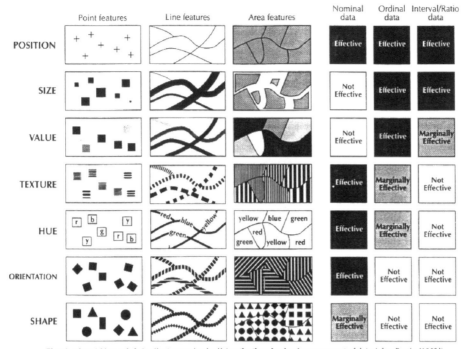

The visual variables and their effectiveness in signifying the three levels of measurement of data (after Bertin [1983]).

Figure 31.4   Visual variables (from DiBiase *et al.* 1992; used with permission from ACSM).

The user must clearly understand the data, and which visual variables/cartographic symbols are appropriate. Misuse of symbolization techniques often results in significant errors in map reading and interpretation.

31.4. Common types of thematic maps

There are certain standard types of thematic maps that cartographers have developed over the past 100 years, including dot, graduated symbol, choropleth, and isarithmic approaches. Each will be briefly discussed.

31.4.1. The dot and graduated symbol methods

As seen in Figure 31.6, two common point techniques include the dot and graduated symbol techniques. Dot mapping provides an excellent visual technique for viewing the clustering, dispersion, linearity, and general pattern of a distribution and is often applied to both population and agricultural data. The user must determine both the size of the dot itself, and the unit value, or how many items equals a single dot (e.g. 100 acres of wheat harvested per dot). The graduated symbol method is excellent for data that are large in number, but close in space. For instance, when mapping the total shipping tonnage brought into ports on the eastern seaboard, the graduated symbol technique is ideal.

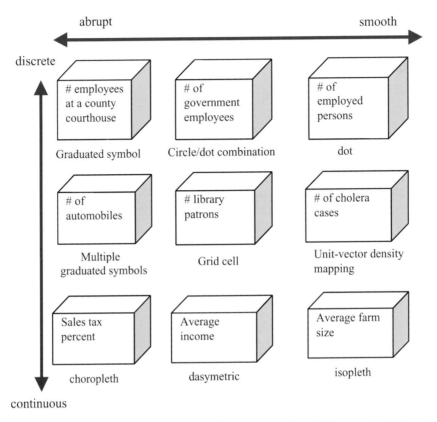

Figure 31.5 A typology of mappable data and mapping types (from DiBiase et al. 1992; used with permission from ACSM).

31.4.2. The choropleth and isarithmic methods

Two additional mapping techniques are the choropeth and isarithmic methods (Figure 31.7). The choropleth method involves applying value or color intensity to enumeration units (census tracts, counties, states, or nations) based on some statistical value.

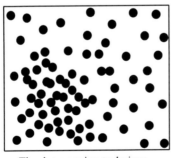

The dot mapping technique

The graduated symbol technique

Figure 31.6 The dot and graduated symbol mapping technique.

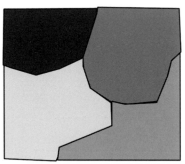

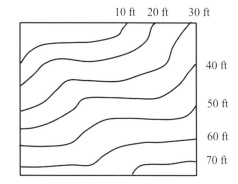

The choropeth mapping technique

Figure 31.7   The chloropleth and isarithmic technique.

The higher the value, the darker the value. Key to the choropleth method are issues of classification and data standardization. Choropleth data must be first be classed into categories for mapping, since the average map reader is unable to differentiate more than seven plus or minus two classes (gray values). All choropleth data must be standardized and not involve raw numbers, thus removing the effect of area. In general, larger enumeration units will have a greater quantity of a phenomenon than a smaller unit. For instance, Texas and California have greater populations than Rhode Island or Connecticut, since they are larger. By calculating a population density, one makes data for the two units comparable.

The isarithmic mapping technique involves lines of constant data value, such as elevation. Such maps are constructed through the 'interpolation' of point distributions, and allow the user to visualize a surface, in this case terrain. Other forms of data, including population densities, rainfall amounts, and even average incomes, can be depicted with the isarithmic technique. Isometric data are those actually recorded at positions on the earth's surface, such as precipitation, barametric pressure, temperature, and depth to bedrock. Isopethic data are 'derived', and include population density, per capita income, and average graduation rates. Isoplethic data require the establishment of some type of centroid, which can be used for interpolation.

This discussion has mentioned some of the very basic mapping forms. Many other symbolization types are possible, including methods for the portrayal of multivariable data. For additional methods, see the basic books by Dent (1996), Robinson *et al.* (1995) or Slocum (1999).

31.5. Data classification

A critical consideration in mapping involves the classification of data. As seen in Figure 31.8, different classifications of the same data can result in significantly different interpretations. When designing a data classification, the user must keep in mind several considerations. These include:

1   Encompass the full range of the data (both minimum and maximum values);
2   Have neither overlapping values NOR vacant classes;

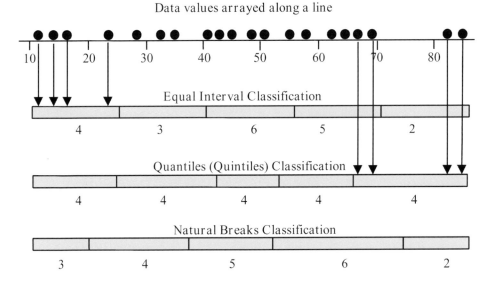

Figure 31.8 Data classification technique using equal intervals, quantiles, and natural breaks.

3 Be great enough in number to avoid sacrificing the accuracy of the data, but not be so numerous as to impute a greater degree of accuracy than is warranted by the nature of the collected observations. One basic technique for determining the number of data classes is Sturges Rule, which states: number of classes x should fall between $2^n < x < 2^{n+1}$. Thus for the 50 states of the US, the two values of n would be:

$2^5 = 32 < 50 < 2^6 = 64$. For 50 observations, which falls between 2^5 and 2^6, either 5 or 6 classes would be needed.

4 Divide the data set into reasonably equal groups of observations;
5 Have a logical mathematical relationship if possible (Robinson *et al.* 1995).

At the top of this illustration a sample data set of 20 numbers is arrayed along a number line ranging from 10 to 85. Numbers in the data set include:

1.	11
2.	14
3.	16
4.	24
5.	28
6.	33
7.	35
8.	41
9.	43
10.	45
11.	48
12.	51
13.	56

14.	58
15.	62
16.	65
17.	67
18.	69
19.	82
20.	84

An equal interval classification assumes an equal distance between the class breaks. In this instance, the equal distance has been set at 15 units, which would yield class breaks of:

Class 1 10–25
Class 2 25–40
Class 3 40–55
Class 4 55–70
Class 5 70–85

Alternatively, a quantiles classification puts an equal number of observations in each class. Thus if there are 100 observations in a data set, and the user desires 5 classes (quintiles), 20 observations will be placed in each category. In the example provided, a quintiles classification (20 data values/5 classes, or 4 observations in each class) is calculated, yielding class breaks of:

Class 1 10–25
Class 2 25–41
Class 3 41–52
Class 4 53–66
Class 5 66–85

Finally, a natural breaks classification may be calculated. Here, the user selects the maximum 'breaks' along the number line, or where the significant gaps appear in the data. The idea is to minimize the internal variation of data set, while maximizing the variation among the classes. Although the breaks are normally determined through graphical methods – number lines, histograms, and frequency curves – George Jenks created an 'optimal classification method' that used an algorithmic approach to determine these ideal breaks. (Slocum 1999). For the sample data set, a possible series of breaks (found by inspecting the number line) might be found at 20, 38, 52. and 75, yielding classes of:

Class 1 10–20
Class 2 20–38
Class 3 38–52
Class 4 52–75
Class 5 75–85

Many other possible classification methods exist, including those using nested means, standard deviations, and an area-under-the curve calculations. The references listed above provide greater details on most of the techniques used by geographic information specialists. The user of modern GISs must be aware of both the classification methods provided by the system, and the limitations of the methods. Simply selecting a classification method based on convenience rarely leads to an effective and meaningful representation of the data.

31.6. Graphic design, a graphical hierarchy, and figure ground

A significant part of the mapping process involves effective design, or a logical graphical presentation of the mapped elements. The map should not be viewed as a uniform graphical plane, but one that visually emphasizes certain features, while de-emphasizing other features. The concept of certain features being prominent (figures) and others being background (ground) is called figure-ground, and is of critical importance in cartographic design. It is Borden Dent in his seminal work on graphical design who systematizes the approach to the creation of effective figure-ground relationships on maps.

The problem involves establishing a logical hierarchy among the features that are to be mapped, and matching this with a commensurate set of visual levels. The first step is to identify a logical organization of mapped elements (Table 31.2).

As seen in Figure 31.9, the actual thematic symbolization (graduate circles, choropleth values) should be positioned at the highest visual level. At this same visual level, the title, legend material, and cartographic labels should be positioned. At a second level, the base information (boundaries, cultural and physical features, and land-water boundaries) should be positioned.

Other cartographic information, such as explanatory information, base map information, and other base elements should be placed at lower levels (visual levels C and D). Figure 31.9 depicts three possible level of a graphical hierarchy.

31.7. Visualization and modern cartography

Cartography is witnessing significant changes as modern computer systems allow more sophisticated forms of visualization, four-dimensional cartographies, and multimedia representations. A new term to describe this expanded capability is scientific, or 'geographic' visualization. For a survey of some recent research activities in this area, see the special issue of the journal *Cartography and Geographic Information Systems* (Vol. 19, Number 4), edited by Monmonier and MacEachren (1992), or the special

Table 31.2 A typical organization of mapped elements in the visual hierarchy (after Dent 1996)

Probably intellectual level	Object	Visual level
1	Thematic symbols (graduated circles)	More prominent
1	Title, legend material, and cartographic labels	
2	Base map, land areas, political boundaries, features	
3–4	Explanatory text	
4	Base map, water features (oceans, lakes, bays)	
5	Other base elements	Less prominent

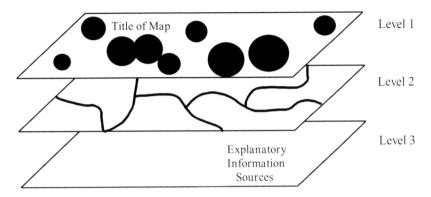

Figure 31.9 Three possible levels of a graphic hierarchy.

issue of the journal Computers and Geosciences (Vol. 23, Number 4), edited by MacEachren and Kraak (1997). What is this new visualization? According to MacEachren and Monmonier (1992):

> In the context of scientific visualization, 'to visualize' refers specifically to using visual tools (usually computer graphics) to help scientists explore data and develop insights. Emphasis is on purposeful exploration, search for pattern, and the development of questions and hypotheses. Advances in scientific visualization are changing the role of maps and other graphics as tools of scientific investigation

One of the more interesting advancements in the area of visualization is the potential for four-dimensional cartography. Unlike the 'static' maps discussed so far, dynamic maps include the temporal dimension, which is normally visualized through animations. In such a dynamic world, the visual variables presented earlier are insufficient. DiBiase *et al.* (1992) propose dynamic mapping offers at least three modes of cartographic expression:

1 Animation – the illusion of motion created from a sequence of still images
2 Sonification – the representation of data with sound
3 Interaction – the empowerment of the viewer to modify a display

They also propose a revised set of dynamic variables that are useful for understanding animation. These include duration, rate of change, and order. Duration is the number of units of time a scene is displayed. In animation, this scene duration may be used as a design variable – longer scenes allow for a more thorough study of a distribution. The rate of change is a proportion, m/d, where m is the magnitude of change in position and attributes of entities between scenes and d is the duration of each scene. A third dynamic variable, order, is the sequence in which the scenes are presented (DiBiase *et al.* 1992). The authors point out that a strict chronological order may not be the ideal method for exploring a spatial distribution, and provide an example of re-ordering the monthly temperature/precipitation differences for five different climate models. The example shows that the most significant differences in model prediction occur in the critical planting months of April to June, a trend not detected with a conventional chronological ordering from January to December

Table 31.3 Uses of dynamic variables

Uses of dynamic variables
Emphasizing location
Emphasizing attribute
Visualizing change
Visualizing spatial change: fly-bys
Visualizing chronological change: time series

(DiBiase *et al.* 1992). These three dynamic variables can then be used for a variety of visualizations, including emphasizing location, emphasizing attribute, and visualizing change (Table 31.3).

Users of GIS will mostly likely see significant advancements in their ability to symbolize spatial data with more sophisticated mapping techniques, including animation in the near future.

31.8. Conclusions

The user of a GIS must be aware of the cartographic capabilities of the system they are using. What classification methods are provided? What forms of symbolization are possible? Can a user customize the map? What operations for generalization are available? Far too often, because of poor education in cartography, users default to customized approaches that are inappropriate for their data, and can easily mislead the user of the analysis. Since more often than not, the final output of a GIS-based analysis is a map, the GIS specialist is well advised to carefully consider the cartographic representation early in the process.

References

Bagrow, L., *History of Cartography*, revised and enlarged by R. A. Skelton, Harvard University Press, Cambridge, MA, 1964.

Bertin, J., *Semiology of Graphics*, translated by W. J. Berg, University of Wisconsin Press, Madison, WI, 1983.

Dent, B. D., *Cartography: Thematic Map Design* (4th ed.), William C. Brown Publishers, Dubuque, IA, 1996.

DiBiase, D., MacEachren, A., Krygier, J. B. and Reeves. C., 'Animation and the role of map design in scientific visualization,' *Cartography and Geographic Information Systems*, 19(4), 1992, 201–214.

MacEachren, A. M. and DiBiase, D., 'Animated maps of aggregate data: conceptual and practical problems,' *Cartography and Geographic Information Systems*, 18(4), 1991, 221–229.

MacEachren, A. M. and Monmonier, M., 'Introduction', Special issue on Geographic Visualization, *Cartography and Geographic Information Systems*, 19(4), 1992, 197–200.

McMaster, R. B. and Stuart Shea, K., *Generalization in Digital Cartography*, Association of American Geographers, Washington, DC, 1992.

Monmonier, M. and MacEachren, A. (eds.), 'Geographical visualization', *Cartography and Geographic Information Systems*, 19(4), 1992.

Muehrcke, P. C., *Map Use: Reading, Analysis, and Interpretation*, JP Publications, Madison, WI, 1983.

Robinson, A. H. and Petchenik, B. B., *The Nature of Maps: Essays Towards Understanding Maps and Mapping*, University of Chicago Press, Chicago, IL, 1976.

Robinson, A. H., *Early Thematic Mapping in the History of Cartography*, University of Chicago Press, Chicago, IL, 1982.

Robinson, A. H., Morrison, J. L., Muehrcke, P. C., Kimerling, A. J. and Guptill, S. J., *Elements of Cartography* 6th ed., John Wiley and Sons, New York, 1995.

Slocum, T. A., *Thematic Cartography and Visualization*, Prentice Hall, Upper Saddle River, NJ, 1999.

Thrower, N. J. W., *Maps and Civilization: Cartography in Culture and Society*, University of Chicago Press, Chicago, IL, 1996.

Chapter 32

Carrying out a GIS project

Rebecca Somers

32.1. Introduction

This chapter provides guidelines for implementing a GIS. It describes the steps, activities, and issues involved in carrying out a GIS project. In some respects, developing a GIS can be daunting – especially for big projects. There are so many choices and decisions to be made and so many things to do. In other respects, GIS implementation can seem deceptively simple – especially if one makes simple assumptions. This discussion will put GIS implementation in perspective and help readers understand the GIS implementation process and get started.

32.1.1. GIS projects and programs

GIS efforts can be classified as projects and programs, ranging from small efforts to large, complex undertakings. A project is a one-time effort, developing a GIS that will serve a specific limited-term project and then be abandoned. Examples of such projects might include the performance of an environmental analysis, the production of maps for a survey, the development of a long-range land use plan, or the design and development of a park. As can be inferred from these examples, however, GIS do not usually disappear when the designated project is over. The GIS developed to plan and construct a facility such as a park or a community could then remain as a long-term management tool for that facility. Ongoing GIS facilities are often termed programs. Although they may start small, and may even start out as a project, the goal of these programs is to develop a lasting facility or asset that will facilitate the organization's work. A GIS project or program may be small and simple, involving limited software, data, and users; it may be large and complex, involving myriad data sets, applications, and users and complex systems and databases; or it could fall anywhere in between.

32.1.2. Typical implementation approaches

Different types of organizations implement geospatial data and technology differently. The GIS components and implementation approach used by any organization depend on its specific needs, but can be characterized by organization type.

Most local government GISs, for example, are multipurpose, comprehensive, enterprise-wide systems. They are designed to serve most of the organization's geospatial data handling needs, integrate all its spatial data, and make the data accessible to all users and departments. The data are usually large scale, based on high-accuracy parcel information, although local governments also use some smaller-scale, generalized data (FGDC, 1997). The implementation approach taken by most local governments is a pro-active, highly coordinated effort that minimizes redundancy and incompatibilities in data and systems. Leveraging GIS efforts and assets is very important to local governments because these systems can be very expensive.

Utilities also develop comprehensive geospatial data systems, and the integration extends beyond GIS technology. These organizations typically interface or integrate their geospatial data and systems with facilities management, work order management, customer information, facilities modeling, and other operational data and systems. Much of their critical information is high-accuracy, high-resolution data, although they also user smaller scale data, particularly because some utilities cover very large geographic areas.

Private sector organizations, on the other hand, take different approaches to GIS development. For example, the GIS resources developed in professional services firms, such as engineering companies, are usually intended to serve specific projects or clients. The data and applications developed for one project may have no relation to those developed for another project and never will. In this type of situation, GISs are implemented independently, focusing only on each project's needs. This independent development of specific GISs within an organization is a business-tools approach. This approach is also used by other types of companies that develop limited GISs in different areas such as marketing, customer service, and facilities planning. In this case, even if the GISs do cover the same geographic areas, the costs of developing them are usually low, business units are independent, and the costs of coordination may outweigh the savings achieved by that coordination. However, in situations where the organization finds that it is starting to spend significant resources developing and supporting independent GISs, it may take a service-resource approach to GIS development. In this model, GIS support services and perhaps data resources are developed and made available to the operating units. This approach benefits the operating units and the company as a whole, but does not involve the extensive coordination, comprehensive planning, and enforced standards that the enterprise approach does (Somers 1998).

32.2. The GIS implementation process

Although organizations differ in their approaches to coordinating their GIS implementation activities and in the resultant components and character of their systems, they follow the same basic GIS implementation process. It is a structured process that ensures that the GIS will meet users' needs.

The process involves five basic phases or tasks:

- Planning – establishing the scope and purpose of the GIS, and developing a general plan;
- Analysis – determining specific GIS requirements;
- Design – developing GIS components and specifications;
- Acquisition and development – acquiring software, hardware, and data, and crafting them into a personalized system; and
- Operations and maintenance – using the GIS and maintaining the system.

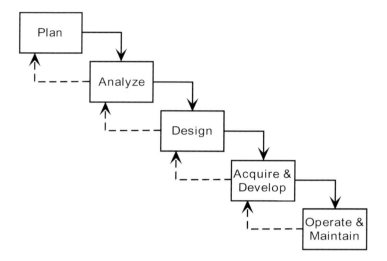

Figure 32.1   The GIS implementation process (Somers 1993–2000).

The process is illustrated in Figure 32.1. It includes feedback loops from each step because information gained in one task may require going back to a previous step and doing more work on that task. For example, analysis of users' needs in step two may necessitate going back to the planning step and re-examining assumptions and goals or adjusting budget plans (Somers 1996).

The following discussion provides an overview of these tasks. It usually takes the viewpoint of a GIS program of moderate to high complexity, so that all points can be raised and discussed. The tasks discussed are crucial for GIS programs of this type. They also apply to simpler and smaller GIS projects and programs, but can be more easily accomplished for those types of efforts.

32.3. Planning a GIS project and getting started

Planning is the first step in developing a GIS project or program of any size or type. Careful planning lays a solid foundation for all the subsequent steps of GIS implementation and helps an individual or organization avoid costly mistakes. Whatever the size of the GIS, it is necessary to address some basic questions before moving ahead.

First, the scope and nature of the GIS must be defined. Will it be a one-time project or an ongoing program? Will it be used for all geospatial data handling in the organization or for only a specific subset of functions, such as mapping? Will most people in the organization use it or will users be limited in number or job function? Will this GIS activity be part of a larger GIS effort? Will geospatial data and technology be integrated with the organization's other data and systems? Will GIS change the way the organization does business or will its impacts be limited? These types of questions help establish the scope and character of the GIS – whether it will be a simple limited work tool confined to specific tasks, an enterprise-wide framework that organizes and integrates spatial data and changes the way the organization operates, or anywhere in between. The scope of the GIS guides all the remaining planning tasks and implementation tasks.

The scope of the GIS determines who should be involved in its development. For example, if the GIS will be an enterprise-wide system serving many different types of uses and applications in many departments, then representatives from those areas must be involved. They represent the end-user perspective and their involvement will ensure that all requirements are met. In addition, large enterprise GISs often require the involvement of top management to provide support. Another perspective to consider is the skills needed to carry out GIS implementation. The scope and nature of the GIS effort will determine the types of background and expertise required to implement it. Required skills might be added through education and training, including specific people on the team, or obtaining outside assistance. For GIS efforts of even modest size, it is usually advisable to form a team and assign a leader. Even if things appear simple at this stage, they are going to get much more complex in the course of developing the GIS and a project manager will be necessary.

In order to work effectively, most GIS implementers and participants require some GIS education. Providing organized GIS educational opportunities in the planning phase is especially important for large projects with many diverse participants. Each participant must acquire an appropriate level of GIS knowledge so they can participate effectively in the planning and analysis activities that lie ahead. This may involve taking short courses, attending GIS conferences or trade shows, reading, or conducting in-house educational events. In any case, it is very important that education be tailored to individuals in terms of their backgrounds, how they will be using the GIS, and how they will participate in the GIS development process.

The third major component of GIS planning is developing resource and benefit estimates. Again, the scope of the GIS largely determines these initial estimates. The nature of the uses, the data, and the system provide indicators of costs. Specific cost details cannot be known yet, but general estimates can be developed based on these factors: whether there will be a few users or hundreds of users; whether the largest portion will be using professional GIS, desktop, or web-based systems; whether there will be a handful of data sets or hundreds; whether the data will be large-scale or small-scale. These factors will provide indications of general cost levels – whether the GIS will cost hundreds, thousands, or millions of dollars. Likewise, the scope and nature of the GIS provide general indicators of benefit levels. Moderate benefits will be gained from map creation applications; higher benefits can be gained from data and map maintenance, depending on the volume; and the highest levels of benefits can be realized from data access, manipulation, and analysis. Together, these estimates can provide a broad picture of the costs, benefits, and time required to implement the GIS.

Planning a large complex GIS may take weeks or months. For these types of projects a formal GIS strategic planning approach is required. Planning a smaller, simpler effort may take only a few days or less and may be done through an informal, yet still analytical method. In any case, however, questions concerning the scope of the GIS, the participants and their roles and skills, and resource requirements must be asked and answered before moving on (Somers 1996).

32.4. Analyzing requirements

Although people usually have general ideas of what they need from a GIS and the planning task addresses those needs, specific requirements analysis is needed to provide the necessary information for effective GIS implementation. In this task, the future uses of GIS and current geospatial data handling situation are examined in analytical detail. The goal is to identify the functional and data needs of the GIS applications, as well as the organizational environment.

32.4.1. Business processes and GIS functions

First, identify all the business processes that involve geospatial data and will use GIS in the future. The GIS scope defined in the planning task guides the identification of these processes and users. If the GIS is being developed to serve only one function, this step will be easy. It is far more complex in large organizations where GIS will be used extensively, and it is common to miss future GIS users when their use of or need for geospatial data is not apparent today. They may be using non-graphic forms of geospatial data and not realize it. It is important to identify as many of the GIS's future users as possible, even if they may not come online for a long time. Identifying their future needs and building those considerations into the GIS design and implementation will ensure that the system can be expanded to accommodate them when the time comes.

A business process analysis approach is very useful in analyzing GIS requirements. This approach examines each current (and planned) operation and maps out the steps and the data involved. The method focuses on the goal, input, output, and steps of the process and the data used to perform it. Tasks and decisions performed by individuals and work units are mapped out. Specific databases and data items, formats, sources, characteristics, and flows are included. Links to other work processes and systems are also identified. In the course of the analysis, desired changes are also included, based on the incorporation of GIS into the process and other changes the organization may be planning. The goal is to define future GIS-based work processes. This business process analysis method provides all the components that describe each user's GIS needs. Another benefit of this approach is that it prevents the mere transposition of current manual operations into GIS. It helps users and organizations make the best use of the power of geospatial data and technology.

Finally, the organizational environment will present conditions, constraints, and opportunities that affect the requirements and business process analysis. For example, the organization may have standards that must be followed, regulatory requirements to meet, set mandates and operating procedures, or politics that affect the way it does business and the way GIS must fit that business. These conditions must be factored in. They may affect individual work processes as well as the overall requirements for the GIS.

32.4.2. Performing the analysis

This analysis is accomplished through user interviews and work sessions. If the planned GIS is small and the end-user is the implementer this task may be relatively simple, but it must still be approached using analytical methods in order to produce the required information. The individual must take a close critical look at how they intend to work and beware of assumptions about data and software. They must also be sure to identify the individual work processes and planned uses of the GIS. They may involve different data and functionality requirements.

It is more common that there are many future GIS users involved in many different business processes in different parts of the organization. They have different viewpoints, different missions and work processes, and different needs in terms of geospatial data and processing tools. In this situation, the GIS implementer or analyst works with each individual user or group to perform the business process analysis. The analyst needs skill and experience in business process analysis to do this effectively. This is also where users' GIS education, accomplished in the planning task, becomes important. They need to understand the basic tools that GIS can provide to their work in order to effectively communicate with the analyst to jointly design their future GIS work process and to ensure that all their requirements are addressed.

32.4.3. Analysis results

The requirements analysis results in a clear, documented specification of the detailed GIS needs of the future GIS users as well as organizational support factors. There is a description and diagram of each future GIS work process and its functional and data needs. Many of these will become applications. Any constraints, opportunities, or problems associated with individual work processes, user groups, or the organization a whole are also identified.

The requirements analysis also provides additional information about the expected benefits and costs of the GIS. Benefits, such as decreases in work time or staff levels, increases in levels of service, and improved data and decision support, are identified in the course of analyzing each work process and the improvements that GIS will bring to it. Analysis of the work processes also provides a better idea of cost information based on data and functionality requirements, as well as the number of users in each group.

Some of this information will contribute to the technical implementation of the GIS; some will be used in the project and GIS management components. All of these products are necessary to build an effective GIS, but they are just pieces – they are not sufficient to go ahead with GIS implementation. They must be put together in the next step.

32.5. Designing system components

The design task involves putting the components together: determining the characteristics and combination of software, hardware, data, processes, and people that will be required to meet the organization's GIS needs. The challenge is to combine the organization's overall goals for GIS and the specific needs of the diverse users and applications, while developing an integrated and effective design. Developing this design is a crucial step prior to obtaining and implementing any GIS components.

32.5.1. Data and database design

Data are the most important aspect of a GIS and should be given primary consideration. Case studies and industry experience indicate that organizations generally spend the largest portion – as much as 80 percent – of their GIS budgets on data. Accordingly, the largest portion of effort and consideration should be spent on the data that will become part of the GIS – rather than the disproportional attention most people devote to technology. A GIS is, after all, just a tool to better use and maintain spatial data, so system design should focus on the data needed to do work and how they are to be handled.

GIS data and databases are discussed in previous chapters (see Chapter 25 on Fundamentals of GIS and Chapter 26 on Geographical data structures). At this point in the GIS development process, it is up to each individual or organization to define their specific GIS database. The data requirements of individual users and work processes are combined into one integrated design. The goal is usually to develop one version of a shared database that meets all users' needs with minimum redundancy and maximum usefulness and accessibility. The design addresses several aspects of the data:

- Data characteristics are defined to suit the combined system requirements. Each data entity is described in terms type and appearance, format, accuracy or resolution, attributes, links to other GIS entities and other databases, volume, source, maintenance responsibility and standards, and distribution and access.
- Data relationships are identified and described through a data model, indicating relationships at the entity level (see Chapter 26).
- Data access and handling requirements are described, ensuring that each user and application will have access to needed data in required form (see Chapter 25).
- Relevant temporal aspects are identified. These will support applications and data management functions such as time series analysis, planning, backup, archive, and retrieval.
- Metadata are identified at the appropriate level and in terms of how they will be used.
- The landbase or basemap is defined, based on users' needs. Content, accuracy, and maintenance procedures for this data set will be crucial to most applications and users. This data set usually involves some of the largest creation or conversion costs, but also may provide some opportunities for cost-effective acquisition and/ or sharing.

For a large organization developing an enterprise GIS, the database design can be a complex process.

32.5.2. Software and applications

The operational needs of the business processes and of the overall database support environment are examined to derive the required functions of the GIS software and applications. GIS software is discussed in Chapter 28 on GIS hardware and software. As with the database design, this is the point at which an individual or organization identifies their specific software needs to support their specific applications and environment. This is not yet the point to select software, but to develop a comprehensive description of what will be needed in terms of functionality. There are several important aspects to consider:

- Applications support. What functions will the applications perform and what basic GIS software tools will be needed to support applications development?
- Data support. What functions and features are needed to support the database design in creation, operation, use, and maintenance activities?
- Data access. What types of data access tools will be needed by users?
- Data integration. What are the data and systems integration support requirements?
- Performance. What are the performance requirements for applications and for other aspects of system operation?

Large organizations implementing enterprise GISs try to minimize the number of different GIS software packages that they implement. Software compatibility is important not just for data sharing, but for system support as well. Therefore, they seek to develop a comprehensive set of system specifications that will help them choose a suite of products that will meet all users' needs. A single user has more latitude in choosing whichever software package will most closely match their specific application needs, but may want to consider outside compatibility and standards for the purposes of future data sharing (and any such needs should have arisen in the planning and the requirements analysis tasks).

32.5.3. Integration with other systems and databases

Historically, most GISs have been standalone systems. Today, GIS users and developers recognize the need to interface GIS with other systems and data. Requirements, uses, and tools for full integration of geospatial data and technologies with other data and systems are becoming more prevalent. An individual's or organization's needs for data and systems interfaces and integration are derived in the planning and requirements analysis tasks. At this point, those needs must be examined and defined in terms of the system design – specifically which data or aspects of the GIS must interface or integrate with other data and systems.

A small GIS may truly be a viable standalone system. Larger organizations, such as local governments and utilities, however, usually need to integrate their GISs with systems and databases that support functions such as appraisal, emergency response, facilities maintenance, or customer support.

32.5.4. Management components

Along with the technical aspects of the GIS design, the management aspects must be designed also. Standards and procedures for database development, data maintenance, data management, system support and management, user support, and project management and coordination must be developed. The particulars will be based on the character and specifics of the GIS design and the environment and users that are to be supported. It is necessary to develop the management design at this stage because it affects other aspects of design as they affect it. For example, if a certain data set, as initially designed, is unsupportable, it must be redesigned now. Another reason that the management components must be designed now is that some of them will be needed soon in the process.

32.5.5. Determining resource requirements

What will it cost to develop a GIS and how long will it take? These are probably the most commonly asked questions. The answer for every organization is different. The resource requirements for any particular GIS depend on the organization's needs. So, as with all other GIS components, the resource requirements are derived from the planning, analysis, and design steps.

Initial cost estimates for the GIS are established in the planning task. Then in the analysis and design tasks, details are collected that provide additional information needed to calculate costs. The combined costs for the development and operation of the GIS include the following components:

- Hardware purchase, upgrade, and maintenance
- Software purchase, development, enhancement, and upgrade
- Software support
- Systems integration
- Data purchase, license, conversion, collection, and creation
- Database development
- Data maintenance and enhancement
- Data preparation
- Quality control and assurance
- Training – initial and ongoing
- On the job learning
- Recruiting and hiring
- System maintenance and enhancement
- Staff
- Consulting and services; and
- Management time

These development and operational costs are offset by the benefits that the GIS will provide. Tangible benefits include costs that can be avoided by using GIS to provide the needed data or functions and costs that can be reduced by performing tasks more efficiently with GIS. Depending on the organization, benefits may also include income and profit. Intangible benefits cannot be measured and, therefore, cannot be factored into the numerical part of the cost-benefit analysis, but they are often some of the most important benefits and should be identified. Such benefits may include better products, better service to citizens, and better planning or analysis results.

Once the costs and benefits have been identified, they must be transformed into an analysis and an evaluation method that can be used by the organization. Some are concerned with total costs vs. total benefits and payback periods. Others are interested in comparative measures such as internal rate of return or net present value that they can use to evaluate GIS investments with respect to other investments.

GIS implementation, operation, and use will also require human resources. The number of people, type and level of skills needed, and how they should be organized and managed to best support the system depends on the type of GIS that will be implemented. Likewise, the skills that users will need depend on their system use. Training, recruiting, and staffing requirements will be derived from these needs.

Time requirements must also be calculated. All the tasks and expenditures necessary to implement the GIS as designed must be scheduled for development of an implementation plan. As mentioned earlier, large enterprise-wide GISs can take years to complete. Small projects may be finished in weeks or even days.

32.5.6. Developing an implementation plan

Although a general plan is developed as part of the planning task, a detailed GIS implementation plan must be developed based on the detailed information identified in the requirements analysis and developed in the design task. The implementation plan spells out all tasks including data development, system acquisition and development, organizational development, and GIS management. It describes the tasks, schedule, and responsibilities for realizing all details specified in the requirements analysis and design tasks, and provides a roadmap and management tool for doing so.

A typical implementation plan would include several key components:

- GIS vision and scope.
- Participants: roles, responsibilities, and organization.
- GIS design: database design, applications, software requirements, hardware, requirements, and integration.
- Tasks: data acquisition, creation, and/or conversion; system acquisition and development; organizational development, including staffing and training; and task responsibilities.
- Schedule: schedule of tasks and milestones.
- Budget.
- Management procedures.

The need for a detailed implementation plan and project manager is evident for large GIS projects that can take many years and millions of dollars to complete. But detailed planning is necessary even for small systems. Without a plan, significant money and time can be lost through mismanagement. Documenting even a simple plan helps ensure that all aspects are covered.

32.5.7. Design results

There are several products relating to the components of the design task:

- Database design, including data descriptions, data model, and metadata specifications;
- Applications descriptions;
- Data/application/user correlations;
- System functionality description, focusing on software requirements;
- Management and organization design and components;

- Cost- benefit analysis and budget; and
- Implementation plan.

All of these components work together, so they must all be developed together and completed together before moving on. If emphasis is given to some aspects while others are neglected, the GIS components can get out of sync, sowing the seeds for later problems.

32.6. Acquiring and developing GIS components

Many people find the task of selecting GIS data and system components daunting – understanding and choosing among the many alternatives. But if the work done in the requirements analysis and design tasks is thorough, then selecting, procuring, and developing GIS components should not be difficult. It may still be complex and time and resource-consuming, but the decisions and tasks should be a matter of following the specifications and plans developed in the design phase.

GIS software packages provide the basic tools for input, editing, storage, maintenance, management, access and retrieval, manipulation, analysis, display and output of spatial data (see Chapter 28). These tools alone may satisfy some users' needs, and most applications will be built with these tools. So the challenge is to select the GIS package or software suite that best meet the organization's needs. A detailed specification is crucial in doing this. It provides the standard against which all alternatives will be evaluated.

Database implementation also presents many options. Depending on the type of data needed, the data sources available, and the costs associated with different sources and methods, an organization may choose to buy, license, collect, or convert the needed spatial data. And among these different sources and methods there are many choices. As with software selection, the challenge is to find the most cost-effective alternative that meets the organization's needs, and the way to do this is to have a detailed data specification against which to compare alternatives. If the organization is developing or converting the data itself, then the detailed specification provides the guidelines for performing the operation.

A variety of system and data development activities may also be required to complete the GIS components. Major items may include applications and other software development, database design, and system installation and integration.

32.6.1. Procurement methods

Depending on the organization and the GIS, any of a variety of methods may be employed for the acquisition of system components. Many public sector organizations must follow formal Request For Proposal (RFP) procedures. The organization may supply required 'boilerplate' content and procedures, and the GIS design and plan components comprise the technical specifications. Other organizations are free to acquire products and services through less formal means. In either case, however, GIS product and service specifications should be thoroughly documented and the alternatives should be evaluated against those specifications.

A large, formal evaluation process usually involves evaluating written responses to the RFP specifications, and then meeting with short-listed vendors to conduct a more detailed evaluation. Evaluation criteria include not only the vendor's response to the specifications and requirements, but their demonstrated ability to meet the organization's needs, experience, track record, costs, and other factors identified in advance. Implementers of small GISs may not have the resources or opportunities to conduct such an in-depth evaluation, but should still screen vendors according to documented requirements, and then spend extra time looking closely at a small set of viable alternatives.

Managing GIS product and service contracts effectively is as important as making the right vendor selection. Clear contract operating procedures must be developed, along with a plan and schedule. Reports must be regular and useful. Vendor responsibilities and deliverables must be clearly specified. And the client's responsibilities must be also specified and agreed upon.

32.6.2. Database development

For many organizations, database development will be the biggest part of their GIS effort. Therefore, to obtain cost-effective and timely data development, many organizations contract out the creation and conversion of their data. Local governments, for example, often contract out their basemap (planimetric, topographic, geodetic control, and digital orthophoto) creation and parcel data conversion. Utilities often do the same for their landbase and facilities data. Other types of organizations acquire GIS as a tool to handle the data they create or obtain in the course of their operations or projects. In these cases there may be little or no basemap development. Still other organizations, particularly those performing 'business GIS' applications buy or license much of their data (Somers 1996). Chapter 25 discusses GIS data development. It is up to the organization to tailor basic spatial data input processes to suit their specifications.

Whatever the method or source for data development, quality assurance must be given prime attention. Quality control requirements should be built into the RFP or data development specifications; the vendor must respond to them, assuring the quality of the data throughout the delivery process; and the user must take responsibility for verifying and maintaining data quality (see Chapter 30 on spatial data quality).

32.6.3. Putting the GIS into operation

Except for limited projects, putting a GIS into operation is a fairly lengthy process that requires careful management. An organization has work that it must continue to do without interruption or slowdown while the GIS is being implemented. And for large GISs, the implementation process can take months or years.

It is usually advisable to conduct GIS implementation in phases. In large GIS efforts there is simply too much data and software to implement all at once. Deciding which data to develop first and/or which users to supply with system access first relies on the analysis done in the design task. That analysis should have revealed factors such as which applications and users need which data sets and software functionality and what the costs and benefits of those applications will be. These factors, in addition to other organizational priorities and opportunities, will reveal advantageous starting points. Organizations often choose to implement those parts of the system that require the least amount of time, money, and effort in order to get early, cost-effective benefits from the GIS. The demonstration of early GIS use can often satisfy many users and managers and build support and resources for further development.

For more complex applications and aspects of the system, it is often advisable to develop a pilot project before proceeding with full implementation. The pilot project comprises a representative, yet relatively small, set of the data, system, applications, and procedures. It gives the GIS implementers and users the opportunity to evaluate the software, data, and procedures and make necessary changes before committing full funding and effort.

In addition to planning phased development and conducting pilot projects, there are other considerations and methods for introducing GIS operations into the organization, including developing test systems, operating parallel systems, and managing switchovers. In most cases, putting a GIS into operation must be viewed and managed as a process, not an event.

32.7. Managing GIS operations

As the GIS project moves from development into operations, the main goals become maintaining the asset and supporting users. A GIS has a lifecycle, as does any system or facility. Throughout the phased development, as well as once the system is in full operation, the system components must be maintained, added to, and phased out when necessary. There may frequently be new technology, data, applications, and users and the cycle of planning, analyzing, designing, and implementing must be continued for effective incorporation of new components and maintenance of existing ones. This process is illustrated in Figure 32.2.

In large GIS environments, many of the GIS operations tasks are similar to Information Technology (IT) management tasks: user support, trouble shooting, training, system management, network management, data management, vendor support coordination, configuration management, system backups, and so forth. And, in fact, many organizations are now incorporating GIS management into their IT operations. However, due to some of the special characteristics of spatial data and technologies, certain aspects of GIS system management require special attention.

While all IT managers are concerned with data management, this responsibility presents additional challenges to GIS managers. Spatial data represent a very expensive and valuable asset. Spatial data and technologies are also powerful tools. Therefore, such matters as data stewardship, data maintenance standards and procedures, metadata, and data access and security are vitally important (Somers 1996). Developing equitable and manageable standards and procedures for funding ongoing GIS system operation, maintenance, development, and data access is also an important aspect of GIS management.

32.8. GIS implementation challenges and success factors

Several factors present GIS development challenges, but are crucial to successfully developing a large GIS project or program:

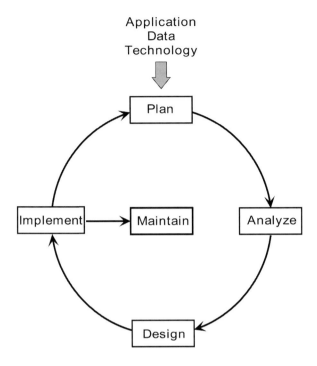

Application
Data
Technology

Figure 32.2 GIS lifecycle (after McDonnell Douglas Corporation 1989).

- Effective planning. A formal strategic planning methodology is required for large complex projects.
- Requirements-based GIS development. GIS software and data will be big investments. It is crucial to understand the organization's entire set of requirements and specifications before selecting or developing data and software. Mistakes caused by hasty or ill-informed decisions or assumptions can be very costly.
- Skillful GIS leadership and management. A large project with many participants requires leadership as well as good project management.
- Upper management support. Large projects that involve many different user groups or departments and will entail significant resources require upper management understanding and support.
- Starting small. Demonstrating meaningful results early in the process and for minimum expenditures has been a key success factor for countless GIS projects.
- User involvement, coordination, and communication. Effective structures, policies, and procedures must be developed for organizing, coordinating, and communicating with the user community. Involving users in the design of the system is also crucial for ultimate GIS acceptance and adoption (Eason 1988).
- Education and training. Ensuring that GIS implementers and users and organizational managers have adequate GIS information and training is crucial to successful GIS development, adoption, and use.
- Change order management. A GIS project can get swamped with constant requests for additions and changes, particularly because as participant's GIS

knowledge and exposure grows, so do their demands. Effective change order management is crucial to keeping it on track.

- Managing risk. Assessing organizational risk at the outset of the project and developing implementation methods that contain and minimize it increase the chances of project success (Croswell 1991).
- Politics. Successful navigation of the organizational political waters is necessary to gain and maintain support for the GIS.

Small GIS projects and programs have fewer or smaller challenges, but still require attention to a few key points in order to be successful:

- Planning and analysis. Many people undertaking small GIS projects feel that planning and analysis are not necessary. They believe they know what they need from a GIS and how to go about building it. By the same token, however, if they do have a clear idea of what they are need, then planning and analysis should be fast and easy. But these tasks are still necessary to ensure that all important aspects are covered and that small, seemingly obvious decisions do not lead to big problems.
- Following the implementation process. Likewise, it is important to follow all the steps of the GIS implementation process. The tasks may be easy and the process may go quickly for small systems, so it should not be a burden. But it is a safeguard in making GIS decisions and investments.
- Data quality. Developers of small systems often have limited resources and end up obtaining or developing inadequate data. Furthermore, they invest additional money and time in those data. Data are an investment and it is important to look ahead and ensure that data that solve immediate needs have longevity. Again, the steps and methods of the GIS implementation process help implementers do this.
- Avoiding unnecessary expenditures. The flip side of under-investing in GIS is over-investing. Limited GIS projects may be able to make due with limited GIS data and software. However, some users spend more money on more sophisticated software and data than they need and more time learning and using those systems than they would have spent on simpler systems. Again, the analyses done in the GIS implementation process will help implementers of such projects select appropriate GIS components (Somers 1993–2000).

References

Croswell, P., 'Obstacles to GIS implementation and guidelines to increase opportunities for success', *Journal of the Urban and Regional Information Systems Association*, 3(1), 1991, 43–56.

Eason, K., *Information Technology and Organizational Change*, Taylor and Francis, London, 1988.

Federal Geographic Data Committee, *Framework Introduction and Guide*, FGDC, Washington, DC, 1997.

McDonnell Douglas Corporation, *Lifecycle Solutions for Infrastructure Management*, McDonnell Douglas Corporation, St. Louis, MO, 1989.

Somers, R., *GIS Implementation and Management Workbooks*, Somers-St. Claire, Fairfax, VA, 1993–2000.

Somers, R., 'How to implement a GIS', *Geo Info Systems* 6(1), 1996, 18–21.

Somers, R., 'Developing GIS management strategies for an organization', *Journal of Housing Research* 9(1), 1998, 157–178.

Part 5

Applications

Chapter 33

Biophysical and human-social applications

Susanna McMaster and Robert B. McMaster

33.1. Applications of GIS

Geographic Information Systems (GIS) have been used to support a wide range of applications. This chapter discusses data sources, analytical approaches and examples of applications in both biophysical and human-social areas. The sophistication of GIS use in the application areas has been increasing over the past decade. Early on, GIS was used primarily for mapping distributions, such as the location of car thefts, or of toxic waste sites. Now, however, GISs are utilized for advanced types of both analysis and display, such as modeling transportation flows and in selecting idealized 'routes' for delivery trucks, or locations for police and fire stations. The chapter is divided into two sections – biophysical applications and human-social applications. Although this division might seem somewhat arbitrary, as explained in Chapter 26 spatial data are either 'field' or 'object'. Natural resource data, such as forest stands, drainage basins, or vegetation covers tends to be continuous (field). Urban-social data, such as population, industry, and transportation tends to be discrete (object). One must be careful, however, as many applications might very well require both types of data – transportation and terrain. An excellent review of applications may be found in a recently published year 2000 issue of the *Journal of the Urban and Regional Information Systems Association* (Vol. 12, Number 2) (Getis *et al.* 2000; Nedovic-Budic 2000; Radke *et al.* 2000; Rushton *et al.* 2000; Wiggins *et al.* 2000; Wilson *et al.* 2000). In this issue are comprehensive reviews of GIS and crime, emergency preparedness and response, public health, transportation planning and management, water resources, and urban and regional planning.

33.2. Biophysical applications

Numerous disciplines in the biophysical arena such as ecology, forestry, soil science, and land use planning have been concerned for some time with the management of natural resources at different scales over space and time. A major goal has been to manage natural areas in a sustainable manner or to preserve remaining wilderness areas. Historically, however, resource planners employed management models that were non-spatial in nature such as the Forest Service's FORPLAN model. More recently, over the past 10–15 years, resource managers have recognized the benefits of using GISs to support spatial analysis of natural resources. Crain and MacDonald (1984) present three evolutionary stages of development of GISs from land inventory to land analysis to land management purposes. The initial phase involves developing a spatial database that can eventually be used in the final phase, involving more sophisticated analytical capabilities as well as predictive modeling and decision-making abilities.

GISs have been used to better understand natural ecosystems and ecological processes in order to conserve or restore them. Additionally, these ecosystems have been impacted to various degrees by human activity as populations grow and economic activity and development increases. Thus, GISs have been instrumental in monitoring, evaluating and resolving environmental impacts. GISs provide planners and managers with the capability to test different alternatives to proposed actions in order to assess the possible impacts and outcomes of an action. 'What if' scenarios can be modeled by spatial analytical capabilities of GIS and help to support improved decision-making. Geographic visualization further enhances our ability to explore potential outcomes of different scenarios.

An important consideration in the development of biophysical applications is the scale of analysis. The issue of scale is a fundamental question for geographers but also to other scientists in the biophysical arena such as ecologists. For example, how does scale influence the understanding of spatial pattern and ecological processes. GISs are able to support analysis at a variety of scales and, depending on the availability of data, can integrate data at disparate scales from a variety of sources (e.g. Global Positioning System (GPS), various governmental data, commercial data, etc.). The purpose of this section is to describe basic data sources and analytical operations useful for biophysical applications and then to discuss more specific applications in order to illustrate the benefits of GIS for such applications.

33.2.1. Data

Biophysical applications involve the use of data from a variety of sources including remotely sensed imagery (e.g. satellite imagery and aerial photographs), field data collected by GPS technology, and digital data and hardcopy maps available from local, state and federal government agencies such as land use/land cover data, species distributions, wetlands, Toxic Release Inventory (TRI) data, forest resource inventories, climatic and weather data, hydrologic and geologic data, geodemographic data, and terrain representations. For example, many natural resource applications involve the use of satellite imagery such as Landsat Thematic Mapper data to map vegetation cover and monitor forest health and landscape change. The Natural Resources Conservation Service uses GPS and GIS to map soils and related data in order to monitor soil and water quality and to recommend soil conservation options to landowners. Many state and local government agencies also have a variety of data useful for biophysical GIS applications (e.g. Minnesota's Land Management Information Center and Department of Natural Resources). A variety of environmental data are becoming very accessible in digital form via the Internet. For example, the US Environmental Protection Agency highlights spatial data and applications at their web site http://www.epa.gov/epahome/gis.htm including an Internet-based mapping application called 'Maps on Demand' (http://www.epa.gov/enviro/html/mod/index.html). This system allows users to map environmental information such as superfund sites, hazardous waste releases, water discharge permits and tax incentive zones for brownfields using the Envirofacts Data Warehouse. Overall, most applications involve the use of a wide variety of data from different sources at different scales and GIS is able to integrate these data sources for use in a variety of applications.

A key characteristic of most natural landscape features such as soils, vegetation and climate is that they exhibit a great deal of natural variation and uncertainty with respect to boundary location. For example, symbolizing soil boundaries as sharp clean lines on a map is inconsistent with the great degree of heterogeneity that may actually occur on a local scale (Burrough and McDonnell 1998). Miller (1994) provides a useful discussion of the challenges faced when mapping very dynamic natural phenomena like plant and animal species distributions. Users of such data must be aware of possible limitations of such data (e.g. data quality, accuracy and/ or precision, sampling density) depending on the purpose of a particular application. The development of data standards and metadata are helping to improve the use of such data, as are advancements in data processing and analysis. Researchers in the area of geographic visualization are also developing techniques for visualizing data accuracy and uncertainty.

33.2.2. Spatial analysis and modeling

One of the fundamental aspects of GIS is the ability to undertake a variety of spatial analytical operations that can assist in modeling complex biophysical processes. The basic analytical capabilities of GIS have already been covered in Chapter 29. In addition to these analytical capabilities (e.g. reclassification and overlay analysis), quite sophisticated analytical methods include the incorporation of spatial pattern analysis, geostatistical techniques and accuracy assessment and error modeling operations. For example, Klopatek and Francis (1999) discuss various spatial pattern analysis methods that are useful landscape ecology, which focuses on the relationship between spatial patterns and ecological processes at a variety of scales. Specific analytical techniques include point and patch analysis such as nearest-neighbor methods, join-count analysis and lacunarity analysis. Griffith and Layne (1999) and Goovaerts (1997) describe various geostatistical techniques useful for natural resource evaluation and biophysical applications such as kriging, the use of stochastic simulation for assessing spatial uncertainty, and understanding the influence of spatial autocorrelation on spatial analysis. Digital Elevation Models (DEM) can be used for a variety of terrain analysis such as developing viewsheds and delinating watershed regions. Understanding the influence of data and modeling accuracy on biophysical applications is another key analytical area in GIS. Mowrer *et al.* (1996), Lowell and Jaton (1999) and Mowrer and Congalton (2000) include examples of GIS-based applications that examine various spatial accuracy assessment approaches in natural resources management and environmental sciences. In order to illustrate how various spatial data sources and analytical techniques can be used to support biophysical applications using GIS, some examples are provided in the next section. It is important to note that these are only a few of many possible applications in the broader range of biophysical applications that exist today.

33.2.3. Examples of biophysical applications of GIS

GIS can support spatial analysis and decision-making in a wide variety of biophysical applications including landscape ecology and conservation planning (Haines-Young *et al.* 1993; Scott *et al.* 1996), natural resource management (e.g. forest and wildlife management) (Morain 1999), and environmental modeling at various scales (e.g. hydrologic, atmospheric, soils and ecological modeling) (Goodchild *et al.* 1996). The need to explore the impact of human activity on the environment also involves the integration of geodemographic data and analysis techniques, and more recently the incorporation of participatory democracy or public participation GIS in an effort to incorporate localized knowledge in order to understand how technology serves society. The examples discussed in this section illustrate how GIS can be used in forest management and bioresource conservation planning.

33.2.3.1. GIS and forest resources management

GISs are being used by the federal, state and local forest agencies as well as private forestry companies as spatial decision support tools to monitor and manage public and private forest lands. More specifically, GIS can be used to manage a variety of activities that fall under the 'multiple use' notion such as timber harvesting activities, wildlife habitat, livestock grazing, watershed management, recreational activities, mineral extraction, and the protection of rare and endangered species, and archaeological sites. GIS management models include those that are descriptive, predictive and prescriptive (Green 1999). Descriptive models are the most common and involve some form of suitability analysis (e.g. determining the optimal site for recreational trail development based on a set of key spatial criteria). Predictive models allow users to test 'What if' scenarios based on a set of spatial variables (e.g. predicting the movement of a toxic gas plume given certain meteorological conditions, etc.) while prescriptive models support a full range of spatial decision-making capabilities that allow managers to assess the different management actions. The following applications include examples of these different management models.

Timber harvest planning and analysis often involves the use of GIS to integrate site specific forest stand inventory data along with terrain data, sensitive species and roadless areas to evaluate the environmental impact of proposed timber sale on national forest lands. GIS can be used to develop a harvest plan, evaluate potential environmental impacts and assess alternatives to the plan. A key aspect of such a study is to protect roadless areas by using data such as Forest Service mapped roads, Roadless Area Review and Evaluation II (RARE II) mapping and roadless areas data produced by the Sierra Biodiversity Institute. Such analysis often incorporates optimization modeling and proximity analysis (e.g. how to most efficiently skid logs to the nearest logging road and minimize erosion and sedimentation of streams and other environmental impacts). Figure 33.1 shows an example of a timber harvest suitability map indicating high, medium and low suitability of stands. The spatial model that generated this map was designed to assess the suitability of stands in the study site for pulpwood management based on certain forest inventory criteria. It consisted of three submodels: pulpwood cover type, biological and economic submodels. Fourteen source maps were used including cover type, physiography, cover size, cover density,

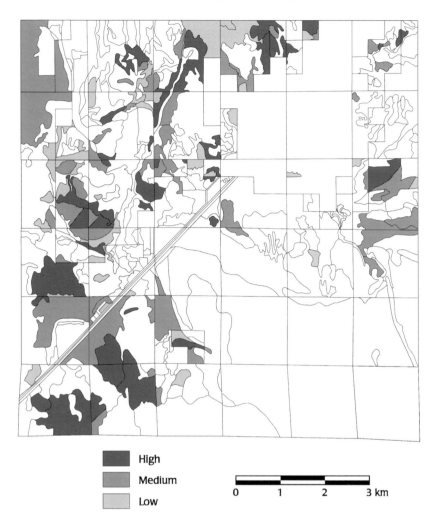

Figure 33.1 The results of a timber harvest suitability map indicating high, medium and low suitability of stands.

High

Medium

Low

0   1   2   3 km

age, diameter at breast height (i.e. the diameter of a tree at 4.5 ft above ground level), basal area per acre (i.e. the sum of basal areas, the cross-sectional area of a tree trunk at breast height, of individual trees on the acre), site index (i.e. an expression of forest site quality based on expected height of dominant trees at a specific age), mortality, damage, volume per acre, cords per acre (a cord is equivalent to a pile of wood measuring 128 cubic feet), stand acreage and roads. A detailed discussion of model development is given in McMaster (1999). Assessing the impact of data quality on the outcome of such analysis is another important concern. For example, forest stand data could be tested against satellite images or orthophotos or field collected data to ground truth the data. The influence of resolution or model specification on harvest decisions can also be assessed. McMaster (1997, 1999) has examined the impact of varying resolution and attribute accuracy on pulpwood harvest decisions.

Management of forested lands can also include watershed analysis, wildlife habitat management (e.g. determining optimal elk calving areas), and forest pest (e.g. predicting areas that are most susceptible to spruce bud worm and gypsy moth infestations) and fire management (e.g. predicting fire hazards and studying the effect of fire on plant and animal species). For example, GIS can be used to assess watershed condition in an effort to facilitate sustainable economic development in natural areas and maintain fisheries and aquatic habitats. This type of analysis, like the timber analysis above, is centered on examining detailed roads data such as road density, road length, type and condition, level of road use, and proximity of roads to streams. A related application area deals with conservation GIS including the habitat and endangered species management.

33.2.3.2. Conservation GIS and gap analysis

Another major biophysical application area centers on preserving and maintaining biodiversity and endemism using GIS. Applications in this area can involve different strategies, e.g. a species approach often focusing on endangered species or a more holistic habitat approach on a broader scale. Gap analysis is a program of the National Biological Service interested in determining the conservation status of land cover types, vertebrates and habitats on a regional scale. Those species not adequately represented in the nation's existing conservation spaces are referred to as conservation 'gaps.' Gap analysis does not focus solely on rare or threatened species but also on 'ordinary' species represented in biodiversity reserves. A key assumption of this approach is that terrestrial vertebrates and vegetation cover can serve as indicators or surrogates for biodiversity in a region. Sources of data used in this approach include Landsat Thematic Mapper imagery, agency and museum collection records, background knowledge of species' ranges, and previously-generated hardcopy and digital maps (e.g. vegetation cover, land use/land ownership data). A key focus of gap analysis is to improve upon the inadequacies of traditional species distribution maps by using a variety of data sources to develop more accurate representations of species distributions. Butterfield *et al.* (1994) illustrate the gap analysis approach to map the breeding distribution patterns of 366 species of vertebrates. At present, the administration and scale of the gap analysis program is at the state or regional level (Scott *et al.* 1996). The gap analysis program website (http://www.gap.uidaho.edu) highlights other research and applications of this approach to preserving biodiversity. GIS is ideal for undertaking gap analysis because it is a useful means of handling large volumes of data from disparate sources.

In some instances, GIS has been used to manage and monitor individual species that are recognized to be rare or endangered and in need of special attention (e.g. old growth redwood forests and grizzly bears in the US). For example, GIS can be used to evaluate the success of grizzly bear recovery zones designated by the US Fish and Wildlife Service (i.e. areas where the population can increase to a more sustainable level) by using grizzly bear siting data, roadless area data and National Park Service area data (The Conservation GIS Consortium 1999). MacKinnon and De Wulf (1994) illustrate the use of GIS to design protection areas for the threatened giant pandas in China.

The future of GIS in supporting biophysical applications is sure to grow and improve as advancements are made in data and analytical procedures and as research continues to improve our understanding of how scale influences ecological process and how they are modeled.

33.3. Human-social applications

Our society is faced with many complex human-social issues that involve the use of large amounts of data in an effort to understand and evaluate the spatial dynamics of a problem so effective policies and solutions can be developed. GIS provide planners and decision makers with the ability to identify at-risk populations using a vast array of geodemographic data and spatial analytical tools. Private businesses also use GIS to support market research and real estate sales. One can find an increasing number of applications of GISs to human-social problems, including geodemographic analyses, crime assessment, environmental justice studies, and concerns in public health.

33.3.1. Data

Human-social applications involve the use of census data, land use/land cover information, transportation networks, public health data, property boundaries and geographic base files like Topologically Integrated Geographic Encoding and Reference system (TIGER), and zoning data. The backbone of much of the spatial analysis for human-social applications are the data sets provided by the census, including the TIGER files and accompanying statistical data. Academic researchers as well as planners, decision-makers and community organizers utilize census data to assess a multitude of population characteristics, including those on housing, poverty, race/ethnicity, income, and transportation, and increasingly wish to analyze temporal trends. In contrast to some biophysical data such as soils and vegetation that tend to be more variable, some human-social data are more precise, such as cadastral data and property boundaries. Users of such data are very concerned with accuracy and precision since errors can be costly and involve liability issues. Increasingly, data from the public health sector are being provided, albeit at rather course resolutions due to confidentiality constraints. For instance, in the State of Minnesota one can obtain public health data only at the county-level, unless confidentiality is maintained; this is normally only possible for research purposes. Thus the type of address-specific data that geodemographic analysts utilize for marketing purposes would be impossible to acquire for health studies. Another key source of data for human-social applications involves transportation information. The Federal Department of Transportation, along with all states, maintain detailed data on transportation routes, flows, and conditions. One specific type of transportation data are the Traffic Analysis Zone (TAZ) datasets for urban areas, where census data on where people work are reaggregated into zones used to predict traffic flows.

33.3.2. Spatial analysis and modeling

GIS has facilitated the rapid growth of geodemographic analysis, including geomarketing, and many forms of population analysis in general, by integrating together the wealth of population data that are now spatially-referenced with the powerful spatial-analytic capabilities of GISs. Examples of these spatial analyses include the assessment of environmental justice/racism at multiple spatial scales (regional, urban, community); the calculation of segregation indices and evaluation of urban poverty, and the indentification of concentrated poverty; the development of neighborhood indicators, including a multitude of economic and social measures based on population data; and the spatial/temporal analysis of census data.

Increasingly, researchers are utilizing GISs and census geographic base files for historical geodemographic analyses. For instance, after the 1990 census it was possible to document the changes in geodemographics between 1980 and 1990, using the 1990 TIGER files. A common application was mapping the change in minority populations between the two periods. It should be noted, however, that researchers are constrained mostly to two or three decades of temporal analysis without the availability of pre-1970 digital files.

Suitability analysis can be useful in both human-social and biophysical applications to determine the optimum location of some phenomena given a set of criteria. For example, where is the best site to locate a school or day care given certain geodemographic and other relevant criteria.

33.3.3. Examples of human-social applications of GIS

33.3.3.1. Racial segregation studies

One specific use of temporal census boundary files and data is in the computation of segregation indices. Although a multitude of segregation indices have been developed over the past 10 years, nearly all work applies such indices to 1990 data. The application of such indices is normally applied for the purposes of identifying areas of concentrated poverty, and to identify the income isolation that many groups experience.

33.3.3.2. Environmental justice research

Environmental justice studies, which rely extensively on census data attempt to determine if certain 'at risk' populations (minority, young, old, disabled, poor) bear a disproportionate exposure to environmental chemicals, are only able to assess the conditions from 1990 onward. Increasingly, however, researchers (Pulido 1999) have argued that examining the historical perspective of toxic sites is crucial in such work. The basic question, of course, is which came first – the toxic site or the 'at-risk' population? Proving or disproving environmental inequity requires 'intent', which can only be determining with careful painstaking historical reconstructions of both the industries themselves and the associated 'at risk' populations. Did, for instance, a subpopulation begin to migrate to certain spaces and a toxic industry then purposefully targeted that space for a factory? Or did the subpopulations themselves, because of a complex series of social-economic issues, cluster around a site due to depressed land values? Intent and causality can only be determined through historical reconstructions. Multiple studies utilize GISs to assess the effect of both spatial scale and data resolution on the results of environmental justice (McMaster et al. 1997). Figure 33.2 shows the results of an analysis of environmental inequity in Minneapolis. On this map, the percent of the population in poverty is depicted in relation to the 38 TRI sites in Minneapolis. One can also perform simple types of spatial analyses here, where

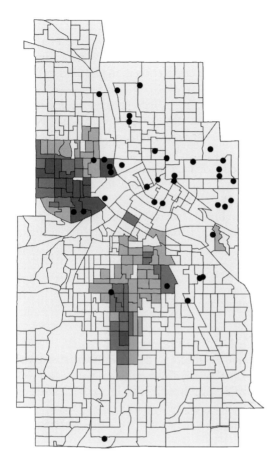

• TRI Site

Percent Non-White Population
1990 Census Block-Groups
(number of block-groups in parentheses)

■	80 to 100	(16)
■	60 to 79	(27)
■	40 to 59	(54)
□	0 to 39	(379)

Figure 33.2   An example of basic geodemographic analysis. The map depicts percent population
in poverty and TRI sites for the city of Minneapolis.

Figure 33.3 illustrates the same data, but where the TRI sites have been buffered to
1,000 yards in order to ascertain whether low-income populations live 'close' to TRI
sites.

33.3.3.3. Public health and epidemiological studies

The field of epidemiology is turning quickly to spatial solutions to both trace the path

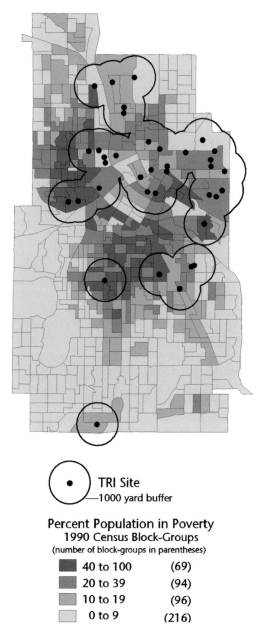

Figure 33.3   An analysis depicting population in poverty and 1,000-yard buffers around TRI sites for the city of Minneapolis.

of diseases and to determine the impact on subpopulations. One can point to a plethora of studies that are utilizing GISs and census data for public health analysis. One common usage (Waller and McMaster 1997) is with standardization of mortality/morbidity statistics based on subpopulations such as age (e.g. younger than five) and ethnicity (e.g. hispanic population).

33.4. Conclusions

There is a growing literature on most of these topics, but is found in dispersed sources. One can often find much of the literature in the cognate disciplines (e.g. public health), and not in the core GIS literature. Readers interested in seeing other examples of GIS applications are encouraged to read the trade magazines of *GeoWorld* and *GeoSpatial Solutions*, as well as the literature published by the vendors themselves.

References

Burrough, P. and McDonnell, R., *Principles of Geographical Information Systems*, Oxford University Press, Oxford, 1998.

Butterfield, B., Csuti, B. and Scott, J., 'Modeling vertebrate distributions for gap analysis', *Mapping the Diversity of Nature*, Miller, R. (ed.), Chapman and Hall, London, 1994, pp. 53–68.

Conservation GIS Consortium, *Conservation GIS Starter Kit: a Workbook for Using Geographic Information Systems for Natural Resource Conservation and Management*, 1999.

Crain, I. and MacDonald, C., 'From land inventory to land management: the evolution of an operational GIS', *Cartographica*, 21(2/3), 1984, 40–46.

Getis, A., Drummy, P., Gartin, J., Gorr, W., Harries, K., Rogerson, P., Stoe, D. and Wright, R., 'Geographic information science and crime analysis', *Journal of the Urban and Regional Information Science Association*, 12(2), 2000, 7–14.

Goodchild, M., Steyaert, L., Parks, B., Johnston, C., Maidment, D., Crane, M. and Glendinning, S., *GIS and Environmental Modeling: Progress and Research Issues*, GIS World Books, Fort Collins, CO, 1996.

Goovaerts, P., *Geostatistics for Natural Resources Evaluation*, Oxford University Press, New York, 1997.

Green, K., 'Development of the spatial domain in resource management', *GIS Solutions in Natural Resource Management*, Morain, S. (ed.), OnWord Press, Santa Fe, NM, 1999, pp. 5–15.

Griffith, D. and Layne, L., *A Casebook for Spatial Statistical Data Analysis: A Compilation of Analyses of different Thematic Data Sets*, Oxford University Press, New York, 1999.

Haines-Young, R., Green, D. and Cousins, S., *Landscape Ecology and GIS*, Taylor and Francis, London, 1993.

Klopatek, J. and Francis, J., 'Spatial pattern analysis techniques', *GIS Solutions in Natural Resource Management*, Morain, S. (ed.), OnWord Press, Santa Fe, NM, 1999, pp. 17–40.

Lowell, K. and Jaton, A., *Spatial Accuracy Assessment: Land Information Uncertainty in Natural Resources*, Ann Arbor Press, Chelsea, MI, 1999.

MacKinnon, J. and De Wulf, R., 'Designing protected areas for giant pandas in China', *Mapping the Diversity of Nature*, Miller, R. (ed.), Chapman and Hall, London, 1994, pp. 127–142.

McMaster, S., 'Examining the impact of varying resolution on environmental models', *Proceedings, GIS/LIS '97*, Cincinnati, OH, October 28–30, 1997.

McMaster, R. B., Leitner, H. and Sheppard, E., 'GIS-based environmental equity and risk assessment: methodological problems and prospects', *Cartography and Geographic Information Systems*, 24(3), 1997, 172–189.

McMaster, S., 'Assessing the impact of data quality on forest management decisions using geographical sensitivity analysis', *Geographic Information Research: Trans-Atlantic Perspectives*, Craglia, M. and Onsrud, H. (eds.), Taylor and Francis, London, 1999, pp. 477–495.

Miller, R., *Mapping the Diversity of Nature*, Chapman and Hall, London, 1994.

Morain, S. (ed.), *GIS Solutions in Natural Resource Management*, OnWord Press, Santa Fe, NM, 1999.

Mowrer, H. and Congalton, R., *Quantifying Spatial Uncertainty in Natural Resources: Theory and Applications for GIS and Remote Sensing*, Ann Arbor Press, Chelsea, MI, 2000.

Mowrer, H., Czaplewski, R. and Hamre, R., 'Spatial accuracy assessment in natural resources assessment and environmental sciences', *Second International Symposium*, USDA, Forest Service, Fort Collins, CO, 1996.

Nedovic-Budic, Z., 'Geographic information science implications for urban and regional planning', *Journal of the Urban and Regional Information Science Association*, 12(2), 2000, 8–91.

Pulido, L., 'Rethinking environmental racism: white privilege and urban development in Southern California', *Annals of the Association of American Geographers*, 90(1), 2000, 12–40.

Radke, J., Cova, T., Sheridan, M. F., Troy, A., Mu, L. and Johnson, R., 'Application challenges for geographic information science: implications for research, education, and policy for emergency preparedness and response', *Journal of the Urban and Regional Information Science Association*, 12(2), 2000, 15–30.

Rushton, G., Elmes, G. and McMaster, R. B., 'Considerations for improving geographic information research in public health', *Journal of the Urban and Regional Information Science Association*, 12(2), 2000, 31–50.

Scott, J., Tear, T. and Davis, F., *Gap Analysis: A Landscape Approach to Biodiversity Planning*, American Society for Photogrammetry and Remote Sensing, Bethesda, MD, 2000.

Waller, L. A. and McMaster, R. B., 'Incorporating indirect standardization in tests for disease clustering in a GIS environment', *Geographical Systems*, 4(4), 1997, 327–342.

Wiggins, L., Deuker, K., Ferreira, J., Merry, C., Peng, Z.-r. and Spear, B., 'Application challenges for geographic information science: implications for research, teaching, and policy for transportation planning and management', *Journal of the Urban and Regional Information Science Association*, 12(2), 2000, 51–60.

Wilson, J. P., Mitasova, H. and Wright, D., 'Water resource applications of geographic information systems', *Journal of the Urban and Regional Information Science Association*, 12(2), 2000, 61–80.

Chapter 34

Local government applications

Zorica Nedović-Budić

34.1. Diffusion of spatial technologies in local governments

Local governments are created to take care of the general health, welfare and prosperity of their citizens and communities. Examples of local government concerns include: regulating physical growth of the community; preventing fire hazards; providing infrastructure for roads, water, and sewer; stimulating economic development; minimizing crime; building schools; and securing publicly accessible recreational opportunities. In doing so, the local governments channel a variety of private interests and goals toward decisions and policies that take into consideration and try to protect public interest.

Computerized information systems have been adopted since the 1950s as a way to improve government performance in serving and protecting the public interest by increasing government's efficiency, effectiveness, and accountability. In the late 1980s, availability of user-friendly and affordable geospatial technologies, including GIS, GPS, and RS, has prompted their intensified adoption by local governments (French and Wiggins 1990; Budic 1993; Masser and Onsrud 1993). The availability of those geographic information technologies has been met with great enthusiasm, given their suitability to handle local government data, over 70 per cent of which can be referenced by location (O'Looney 1997).

Since the late 1980s, local governments have invested heavily in geospatial technologies, GIS in particular (Huffman and Hall 1998; Warnecke *et al.* 1998). By 1997, almost one third (27.1 per cent) of all cities, about one half (43.3 per cent) of all counties, and 77 per cent of large cities and counties in the US used GIS. RS and GPS have not been as prevalent, but they are important sources of spatial data for local government applications. According to the same sources RS technologies and data are used by 2.1 per cent cities and 32.5 per cent counties, and GPS is used by 7.7 per cent cities and 22.9 per cent counties. Digital orthophotography is more common source of remotely sensed local data than satellite images. It is used by 44 per cent of the large cities and counties. Availability of RS images of 1 m and higher resolution may change this practice, and increase the reliance of local governments on satellite data.

34.2. Examples of local government applications

Geospatial technologies (GIS/GPS/RS) are applied to a wide range of local government functions. The scope of those functions varies across more than 80,000 of the local government entities accounted for in the US (over 3,000 counties; over 35,000

Table 34.1 Functions of local government (Huxhold 1991)

Health and safety	*Public works*	*Recreation and culture*
Building codes	Building and structures	Cable television services
Disaster preparedness	Forestry services	Community events
Fire services	Transportation	Historic/cultural preservation
Health services	Utilities	Recreational facilities
Police services	Planning	Education
Safety services	Engineering	
Sanitation	General	
Urban development	*Administration*	*Finance*
Business regulation	Equipment maintenance	Budget administration
Economic development	Information services	Property assessment
Environmental amenities	Records management	Revenue collection
Harbor management		
Housing	*Management*	
Land-use control/mapping	Coordination of services	Districting
Neighborhood preservation	Elections	Land-records management
Planning	Legislative activity	General
Demographic analysis	Long-range planning	
General	Policy development	

municipalities, townships and towns; about 15,000 school districts, and about 30,000 special districts) (US Bureau of the Census 1992). The main local government functions are: health and safety; public works; recreation and culture; urban development; administration; finance; and management (Huxhold 1991) (Table 34.1).

Health and safety applications rely primarily on emergency services – fire and police protection. Those two are the key local government functions, and in many cases, the backbone of local government usage of geospatial technologies. Street network files coupled with accurate addresses and routing software allow for prompt response to emergency calls. With increased incidences of natural disasters (e.g. flooding, earthquakes, and storms), geospatial technologies have also become a prominent tool for disaster mitigation and evacuation planning. Figure 34.1 displays results of a spatial query on the properties that may be affected by 100-year flooding. General community health applications are also common. Figure 34.2, for example, shows the hazardous waste sites in the city of East St. Louis, IL. The information about the sites is derived from the US Environmental Protection Agency (EPA) database developed and made accessible by the 1980 Comprehensive Environmental Response, Compensation, and Liability Act (CERCLA). Further analysis about the exposure of various population groups and the suitability of former industrial sites (so called 'brownfields') for re-use can be performed with these data.

Urban development applications span most of the planning departments' functional areas including neighborhood and economic development, land use and comprehensive planning, and environmental protection. Figure 34.3 shows a land use map draped over a three-dimensional model of earth surface representing the physical characteristics and features in an area of interest. Figure 34.4 shows population data for Champaign-Urbana, in Illinois, US, providing the basis for further socioeconomic analysis.

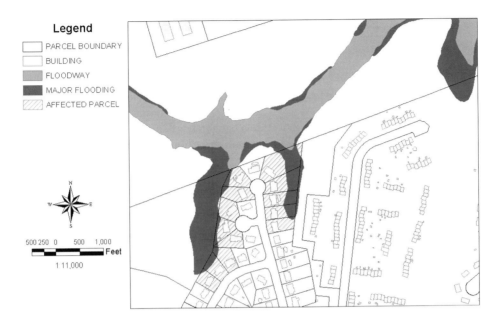

Figure 34.1   Properties affected by flooding (digital data source: ESRI, Inc., PC AUC Info Training Kit).

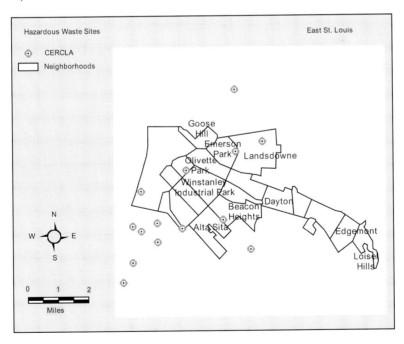

Figure 34.2   Hazardous waste sites (digital data source: University of Illinois @ Urbana-Champaign (UIUC), East St. Louis Action Research Project (ESLARP).

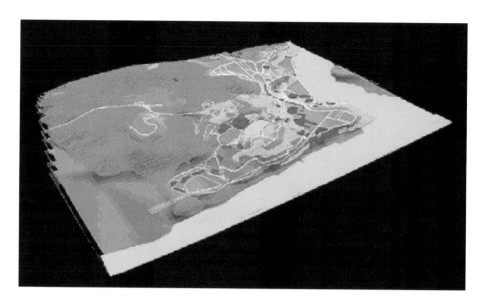

Figure 34.3   Land use and roads draped over topography (Source of digital data: UIUC, UP 419 GIS class project).

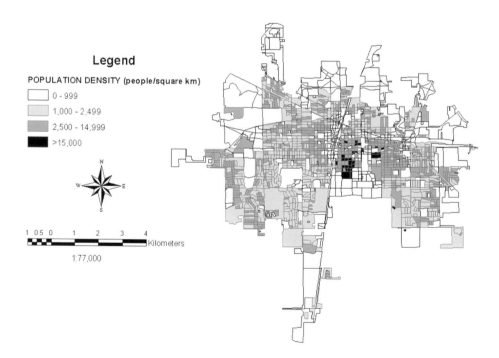

Figure 34.4   Population density by census block group (source of digital data: UIUC, UP318 GIS class exercise).

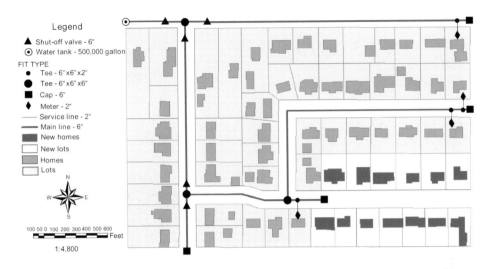

Figure 34.5   Utility mapping (source of digital data: ESRI, Inc., ArcInfo 8.0 sample/exercise dataset).

Public works applications are usually performed by engineering units and often entail datasets with higher degree of positional accuracy then those used by planners. Maintenance of water utility features (Figure 34.5) and display of street network characteristics (Figure 34.6) are two examples of engineering base data.

Administration and *management* applications focus on inventorying and allocating local government resources, providing public access to those resources, performing political functions, and creating long-range plans and policies. Figure 34.7 is an example of using geospatial technologies for creating voting districts.

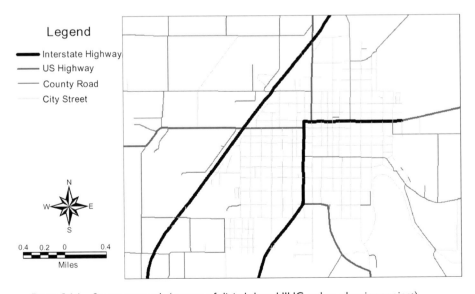

Figure 34.6   Street network (source of digital data: UIUC, urban planning project).

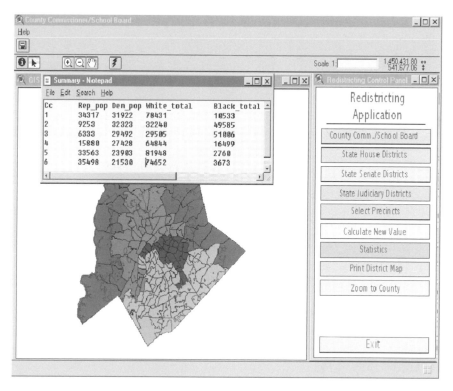

Figure 34.7 Election districts generated by GIS districting function (source: Macklenburg County, North Carolina, GIS Department)

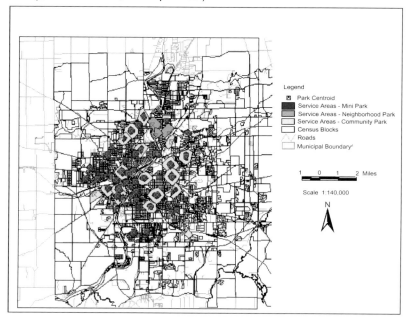

Figure 34.8 Delineation of service areas for mini, neighborhood and community parks (source of digital data: UIUC, GIS-based Illinois Recreational Facilities Inventory Project (G-IRFI)).

Recreation and cultural related applications focus on providing public amenities and facilities in the areas of recreation, education, and culture. Geospatial technologies can be useful in planning and managing those recreational amenities and facilities. For example, street network files can be used to derive service areas of each recreational area in a city (Figure 34.8). Land cover data derived from satellite images are used to understand the vegetation patterns between existing recreational sites and to explore the landscape patterns that may provide for connectivity between the sites. Streams, abandoned railroads, bike trails, and utility corridors are targeted as potential connections (Figure 34.9, see plate 22).

Finally, property assessment is probably the most fundamental local government function in the area of *finance* for which geospatial technologies are becoming extensively applied. Deriving from various legacy land records management systems and computerized mapping software, the geospatial technologies applications in property assessment allow already sophisticated Computerized Assisted Mass Appraisal (CAMA) systems to take advantage of spatial representation and visualization tools (see Figures 34.1 and 34.5).

34.3. Building a local database

As reported in a recent national survey (Warnecke *et al.* 1998) over 40 per cent of the local governments sampled have the following components in their geospatial database:

1 Roads
2 Hydrology
3 Political/administrative boundaries
4 Cadastral/land records
5 Land use/zoning
6 Elevation
7 Digital imagery, and
8 Geodetic control

Geodetic control, parcels, street centerlines, and planimetrics (e.g. hydrography, building outlines) are the most common base maps.

Database development consumes substantial resources, particularly in the initial stage of system development. Database investments often exceed a million dollar figure, and raise exponentially with an increase in accuracy of the map (Antenucci *et al.* 1991). The database costs represent a considerable portion – up to 80 per cent – of the total system development.

Geographic information databases are developed from multiple sources and by a variety of methods, including (Ngan 1998):

1 Digitizing
2 Scanning and automated vectorization
3 Coordinate geometry (COGO)
4 Third party data (Digital Line Graphs (DLG), Topological Integrated Geographic Encoding and Reference (TIGER)/Line)
5 GPS/surveying
6 Photogrammetric surveys

7 RS; and
8 Digital orthophotographs

Whenever possible, it is most efficient to use existing ('third party') data already available in digital form. These data come in a variety of formats including drawing, proprietary GIS formats, and image files. Table 34.2 summarizes the information of various geospatial data formats. For further discussion of data acquisition and development methods see Chapters 17 and 26.

34.4. Managing local government geographic information resources

Local government adoption of geospatial technologies requires various organizational and management activities. The targets of geospatial technology transfer are various local government entities, including organization as a whole and its units, subunits, and individual employees (Budic and Godschalk 1994). Geospatial technology implementation activities may be pursued in a single department, shared by multiple departments, or corporate (enterprise-wide). Huxhold and Levinsohn (1995) differentiate three types of organizational placements: (a) operational – within an existing functional unit; (b) administrative – as a separate functional unit focused on managing information resources, directly responding to top management; and (c) strategic – as a service unit to government top management and administration. Among the operational units, geospatial technologies are most commonly placed in planning, tax assessment, or public works/utilities departments; information systems, data processing or GI/GIS departments – the units specifically created to deal with government information resources – often house and manage the geospatial technologies (Budic 1993; Warnecke *et al.* 1998).

Geospatial technologies and information resources can also be distributed across organizational boundaries to include multiple local governments and non-profit groups, or to involve private sector partners (O'Looney 1997). Interorganizational geographic information activities are stimulated by existing interdependencies between various organizations, but are also challenged by the complexity of the relationships (Nedović-Budić and Pinto 1999a). To illustrate this complexity, Fletcher (1999) proposes four levels of interoperability, global, regional, enterprise, and product; three types of interoperability: institutional, procedural, and technical; and three dimensions of interoperability: horizontal, vertical, and temporal (Figure 34.10).

An important factor for achieving interoperability and multi-participant developments of geospatial technologies is sharing and easy access to geospatial information. Sharing geospatial information is believed to promote more effective use of organizational resources and cooperation among involved organizational entities (Brown *et al.* 1998; Nedović-Budić and Pinto, 1999b, 2000). Obstacles to data sharing are numerous, including both technical and non-technical factors. On the data side, for example, it is very hard to resolve the varying needs for scales and accuracy that local governments located in the same region may have. The scales of geographic data layers used by local governments range from 1:500 to 1:24,000. The accuracy requirements range from subcentimeter to dozens of feet (20 m). Finally, the temporal scale – the frequency of update – can also vary greatly across the local organizations.

Huxhold (1991) argues that effective government information systems integrate data at operational, management, and policy levels, visualized as the government business

Table 34.2 Digital geographic data formats (Lazar 1998)

Format	Description
AutoCAD drawing exchange format (DXF)	A vector format that has become the de facto standard for transfer of data between different CAD systems. Often used to transfer geometry into and out of GIS systems, DXF is not well-suited for transferring attribute data.
ARC/INFO export format (e00)	A vector format intended to transfer data, including attributes, between different ESRI systems. Despite being a proprietary format, there are other GIS systems and third party products that can read or write e00 formats.
ArcView shape file format	Openly published, this vector format is available for use by other GIS vendors. It consists of three types of files: main files (SHP), index files (SHX), and dBASE tables (DBF) for storing attributes.
MapInfo interchange format (MIF/MID)	This format, also formally proprietary, has nonetheless been widely implemented in other GIS systems. MIF files store vector graphical information, while MID files store attribute data.
MicroStation design file format (DGN)	An openly documented (except for some newer and product-specific extensions) vector format used by Bentley's MicroStation CAD software. MicroStation is the platform on which the Modular GIS Environment (MGE) and MicroStation Geographics GIS packages are built. The format does not store attribute data, but can store links to relational database records. MGE and Geographics also have export formats that transfer all files and database tables associated with a project.
DLG format	A vector format used y the US Geological Survey (USGS), and to a lesser extent, other federal and state government agencies. There are several varieties of the DLG format; DLG optional is most common. This format and USGS data products that use it are often called 'DLG-3'; '3' refers to the topological (or 3D) nature of the data. DLG format only supports integer attribute information for spatial objects.
TIGER/line format	The format used by the Census Bureau to distribute vector and attribute data from its TIGER database.
Spatial data transfer standard (SDTS) format	A standard format (used by the USGS, other federal agencies in Australia and South Korea) designed to support all types of vector and raster spatial data, as well as attribute data. The Topological Vector Profile (TVP) and the Raster Profile are implementation of subsets of the SDTS.
Spatial Archive and Interchange (SAIF) format	The national spatial data exchange standard in Canada, supporting vector, raster, and attribute data.
National transfer format (NTF)	The national spatial data exchange standard in the UK, supporting vector, raster, and attribute data.
Vector product format (VPF)	A vector format with attribute data that is part of the NATO DIGEST standard. It is used by the Digital Chart of the World (DCW).

Table 34.2 (continued)

Format	Description
Tagged image file format (RIFF or TIF)	A raster format frequently used for imagery, GeoTIFF is an extension of TIFF that includes georeferencing information.
Joint photographic experts group (JPEG or JPG) format	Another raster format commonly used for imagery.

pyramid – with operations representing the base and policy representing the tip of the pyramid. Policy level information is the most general, used to address organization-wide issues; management information is used for efficient and effective allocation of government resources; and the most detailed operational level information supports the daily delivery of services to the public, e.g. issuance of building permits, pavement repairs, and tax assessment. The key to successful use and management of geospatial technologies and information resources is that database development and maintenance becomes internalized into daily organizational processes (Budic and Godschalk 1994). While outsourcing can be used for initial data development, building an in-house capacity for the use of geospatial technologies in daily organizational functions secures their full incorporation in local government settings. Both Huxhold and Levinsohn (1995) and Brown and Brudney (1998) encourage public information managers to balance contracting with development of in-house expertise.

34.5. Conclusion

This chapter presented the trends in adoption of geospatial technologies in local government settings, a cross-section of local government functions and their corresponding applications, and the major sources and issues in building a local government geospatial information database. The local governments are the most massive user community for geospatial technologies. At the same time, they are the most challenging environments for introducing the new technologies. Their implementation in local government settings is ridden by obstacles of both technical and organiza-

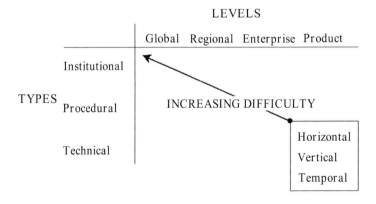

Figure 34.10 Interoperable system dimensions (created by David R. Fletcher for the URISA publication Enterprise GIS, 1999).

tional nature (Croswell 1991; Brown 1996). Finances, staffing, and organizational conflict are the strongest impediments to successful introduction of geospatial technologies. On the technical side, data development costs, data formats, and integration, are the main implementation challenges.

Local governments, however, are also the most important community of current and potential users of geospatial technologies and the investments into their modernization are worthwhile. Most decisions affecting the quality of life and the livability of local settlements are made at the local level, where the new geospatial technologies can make the largest impact.

References

Antenucci, J.C., Croswell, P.L. Kevany, M.J. and Archer, H., *Geographic Information Systems, A Guide to the Technology*, Van Nostrand Reinhold, New York, 1991.

Brown, M. M., 'Geographic information systems and the local government setting: an empirical assessment of the hurdles to success', *State and Local Government Review*, 1996.

Brown, M. M. and Brudney, J. L., 'A smarter, better, faster, and cheaper government? Contacting and geographic information system adoption', *Public Administration Review*, 1998.

Brown, M. M., O'Toole, L. J. Jr. and Brudney, J. L., 'Implementing information technology in government: an empirical assessment of the role of local partnerships,' *Journal of Public Administration Research and Theory*, 8, 1998, 499–525.

Budic, Z., 'GIS use among southeastern local governments,' *Journal of Urban and Regional Information Systems Association*, 5(1), 1993, 4–17.

Budic, Z. D. and Godschalk, D. R., 'Implementation and management effectiveness in adoption of GIS technology in local governments', *Computers, Environment, and Urban Systems*, 18(5), 1994, 285–304.

Croswell, P. L., 'Obstacles to GIS implementation and guidelines to increase the opportunities for success', *Journal of the Urban and Regional Information Systems Association*, 3(1), 1991, 43–56.

Fletcher, D. R., 'The interoperable enterprise', *Enterprise GIS*, Nancy R. VonMeyer and Scott Oppman (eds.) Ch. 2, Urban and Regional Information Systems Association, Park Ridge, IL, 1999, pp. 7–16.

French, S. and Wiggins, L., 'California Planning Agency experiences with automated mapping and geographic information systems', *Environment and Planning B*, 17(4), 1990, 441–450.

Huffman, L. A. and Hall, G., 'Use of innovative technology applications in local government', *Special Data Issue*, No. 1, International City/County Management Association, Washington, DC, 1998.

Huxhold, W. E., *An Introduction to Urban Geographic Information Systems*, Oxford University Press, Oxford, 1991.

Huxhold, W. E. and Levinsohn, A. G., *Managing Geographic Information System Projects*, Oxford University Press, New York, 1995.

Lazar, B., 'External Data Sources and Formats', in Pat Hohl (Ed.), *GIS Data Conversion Strategies, Techniques, and Management*, Chapter 9, OnWord Press, Albany, New York, 1998, pp. 179–204.

Masser, I. and Onsrud, H., *Diffusion and Use of Geographic Information Technologies*, Kluwer Academic Publishers, Dordrecht, 1993.

Nedović-Budić, Z. and Pinto, J., 'Understanding interorganizational gis activities: a conceptual framework', *Journal of Urban and Regional Information Systems Association*, 11(1), 1999a, 53–64.

Nedović-Budić, Z. and Pinto, J., ' Interorganizational GIS: issues and prospects', *Annals of Regional Science*, 33, 1999b, 183–195.

Nedović-Budić, Z. and Pinto, J., 'Information sharing in an interorganizational GIS environment', *Environment and Planning B: Planning and Design*, 27(3), 2000, 455–474.

Ngan, S., 'Data models, collection considerations, and cartographic issues', *GIS Data Conversion: Strategies - Techniques - Management*, Ch. 8, Hohl, P. (ed.), Onword Press, Santa Fe, NM, 1998, pp. 161–178.

O'Looney, J., *Beyond Maps: GIS and Decision making in Local Government*, International City Management Association, Washington, DC, 1997.

US Bureau of the Census, *Statistical Abstract of the United States 1992* (112th ed.), Table 448, US Government Printing Office, Washington, DC, 1992, p. 278.

Warnecke, L., Kollin, C., Beattie, J., Lyday, W. and French, S., *Geographic Information Technology in Cities and Counties: a Nationwide Assessment*, American Forests, Washington, DC, 1998.

Chapter 35

Geographic information technology in state governments of the US

Lisa Warnecke and Drew Decker

35.1. Introduction

Governing roles in the US are uniquely shared by a combination of local, state, and federal governments. Rather than all public sector roles assigned by the federal government as in many unitary forms of government throughout the world, America's governmental system – known as federalism – generally provides that states have sovereign rights, and that states and localities can determine many of their own roles and activities. Over time, states have had an increasingly important role in American government, both in general, and regarding Geographic Information (GI) and GI Technology (GIT). *These terms are used in this chapter because related state policy and institutional attention and direction may be broader than geospatial technology.* This chapter provides an overview of GI/GIT applications, factors, institutionalization, activities, and trends among the nation's 50 states.

35.2. Applications

The nation's 50 state governments generally mirror the federal government in that they each have three similar branches, including the executive branch (governor), legislative branch (a legislature or general assembly, with two bodies, except Nebraska), and the judicial branch. While most GIT use to date has occurred in the executive branch, the other two branches are beginning to apply this technology as well. For example, an important responsibility of state legislatures is to conduct reapportionment for many elected officials at the turn of each decade. Since 1980, states have used increasingly sophisticated GIS tools to conduct this work. By exposing this technology to elected officials that make these often crucial decisions, they can see and sometimes extend the use of this technology to other applications. The judicial branch also is becoming more exposed to GIT as a tool used in testimony to render decisions in legal proceedings, and also to reveal inequities in terms of government decision making. There are no known investigations of GIT use in the legislative and judicial branches of the nation's state governments.

Research has been conducted about the use of GIT, and particularly GI Systems (GIS) and satellite data, in the executive branches of the nation's 50 states for almost 20 years (Warnecke 1987). Each of the 50 state governments used GIS in at least one agency, and a few states had five or more agencies with GIS capabilities in 1990 (Warnecke *et al.* 1992). As indicated in Table 35.1, GIS had been applied to virtually

Table 35.1 GIS use in state government (1995)

No. of states	Function of state government
General state government: administration, planning, finance, revenue, asset management	
13	Revenue, including property taxation
13	Census data center
12	State planning
9	Budget, finance, comptroller, state property management
4	State surveyor, cartographer, geographer
3	Library
1	Banking regulation
Environmental/natural resources	
49	Water – quantity, quality, rights, or drinking
42	Wildlife, game, fish or biological resources
39	Geological survey
30	Waste management, including solid, low level
29	Air quality
27	Forestry
27	Agriculture
24	Oil/gas/mining regulation and reclamation
22	Public lands management
22	Parks management
20	National heritage program
18	Coastal resources
12	Energy
Cultural resources	
19	Historic preservation
14	Archaeology
1	Other – museum
Infrastructure	
50	Transportation
9	Utility regulatory commissions
Human services	
25	Health (primarily epidemiology)
6	Social services
5	Employment security and labor
3	Education
Other	
24	Public safety, emergency management and military
20	Economic development
20	Community and local affairs

every function of governance in states by 1995 (Warnecke 1995). Society and government at all levels is experiencing an explosion of GIT usage in a growing variety of applications. As described below, GIS use is now widespread among the states in early application areas, such as natural resources and transportation. State government applications also are expanding in human and social services, public health, public safety, criminal justice, emergency management, economic development, and growth management. However, the rate of GIS adoption seems to vary by function. While GIS represents most GIT use, more recent states investigation identified satellite data use in one or more agencies in 42 states (Warnecke 1997).

States initially used GIT independently – and often experimentally – to automate mapping processes or conduct rudimentary analyses. More recently, GIT has been applied to all levels of activity within state organizations, from policy analysis and development to planning, management, operations, regulation, adjudication, licensing, leasing, and other applications. GIS also supports business and financial functions in government, including analysis and optimization of revenues and expenditures, and particularly in collections and resource allocations. More and more state agencies apply GIT as an essential tool to conduct several business processes and deliver public services in a more efficient, effective, and equitable manner. GIT is becoming an important component of government Information Technology (IT) modernization and 'reengineering' initiatives within and among individual agencies. While GIT benefits are typically measured in terms of individual agencies and governing functions, the trend is clearly toward multiple agencies within a state government now using GIT within an enterprise wide approach. As of 1999, some larger states, such as New York and California, have over 20 separate agencies using some form of GIT. As stated by Wendy Rayner, New Jersey's Chief Information Officer, 'every agency, if they know it or not, needs GIS capability to make decisions.' A review of some key state government applications and experience is provided below.

35.2.1. Planning, growth management and economic development

The planning function has long used GI/GIT because it inherently requires the integration and analysis of many, often disparate types and formats of data. However, political support for planning is often cyclical, and can directly impact the resources available, support for and use of GIT. Some of the earliest state GIS applications can be traced back over 30 years by state planning agencies to assist in land use planning, often stimulated by federal funds. For example, federal support helped develop the Maryland Automated GI (MAGI) system, one of the first state GIS initiatives in the country, that was used in statewide land use planning, and specifically, to prepare the state comprehensive land use and development plan. Mississippi's State Data Base (now the Mississippi Automated Resource Information System (MARIS)) used similar funding and became one of the earliest users of satellite imagery and one of the first state GIS programs to be authorized by statute (Warnecke in Foresman 1998). Early state planning statutes and initiatives in Connecticut, Minnesota, New York, North Carolina and Ohio also stimulated land use data systems, and became leading drivers for GIS.

Several land planning applications have emerged over time. Minnesota, Ohio and South Dakota conducted land suitability and capability analyses for specific regions or counties using early geoprocessing tools. States have identified 'critical' areas warranting protection using GIS since the 1970s. Farmland preservation and protection of shore lands, rivers, or wetlands were early applications, such as by New York's Adirondack Park Agency and the California Coastal Commission. Recent attention has been given to protecting drinking water supply and wildlife habitat, such as through the Gap Analysis Program (now managed by the US Geological Survey's (USGS) Biological Resources Division), which was initiated in 1988 as a nationwide, state-by-state project to inventory, evaluate and preserve biological diversity. In addition, The Nature Conservancy's Natural Heritage Program, initiated in 1974, provided an early focal point in each state for the management of biological data inventories that can be used with GIS. As a result of more public referenda to use state funds to acquire land for public use, states increasingly use GIS to prioritize areas for purchase. For example, voters in Florida recently approved $3.2 billion for land acquisition, with GIS used to inventory potential sites, set criteria for site selection, and monitor acquisition efforts.

Among the strongest state GI/GIT legislative directives to date are state planning or growth management laws. At least 25 per cent of the states adopted planning laws that have legitimized and increased such usage within state governments, and also often in local governments, such as Georgia, Maine, New Jersey, Vermont, Rhode Island and Washington. State funding and assistance were sometimes provided to develop or assist in preparing local plans in accordance with state goals, often including GIS and related data development. For example, Vermont's Regional Planning Councils were equipped with GIS hardware and software to help localities, and Georgia's Regional Development Centers were given significant data roles for GIS. Other states have also provided funding for GIS use for local and regional planning, including New Hampshire and Utah.

Some states also use GI/GIT for economic development, which is a slightly different purpose, but generally requires similar data as planning applications. Some states receive significant political support for this application; such as to attract businesses and optimize siting decisions. For example, Illinois, North Carolina and others used GIS to develop proposals for the superconducting super collider project in the late 1980s. South Carolina is long recognized as a leading GIS user for economic development. The state developed natural resources, transportation, utilities, and socio-economic data with GIS specifically to attract businesses, with successes such as the Bavarian Motor Works (BMW) manufacturing plant. As shown in Table 35.1, 20 states indicated some use of GIS for economic development in 1994 (Warnecke 1995).

Fewer states indicated use for planning (12) than economic development (20) in 1994, due in part to a political climate lacking support for government planning. However, it is postulated that a similar number or more states now use GIS for planning due to growing concern about urban sprawl and traffic congestion, and the resulting resurgence of support for land use planning at a policy level. Several states have adopted 'smart growth' initiatives lately, with expectations for increased GIS use accordingly. In addition, the US Department of Housing and Urban Development recently provided GIS software and support for use by its 50 state counterparts and grantees. This is an important trend for the future because almost by definition, planning encourages coordination and integration of often disparate and sometimes conflicting data to analyze conditions, develop scenarios, and plan for the future. This can, in turn, encourage multiple agencies to develop compatible data. This attention to data can further stimulate coordination and commonality at a 'horizontal' level among agencies within individual governments, and 'vertical' coordination with others. Moreover, policy support for coordinated data activities is growing within states across the country, and GIS is increasingly viewed as a way to manage, present and use data from several sources.

35.2.2. Environment and natural resources

Authority, roles and responsibilities for Environmental and Natural Resources (ENR) are traditionally shared by federal, state and local governments, with each level having a leading role for specific ENR functions. States have numerous ENR responsibilities, some of which were among the earliest and currently are some of the most developed applications in state government today. Early uses were initiated to better manage various individual natural resources for which states have stronger official responsibility than the federal government, such as forestry, water and wildlife. As reflected in Table 35.1, the most recent inventory of state GIS uses in 1994 identified applications in virtually all ENR functions, including: water resources (49), wildlife (42), geology (39), waste management (30), air quality (29) and forestry and agriculture (27 each), with additional uses such as energy, public lands and parks management, and others (Warnecke 1995).

Water resources management was one of the earliest and is one of the strongest uses of GIS in states. For example, the Texas Water Oriented Data Bank that was authorized by the legislature in the 1960s to coordinate response to droughts and floods was responsible for stimulating some of the earliest uses of GIS among the states. Water usage and analysis was also one of the first GIS uses in other states, such as Florida, Idaho, Iowa, Massachusetts, Utah and Wyoming. For example, Idaho's Department of Water Resources (DWR) began to use satellite imagery and geoprocessing to analyze and monitor irrigated agriculture as early as the mid 1970s, later extending to other applications. Other states similarly used GIT for water quantity, a critical function in western states, including appropriation and use management. However, water quantity also has become an important focus, both for ground and surface water.

The second most frequent ENR application in 1994 was wildlife management, another strong state function as compared to the federal government. Several states initiated and continue to have strong GIS activities in this regard, such as Colorado and Tennessee, with more recent activities stimulated in part through the Gap Analysis Program.

State Forestry Organizations (SFOs) were among the earliest GIS users to help accomplish their missions, including to plan for and manage state owned forests, conduct statewide forest inventories, mitigate and control fires, manage pests, and provide technical assistance for and conduct oversight over private forested lands. One of the largest SFOs, Washington's Department of Natural Resources (DNR) uses GIS to help manage over three million acres of land held in trust to support public institutions including schools. DNR has one of the largest GIS installations of any state agency in the country. Other states with strong forestry GIS usage include Alaska, California, Maine, and Minnesota. GIS also has been used extensively to manage non-forested public lands and other natural resources. Agricultural GIS applications also are expanding in states, including evaluation of various conditions, crop inventories and yield forecasting, and pest infestation and crop disease monitoring.

Environmental protection is a growing application in states, due to both state initiative and federal direction and funding. Both state and federal environmental legislation has stimulated the development of GIS and data for many applications over the years. For example, Illinois and Kentucky used federal funding for some of their earliest GIS activities to determine lands unsuitable for coal surface mining. Ohio's Environmental Protection Agency (EPA) was the first state environmental agency to start using GIS in 1975, with both state and US EPA (USEPA) funding. USEPA and state support for GIS has grown across the country during the last two decades. GIS use has grown for individual media such as water, air and waste management, but also to integrate use of data for 'multi-media' decision support and management. Recent environmental GIS applications help integrate and streamline processes, such as 'one-stop permitting,' and public access to data about environmental conditions, increasingly available through the Internet.

35.2.3. Infrastructure: transportation and utilities

State governments have several important infrastructure responsibilities, including management of highway, transit and other transportation facilities and services, and regulation of utilities. Most early transportation and utility users initiated use of Computer Aided Drafting or design (CAD) or mapping (CAM) systems to automate preparation of facility or highway construction drawings and maps. Among utilities, they became known as Automated Mapping/Facilities Management (AM/FM) systems.

State Departments of Transportation/highways (DOTs) were among the earliest, and continue to be some of the largest state government users of GIT, initially to automate various large-scale manual drafting and mapping processes (CAD/CAM), and more recently, to apply analytical capabilities for analysis of conditions and to improve planning and management processes. DOTs initiated CAD/CAM systems beginning in the 1960s, as in New York and Pennsylvania, with most penetration in other states during the 1980s. A 1984 survey found that 19 state DOTs had Intergraph systems and four had other systems (Arizona DOT 1984), while virtually all state DOTs had some type of CAD/CAM activity by 1990. Some mapping installations in DOTs have been the largest technology configurations of any GIS-related systems in state governments, such as in Wyoming where the DOT had over 60 workstations in 1991.

Highway safety and pavement management were early areas benefiting from analytical GIS capabilities, though these programs often were developed and sometimes continue to be organizationally separate from CAD/CAM operations within DOTs. Early examples include Arizona's accident location identification and surveillance system, which was developed in 1970 and was the first statewide digital spatial database in Arizona, and Ohio's state accident identification and recording system. It retrieved and displayed accident-related information for over 112,000 miles of roads in the state, and GIS capabilities were later expanded for use in road inventory, pavement and bridge management, skid resistance and other applications. Wisconsin's DOT was perhaps the first in the US to apply GIS as part of a comprehensive information management approach to fulfill operational, management and planning needs, and several other DOTs have adopted similar approaches.

Interagency GI/GIT coordination has grown in state governments over time, but DOTs often operate separately from other agencies. However, DOTs often work with counterparts in other states to address similar needs, with more attention to common GIT issues across states. For example, various interstate DOT committees address related issues, the Transportation Research Board has sponsored studies about GIS in the 50 DOTs, and a national GIS in Transportation (GIS-T) conference has been held annually since 1987. Federal transportation direction, funding and technical assistance has stimulated GIT use in DOTs and other transportation organizations, including the nation's over 300 Metropolitan Planning Organizations (MPOs). In fact, 1998 legislation included specific language concerning remote sensing. Today, it is generally known that all state DOTs and almost all MPOs use GIS in addition to CAD/CAM.

While states do not traditionally operate utilities, it is important to recognize that they have regulatory authority over those operating within their borders. Some Public Utilities Commissions (PUCs) use GIS for both internal and regulatory purposes, such as to monitor service areas and analyze demographics. GIS has recently been applied to help manage competition among utilities to ensure the public is adequately served. Early innovators included North Dakota and Ohio, with other states emerging as active users including Kansas, Mississippi and New Hampshire. Ten PUCs were identified as GIS users in 1990, however, four of them ceased efforts while another five began using GIS in the early 1990s (Warnecke 1990; Demers *et al.* 1995). A growing trend is for PUCs or other related state agencies to use GIS to develop and manage addressing data for statewide emergency (E911) communications systems or to provide data or technical assistance for localities in this regard, such as Maine, Oregon and Vermont.

35.2.4. Other Applications

Planning, ENR and transportation were the early and continue to be the dominant uses of GIS in states. Other applications emerged in innovative agencies during the 1980s, but did not penetrate most states until recently. Growth in the quality and availability of data and technology has stimulated additional state government functions and agencies to apply GIS. As shown in Table 35.1, several states now apply GIS for public safety and emergency management; health, human social services; cultural resources, and other uses.

While first initiated as long ago as some of the other application areas described above, public safety and emergency management applications grew most significantly during the 1990s. This application is an important driver for GIS and integration of data from other sources similar to planning, but with the additional need for real-time use. Disaster response has been an important impetus for GIS activities and funding in some states. For example, while GIS efforts were previously underway in their states, agencies expanded their GIS efforts in response to major disasters including the Exxon Valdez oil spill, hurricanes Andrew and Hugo, the Midwest floods, and the west coast fires. These disasters, particularly Alaska's oil spill, caused other states to expand GIS efforts to better prepare for emergencies, such as Maine, New Jersey and Rhode Island. GIS use has also expanded to other types of emergencies, such as flooding and floodplain management (34 states), hazardous materials/radiological incidents (31), earthquakes (12), fires (14), and hurricanes (9) (Warnecke 1996). Other public safety functions also use GIS, such as the California Department of Justice, which uses GIS for narcotics tracking. In addition, as mentioned above, modernized communications systems with (E911) capability increasingly use GIS to optimize response.

The growing severity and cost of disasters during the 1990s resulted in an increased need for and use of better technological capabilities to respond and recover from emergencies, but also to better prepare for and mitigate their impact. State Emergency Management Agencies (EMAs) and others are broadening their GIT applications as a result. Similar to the DOTs, many of these activities have been conducted in coordination with the Federal EMA (FEMA), which has provided GIS software and other assistance to each state EMA.

While adopting GIS later than other parts of state governments, various state human and social services, labor and education agencies started using GIS in the 1980s. Table 36.1 shows that most state GIS use has been in epidemiology, largely because ENR data can be analyzed with data about the occurrence of diseases to understand environmental implications on people. Additional applications are emerging as well. For example, New York's Department of Social Services pioneered social services GIS applications to support the state's neighborhood-based strategy for services delivered to children. South Carolina has used GIS to target services for children based on the locations of poverty, crime, teenage pregnancy, infant morbidity and other conditions, while Illinois and Washington use demographic and migration data with GIS to help locate health care and other facilities and services. Some state labor developments also use GIS, such as Alaska, Arizona and Washington. Labor statistics and trends are increasingly analyzed geographically to target and locate services, such as for job search and placement for welfare recipients, as well as input for transportation and land use planning. Some state education departments have used GIS for similar program planning and management applications, particularly to ensure the equitability of resource allocations. GIS applications are expected to grow dramatically in the twenty-first century, particularly in public safety, and human and social services, because these are among the most costly functions of state governments, and because continually improving data and technology can be deployed for these applications in a very cost-effective manner.

35.3. Factors impacting GIT in state governments

Several internal and external factors have influenced the development and use of GIT in states over time. Public and governmental needs and actions increasingly seem to drive GIT and related data development. Practitioners and researchers alike recognize that GIT support, use and utility depend on political, institutional, financial and human factors in addition to technical capabilities. Depending on conditions within individual governments, these factors may be obstacles, facilitate efforts, or in some cases, result in a combination of both influences.

Societal and government trends affect GIT in several ways. Early experimentation and development was facilitated by increasing public consciousness, societal concern and particularly, federal government activism, though such action is cyclical and varies in direction and emphasis over time. Strong federal government involvement in domestic matters during the 1960s led to legislation and programs providing incentives, assistance, and funding for state governments, such as land use planning discussed above. Governmental activism also was reflected in the 'environmental movement' as numerous significant and long-lasting federal and state environmental laws and institutions were established. While funding support for planning has diminished, many state ENR GIS applications continue to be aided by federal ENR agencies in the Departments of Agriculture, Commerce and Interior, and the USEPA as discussed above. Federal government activism also is exhibited in direct transfer of technology, data and assistance to states. The National Aeronautics and Space Administration (NASA) funded investigation and use of satellite imagery and image processing in states during the late 1970s and early 1980s. The USGS and the Federal Geographic Data Committee (FGDC), (see Chapter 36) provided aid to states and others during the 1990s. NASA readdressed state needs in the mid 1990s (Warnecke 1995), and is establishing a new state and local government program. Four federal agencies funded a nationwide study of GI/GIT whilch encouraged greater federal attention to state and local governments (National Academy of Public Aministration 1998). Federal activism continues to aid state GIT efforts in various respects.

State government direction and actions also stimulated GIT development and use. States and localities responded well to federal funding, but during the 1980s began to resist federal preemption's, regulations and 'unfunded mandates' that accompanied some legislation. At the same time, many new federal roles were transferred or 'devolved' to states, which helped stimulate states to improve their institutional capacity and technical capabilities. States have become increasingly active in many areas addressed by the federal government, as well as additional areas such as economic development. State initiative in these areas created opportunities for new GIT applications as described above. Improved state capacity and capabilities also spread to local governments, with states funding new programs to improve local financial and personnel systems among others which in turn helped increase local government use of GIT (Warnecke *et al.* 1998). At the same time, as governing issues continue to cross local government boundaries, state governments are increasingly required to address local issues, such as land use planning. These trends and state activism create fertile ground for new GIT applications at the state level as described above.

GIT conditions also are influenced by political, institutional and financial factors within individual governments. Moyer and Niemann concluded that these factors are 'one of the least understood, least discussed and most important aspects of GIT (Moyer and Niemann, 1994). There are inherent difficulties in modernizing government and adopting innovation, such as the diffusion of GIT and other integrating information systems. Leadership support, policy direction, new institutional arrangements and stable funding are typically required to develop enterprise approaches and to maximize benefits across multiple agencies. However, implementation of internal organizational, personnel and financial changes are often resisted even when leadership direction and support exists. These factors increasingly impact GIT over time. Though limited investigation has been conducted about these factors, it is clear that several agency level and statewide GI/GIT initiatives have been abandoned over the years due to these factors.

Funding is a particularly crucial factor in states, since limited resources can severely stymie GIT development, use and coordination. Justifying GIT expenditures is difficult because anticipated benefits are often hard to document and long term in nature, and not necessarily quantifiable in economic terms because savings are often created by reducing duplication of efforts and improving cooperation between agencies and governments. This problem has diminished to some extent in recent years, but remains in many government organizations. But even when accepted, if GIT is not funded in the long term as a regular component of an agency's business processes and budget, then continual efforts are required to justify work and expenditures, thus detracting from actually applying the technology to an agency's needs and effectively coordinating plans, efforts and data with others.

While internal governmental factors, and particularly bureaucratic and financial conditions, can stymie state adoption and hinder full realization of GIT potential and benefits, it is well recognized that technological developments stimulate such usage. Better and cheaper GIT tools, and more available and useful data, are strong drivers for technology adoption. However, government data are highly influenced by the internal institutional conditions discussed above – resulting in most databases being highly fragmented. Data are usually contained within individual programs and agencies, and not managed or financed as a strategic enterprise wide resource. Combining and integrating data from differing sources is challenging at best due to data incompatibilities and other technical factors, but also because data coordination usually is not a high priority of political or administrative leaders. As a result, there is limited support and funding to address these issues or to develop standards, data coordinating centers, or other mechanisms to provide access and share data resources.

35.4. Institutionalization of statewide GI/GIT approaches

Despite often pervasive institutional and data challenges, states began to coordinate GIT activities in the 1980s (Warnecke 1987). Typically initiated from the 'ground up' rather than the 'top down,' entrepreneurial GIS users can make a big difference. Many stories reveal how commendable GIS advocates broke barriers between agencies to coordinate efforts and share data, often without the knowledge or support of their superiors. Previous efforts coordinated hard copy mapping through State Mapping Advisory Committees (SMAC), but GIS use often emerged in different agencies. Concern with duplicative GIS activities led to broader attention to GI in the 1990s. In general, maturation of coordination approaches in states has meant an evolution in the conceptualization, use and range of technologies, data, applications and participants. As a result, state government authorizing direction, interorganizational groups, and coordinating staffs are beginning to address GI/GIT in a holistic manner. While most attention is on GIS, focus is growing to coordinate remote sensing and Global Positioning System (GPS) activities, as well as GI residing in tabular databases. Through time, previous SMAC and other related interagency efforts are being subsumed conceptually and organizationally within broader GI/GIT approaches. Enterprise wide GI/GIT approaches and institutionalization is evidenced in several ways in state governments, and provides evidence of continuing maturation of statewide GI/GIT approaches.

Authorizing directives are increasingly adopted in the form of statutes and executive orders which directly or indirectly promote GI/GIT development and coordination. While thorough analysis has not been conducted in this regard, an inventory of state GI/GIS directives conducted by this author identified 100 state directives among the 50 states, with 49 of these directives authorized from 1991 to 1993 (National Research Council 1994). The most common purpose of these directives is to sanction GI/GIT coordination, and often to authorize GI/GIS coordination groups, or studies. Some directives authorize related offices, databases or funding for statewide or narrower missions, usually such as natural resources management, environmental protection or growth management. Some directives specifically address GI or 'GIS data' in a modification of open record laws to provide cost recovery, while others authorize GI/GIT assistance to local governments. While more GI/GIT directives are adopted over time,

Table 35.2 Incidence and authorization of state GI/GIS coordinators

Authorized			Unauthorized		
Year	Number	% of total authorized	Number	% of total unauthorized	Total coordinators
1985	10	59	7	41	17
1988	15	52	14	48	29
1991	30	75	10	25	40
1994	31	77.5	9	22.5	40
1995	33	80.5	8	19.5	41

few are comprehensive, such as to fund a coordination program, or establish oversight to ensure multiagency data commonality or compliance with direction.

States began establishing broadly-focused interagency GI/GIT coordination groups in the mid 1980s, now generally known as 'GI Councils' (GICs). At least one such group existed in each of the 50 states in 1990 (Warnecke *et al.* 1992), though almost 90 independent GI/GIT groups were identified among the 50 states in 1993 (National Research Council 1994). The degree to which a group operates officially, and its level of activity varies considerably by state and over time. Some states having very cyclical GI/GIT coordination histories, including the establishment of new groups after others have ceased operations. However, state GI/GIT groups have generally experienced an increase in authorization, membership participation, strength, resources, and direct or implied responsibility and influence over the direction of GI/GIT among their agencies.

Today, it is estimated that over 40 states have at least one group with some degree of official stature. Some states have a policy level group advised by one or more sectorial or technical groups, sometimes focused on GIS, GPS, base mapping, standards or other issues. Group participants can include representatives of virtually all state government functions, sometimes with legislative participation. Members increasingly represent additional sectors, particularly localities, but also federal agencies, regional organizations, academic institutions, Indian tribal governments, utilities, and others. Broadening participation typically expands focus from interagency coordination to also address the needs of other GI/GIT users operating within a state. Moreover, increasing involvement by multiple sectors often facilitates many forms of GI/GIT partnerships.

Increasing incidence of statewide GI/GIT coordination entities is a third example of state government institutionalization. These entities serve as a focal point within the state bureaucracy, and can range from less than the full time effort of one individual to an office with over 30 staff. They typically complement, chair, and/or staff GI/GIT groups discussed above, and can have several GI/GIT development and coordination responsibilities. A few states have designated some official related roles; such as cartographers, geographers, and surveyors, but a larger number have established broader GI/GIT coordination entities in the 1980s. As indicated in Table 35.2, the greatest increase in GI/GIT coordinators occurred in the late 1980s, from 17 in 1985 to 40 in 1991. The number of authorized coordinators increased at a greater rate than coordinators in general. GI/GIT coordination responsibilities were shared by two organizations in nine states. While many early coordinators were located in agencies with ENR responsibilities, the trend is toward locating them in agencies with central or government wide roles, and most often in information or IT entities.

35.5. State GI/GIT coordination roles, responsibilities and activities

Most statewide GI/GIT coordination efforts are conducted by a combination of groups, coordination entities, and others' work. They have varying roles, and conduct several different activities according to the nuances of individual states, though facilitating communication and coordination among GIS users is an initial and continuing goal in virtually all states. One of the most significant differences is that some statewide GI/GIT entities primarily coordinate activities while GIT work is housed in individual agencies. In other states, a statewide entity may provide GIT services for agency or external clients, acting as a 'service bureau.' With either approach, more and more groups and entities serve in statewide policy and planning roles, with or without official direction or oversight in this regard. For example, mature groups and staffs may lead, develop and adopt plans, policies, procedures, guidelines and standards; prioritize and implement statewide data layers; monitor agency GI/GIT activities; and/ or provide access to statewide data resources. Information about GI/GIT coordination activities are increasingly available on web sites, but also in annual reports, directories, newsletters, and other media. State GI/GIT groups and/or entities can have one or more of the following roles and activities:

- Serve as a clearinghouse concerning activities, projects and plans about GI/GIT in state agencies and possibly other entities, including provision of directories, guides, annual reports, newsletters and other materials with regularly-updated information.
- Provide data clearinghouse, access and dissemination functions for data indexed and possibly maintained in a state GI/GIS database, and perhaps provide customized data searches, manipulation and interpretation to meet user needs.
- Develop and implement data and metadata policies, guidelines, standards and procedures to encourage data commonality and sharing, including accuracy and scale requirements to meet overall state needs.
- Promote collaborative planning for future data development and other work, including helping prioritize, coordinate and gather resources to develop and maintain data that is conducted by multiple organizations.
- Synthesize input from various entities to prioritize common data and other needs, gather resources to accomplish these needs, and carryout data development and/or acquisition plans.
- Develop data, sometimes with general appropriation or collaborative interagency funding to ensure data is useful for more than one purpose, project or agency.
- Provide contract GIT services for state agencies and others.
- Staff GI/GIT coordination and user groups.
- Hold GI/GIT conferences and meetings to facilitate information exchange.
- Provide GIT educational services for state agencies and others.

In many respects, state coordination groups and entities serve as 'cheerleaders' for GI/ GIT across several sectors. For example, most of the states have been active supporters of the annual 'GIS Day' first held on November 19, 1999, which was sponsored by National Geographic Society, the Association of American Geographers, and Environmental Systems Research Institute. GIS Day was initiated to increase awareness about how GIS has made substantial contributions to society, with particular focus on educating children. Through the efforts of state GI/GIT coordinators, about 30 Governors signed proclamations officially designating GIS Day in 1999. Many open houses, demonstrations and other activities have been held on these days in various places around the country.

35.6. Statewide data initiatives

Effective use of GIT increasingly requires accurate and quality data. Statewide data are not only a very valuable resource for state agencies, but also for substate entities, such as local governments and regional organizations, federal agencies, private companies and other national organizations. State GI/GIT direction, roles and approaches vary considerably by state, particularly concerning data initiatives.

Most states are, to some degree, developing clearinghouses and standards to provide access to data or metadata about data that can be used with GIS. To date, about 30 of the states have clearinghouses linked to the FGDC and the National Spatial Data Infrastructure (NSDI). State approaches differ in that some focus on providing access to metadata, while others also provide access to actual data resources in an integrated format. For example, California, Florida, and Virginia are developing distributed clearinghouses with metadata available from a central location, and state agencies are responsible for managing and maintaining their own data holdings. Alternatively, states such as Maryland, North Carolina, and Utah provide direct access to data often developed by several agencies and projects.

While most state coordination efforts strive to provide access to data or metadata, states differ significantly concerning their data roles, specifically whether or not, and the degree to which they act at a statewide, multiagency level to determine, manage, fund, develop, implement, maintain, and/or distribute statewide data. Most data roles in the past, and in some states today, are conducted at a programmatic level to meet individual functional or agency needs, such for one of the many applications discussed above. However, state GI/GIT coordination groups and entities increasingly recognize that independent agency data activities often result in duplicative, incompatible, costly, and sometimes conflicting data. States vary considerably in the direction, approach, long term commitment, and programs developed to address increasing demand for multi-purpose, statewide data resources for use with GIS. Some options include:

State as Owner: Certain datasets are needed by many agencies and are too complicated and costly for one agency to manage. In these cases, a state entity with central GI/GIT roles may officially manage the production and maintenance of a certain dataset, and 'owns' and disseminates the data. Data requiring large capital investments and strict standards are candidates for state ownership, such as digital orthophotos and land cover (see framework discussion below).

State as Coordinator: A common role is to coordinate and help manage data that are primarily the responsibility of others. A central GI/GIT entity may initially create a dataset, and then oversee future data development and maintenance by one or more other entities. For example, the state may develop a seamless transportation dataset that is maintained by local governments and/or regional organizations using state standards and guidelines. In this case, the state collects updates at predetermined intervals and integrates them at the state level. The state may financially support various aspects of this data role, including distribution.

State as Catalyst: State government can encourage and/or support one-time data development, and provide standards and guidance for future work, but have a diminished role regarding maintenance and updates. This is applicable for specialized data that are not a leading statewide, multiagency data priority.

States with GI/GIT entities conducting data development and management roles differ in terms of their choice of datasets, scales, and currency. For example, states with small areas but relatively large populations may develop data at the 1:5,000 scale, while larger, rural states may focus efforts at the 1:24,000 scale. Most data efforts typically focus on 'base mapping' or 'foundational' or 'framework' data that can be used with GIS for several purposes. FGDC worked with representatives of several sectors during the late 1990s to determine that certain data sets are most often needed for GIS applications and to derive additional data (http://www.fgdc.gov/framework/overview.html). Seven discrete data themes were identified as 'framework' data, including geodetic control, digital orthoimagery, elevation, highway center lines, hydrography, governmental units, and cadastral data. States have generally adopted this approach, but with unique variations such as adding or further specifying categories of data according to individual state priorities.

All seven data themes are useful, but digital orthoimagery has been a recent and significant data resource for use with GIS. Digital Orthophoto Quadrangles (DOQs) are generally scanned and correct aerial photographs that combine the detail of a photograph with the spatial properties of a map. DOQs typically have spatial resolutions of 1 m (1:12,000 or better) and are produced from either black-and-white or color infrared photos. With the rapid adoption of DOQs as base imagery by states, DOQs are assuming roles similar to those of the original 1:24,000 scale, 7.5-minute quadrangle maps produced by the USGS. DOQs are being produced nationwide under the auspices of the National Digital Orthophoto Program (NDOP) and other independent efforts. Several federal agencies participate in NDOP, which is coordinated by a federal interagency body that assigns priorities, obtains funding from contributing agencies, and organizes input data for DOQ development. NDOP endeavors to complete DOQ production in the US and continue to support the products through maintenance and data updates, including adaptation to new data sources, products and production methodologies as they become available.

While NDOP provides standardized data across states, state progress, resources, and approaches at developing other data resources vary considerably, depending on both statewide and agency-specific efforts. A state's overall data resources for use with GIS depend on applicable statewide data initiatives, but also data developed by functional agencies to meet their individual missions, such as natural resources or highway management. However, an increasing number of states have received funding directly from their legislatures for statewide data development efforts, such as Florida, Kansas, Texas and Utah. One of the most 'data rich' states is Minnesota, particularly because its legislature has specifically committed up to $5 million per year for development of natural resources data for use with GIS through a dedicated fund.

Some examples of statewide data development programs are described below. These examples illustrate the variety of tools states employ to better distribute their data. While the role of states in GIS data development and services is rather straightforward, the degree of creativity applied to these services is very high. States have shown they have the ingenuity to match their data resources with the needs of their citizens. The examples cited above are not meant to illuminate the 'best' programs or describe what should be done with spatial data. They are intended to show what solutions states apply in developing and distributing spatial data.

Florida: Maximizing data availability and access. Emphasis in this high growth state is on making access to data as organized and easy as possible, with strong use of standards and metadata. GIT is distributed among several agencies represented and a wide range of geographically-referenced data are available through the Florida Data Directory (FDD). Florida has had a statewide approach to GI/GIT since 1985 when the Growth Management Data Network Coordinating Council was formed to help meet the state's growth management needs. It evolved to become the Florida GI Board (GIB), authorized by statute in 1996 (http://als.dms.state.fl.us/). Like most states, GIB is aided by technical groups. However, Florida is one of the few states also having a GI/GIT group comprised of federal agency representatives; the Federal Intrastate Interagency Coordinating Committee. This group helps strengthen GI/GIT coordination between Florida and the federal government. Florida's approach to data access is also unique in that FDD more closely follows the search tools available in library referencing schemes. Primarily metadata driven, entities contributing data provide metadata to the metadata collector portion of FDD where it is cataloged and verified. The automated library system cataloging server then provides access to contributed data through direct data downloads or links. Users search for data via catalog indices or through text queries. Searching the category list, for example, the user selects the theme of interest (i.e. forestry) and the server provides information derived from the metadata. Detailed summary metadata is provided, including data description, citation, dates, format information, and a graphic showing the regions covered. FDD is an excellent example of proper metadata application and ease of data access. Planned FDD enhancements include an Internet-based interactive map interface through which users can define regions and select data. In addition to FDD, the University of Florida's Department of Urban and Regional Planning operates another type of data distribution hub, which is based on spatial data packaging. The Geo-Facilities Planning and Information Research Center (GeoPlan) manages and distributes the Florida Geographic Data Library (FGDL) (http://www.geoplan.ufl.edu/index.html). FGDL distributes satellite imagery, aerial photographs, tax data, and other data for use with GIS. Data are organized by county and other political boundaries, distributed via CD-ROM, and are updated frequently.

Texas: Building foundational data. As one of the nation's largest states with 254 counties – the only one with over 160 counties – Texas has many GI/GIT challenges. One of the earliest issues identified by statewide GI/GIT coordinators was the lack of statewide data resources. As a result, Texas's coordinating group (now named the Texas GI Council (TGIC)), organized a multi-year, seven layer, cost-sharing effort to develop much of the framework data at 1:24,000 scale in one large program; the Strategic Mapping Program (StratMap – http://www.stratmap.org). The Legislature provided funding for the 4 year program to develop DOQs, elevation contours, hydrography, transportation, boundaries, digital elevation models (DEMs), and soil surveys statewide where possible. TGIC also developed partnerships with federal, local and regional entities to raise required funds, providing many leveraging opportunities and benefits. Limited availability of state funds combined with the goal of completing seven data layers made the success of the multi-agency and multi-level partnerships even more imperative. TGIC will regularly maintain and provide access to the data. StratMap, while still on-going, shows that states can form partnerships with entities ranging from small local governments to huge federal agencies interested in common mapping goals and regions. The unique and effective position of states between local and federal entities is strongly evidenced in Texas in this effort to develop and implement such a large cooperative data development program.

Vermont: A large scale solution for a small state. This New England state covers only 9,609 sq. miles, and has a history of state government activism in land planning issues, coupled with strong municipal governance and few county government roles. The General Assembly (legislature) first authorized statewide GI activities in 1988 through the state's Growth Management Act, and has provided direction since then. It created the Vermont Center for GI (VCGI) in 1992, and established it as a public, not-for-profit corporation in 1994 (http://geo-vt.uvm.edu/). VCGI is developing and implementing a comprehensive strategy to develop and use GI/GIT, including statewide base layers. A statewide digital database based on 1:5,000 scale orthophoto coverage was initiated in 1988, before the federal 1:12,000 orthophotos program described above (NDOP) was developed. VCGI's web site provides access to the state's many on-line data holdings, and a recent CD-ROM also facilitates data use. Available data include building locations, elevation contours, soils, feature codes, well locations, electric transmission corridors, and 1:5,000 roads which can be used for many applications. Such rich data resources often are not available at the statewide level, and particularly at this scale. In addition, VCGI is helping the Vermont Enhanced 9-1-1 Board implement statewide E911 service (http://www.state.vt.us/e911/), (http://geo-vt.uvm.edu/cfdev2/VCGI/proj/e911.cfm). A spatially-referenced statewide address database was developed using GIS and GPS, and now is used for additional applications (Westcott 1999). VCGI also uniquely developed an approach for the US Census Bureau to incorporate Vermont's street centerline network data into the Bureau's Topological Integrated Encoding and Referencing (TIGER) database, thus improving both the positional and attribute accuracy of TIGER data in Vermont (Sperling and Sharp 1999).

Wisconsin: State-sponsored statewide and local database development. A unique legacy of support for parcel-based land information systems and expertise in satellite remote sensing has enabled Wisconsin to have unparalleled programs and data resources. Components of Wisconsin's statewide GI/GIT approach include the Wisconsin Land Information Program (WLIP) and a GIS service center located in the Department of Administration, a State Cartographer's Office, the Wisconsin Land Information Clearinghouse (WISCLINC) and the Wisconsin Initiative for Statewide Cooperation on Land Cover Analysis and Data (WISCLAND). Unique among the states, WLIP provides grants to counties to facilitate land information mapping and record keeping, funded by fees assessed for all land deed transfers in the state (http://www.doa.state.wi.us/olis/wlip/index.asp). The Legislature created the WLIP and the Wisconsin Land Information Board (WLIB) to govern it in 1989. Most collected funds (two thirds) are retained by counties for land record documents, mapping, and related salaries, while the remaining one third are state-managed, including provision of local grants and administrative support. The result is a local/state partnership that mixes local knowledge and responsibility with state standards and support. WISCLAND is a separate public/private cooperative effort to develop, maintain, and distribute statewide GI, including several 'framework' datasets including DOQs, elevation models, and hydrography (http://feature.geography.wisc.edu/sco/wiscland/wiscland.html). Unique compared to other states, WISCLAND includes a multipurpose land cover database created by processing satellite data derived from 30 m Landsat imagery. The data are classified according to a schema designed to meet the state's needs, yet are compatible with established systems such as the Anderson classification. Additional data often not available in states include wetlands, land use, digital soil surveys, and floodplains. WISCLINC is the state's clearinghouse that provides access to much of the state's GI.

35.7. Conclusion

As reviewed in this chapter, states have expanded, coordinated and institutionalized GI/GIT approaches and activities within their organizations over time. States employ more than just 'smaller' versions of federal GIT; in fact, in some respects, states have a wider range of applications and more advanced institutional arrangements than other sectors. In addition, state GIT activities have an increasing influence upon federal, local and other organizations. This is largely because overall state activism has increased, while much federal direction and funding that supported state and local GIT activities in past years has decreased. Another important trend is that the notion of a 'state' is expanding beyond state government as an entity to also encompass other organizations with interests and efforts for the area defined by a state's geographic borders. States increasingly provide external organizations with a focal point and opportunities to help develop and conduct GIT coordination and activities across several sectors. These initiatives promise to enhance GIT usage, maturation, and coordination, as well as data development, quality, maintenance and access within states and other sectors. As a result, improved government efficiency, effectiveness and equity will surely result – along with multiple benefits for the public they serve.

References

Arizona Department of Transportation, *Feasibility Report and Implementation Plan for Application of a Computer-Aided Design and Drafting (CADD) System*, Arizona Department of Transportation, Phoenix, AZ, 1984.

Demers, L. *et al.*, *Survey of GIS Use by Public Utility Commissions*, New York State Department of Public Service, Albany, NY, 1995.

Moyer, D. D. and Niemann, B., 'Institutional arrangement and economic impacts.' *Multipurpose Land Information Systems: the Guidebook*, vol. 17, Federal Geodetic Control Subcommittee, Coast and Geodetic Survey, National Oceanic and Atmospheric Administration, Silver Spring, MD, 1994, pp. 1–28.

National Academy of Public Administration, *Geographic Information for the 21st Century: Building a Strategy for the Nation*, National Academy of Public Administration, Washington, DC, 1998.

National Research Council, *Promoting the National Spatial Data Infrastructure Through Partnerships*, National Academy Press, Washington, DC, 1994.

Sperling, J. and Sharp, S A., 'A prototype cooperative effort to enhance TIGER.' *Journal of the Urban and Regional Information Systems Association*, 11(2), 1999, pp. 35–42.

Warnecke, L., 'Geographic information coordination in the states: past efforts, lessons learned and future opportunities', *Proceedings of Conference: Piecing the Puzzle Together: A Conference on Integrating Data for Decision-making*, May 27–29, National Governors Association, Washington, DC, 1987.

Warnecke, L., *Report on the Survey of State Regulatory Commissions' Interest and Experience in AM/FM/GIS*, AM/FM International, Aurora, CO, 1990.

Warnecke, L., *Geographic Information/GIS Institutionalization in the 50 States: Users and Coordinators*, National Center for Geographic Information and Analysis, University of California, Santa Barbara, CA, 1995.

Warnecke, L., *Geographic Information System Technologies Coordination and Use by States for Emergency Management Functions*, Prepared for the Federal Emergency Management Agency, 1996.

Warnecke, L., *NASA as a Catalyst: Satellite Data in the States*, National Aeronautics and Space Administration, Washington, DC, 1997.

Warnecke, L., 'State and local GIS initiatives', *The History of Geographic Information Systems: Perspectives from the Pioneers*, Ch. 14, Foresman, T. W. (ed.), Prentice Hall, Upper Saddle River, NJ, 1998.

Warnecke, L., 'Geographic information technology institutionalization in the nation's states and localities,' *Photogrammetric Engineering and Remote Sensing*, 65(11), 1999, pp. 1257–1268.

Warnecke, L. *et al.*, *State Geographic Information Activities Compendium*, Council of State Governments, Lexington, KY, 1992.

Warnecke, L., Kollin, C., Beattie, J. and Lyday, W., *Geographic Information Technology in Cities and Counties: a Nationwide Assessment*, American Forests, Washington, DC, 1998.

Westcott, B., 'GIS and GPS - the backbone of Vermont's statewide E911 implementation,' *Photogrammetric Engineering and Remote Sensing*, 65(11), 1999, 1269–1276.

Chapter 36

National, international, and global activities in geospatial science

John J. Moeller and Mark E. Reichardt

36.1. Introduction

In recent years, a growing number of countries have recognized the importance of geographic information to improve decisions, and to address the needs for sustainable development highlighted by international activities such as the Kyoto Protocol and the Rio Agenda 21 Summit. The technology revolution of the 1990s produced new capabilities in Geographic Information Systems (GIS), information technology and communications technology, which called for improvement in the way geographic information was collected, managed, made accessible and used. At the same time, governments were decentralizing, and there was a growing awareness of the need to address the social, economic and environmental dimensions of issues as part of decision-making. These developments helped focus spatial data coordination efforts in the US and elsewhere towards the development of spatial data infrastructures to provide a set of policies, standards, practices, technologies, and relationships to facilitate the flow of geospatial data and information at all levels across government organizations and between sectors. Infrastructure activities in a dozen or more nations and within several regions of the world spurred the idea of a global infrastructure that could provide the means for a worldwide linkage. This chapter provides an overview of national infrastructures and will discuss the emerging Global Spatial Data Infrastructure.

36.1.1. National security

National security issues are high priority considerations for many nations, and in the past have often been the primary drivers for mapping and geospatial data activities. In the US as in many other countries, defense related programs are paying benefits far beyond their initial purposes by establishing the foundation for non-defense related developments and for commercial activities that generate jobs and economic opportunities around the globe. The Internet and the World Wide Web as media for rapid exchange of data and information are providing a technology base for spatial data clearinghouse networks and web-mapping services. The global positioning satellite system is providing the capability of obtaining precise geodetic positions for all kinds of natural and man-made features. Satellite imagery of 1-m resolution is becoming more readily available through commercial sources. These and other technological advances are giving us the capability to collect, process and use vast amounts of spatial information to address and solve a wide variety of issues.

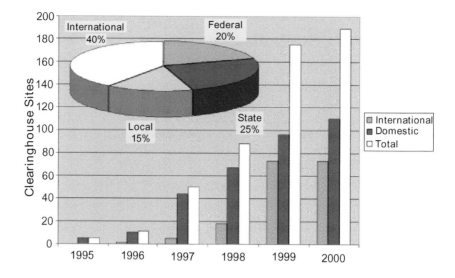

Figure 36.1   NSDI clearinghouse growth 1994 – January 2000 (source: FGDC).

36.1.2. Decisions and place-based information

Throughout the world decision-makers are faced with tough choices. They are asked to make decisions that affect the future wellbeing of nations, business, individual people, the environment, and communities. Too often these decisions are made without the knowledge of all the facts or alternatives. Consequently, decision makers must rely on traditional assumptions that trade-offs are between the environment and the economy, between the interests of individuals, or between one geographic area and another and are situations in which one side wins and the other loses. This approach has often led to stalemates and to actions that do not benefit more than a few special interests.

In many places, people now believe in a different and better way to approach issues. We see the implementation of concepts of sustainable development that call for a balancing of interests to achieve healthy ecosystems, strong economies and social well being. These approaches are building on the idea that by focusing on a geographic area or place, citizens and decision-makers can develop a more comprehensive understanding of the relationships between the various elements of sustainability. This focus on place-based information enables decision-makers to involve scientists and government organizations, and increasingly the citizens most likely to be affected by the decision. This ability to integrate various types of information in a place-based decision-making approach is only becoming feasible by the development of geospatial technologies and by cooperative efforts to establish spatial data infrastructures.

It is now possible to obtain, integrate, and analyze large amounts of information to visualize alternative scenarios that facilitate greater knowledge of future outcomes. These approaches are helping to reduce 'win-lose' situations and create the desire to find alternative solutions that serve the long-term interests of the community as a whole. In the US, this approach represents a return to community-based decision-making that in many cases was lost when computers and information technology favored the centralization of information. Spatial data infrastructures that capitalize on Internet technology and the development of interoperable geoprocessing systems and technology are helping communities and nations actually implement place-based decision processes. The modern world has produced many marvelous capabilities, but these capabilities have also led to profound changes in the environment, economic forces, and social interaction. These changes must not occur in a vacuum without consideration of their impact on the well being of our citizens and on future generations. The improved understanding of place and the effects of actions within that place hold the promise that communities and societies will use geospatial science to provide information in a way that that addresses issues to achieve long-term sustainability.

36.1.3. Local to global and the role of national efforts

We have often heard the phrase 'Act locally – think globally'. Until recently, this phrase has been treated as just an idea. Today however, we exist and work in both a local and a global context. People are linked across great distances by communications and information in such an extensive fashion that a person can be part of a local group at one moment and then part of a global group in the next without leaving their seat. The concept of place and the interconnectivity of our actions take on new dimensions. While communities often do not need to consider the ramifications of an action at the global level, they often need to consider impacts of actions in relation to neighboring communities. Information technologies give us the ability to tie into larger networks and to see and better understand these interconnections. Local areas are part of larger regional and global communities and all components of this connected system must be healthy for the whole system to be at its best.

It is important to consider how this connectivity can be built and maintained from the local to global levels in ways that preserve the sovereignty and dignity of local populations while advancing the sustainability of the world population. The key role in achieving this balance of interdependence is through the efforts of nations in the development and use of geospatial information. Many countries are developing or planning to develop spatial data infrastructures (Onsrud 1998). As these infrastructures develop, it is becoming apparent that many types of geographic information are important for addressing global issues. Science and technology are helping create new knowledge and understanding, which can advance our collective ability to solve local and ultimately global problems if we can find a way to share our knowledge. Data and information are an infinitely sharable commodity. Sharing enriches the collective storehouse of knowledge. National spatial data infrastructures are being enhanced to increase the access and availability of spatial data and to enrich the data and knowledge wealth of the nation. These national efforts, while all tailored to meet the needs of individual countries, contain a number of common elements. If these national efforts continue to coalesce around a few critical linking mechanisms, we will likely see great strides forward in our ability to 'Act locally – think globally'.

36.2. National needs and requirements

36.2.1. Policy perspectives

The needs of nations vary across the globe as does their legal, cultural and policy background. Therefore while there are opportunities for collaboration and for agreement on a fundamental infrastructure for transferring and using spatial data, the areas of common interest and agreement are still emerging. A recent survey of National Spatial Data Infrastructure (NSDI) activities around the world (Onsrud 1998) provides some substantive data on national needs and requirements. Around the globe we see nations rethinking their role with respect to geospatial data and attempting to understand recent geoprocessing technological developments. Technology is driving many of the national efforts in geospatial data and information. However, the most successful endeavors appear to also have a well-defined policy base.

From a technology perspective, nations see the benefits of new capabilities that allow the collection, storage, processing, and display of vast amounts of geographically referenced information. While much of the desired functionality is still being developed, current technologies along with those available in the immediate future hold great promise for improving the cost efficiency and ease of use of GIS systems, decision support technologies, and spatial data collection and management technologies. In the near future, all nations may be able to establish at least a basic capability of geospatial technology tools.

From a policy perspective there are several key issues that drive the development of national efforts to improve the use of geographic information. The key policy drivers appear to be: technology modernization; the need to coordinate among currently disparate government functions or levels of government; the desire for better government; and the desire to implement principles of sustainable development. Countries may have differences in their approach for addressing these drivers, but there are some fundamental common features. Most of these efforts are focused on establishing national strategy or plans. All efforts address a growing role for the use of geographic information; and all are attempting to establish a way of linking information providers and users.

36.3. The role of national data programs

Most nations have programs that provide geospatial data. Many of these are national mapping programs whose mission has traditionally been to provide national level map products. As mapping capability improved in the late 1900s, more and more nations have included national programs for environmental and natural resource data and information or other nationally important data. These programs have served critical needs for both military and civilian users and have laid the foundation for the distributed spatial data infrastructures existing today and being planned for the future. Technology developments and the growth of the Internet along with the devolution of government from central structures are leading mapping agencies away from traditional products towards open geoprocessing and information/services distribution as critical national programs. The ability of national programs to move quickly from a traditional product paradigm to a focus on information and services may indeed determine their continued relevance in the new era of the Internet and spatial data infrastructures.

National programs must maintain or establish a connection to local requirements and needs, and address broader regional and global efforts. Nations provide the fabric that holds all of the elements of the political and governmental infrastructure together. They can provide the connections and establish or facilitate operating climates that link local to national and national to regional and global. The NSDI development in the US is an example of the national government and its key spatial data programs moving to play this new role as a change agent and as a the transition point from community views to a worldwide reach.

36.4. Case examples of some leading nations

The following sections discuss the progress of spatial data infrastructure development and implementation around the globe. These are just a few case examples that provide some perspective regarding the similarities and differences between initiatives, and the implications associated with establishing a global spatial data infrastructure necessary to collaborate on issues of great environmental, social, and economic importance.

36.4.1. The United States of America

The genesis of today's US NSDI can be traced back to 1953, when the US Office of Management and Budget issued Circular A-16 on Surveying and Mapping to create a federal process for the assessment and coordination of surveying and mapping functions being performed across the federal government. Through a revision of this circular in 1990, the Federal Geographic Data Committee (FGDC) was established as a formal coordination mechanism with the objective of building a 'national digital spatial data resource' (United States Office of Management and Budget 1990).

A major report (National Research Council 1993) solidified the concept of the NSDI. This report, along with strong interest in federal government reform by the Vice President's National Partnership for Reinventing Government, resulted in the endorsement of a NSDI. This ultimately led to Presidential Executive Order 12906 in April 1994 (United States Executive Office of the President 1994), calling for:

1 A *NSDI* as a key component of the national information infrastructure;
2 The development and use of a national geospatial data clearinghouse;
3 Use of a national distributed *framework of data* for registering and referencing other themes of geospatial data;
4 *FGDC-endorsed standards* for data content, classification and management for use by Federal and available to all other geospatial data producers and users.

36.4.1.1. US progress towards a NSDI

The Executive Order provided the basis for more aggressive national efforts to advance the NSDI. In the 1990s, FGDC stakeholders established a core set of essential NSDI components comprised of: data content and metadata standards; a nationwide set of framework data; metadata to help inventory, advertise, and intelligently search geographic data sets; a NSDI clearinghouse that allows for catalog searches across multiple geospatial data servers on the Internet; and strong partnerships with the public and private sectors to advance the NSDI. Through advancements in each of these components, and a strong US policy for low cost access to federal information, significant progress has been achieved toward fulfillment of the NSDI Executive Order.

Significant progress has been made in the promotion of data content standards and medadata practices. As of December 1999, 14 data content standards had been endorsed by the FGDC for use nationwide, with another 19 standards in review (see http://www.fgdc.gov/standards/standards.html). The FGDC has worked with stakeholders to implement a national metadata standard, which is now in broad use at the federal level and by other communities throughout the nation to help inventory, advertise and access geospatial data. Several states have formed pacts with local, and federal organizations to use FGDC compliant metadata statewide to maximize geospatial data sharing. The International Standards Organization Technical Committee 211 is drafting an international metadata standard. When approved, the FGDC will support the US transition to this ISO standard.

> US NSDI vision: 'Current and accurate geospatial data will be readily available to contribute locally, nationally, and globally to economic growth, environmental quality and stability, and social progress' (FGDC 1997).

The generation of framework data is occurring throughout the nation. Framework data provides a minimum critical set of geospatial data from which other data can be registered and added to meet specific user needs. Based on a recent framework data survey (see National States Geographic Information Council 1999), approximately half the counties surveyed were actively producing one or more of the seven components of framework data as a vital part of their geospatial infrastructure.

The US NSDI clearinghouse provides a discovery and access service for geospatial data, based on FGDC compliant metadata and internationally accepted cataloging processes. The clearinghouse has grown from its first operational site in 1994, to 183 clearinghouse sites now accessible from six NSDI clearinghouse gateways (see www.fgdc.gov/clearinghouse/clearinghouse.html). Of the 183 sites, 20 per cent represent federal interests, 25 per cent have statewide or university scope, 15 per cent are local communities, and 40 per cent of sites are international in scope or location. International sites have been a rapidly growing component of the clearinghouse (Figure 36.1).

The FGDC has been successful in building critical partnerships needed to advance the NSDI. Since its inception, the FGDC has expanded its federal stakeholder representation from 12 to 17 federal departments. This increase in participation has resulted in a large part from the rapid growth of geospatial programs within social and environmental areas of governance. By the end of 1999, the FGDC had established a broad stakeholder network of federal, regional, state, local and tribal organizations and communities in the US. In 1998, a series of six NSDI Community Demonstration Project sites were implemented (see www.fgdc.gov/nsdi/docs/

cdp.html) (FGDC 1998). These projects were designed as a cooperative effort between the federal government and local communities to provide a detailed illustration of the value of the NSDI to critical community decision-making in the areas of citizen-based land use planning, environmental restoration, crime management, and flood mitigation; and to help build support for broader application of NSDI in communities across the US.

36.4.1.2. Evaluating the NSDI – 1999 national geodata forum

While significant progress has been made towards many of the US NSDI goals, there remain many significant issues that are hampering further advancement. At a 1999 Geodata Forum held in Washington DC, over 460 members of the private sector, academia, government at all levels, and international representatives gathered to examine the NSDI in the context of 'Making Livable Communities a Reality' (FGDC 1999). The results of this forum and a major Congressional Hearing that immediately followed the Geodata Forum, helped to solidify a number of recommendations for change to the FGDC and NSDI strategic direction (see http://www.fgdc.gov/99Forum/). Recommendations focused on the need to create a new and more flexible national organization composed of a more broadly represented mix of private and public sector organizations to steer the NSDI as a national resource. Recommendations also urged the FGDC to consider increased emphasis on education and outreach and community applications of NSDI to enhance decision-making. Finally, geodata recommendations highlighted the need to identify new financing strategies to continue and accelerate the growth of NSDI.

36.4.2. Canada

The Canadian Geospatial Data Infrastructure (CGDI) is an effort to establish an NSDI for Canada. In August 1999 the Canadian government launched GeoConnections – a $60 million (Canadian) national initiative to accelerate the development of the CGDI through partnerships and matching investments from provincial and territorial governments, and the private and academic sectors. The CGDI is the mechanism by which Canadian governments will significantly enhance access to geoinformation and in turn accelerate development of knowledge based economic activities. Coordinated through the GeoConnections initiative, which reports through Natural Resources Canada, development of the CGDI is well underway through the efforts of many partners in government, and the private and the academic sectors. GGDI has five technical components (excerpted from Labonte *et al.* 1998):

1. Access: national electronic access to digital geospatial or geographic information held by public agencies, greatly expanding access and usage.
2. Data framework: a common national framework for geospatial information, making information easier to use.
3. Standards: participation in the development of and adoption of international standards, making information easier to use.
4. Partnerships: federal-provincial partnerships to improve federal-provincial relations, and leading to greater efficiencies in cost-sharing activities (production, management, and distribution).

5   Supporting policy environment: foster a policy environment that promotes the broadest use of information by the public, enabling business and private sector industry to acquire, add-value, commercialize and use this government information.

CGDI vision: 'to enable timely access to geo-info data holdings and services in support of policy, decision-making and economic development through a cooperative interconnected infrastructure of government, private sector and academia participants'.

Source: GeoConnections Canada at
http://www.geoconnections.org/english/partnerships/index.html

---

CGDI development is proceeding with results on all fronts. The GeoExpress-Access program is providing ready access to government spatial information via the Internet. As with the US, this on-line catalog service is based on international catalog standards and is modeled after FGDC and ISO metadata standards. Thus, the Canadian and US clearinghouses, as with other catalog and metadata compliant clearinghouses in South Africa and Australia offer the makings of a global catalog service as more nations register conformant catalog servers.

Development of nation-wide framework data is underway, with the completion of a 'Data Alignment Layer' from which other spatial data can be rapidly and accurately integrated. Canada is also very active in the development and application of international data and technology standards to assure consistent data generation, inventory, discovery, and access to address national and international activities. As with the US, pilot projects geared toward sustainable communities is a major portion of the CGDI program thrust to help communities better plan and manage economic, environmental, and social programs. Pilots will be focused on helping communities to build geospatial capacity. The CGDI program also provides focus on education and outreach through the National Atlas of Canada. This on-line service provides perspectives on Canada for students and citizens, and will help place local (provincial and territorial) data in context with national data. Finally, CGDI programs will also focus on building the skills network needed to sustain and grow jobs and industry sectors.

36.4.3. The Netherlands

As with many other national SDI programs, the evolution of the NSDI in the Netherlands was the result of interest by the Dutch government to better coordinate state geographic information activities. In 1984, the Netherlands Council for Geographic Information (RAVI 2000) was established as an advisory Council for the Secretary of State for Housing, Spatial Planning and the Environment who acted as Coordination Minister for geographic information. An Information Provision Decree of 1990 formalized the national coordination role of the Coordinating Minister and RAVI. In 1992, the RAVI plan and schedule was approved by the Dutch Parliament with the following three major thrusts: a description and plan for a NSDI in the Netherlands, a plan to coordinate the interests of geographic information users, and a proposal on how to realize the NSDI through joint efforts of the Minister and RAVI.

In 1993 RAVI was restructured into a public/private foundation to steer the NSDI through self-regulation of geographic information. Primary funding sources for RAVI were established from the Coordination Minister and the other major government agencies. Over the next few years, a business plan was developed, and members of the private sector were encouraged to form a 'business platform' to directly address RAVI goals, objectives and projects. The business platform was originally comprised of 25 members of the geospatial data and service providers, and has recently grown to 40 members, with broadening representation being encouraged from other key business sectors including real estate, insurance, etc.

In 1996, a major grant was awarded to RAVI from the Dutch Cabinet's National Action Plan to commence development of an electronic highway for the National Clearinghouse of Geographic Information (NCGI). By 1997, the first prototype NCGI became available on the Internet. A second prototype was completed during the following year, and an additional grant program was implemented in 1998 to further develop the NCGI over the following 4 years. Additional actions are being taken to define and implement 'authentic registers' (or trusted sources) for geospatial data, and to publish a Policy Framework for Geo Information. Today, the major areas of focus for RAVI are in the area of public access to information (freedom of information), and the development of an overall information policy framework for the nation.

36.4.4. Colombia

The following Colombia Spatial Data Infrastructure (ICDE) overview is summarized from a spatial data infrastructure 'Cookbook' developed by the Global Spatial Data Infrastructure Technical Working Group (Global Spatial Data Infrastructure (GSDI) Technical Working Group 2000). Spatial data infrastructure efforts in Colombia have evolved over the years, largely as a result of government mandates to respond to a variety of municipal to national programmatic needs. In the early 1990s a series of laws and decrees were issued for Colombian agencies to develop information systems to manage the environmental, geologic, demographic, and mapping data for the nation. As an example, recent legislation (Law 388 of 1997), required municipalities to develop a territorial ordering plan to define and regulate land use. Geographic data are the key to ensure compliance with the law. Additionally, the Colombian Coffee Growers Federation and the National Oil Company ECOPETROL developed information systems to address the geospatial data needs of these industries.

A high level team drafted a set of government policies on information in 1996, producing policies that emphasized the need to manage information like a strategic national resource. These policies viewed the use of information technology as a means to promote social welfare and citizen service, and to link government agencies with outside sectors. The concept of the ICDE formed from the efforts noted above, with the purpose of integrating geographic information systems into a consistent framework for the nation.

The Colombian Spatial Data Infrastructure (ICDE) is defined as the set of policies, standards, organizations, and technology working together to produce, share, and use geographic information on Colombia in order to support national sustainable development.

In 1998, the Colombian government defined as a priority the establishment of a long-term multilateral alliance between Colombia and the US, the 'Environmental Alliance for Colombia' (Pastrana 1998), aimed at the promotion of technical, scientific, managerial, informational, financial and political cooperation for the knowledge, conservation and sustainable development of Colombian natural resources.

An Inter-Institutional Committee was set up in November 1998 to create consensus on number of topics related to geographic information. The government agencies in charge of geographic information production agreed to work jointly to define policies, guidelines and strategies to foster the production and publication of geographic data in Colombia and facilitate data integration, use and analysis by the agencies' information systems. The committee also decided to promote carrying out actions to develop autonomous information systems in a coordinated and harmonized way as integral part of a national geographic information system. The Committee agreed to coordinate actions including: the development of guidelines for geographic information production and management, strategies for standardization of products to best meet customer needs, legal and business strategies, and strategies to strengthen Colombian telecommunications and information technology infrastructures.

By early 2000, the formalization of commitment to build the Colombian NSDI had been formalized between the Colombian government and key members of the private sector. The main Colombian producers of geographic data and two of the most relevant information users, the National Oil Company and the Coffee Growers Federation signed an agreement to jointly develop and implement the ICDE. In addition, the Colombian Presidents National Council for Social and Economic Policy recently approved a policy called the 'Colombian Connectivity Agenda', focused on the development of telecommunication infrastructure and to strengthen the use of the Internet by the community. This policy incorporates the development of the SDI for Colombia, and assigns responsibility for spatial content to the main geographic information producers.

36.4.5. Australia

Since 1996, the Australia New Zealand Land Information Council (see ANZLIC 2000) has been promoting the concept of an Australian Spatial Data Infrastructure (ASDI). The ASDI provides a technical and administrative framework that will allow spatial data of all types to be correlated and analyzed, thereby extending its usefulness and maximizing the community benefit from the government investment in the data.

A key part of the ASDI is the Australian Spatial Data Directory (ASDD), which currently provides a mechanism for discovering existing spatial data, thereby avoiding duplication. In due course, that ASDD will evolve into a Clearinghouse that will provide on-line access to the ASDI; it may also include product development tools. To make the ASDD work, metadata standards have been developed and are being widely implemented.

ANZLIC has developed a national policy statement on data management, including access and pricing. Member jurisdictions are striving to develop their policies towards the goals set by ANZLIC. ASDI compliance criteria have been developed and evaluated in jurisdictions. They will be refined in 2000. ASDI compliance requires more than adoption of technical standards. It requires that metadata be compiled and submitted to the ASDD. It also requires that the data are accessible under conditions established by the relevant jurisdiction.

ANZLIC has also developed custodianship guidelines as a means of ensuring accountability for the care and maintenance of information within the public sector. The principle of custodianship assigns to an agency certain rights and responsibilities for the collection of spatial information and the management of this on behalf of the community. The rights of a custodian include certain rights to license and market the information while the responsibilities include maintenance and quality of the information. The principle of custodianship also ensures access to the information and provides a recognized contact point for the distribution, transfer and sharing of the information. In 2000, ANZLIC will be determining the fundamental datasets that comprise the ASDI and will identify custodians and sponsors for them.

Beyond the ASDI, there are ranges of access and equity issues, regarding both spatial and non-spatial information, that need attention. They include: the need for a communication network with the coverage and capacity to give the regional and rural population access to on-line information; the need to identify a process for capturing and managing the information compiled by land care and other community-based groups; and the need to develop interfaces and expert systems for easy use by landholders and others in rural and regional Australia.

As described above, spatial data infrastructures have become key needs in overall national information infrastructures. The cases of the US, Colombia, the Netherlands, Canada, and Australia represent a few of the SDI initiatives now underway. Many of these efforts have common characteristics, a strong national priority for spatial data as a national resource, the desire to reduce duplication of effort, the need to promote policy to address collaboration and sharing, and the need for unique public/private partnerships at all levels to advance spatial data infrastructures. Many nations are establishing spatial catalogs based on internationally accepted practices and standards to inventory, advertise, and promote access to spatial information.

36.5. Beyond borders – regional and global spatial data infrastructures

NSDI programs are designed to encourage a level of consistency for nations and their localities to share critical geospatial information vital to improved decision making. Similarly, the concept of Regional and Global Spatial Data Infrastructures has arisen as a result of the need to address the broad issues that tend to transcend national borders. The United Nations (UN) Conference on Environment and Development held in Rio de Janeiro in 1992 called for action to reverse environmental deterioration, and for the establishment of a sustainable way of life to carry us into the twenty-first century. To achieve these goals, Agenda 21 of this conference called out measures to help reduce pollution, deforestation and to address reversing other adverse environmental activities. Geospatial data was identified as a critical resource for decision-making.

In response to the need for broader multinational collaboration on environmental, economic and social issues, Regional Spatial Data Infrastructures have emerged, with a focus on the issues unique to of a broad area. In the case of the Permanent Committee on Geospatial Infrastructure for Asia and the Pacific, 55 nations have agreed to cooperate on the establishment of a common regional datum, the sharing of best practices, and other matters to further promote collaboration on issues of mutual concern (see http://www.permcom.apgis.gov.au/). Similarly, the European Umbrella Organization for Geospatial Infrastructure (EUROGI) is dedicated to the promotion of regional collaboration for European nations (see http://www.eurogi.org/) (European Umbrella Organization for Geographic Information 2000). In March 2000, 21 nations in the Americas, including the Caribbean, have signed a provisional agreement to form a Permanent SDI Committee for the Americas by 2001.

Some issues, such as climate change are truly global in nature. To meet the challenge of global collaboration for spatial issues, a group of representatives from various nations held the first Global Spatial Data Infrastructure Conference in Bonn, Germany in 1996. Since that time, three additional conferences have been held, with the results of each conference helping to provide the additional detail and focus needed to promote the GSDI as an umbrella organization working to create a global infrastructure capable of supporting transnational and global issues without impeding national and local objectives. Membership in the GSDI *ad hoc* Steering Committee now includes representatives from all continents.

GSDI is defined as '... the policies, organizational remits, data, technologies, standards, delivery mechanisms, and financial and human resources necessary to ensure that those working at the global and regional scale are not impeded in meeting their objectives...' (http://www.gsdi.org/).

The GSDI Technical Working Group has released a 'GSDI Implementation Guide', which will serve as a global reference guide for SDI developers (GSDI 2000). This document will provide significant detail regarding core data categories, accuracy, and resolution; the standards and best practices for data documentation, discovery, and access; case studies and other services necessary for nations to assure globally compatible spatial data infrastructures. This document will highlight internationally agreed upon specifications and proven practices to promote consistency and ease of data sharing across national borders.

The GSDI offers significant potential to guide the development of compatible national and regional infrastructures that offer nations opportunity to collaborate on the significant issues that affect the broader transnational and global community. To better understand the GSDI, it is appropriate to look at the advancements that have been achieved to date by nations and organizations working collaboratively to achieve international consensus on data, standards, dissemination, policy, and resources. This section summarizes some of the major programs that have contributed to the Global Spatial Data Infrastructure. This list is by no means exhaustive, and is offered only to provide examples of the work that is being accomplished towards a GSDI.

Geospatial data – The International Steering Committee for Global Mapping (ISCGM) (http://www1.gsi-mc.go.jp/iscgm-sec/index.html) was created as a response to Agenda 21 from the UN Conference on Environment and Development (UNCED) held in Rio de Janeiro in 1992. As a result of a call from Agenda 21 to create global environmental data, the Japanese Geographical Survey Institute/Ministry of Construction formed the ISCGM and set forth to create Global Map. The goal is the production of a global data set (Global Map) containing elevation, vegetation land use, drainage systems, transportation networks, and administrative boundaries at a scale of 1:100,000,000. As of February 2000, 75 nations representing every continent are participating in this initiative.

Standards – Internationally recognized data content, metadata, technology, and process standards are essential in building a global SDI:

1 Technology Standards – The OpenGIS Consortium (Open GIS Consortium at http://www.opengis.org) is an organization 'whose mission is to promote the development and use of advanced open systems standards and techniques in the area of geoprocessing and related information technologies.' OpenGIS concentrates on the creation of technology specifications that ease the sharing of geospatial data and processing between vendor products on the Internet.

2 Data Standards – The International Standards Organization Technical Committee 211 (ISO/TC211) goal is the 'standardization in the field of digital geographic information.' Specific goals are to publish a structured set of standards for information concerning 'objects or phenomena that are directly or indirectly associated with a location relative to the Earth.' (ISO web site), and to establish the relevant links to other technical standards that may apply.

Delivery mechanisms – To provide access to data on a global scale, several GSDI member nations are linking to a GSDI Catalogue Registry being maintained by the US for the GSDI community. By linking to this catalogue registry, spatial data Gateways in the US, Canada, South Africa, and Australia offer access not only to their national geospatial data servers, but to a growing list of international servers compliant with internationally accepted standards for cataloguing and metadata (an approved ISO standard for metadata is anticipated by the end of the year 2000).

Funding/human resources – Member nations have been working internationally to help educate, train, and develop NSDI's consistent with internationally recognized standards and practices. Often, this is a result of 'in kind' support through the provision of training materials, technical expertise and training from member nations. Funding for national, regional and global SDI initiatives can potentially come from a number of sources: National Mapping Agencies, the UN, the World Bank, private funding concerns, and foundations. Adequate funding in the developing nations is a particularly important issue if SDI activities are to be advanced to enable regional and global collaboration on spatial issues.

Policies – As NSDI development progresses, undoubtedly there will be situations where data and service sharing between nations are inhibited by policies that constrain data access, protect privacy/security, or promote use restrictions (copyright, pricing, etc.) at varying levels. It is not the intent of the GSDI to drive policy, rather to offer a collaborative environment for member nations to discuss, understand, and potentially resolve differences so that appropriate action can be sought nationally. This is an area that may potentially become a major factor in determining the degree to which data can be shared on a global basis.

Indeed, the contributions of government and private programs have yielded key elements of the GSDI, many of which have become part of the overall GSDI reference environment needed to help gain compatibility at a transnational and global level. However, much more work needs to be accomplished to address the remaining technology, policy, and resource issues that are limiting the implementation of the GSDI. The vision of a Digital Earth voiced by US Vice President Gore in January 1998 (Gore 1998), has become a key programmatic effort led by National Aeronautics and Space Administration (NASA) in cooperation with other US agencies, the private sector, and academia to find ways to focus research and development resources on the areas of SDI in need of attention. Digital Earth is gaining an international momentum, with an International Symposium on Digital Earth (see http://159.226.117.45/de99.htm) held in Beijing, China in 1999, and a second international symposium to be hosted by Canada in the year 2001.

In closing, Spatial Data Infrastructures are advancing at the local, national, regional, and global levels to encourage collaboration and sharing for place based decision making. An emerging structure is developing through global and regional SDI organizations to help nations establish and implement spatial data infrastructures that serve not only national needs, but also those broader issues that will require collaboration across borders.

Acknowledgements

The authors wish to thank the following contributors for their work in providing content to this chapter: Peter Holland, General Manager, Australian Surveying and Land Information Group, Department of Industry Science and Resources. Postal address: PO Box 2, Belconnen ACT 2616, Australia. Street address: Scrivener Building, Dunlop Court, Fern Hill Park, Bruce ACT 2616, Australia. Tel: +61-2-6201-4262, Fax: +61-2-6201-4368, Mobile: +61-(0)412-620-132. Email: *peterholland@auslig.gov.au* Internet: *www.auslig.gov.au*. Jeff Labonte, Program Coordinator, GeoConnections Secretariat, 615 Booth Street, Room 650, Ottawa, Canada ON K1A 0E9. Tel: (613) 992-8609, FAX: (613) 947-2410. Email: *labonte@nrcan.gc.ca*. Dora Inés Rey Martínez, Advisor for SDI Projects, Agrologist, MBA, Instituto Geográphico Agustín Codazzi, Carrera 30 No. 48-51 Bogotá, Colombia. Tel: (571) 3681057, Fax: (571) 3680950.

References

ANZLIC, *ANZLIC: The Spatial Information Council*, 2000, available at http://www.anzlic.org.au.

Document, *Proposal for the Design and Implementation of a Colombian Geospatial Information System*, Cartagena, Colombia, May 1999.

EUROGI, *About EUROGI*, European Umbrella Organization for Geographic Information, 2000, available at http://www.eurogi.org.

Federal Geographic Data Committee, *Executive Summary: 1999 National Geodata Forum Making Livable Communities a Reality*, 1999, available at http://www.fgdc.gov/99Forum/.

Federal Geographic Data Committee, *Community Demonstration Project Overview*, 1998, available at http://www.fgdc.gov/nsdi/docs/cdp.html.

Federal Geographic Data Committee, *A Strategy for the National Spatial Data Infrastructure*, 1997, available at http://www.fgdc.gov/nsdi/strategy/index.html.

Global Spatial Data Infrastructure (GSDI) Technical Working Group, *Developing Spatial Data Infrastructures: The SDI Cookbook (Draft)*, March 2000, available at www.gsdi.org.

Gore Jr., A. M., 'The Digital Earth: Understanding our planet in the 21st Century', given at the California Science Center, Los Angeles, California, 1998, available at http://www.digitalearth.gov/VP19980131.html.

National Research Council, *Toward a Coordinated Spatial Data Infrastructure for the Nation*, National Academy Press, Washington, DC, 1993.

National States Geographic Information Council, *Framework Data Survey Preliminary Report*, a supplement to *GeoInfo Systems*, 1999, available at http://www.fgdc.gov/framework/survey_results/readme.html.

Onsrud, H., Survey of National and Regional Spatial Data Infrastructure Activities around the Globe, 1998, available at http://www.spatial.maine.edu/~onsrud/GSDI.htm.

Open GIS Consortium, *Catalog Interface Implementation Specification (Revision 1.0)*, undated, available at http://www.opengis.org.

Pastrana, A., *Formal Announcement of Alianza Ambiental por Colombia*, meeting held in Washington, DC, October 1998.

RAVI, *Dutch Council for Geographic Information homepage*, 2000, available at http://www.euronet.nl/users/ravi/english.html.

United States Executive Office of the President, *Coordinating Geographic Data Acquisition and Access: the National Spatial Data Infrastructure*, Executive Order 12906, April 1994, Available at http://www.fgdc.gov/publications/documents/geninfo/execord.html.

United States Office of Management and Budget, *Circular Number A-16 Revised: Coordination of Surveying, Mapping, and Related Spatial Data Activities*, Washington, DC, 1990. Available at http://www.whitehouse.gov/OMB/circulars/a016/a016.html.

Chapter 37

Private sector applications

Robin Antenucci and John Antenucci

37.1. Technology and information integration yield profits

Profits generated through higher revenue, reduced costs, accelerated earnings and competitive positioning and market share are a basic motivating influence of the commercial sector to embrace a new 'workflow.' Changes to workflow or the business process within an organization are frequently met with skepticism and even hostility. Commercial ventures are frequently hesitant to tread on new ground if there is not a precedent and an experience set that establishes the economic value of such a change.

Despite the fact that for the better part of 30 years the commercial sector has been constantly refreshed with more potent and less expensive technology, organizations continue to demonstrate hesitancy when approached with the idea of changing or 'reengineering' their business process to leverage a new technology or the adoption of one. The introduction of the Global Positioning System (GPS), Geographic Information Systems (GIS) and Remote Sensing (RS) techniques have not been an exception to the rule.

GIS may have had, perhaps, the higher barrier to entry in the commercial market space given its parentage in academia and in the government sector. GPS, developed for the national defense by commercial contractors, had high costs to overcome but a body of commercial entities capable of promoting its broad based utility. And RS technologies, developed from the feedstock of the US space program has, until recently, been more of an exotic source of information.

Coupling two or more of these technologies into a commercial workflow represents a significant challenge to organizational dynamics in the private sector. The integration requirements associated with the technologies have not always been trivial from a technical standpoint. Likewise, the integration of the derivative information has been equally a challenge – both from a technical viewpoint and from a perspective of absorption into the analytical needs and information flow of organizations.

Yet, as the case studies below exemplify, the successes are real – and not so unusual once an appreciation of the individual technologies is developed. GPS is frequently relied upon as a baseline and as a vehicle for accumulating and registering data for a particular geography. Remotely sensed data – in the broadest definition of that phrase – provides a broad range of techniques for the collection, analysis and portrayal of data. GIS finds itself used as a source of collateral data, as a mechanism for aggregating data (and their respective positions) and serving as both an analytical tool and a visualization media.

The technologies seem to have a, perhaps temporary, stronghold on resource development and exploitation ventures: whether the cropping of bananas, the insurance of crops or the development of oil fields. On the other hand, the benefits are more than monetary and the utility nearly unconstrained, where a feature can be associated with some referencing system (Ardila 1996).

37.2. Commercial applications

37.2.1. GPS and digital mapping benefits banana production in South America

In Columbia, South America, banana production is an important part of the economy, accounting for 21 per cent of all agricultural products. Banana exports to the US are valued at over $371 million each year. Only coffee surpasses this product in exports.

The banana crop is vulnerable to a damaging fungi known as the Sigatoka fungus which can destroy a major portion of each annual harvest if not controlled properly.

Banana plantations managed by growers such as Dole and Chiquita have traditionally used ground based flagging crews to guide crop dusting aircraft. These crop dusters spray more than 100,000 acres of bananas several times each season to prevent the fungus infestation. The flag crews establish flight track lines for the pilots in an attempt to accurately spray each banana field while avoiding power lines and water bodies. After an application, the ground crews must move to the next flight line and position the flags above the vegetation. While this occurs, the pilot circles in the air awaiting the next signal to spray.

Social and environmental scientists find this method hazardous to the flagging crews, particularly from physical contact and inhalation of fungicide. In addition the technique is imprecise, environmentally unsound and an inefficient use of costly aircraft and pilot time.

In 1995 a group of growers learned about GPS technology for agricultural applications and decided to invest in the technology to improve their spraying programs. The growers hired a surveying company from Medillion, Columbia to implement a differential GPS navigation system (Ardila 1996).

The components of the system include a Differential GPS (DGPS) receiver, GPS and differential correction antennas mounted above the cockpit, an indicator light bar mounted in the cockpit, and a moving map display. Spray boundaries were mapped and digitized for each banana crop lot in each grower's plantation. Hazards such as power lines and water bodies were also mapped. A GIS database was developed to track data on each lot sprayed such as the chemical used, the per cent of coverage, etc.

With the new system, as the pilot approaches the lot to be sprayed, the indicator light bar displays the planes location relative to the path to be sprayed through an interconnection between the DGPS and the digital map. LED's on the indicator light bar change color as the area to be sprayed is approached. Red indicates "do not spray", changing to orange to indicate "prepare to spray", and when the plane is over the target lot, a green light on the display tells the pilot to "release the chemical spray". At the end of the flight line, the display changes back to red. If the pilot nears a power line or water body, a warning is flashed on the light bar notifying the pilot to change course.

The system was implemented operationally in 1996 and has provided a number of significant benefits to the growers' operations:

Flexibility: the pilots no longer have to wait on flaggers to relocate when ground conditions are such that a planned flight line cannot be sprayed. This often occurs due to changing wind and weather conditions. With the aerial guidance system, the flaggers are no longer needed, reducing operational costs. Pilots can also review their flight lines prior to landing, and make course or spraying corrections without having to wait until the next mission. If the pilot has to refuel, the DGPS guidance system allows them to navigate back to the point where they discontinued the spraying operation.

Precise control: data stored in the moving map display accurately records information about the spraying operation which is used to determine if the spraying missions are being carried out properly by the contracted pilots. Key information recorded provides data that is subsequently analyzed for cost control purposes and to determine the effectiveness of specific fungicide 'lots' or types.

Flaggers: One of the most important and dramatic benefits is that the flaggers are no longer exposed to dangerous chemicals. Most of the flaggers have been assigned to positions in the packinghouses where production has increased.

Environmental: The digital cockpit map display allows the pilot to know if a no spray area such as a water body or a hazard such as a power line is being approached. Prior to this the pilot had only visual clues as to whether a no spray area or hazard was near.

Costs: The traditional, manual system cost growers more than $150,000 US per year. The incorporation of digital cockpit maps and GPS reduced system costs to approximately $23,000 US per year, a saving of over $126,000 US or more than 80 per cent per year. The system investment cost was approximately $109,000 US and the growers achieved recovery of this cost in the first 10 months of operation.

37.2.2. Integration of GPS and GIS within the insurance industry

Fireman's fund insurance company of Overland, Kansas is using mobile GPS technology integrated with GIS to provide a competitive advantage over other insurance providers in the agricultural field (Kozero 1999). The advantage stems from both the use of the technology for accurate underwriting of crops and more accurate assessments of damages as well as a 'value-added' service of providing their customers with high accuracy maps at no cost.

Essentially, Fireman's fund has developed a proprietary GIS system that relies on GPS measurements of farm fields for determining the insurance premiums for various crops. The deployment of GPS technology on All Terrain Vehicles (ATVs) and the integration of GPS and GIS system components provide a capability to quickly and accurately map cropped areas while agents are in the field.

Insurance agents use ATV's to drive the perimeter of each of the customer's fields using a GPS receiver to collect coordinates defining the boundary with an accuracy of 1 m (Figure 37.1, see plate 23). The coordinates are then downloaded to the GIS system, which is accessed through a PC housed in the agents van (Figure 37.2, see plate 24). Within a few minutes agents are able to generate a map of each field. Each cropped field is mapped in this manner to develop accurate premium assessments. In addition, hardcopy and digital maps are produced as a derivative product and are provided to the customer at no cost.

Fireman's fund benefits from this innovation because their clients are now charged accurately for their policies and the fund is accurately assessed for losses. Since most crop insurance premiums are computed by the acre, accurate acreage data is important. Traditional insurance rate calculations are based on measuring fields using approximations obtained by measuring the perimeter of roadways surrounding them and estimating acreage of subdivided fields. These techniques almost always overestimated the amount of acreage. By having this accurate mapping system, the company has a unique advantage over their competition both in assessments of premiums and also when required to underwrite crop losses. Another benefit to the company is that agents can process client's claims at a faster rate and more consistently. The map information is stored in the company's GIS database, so the agent has the key information readily available to begin the work to process claims. Faster claims processing means more satisfied customers; adding to the competitive advantage.

In addition, the customer benefits in a number of ways: First, they have immediate access to accurate maps of each field. That translates, in over 90 per cent of the cases, to savings in their insurance premiums. Since the GPS measurements pinpoint field boundaries, most farmers have found that they have fewer actual acres cropped than other insurance companies had estimated and charged them. Secondly, many farmers obtain the accurate field maps in a digital form and use this information as a base to better manage their fields for fertilizer application, pest control, rotation planning, and other good farming practices. Hard copy versions of the digital maps are useful to the farmers for meeting the requirements of federal government in reporting production and filing claims, adding to customer benefits.

37.2.3. Lumber company integrates GIS, GPS and RS to yield savings and stay competitive

Weyerhaeuser Corporation of Tacoma, Washington has integrated GIS technology with GPS and RS technologies in identifying potential stands of merchantable timber throughout the Pacific Northwest US (Needham 1999).

With millions of acres of private forestlands to assess for potential procurement of softwood and hardwood logs, Weyerhaeuser is using RS techniques to rapidly identify tree types and stand age characteristics to avoid the costly process of using field crews to complete this work. In the Pacific Northwest, foresters use classified Landsat Thematic Mapper imagery to identify tree stands consisting of primarily mature deciduous and coniferous hardwoods. These stands are mapped and, using the company's GIS system combined with land ownership data, roadways, and other base data. The information is downloaded to laptop PCs that are installed in company trucks. The laptops are equipped with GIS software and integrated with GPS units. Procurement foresters use the GPS unit to navigate to potential procurement sites and investigate the area from the ground to obtain more detailed information about the timber stand and site.

These technologies have enabled Weyerhaeuser to more quickly and accurately prioritize areas for procurement. This means that they can develop their pricing strategies more quickly and begin negotiations with the landowners sooner than their competitors. By having tree stands identified and mapped, the company is also able to develop better strategic plans for future purchases and react more quickly to changing market demands for various types of timber. The procurement foresters in the field are also able to perform their work more effectively and accurately. With the GIS provided mapped tree stands viewed on the truck installed PCs, these staff can record more detailed information about particular stands and using the GPS devices, can more quickly navigate to the potential sale sites.

With these techniques proving successful in the Northwest, the company is using these systems in other procurement activities and in the management of pulpwood plantations. Remotely sensing data is being used in the Southeast to map pine plantations and provide a base of information for the management of these sites. From pest control to harvest schedules, the use of GIS integrated with remotely sensed data is giving Weyerhaeuser the tools needed to stay competitive and maintain profit margins.

37.2.4. Forestry industry in British Columbia uses GIS, RS, and visualization technologies to save money

Forestry practices in British Columbia have become increasingly complex over the past several decades. Environmental concerns and public sentiment have taken a forefront in determining how logging activities take place. The British Columbia government has implemented strict regulations in their Forest Practices Code that logging companies must meet in order to attain permission to harvest a given site. Logging companies must submit detailed designs for clear cuts that meet visual impact regulations such that minimal alterations to the landscape occur. In some areas, only partial cutting is allowed so that no obvious change to the landscape is apparent.

The logging companies must submit detailed proposals to the government to attain permission for harvesting timber (Allen 1998). The proposals are complex documents that must include images of the site in its present state and renderings of its post cut state. Rendered images must be submitted that include views from several different points or perspectives in order to show what the site will look like after the logging is completed. The more realistic the renderings are, the more likely that approval for the cut will be granted. If the images of the post logging landscape do not meet the criteria of the government or are deemed unacceptable as a result of public comment, the logging company must redesign and resubmit their proposal. This costs the company value time in the approval process, more money in the revision of the proposal, and it opens the door for competitors to gain approval for the same site.

In the early 1990s, the rendered images were typically developed using GIS technology and 3D wire frame images that showed perspective views of the cut areas. Most often, the logging companies hired local consulting firms to produce these images since they did not have the in house expertise to perform this work. The fees charged to the logging companies could be very high – over $10,000 US for a single proposal – and over time, as software tools improved and image production software developed, the charges became even higher.

The time requirements for development of the harvesting proposal including the post logging site renderings were likewise very high. The renderings alone often took over a week to generate! If a proposal needed to be revised, and new images prepared, the time requirements and costs were even more extensive and could be detrimental to the approval process.

In 1998 the forest industry adopted a software animation tool that was developed by the film industry. The tool is now used widely within the forest industry to perform terrain modeling and landscape visualization. This visualization software integrates data derived from GIS and RS technology, and GPS data can also be imported into the system. Many of the logging companies are implementing this tool and integrating it with their GIS systems in an effort to save costs in the development of clear-cut proposals and their associated post cut renderings.

Digital Elevation Models (DEM) data is used first to develop an image of the topography or terrain (Figure 37.3). The data is imported from a company's GIS or developed directly in the visualization system. Aerial photographs or satellite imagery provides images of the logging area showing species specific trees. The imagery is orthorectified and interpreted for species type or class using commercial RS software. The data is subsequently stored within the company's GIS and, then imported into the visualization system to be used as a color index for the generated image. This results in a much more realistic image than the past technique of draping a photo or satellite image over a 3D wire frame terrain model (Figure 37.4, see plate 25).

The visualization software generates an image of the site with realistic landscape features such as trees. Other surface features such as roads, lakes, power lines, etc. can be incorporated into the image from digital data within the GIS. Administrative and

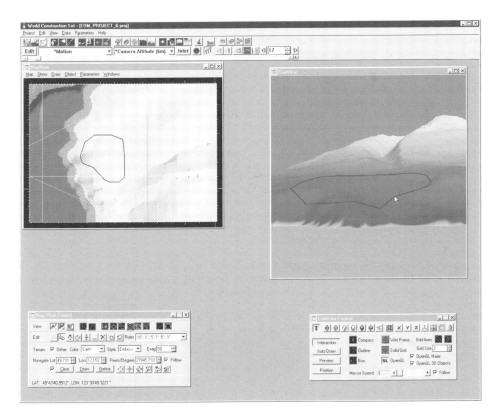

Figure 37.3 Developing a 3D terrain model using visualization software.

political boundaries as well as demarcations of property ownership can also be imported from the GIS and superimposed on the rendering.

One of the significant benefits that the visualization tools provide is the ability to generate alternative harvest scenarios for the proposal area showing only a portion of the trees removed. If the site that the logging company is bidding on for harvest is in a highly visible area, such as along a major roadway, the government may only allow partial clearing and in some instances staged or sequenced harvesting.

The visualization software, using data derived from the GIS and RS tools, can create an image of the site that accurately reflects what the area will look like with a percentage of the trees removed through selective harvesting techniques. Within the visualization software, a density value is set for the site and a species image is selected. When the rendering is complete, the image resembles an actual photograph showing the site with the selected tree density that would remain following a partial clearing operation (Figure 37.5, see plate 26).

Subsequent to the logging operation, field crews equipped with GPS walk and map the boundaries of recent clear-cuts. The resultant data is imported to the GIS as polygons reflecting the specific location of the timber cuts. The visualization software for deriving comparisons between planned cuts and actual cuts can then use these data. In addition, the information is available for later harvest proposals in adjacent areas.

37.2.5. Texaco integrates GIS and RS to improve oil exploration techniques

In its constant efforts to explore the earth for petroleum resources, the US based company, Texaco, has implemented GIS and RS technologies to assist in their exploration activities (Lyle 1999). These technologies are proving beneficial to Texaco in their management of overseas mapping and exploration projects where base maps and other key data are unavailable or unreliable.

Landsat and Spot imagery is combined with surface and subsurface geology data, well locations, and topographic data to identify viable sites for exploration. The company acquires imagery for huge areas of land throughout the world as part of the exploration process. The imagery is analyzed and used to map lineaments and surface feature characteristics that may indicate the presence of hydrocarbons. These tools help the company narrow the focus from huge land areas with potential oil fields to smaller areas that, through the analyzed data, indicate viable sites for on the ground exploration and sampling. This technique of analyzing imagery saves the company hundreds of hours of staff time by enabling geologists to more rapidly identify potential exploration sites across large land areas. The wide spectrum of collected data types and imagery allow staff geologists to 'see' the site and gain perspective on the relationships of surface and subsurface features. This greatly reduces the time spent determining the potential of a site and saves the company money in the development of data needed to analyze potential sites.

Once an area has been targeted for exploration, the company uses GIS to develop models for the drilling sites reflecting not only geological but also social and environmental considerations. In the field, workers collect detailed information about the site such as digital photos of outcrops, video strips of topography, and field observation notes. GPS units are also used in the field to identify and precisely locate seismic lines and core sample locations. This information is combined with base information in the

GIS and is used to form a complete picture of the potential exploration target area. In the GIS, 3D representations are generated for the potential drill sites. These generalized data are subsequently imported to Computer Aided Design (CAD) tools and used to build detailed site plans for the location of the well rig machinery, roadways, and other drill site infrastructure.

The GIS tools in use at Texaco have given the company improvements in the integration and coordination of activities at potential drill sites, better and substantive data for evaluation and analysis of sites, savings in time and money, and a means to collect and store critical corporate information that remains in the corporate data base and is unaffected by changes in key personnel. By optimizing their operations with technology driven processes, Texaco has realized benefits in both dollar savings and staff efficiencies. The company believes its use of GIS and satellite imagery has provided a competitive edge over its competition.

37.3. Financial advantage favors integration

The adoption by a commercial enterprise of any or all of the three technologies, GPS, GIS and RS, would be driven by an expectation of both tangible and intangible benefits. As characterized in (Antenucci et. al 1991) one formal structure for examining benefits of technology implementation identifies five distinct benefits:

- Type 1 – Quantifiable efficiencies in current practices, or benefits that reflect improvements to existing practices.
- Type 2 – Quantifiable expanded capabilities, or benefits that offer added capabilities.
- Type 3 – Quantifiable unpredictable events or benefits that result from unpredictable events.
- Type 4 – Intangible benefits or benefits that produce intangible advantages.
- Type 5 – Quantifiable sale of information, or benefits that result from the sale of information service.

In the preceding case studies at least three benefit types are apparent. Type 1, 2 and 4 are apparent and the potential for achieving Type 3 and 5 benefits exists. Clearly, the actions of a commercial enterprise to take on a new technology, to undertake the integration of several technologies and to change work flows and business practices to leverage the potential for cost savings, higher productivity and higher profits, are typically quantifiable.

The use of GPS and on board GIS to guide spraying operations reduced costs in terms of reduced fuel consumption, lower costs of fungicide, and lower costs of ground staff. And, it does not take much imagination to reflect on the intangible sociological and human health benefits that accrue when the operation eliminated exposing 'flag bearers' to direct aerial spraying. Its not much of a stretch to suggest that the companies may have avoided a currently un-quantified liability in the event that the exposure to the fungicide results in financial claims for illness and death. Carrying this example one step forward, the fact that the companies can systematically document the effectiveness of specific fungicides in specific fields under known conditions-creates the opportunity that they may sell the resultant information to others (a Type 5 benefit).

In a similar vein, the information collected by the Fireman's fund could be a valuable information asset to the farmer – or more likely, the agra-industry that has a financial stake in the yield of the insured land. Again, the derivative information may be sold for secondary purposes favorably affecting both gross revenues and net income.

Texaco might see secondary income from the information collected as well fields are jointly developed and production and regulatory compliance information is shared. Similarly, an oil company may find the information base created for one purpose to be invaluable in the event of a catastrophic failure of equipment, fires or natural disasters.

Returning to the initial premise, justification for the technology in the commercial sector will more often than not be established against expectations of productivity enhancements, new capabilities and the generation of information from, heretofore, disjointed sets of data. As a consequence, commercial enterprise can establish measurement systems that facilitate a determination of return on investments made in the introduction and integration of the selected technologies.

References

Allen, B., 'Forestry in the third dimension', *EOM*, 7(11), 1998.

Antenucci et. al., *Geographic Information Systems: a Guide to the Technology*, Chapman & Hall, New York, 1991.

Ardila, M. J., 'Precision farming', supplement to *GPS World Magazine*, Advanstar Communications, Cleveland, OH, 1996.

Kozero, J., *Press release*, Fireman's Fund Insurance Company, Overland Park, KS, 1999.

Lyle, D., 'GIS in exploration', *Harti̇s Oil and Gas World*, June 1999.

Needham, S., *Personal communication*, Weyerhaeuser Corporation, Tacoma, WA, 1999.

Index

absorption 63, 64, 69–72, 74, 253, 256, 261, 266–268, 306, 337, 365–378, 380, 381, 608
accuracy
  assessment 338, 349–359, 361–363, 368, 373–375, 553, 562
  attributes 450, 504, 510, 519, 555, 590
  change detection 361, 395
  classification 340, 355, 361–363, 368
  degradation 203
  ephemeris 225
  geometric 336, 350–352, 396
  high level 319
  levels 5, 19, 82, 83, 86, 87, 89, 90, 93, 100, 123, 125, 147, 155, 166, 167, 169–171, 192, 202, 218, 222, 256, 258, 319, 372, 488
  measures 168, 174, 358, 359
  positioning 77, 78, 84, 87, 89, 90, 119, 137, 139, 156, 162, 163, 165, 204, 219, 222, 226
  predictable 202
  radiometric 351, 352
  range 171
  repeatable 201, 202
  requirements 92, 135, 170, 185, 187, 192, 570
  semantic 510
  standards 139, 168, 174, 184, 186, 187, 370
  survey 92, 168, 172, 184, 192
  temporal 509
  thematic 244, 350, 352, 359, 510
ADAR 281, 295
adjustment 18, 26, 56, 168–170, 175, 186, 187, 196, 197, 199, 305, 396
AHS 282
Airborne Data Acquisition and Registration Systems See ADAR
Airborne Hyperspectral Scanner See AHS
Airborne Imaging Spectroradiometer for Applications See AISA

Airborne Multispectral Digital Camera See AMDC
Airborne Multispectral Scanner See AMS
Airborne Oceanographic LIDAR See AOL
Airborne Thematic Mapper See ATM
Airborne Visible Infrared Imaging Spectro-meter See AVIRIS
AISA 283, 296
alignment 85, 132, 133, 456, 600
altitude 79, 106, 226, 236, 237, 271, 279, 280, 284, 287, 288, 290, 292–295, 299, 307, 315
AMDC 281, 282, 295
AMS 261, 282, 295
angle of incidence 265
angle of reflection 71
ANZLIC 602, 603, 606
AOL 284, 296
apex 40
Arc/Info 339, 341, 342, 375, 385, 387, 438, 468, 490, 571
atmospheric absorption 63
atmospheric attenuation 306, 338, 394
attitude 219, 293, 307
Australian New Zealand Land Informatiion Council See ANZLIC
availability 83, 87, 93, 102, 118, 135, 157, 159, 162, 176, 179, 180, 201–203, 205, 219, 222, 244, 304, 318, 366, 391, 422, 501, 510, 512, 552, 558, 563, 581, 589, 595
AVIRIS 238, 239, 252, 261, 264, 272, 283, 284, 296, 315, 323, 326, 329, 330, 393
azimuth 47, 130, 180, 194, 220

band
  absorption 70, 267
  infrared (IR) 62–64, 68, 233, 236–240, 249, 253, 255, 261, 264, 267, 268, 270–272, 275–277, 280, 292, 296, 300, 315, 320, 323, 335, 340, 365–367, 379, 380, 382, 588